REGULATED BIOANALYTICAL LABORATORIES

REGULATED BIOANALYTICAL LABORATORIES

Technical and Regulatory Aspects from Global Perspectives

MICHAEL ZHOU Ph.D.
Synta Pharmaceuticals Corporation

A JOHN WILEY & SONS, INC., PUBLICATION

Published by John Wiley & Sons, Inc., Hoboken, New Jersey
Published simultaneously in Canada

For general information on our other products and services or for technical support, please contact our Customer Care Department within the United States at (800) 762-2974, outside the United States at (317) 572-3993 or fax (317) 572-4002.

Wiley also publishes its books in a variety of electronic formats. Some content that appears in print may not be available in electronic formats. For more information about Wiley products, visit our web site at www.wiley.com.

Library of Congress Cataloging-in-Publication Data:

Zhou, Michael.
Regulated bioanalytical laboratories : technical and regulatory aspects from global perspectives / Michael Zhou.
p. cm.
Includes index.
ISBN 978-0-470-47659-8 (cloth)
1. Medical laboratories–Qualtiy control. 2. Biological laboratories–Qualtiy control. 3. Pharmaceutical technology–Qualtiy control. I. Title.
R860.Z56 2011
610.28'4–dc22

2010022398

Printed in Singapore

10 9 8 7 6 5 4 3 2 1

CONTENTS

Preface xiii

Acknowledgment xvii

Contributors and Advisors xix

1 Introduction, Objectives, and Key Requirements for GLP Regulations 1

1.1 Introduction 1

1.1.1 Good Laboratory Practices 1

1.1.2 Bioanalytical Laboratories—Bioanalysis 4

1.1.3 Good Laboratory Practices Versus Bioanalytical Labs/Bioanalysis 7

1.2 Objectives and Key Requirements for GLP Regulations 8

1.3 Fundamental Understanding of GLP Regulations and Principles 10

1.3.1 Elements of Good Laboratory Practices 11

1.4 Key Elements of Bioanalytical Methods Validation 16

1.4.1 Reference Standards 19

1.4.2 Method Development—Chemical/Chromatographic Assay 20

1.4.3 Calibration/Standard Curve 21

1.4.4 Stability 21

1.4.5 Reproducibility 23

1.4.6 Robustness or Ruggedness 23

1.5 Basic Principles of Bioanalytical Method Validation and Establishment 23
1.5.1 Specific Recommendations for Method Validation 24
1.5.2 Acceptance Criteria for Analytical Run 29
References 33

2 Historic Perspectives of GLP Regulations, Applicability, and Relation to Other Regulations 35
2.1 Historic Perspectives of GLP Regulations 35
2.1.1 Economic Assessment 39
2.1.2 Environmental Impact 40
2.2 Applicability and Relations to Other Regulations/Principles 42
2.2.1 GLP, GCP, GMP, and Part 11 42
2.2.2 General Terminologies and Definitions of GxPs (GLP, GCP, and cGMP) 47
2.3 Comparison of FDA GLP, EPA GLP Regulations, and OECD GLP Principles 47
2.3.1 US and OECD GLP Similarity and Differences 53
2.4 Applications of GLP to Multiple Site Studies 55
2.4.1 Roles and Responsibilities 57
2.4.2 Performance of the Studies 61
2.4.3 Applications of GLP to *In Vitro* Studies for Regulatory Submissions 64
2.5 21 CFR Part 11 in Relation to GLP Programs 66
2.5.1 A New Risk-Based Approach 67
2.5.2 Understanding Predicate Rule Requirements 67
2.5.3 21 CFR Part 11 Best Practices 68
2.5.4 Use of Electronic Signatures 71
2.6 GLP, cGMP, and ISO Applicabilities, Similarity, and Differences 74
2.6.1 GLPs, cGMPs, ISO 17025:2005: How Do They Differ? 74
2.6.2 GLPs Versus GMPs 74
2.6.3 GLPs Versus ISO/IEC 17025:2005 75
2.6.4 ISO Versus GLPs 76
2.7 Good Clinical Practices and Good Clinical Laboratory Practices 78
2.8 Gap and Current Initiatives on Regulating Laboratory Analysis in Support of Clinical Trials 80
References 84

3 GLP Quality System and Implementation 87
3.1 GLP Quality System 87
3.1.1 Regulatory Inspection for GLP Quality System 95
3.1.2 Good Laboratory Practice Inspections 99
3.1.3 GLP Quality System Objectives 103
3.2 Global GLP Regulations and Principles 106
3.2.1 General 106

3.2.2 Responsibilities and Compliance 107
3.2.3 Statement of Compliance in the Final Report 107
3.2.4 Protocol Approval 108
3.2.5 Assignment of Study Director 108
3.2.6 Laboratory Qualification/Certification 108
3.2.7 Authority Inspections 108
3.2.8 Archiving Requirements 108
3.3 Implementation of GLP Regulations and OECD Principles 109
3.3.1 Planning (Master Schedule) 114
3.3.2 Personnel Organization 115
3.3.3 Curriculum Vitae 115
3.3.4 Rules of the Conducts of Studies 116
3.3.5 Content of Study Protocol 116
3.3.6 Approval of Study Protocol 118
3.3.7 Distribution of Study Protocol 118
3.3.8 Protocol Amendment 118
3.3.9 Standard Operating Procedures 119
3.3.10 SOP System Overview 119
3.3.11 Characterization 121
3.3.12 Test Item/Article Control before Formulation 121
3.3.13 Preparation of the Dose Formulation 123
3.3.14 Sampling and Quality Control of Dose Formulation 125
3.4 Initiatives and Implementation of Bioanalytical Method Validation (Guidance for Industry BMV—May 2001) 126
3.4.1 Summary 127
References 128

4 Fundamental Elements and Structures for Regulated Bioanalytical Laboratories 131
4.1 Introduction 131
4.2 Fundamental Elements for Bioanalytical Laboratories 133
4.2.1 Document Retention and Archiving 136
4.3 Basic Requirements for GLP Infrastructure and Operations 139
4.4 GxP Quality Systems 143
4.4.1 Laboratory Instrument Qualification and Validation 149
4.4.2 Procedural Elements and Function that Maintain Bioanalytical Data Integrity for GLP Studies 150
References 166

5 Technical and Regulatory Aspects of Bioanalytical Laboratories 167
5.1 Fundamental Roles and Responsibilities of Bioanalytical Laboratories 167
5.1.1 Technical Functions of Bioanalytical Laboratories 168
5.1.2 Basic Processes in Bioanalytical Method Development, Validation, and Sample Analysis 173

5.2 Qualification of Personnel, Instrumentation, and Analytical Procedures 178
5.2.1 From Regulatory Perspectives: Personnel, Training, and Qualification 183
5.2.2 Facility Design and Qualifications 186
5.2.3 Equipment Design and Qualification 186
5.2.4 Analytical/Bioanalytical Method Qualification and Validation along with Related SOPs 197
5.3 Regulatory Compliance with GLP Within Bioanalytical Laboratories 204
5.4 Joint-Effort from Industries and Regulatory Agencies 206
5.4.1 Ligand-Binding Assays In-Study Acceptance Criteria 213
5.4.2 Determination of Metabolites during Drug Development 216
5.4.3 Incurred Sample Analysis 216
5.4.4 Documentation Issues 217
5.4.5 Analytical/Validation Reports 218
5.4.6 Source Data Documentation 218
5.4.7 Final Report Documentation 219
5.4.8 Stability Recommendation 219
5.4.9 Matrix Effects for Mass Spectrometric-Based Assays 221
5.4.10 System Suitability 222
5.4.11 Reference Standards 222
5.4.12 Validation Topics with No Consensus 222
5.4.13 Specific Criteria for Cross-Validation 223
5.4.14 Separate Stability Experiments Required at − 70°C if Stability Shown at − 20°C 223
5.4.15 Stability Criteria for Stock Solution Stability 224
5.4.16 Acceptance Criteria for Internal Standards 224
5.4.17 Summary 224
References 226

6 Competitiveness of Bioanalytical Laboratories—Technical and Regulatory Perspectives 229
6.1 Technical Aspect of Competitive Bioanalytical Laboratories 229
6.2 Bioanalytical Processes and Techniques 232
6.2.1 Sample Generation, Shipment, and Storage 232
6.2.2 Sample Preparation 233
6.3 Enhancing Throughput and Efficiency in Bioanalysis 243
6.3.1 Chromatographic Separation 244
6.3.2 Selective and Sensitive Detection 251
6.4 Technical Challenges and Issues on Regulated Bioanalysis 254
6.4.1 Matrix Effect 254
6.4.2 Method Validation and Critical Issues during Sample Analysis 256
6.4.3 Method Transfer 258

6.5 Regulatory Aspects of Competitive Bioanalytical Laboratories 264
6.5.1 General Consideration 264
6.5.2 Historical Perspective 265
6.5.3 Personnel—Training and Qualification 267
6.5.4 Facility—Design and Qualifications 269
6.5.5 Equipment Design and Qualification 270
6.5.6 Standard Operating Procedures 272
6.5.7 Laboratory/Facility Qualification Perspectives 272
6.6 Advanced/Competitive Bioanalytical Laboratories 277
6.6.1 Strategy Versus Tactics 278
6.6.2 Bioanalytical Laboratory Assessment 279
6.6.3 Capacity 279
6.6.4 Experience 280
6.6.5 Quality 281
6.6.6 Performance and Productivity Measures 281
6.6.7 Information Technology and Data Management 282
6.6.8 Communication 282
6.6.9 Financial Stability 283
6.6.10 Ease of Use 283
6.6.11 Contracting Bioanalytical Services 284
6.6.12 The Contracting Process 284
6.7 Applications and Advances in Biomarker and/or Ligand-Binding Assays within Bioanalytical Laboratories 286
References 290

7 Sponsor and FDA/Regulatory Agency GLP Inspections and Study Audits 297
7.1 GLP versus Biomedical Research Monitoring and Mutual Acceptance of Data for Global Regulations and Inspections 298
7.2 Purposes and Benefits of Regulatory Inspections/Audits 303
7.2.1 Criteria for Selecting Ongoing and Completed Studies 304
7.2.2 Areas of Expertise of the Facility 305
7.2.3 Establishment Inspections 305
7.2.4 Organization and Personnel (21 CFR 58.29, 58.31, 58.33) 305
7.2.5 Quality Assurance Unit (QAU; 21 CFR 58.35) 307
7.2.6 Facilities (21 CFR 58.41–58.51) 308
7.2.7 Equipment (21 CFR 58.61–58.63) 309
7.2.8 Testing Facility Operations (21 CFR 58.81) 310
7.2.9 Reagents and Solutions (21 CFR 58.83) 311
7.2.10 Animal Care (21 CFR 58.90) 311
7.2.11 Test and Control Articles (21 CFR 58.105–58.113) 312
7.2.12 Test and Control Article Handling (21 CFR 58.107) 313
7.2.13 Protocol and Conduct of Nonclinical Laboratory Study (21 CFR 58.120–58.130) 314

7.2.14 Study Protocol (21 CFR 58.120) 314
7.2.15 Test System Monitoring 314
7.2.16 Records and Reports (21 CFR 58.185–58.195) 314
7.2.17 Data Audit 316
7.2.18 General 316
7.2.19 Final Report Versus Raw Data 317
7.2.20 Specimens Versus Final Report 318
7.2.21 Refusal to Permit Inspection 318
7.2.22 Sealing of Research Records 318
7.2.23 Samples 319
7.3 Typical Inspections/Audits and Their Observations 320
7.4 Regulatory Challenges for Bioanalytical Laboratories 321
7.4.1 Introduction 321
7.4.2 Analysis of Current FDA Inspection Trends 324
7.4.3 Discussion and Analysis of Specific Potential FDA 483 Observation Issues 325
7.4.4 Method Validation Issues 325
7.4.5 Batch Runs Acceptance Criteria Issues 329
7.4.6 Events/Deviations Investigation/Resolution Issues 331
7.4.7 Test Specimen Accountability Issue 333
7.4.8 Recommendations to Support an Effective FDA Inspection Readiness Preparation 334
7.5 Handling and Facilitating GLP or GxP Audits/ Inspections 334
7.5.1 General Preparation for an Inspection 336
7.5.2 Why Are Audits/Inspections Needed and Conducted? 342
7.5.3 Written Policy in Place 342
7.5.4 Positions on Controversial Issues 343
7.5.5 The Inspection Coordinator 344
7.5.6 Follow-Up Procedures 348
7.5.7 Summary 349
References 351

8 Current Strategies and Future Trends 353
8.1 Strategies from General Laboratory and Regulatory Perspectives 354
8.2 Strategies from Technical and Operational Perspectives 356
8.3 Biological Sample Collection, Storage, and Preparation 360
8.3.1 Sample Collection and Storage 360
8.3.2 Sample Preparation Techniques 361
8.3.3 Off-Line Sample Extraction 364
8.3.4 On-Line Sample Extraction 364
8.4 Strategies for Enhancing Mass Spectrometric Detection 366
8.4.1 Enhanced Mass Resolution 368
8.4.2 Atmospheric Pressure Photoionization 369

8.4.3 High-Field Asymmetric Waveform Ion Mobility Spectrometry 370
8.4.4 Electron Capture Atmospheric Pressure Chemical Ionization 370
8.4.5 Mobile Phase Optimization for Improved Detection and Quantitation 371
8.4.6 Anionic and Cationic Adducts as Analytical Precursor Ions 372
8.4.7 Derivatization 372
8.5 Strategies for Enhancing Chromatography 374
8.5.1 Ultra-Performance Chromatography 375
8.5.2 Hydrophilic Interaction Chromatography for Polar Analytes 376
8.5.3 Specialized Reversed-Phase Columns for Polar Analytes 377
8.5.4 Ion-Pair Reversed-Phase Chromatography for Polar Analytes 378
8.6 Potential Pitfalls in LC–MS/MS Bioanalysis 378
8.6.1 Interference from Metabolites or Prodrugs due to In-Source Conversion to Drug 378
8.6.2 Interference from Metabolites or Prodrugs due to Simultaneous $M + H^+$ and $M + NH_4^+$ Formation or Arising from Isotopic Distribution 379
8.6.3 Pitfall in Analysis of Two Interconverting Analytes due to Inappropriate Method Design 383
8.6.4 Matrix Effect 383
8.7 Trends in High-Throughput Quantitation 386
8.7.1 System Throughput 386
8.7.2 High-Speed HPLC 386
8.8 Trends in Hybrid Coupling Detection Techniques 388
8.9 Trends in Internal R&D and External Outsourcing 388
8.10 Trends in Ligand-Binding Assays and LC–MS/MS for Biomarker Assay Applications 397
8.11 Trends in Study Design and Evaluation Relating to Bioanalysis 399
8.12 Trends in Applying GLP to *In Vitro* Studies in Support of Regulatory Submissions 403
8.13 Trends in Global R&D Operations 404
8.14 Trends in Regulatory Implementations 407
8.14.1 Calibration Range and Quality Control Samples 407
8.14.2 Incurred Sample Reproducibility (Duplicate Sample Analysis) 408
8.14.3 LIMS and Electronic Data Handling, Security, Archiving, and Submission 409
8.15 Trends in Global Regulations and Quality Standards 412

8.16 Trends in Compliance with 21 CFR Part 11 414
8.16.1 21 CFR Part 11 Software Requirements 415
8.16.2 Building a Roadmap for Compliance with 21 CFR Part 11 415
8.16.3 Low Hanging Fruits in the Roadmap for Compliance with 21 CFR Part 11 416
8.17 Summary 419
References 421

9 General Terminologies of GxP and Bioanalytical Laboratories 431
9.1 General Terminologies for GxP and Bioanalytical Laboratories 431
9.2 GLP Basic Concepts and Implementation 469
9.2.1 The Study Protocol 470
9.2.2 Raw Data 471
9.2.3 The GLP Archive and the Archivist 472
9.2.4 Expansion of GLP Scope 473
9.2.5 OECD GLP 473
9.3 GLP Guidance Documents 474
9.3.1 FDA Guidance for Industry on Bioanalytical Method Validation 474
9.3.2 OECD GLP Guidance Documents 474
9.3.3 Swiss GLP Guidance Documents 475
References and Sources for Above Terminologies 475

Appendix A Generic Checklist for GLP/GxP Inspections/Audits 479

Appendix B General Template for SOP 489

Appendix C Typical SOPs for GLP/Regulated Bioanalytical Laboratory 493
Quality Assurance—GLP 493
Bioanalytical—GLP Laboratories 494

Appendix D Basic Equipment/Apparatus for Bioanalytical Laboratory 497

Appendix E Website Linkages for Regulated Bioanalysis 499

Index 503

PREFACE

The Good Laboratory Practice (GLP) regulations were established in the 1970s by the United States Food and Drug Administration (FDA), and published in the Code of Federal Regulations (21 CFR Part 58). The Organization for Economic Cooperation and Development (OECD) established the Principles of GLP in 1981. United States Environmental Protection Agency (EPA) adopted its own set of GLP regulations shortly thereafter, governing the research surrounding pesticides and toxic chemicals. Bioanalytical laboratories have increasingly become critically important in data generation for discovery, preclinical, and clinical development in life science industries. Bioanalysis, employed for the quantitative determination of drugs and their metabolites in biological fluids, plays significant roles in the evaluation and interpretation of bioequivalence (BE), bioavailability (BA), pharmacodynamic (PD), pharmacokinetic (PK), and toxicokinetic (TK) studies. The quality of these studies, which are often used to support regulatory filings, is directly related to the quality of the underlying bioanalytical data. It is therefore important that guiding principles for the validation of these analytical methods be established and disseminated to the pharmaceutical and life science communities.

The focus and working groups from American Association of Pharmaceutical Scientists (AAPS), European Bioanalysis Forum (EBF), Food and Drug Administration, European Medicine Agency (EMEA), and other related organizations have held a series of workshops and seminars focusing on key issues relevant to bioanalytical methodology and provided a platform for scientific discussions and deliberations. As bioanalytical tools and techniques have continued to evolve and significant scientific and regulatory experiences have been gained, the bioanalytical community has continued its critical review of the scope, applicability, and success of the presently employed bioanalytical guiding principles. Life science products including foods, nutritional supplements, medicine/drug discovery, research, and development expand to wide scope of life sciences meaning varieties of industries such as environmental toxicology, food nutritional analysis, biotechnology, biopharmaceuticals, pharmaceuticals, hospitals, diagnostic/medical device industries, and so on. Bioanalytical sciences basically support above-mentioned areas under FDA GLP and other related regulations, principles, and guidelines. Below are some example elements/infrastructure that are required for regulated Bioanalytical Laboratories:

(1) Responsible management team with quality system
(2) Qualified personnel selection, staffing, and training
(3) Standard operating procedures (SOPs)
(4) Installation, operational, and performance qualification (IQ, OQ, and PQ, respectively) of facilities, instrumentation, and software
(5) Quality control (QC) procedures and staffing
(6) Quality assurance unit (QAU)
(7) Data generation and security assessment
(8) Documentation and archival process
(9) Laboratory information management system (LIMS)
(10) Final gap analysis

This book provides useful information for bioanalytical/analytical scientists, analysts, quality assurance managers, and all personnel in bioanalytical laboratories through all aspects of bioanalytical technical and regulatory perspectives within bioanalytical operations and processes. Readers will learn how to develop and implement strategies for routine, nonroutine, and standard bioanalytical methods and on the entire equipment hardware and software qualification process. The book also gives guidelines on qualification of certified standards and in-house reference material as well as on people qualification. Finally, it guides readers through stressless internal and third party laboratory audits and inspections. Highly comprehensive content with specific chapter by chapter is elaborated making it easy not only to learn the subject but also to quickly implement the recommendations.

It takes account to most national and international regulations and quality and accreditation standards such as GLP, cGMP, GCP, and GCLP from US FDA, ICH, WHO, and EU, accreditation standards such as ISO17025 and to corresponding interpretation and inspection guides. The text begins with an introductory overview of the roles of bioanalytical laboratories in pharmaceutical and biotechnology drug development. Regulatory wise, it describes some fundamental understanding of regulatory aspects within bioanalytical laboratories—current and future requirements as far as GLP and/or GxP quality systems, facility, and personnel infrastructure and qualification along with continuing improvement on a daily basis. From technical standpoints, the book also elaborates the strategies for sample preparation, along with essential concepts in extraction chemistry. Particular strategies for efficient use of automation within bioanalytical laboratories are also presented. With regards to instrumental analysis, fundamental approach is presented within the areas of LC–MS/MS and other hyphenated analytical techniques. Ligand-binding assays are also discussed to recognize its increasingly crucial applications within bioanalytical laboratories. Important objectives that can be accomplished when the strategies presented in this book are followed include: improved efficiency in moving discovery compounds to nonclinical and clinical status with robust analytical methods; automation for sample preparation; modern analytical equipment, and improved knowledge and expertise of laboratory staff. It has been widely accepted that good sciences are not enough to meet regulatory requirements. In author's opinion, good laboratory practices may not improve any "poor" sciences, but indeed make "good sciences" better as to ensure the quality, integrity, and reconstructability of data. GLP or GxP is all about documentation. In another word, nothing has been properly done without any documentation. GLP principles may also enhance the opportunity of *L*imiting waste of resources, *E*nsuring high quality of data, *A*cquiring comparability of results, and *D*eriving to mutual recognition of scientific findings worldwide and ultimately *S*ecuring the health and well-being of our societies, as being *LEADS* concept per author's perspectives.

ACKNOWLEDGMENT

The need of this book has been apparent as noted that Good Laboratory Practice must be followed while generating data for regulatory consideration. I am grateful to peer-reviewers for their positive feedback and encouragement since this project started. The staff at John Wiley and Sons has provided me with tremendous support. In particular, I would like to thank my editor, Jonathan Rose, who has been an invaluable resource during the project as well as other advice. I am truly indebted to all of the contributors for their willingness of sharing their experiences, knowledge, and perspectives on bioanalytical technical and regulatory aspects. The contributions as well as the many discussions and interactions are worth noting! My special gratitudes extend to Dr. Vinod P. Shah for his expert advice and inputs. The growth of bioanalytical laboratory operations and contributions in life science industries has been truly remarkable. I am thankful to have had the opportunity to interact with so many people (contributors, reviewers, and advisors) who shared a common passion for the analytical/bioanalytical sciences and regulatory compliance for the betterment of the world and health.

Finally, I thank my wife, family, and friends for their courage and continued support for everything I do.

CONTRIBUTORS AND ADVISORS

Author sincerely appreciates all of the advice, review, suggestions, and edits from following field experts on respective chapters and sections.

Frank Chow Lachman Consultants
1600 Stewart Avenue, Suite 604
Westbury, NY 11590, USA

Howard M. Hill Huntingdon Life Sciences
Huntingdon, Cambridge, PE28 4HS, UK

Mohammed Jemal Bristol-Myers Squibb
Route 206 and Province Line Road
Princeton, NJ 08543, USA

Marian Kelley MKelley Consulting LLC
1533 Glenmont Lane
West Chester, PA 19380, USA

Jean Lee Amgen Inc
100 Amgen Center Drive
Thousand Oaks, CA 91320, USA

Raymond Naxing Xu Abbott Laboratories
100 Abbott Park Road
Abbott Park, IL 60064, USA

1

INTRODUCTION, OBJECTIVES, AND KEY REQUIREMENTS FOR GLP REGULATIONS

1.1 INTRODUCTION

1.1.1 Good Laboratory Practices

Good laboratory practices (GLPs 21 CFR PART 58) is a standard by which laboratory studies are designed, implemented, and reported to assure the public that the results are accurate/reliable and the experiment can be reproduced accordingly [1], at any time in the future. In less technical terms, GLP is the cornerstone of all laboratory-based activities in any organization that prides itself on the quality of the work it performs. And, despite its immediate association with the pharmaceutical sector, GLPs can (and should) be applied to virtually all industries in which laboratory work is conducted, including companies involved in drug development, manufacturing, foods, pesticides (agrochemicals), drink production, and engineering testing. In addition, commercial testing laboratories (for toxicology, metabolism, materials, and safety, for example), research establishments, and universities—in fact, all laboratories engaged in product or safety testing or research and development—should adopt and apply the doctrines of GLP.

GLP is not a luxury. It is a necessity for any professional laboratory wishing to gain and retain the respect of its employees, clients, regulators, and perhaps most importantly, its competitors. If a company is seen to be applying and adhering to the highest standards of laboratory practice, it will gain significant competitive advantage and will compete successfully for business and recognition within its

Regulated Bioanalytical Laboratories: Technical and Regulatory Aspects from Global Perspectives,
By Michael Zhou

operational environment. Conversely, without rigidly enforced GLPs, good clinical practice (GCP) [2], good manufacturing practices (GMPs) [3], or GxPs—a scientific organization will not achieve the commercial success and respect that its products and personnel deserve.

Published GLP regulations and guidelines have a significant impact on the daily operations of analytical and/or bioanalytical laboratories. GLP is a regulation that enhances good analytical practice. Good analytical/bioanalytical practice is important, but it is not enough. For example, the laboratory must have a specific organizational structure and procedures to perform and document laboratory work. The objective is not only quality of data but also traceability and integrity of data. However, the biggest difference between GLP and non-GLP work is the type and amount of documentation. GLP functions as a regulation, which deals with the specific organizational structure and documents related to laboratory work in order to maintain integrity and confidentiality of the data. The entire cost of GLP-based work is about 40% or more additional (from case to case) when compared to non-GLP operations. For a GLP inspector, it should be possible to look at the documentation and to easily find out the following:

- Who has done a study
- How the experiment was carried out
- Which procedures have been used, and
- Whether there has been any problem and if so
- How it has been addressed and solved where applicable

And this should not only be possible during and right after the study has been finished but also 5–10 or more years later.

From worldwide perspectives, good practice rules govern drug/product development activities in many parts of the world. World Health Organization (WHO), which has published documents on current good manufacturing practices (cGMPs) and GCPs, has not previously recommended or endorsed any quality standard governing the nonclinical phases of drug/product development. GLPs are recognized rules governing the conduct of nonclinical safety studies, ensuring the quality, integrity, and reliability of their data. To introduce the concepts of GLP to scientists in developing countries, workshops on GLP have been organized in these regions. As an outcome of the workshops (industries and regulatory bodies), it became apparent that some formal guidance would be needed for the successful implementation of the GLP regulations.

The first scientific working group on GLP issues was convened on November 25, 1999, in Geneva, to discuss quality issues in general and the necessity for a WHO guidance document on GLP in particular. The working group concluded that it was important to avoid the coexistence of two GLP standards, the Principles of good laboratory practice of the Organization for Economic Cooperation and Development (OECD) [4] being the internationally recognized and accepted standard, and recommended that the OECD Principles be adopted by WHO for Research & Training in Tropical Disease (TDR) as the basis of this guidance document. The experts also recognized the need to address quality issues in areas other than the strictly regulated safety studies for regulatory submission, and recommended that some explanation be included in this guidance document. The working group further recommended that WHO/TDR should request OECD's permission to publish the existing OECD GLP text with a WHO endorsement, and to supplement it with an explanatory introduction. Classical drug development (drug life cycle) is characterized by four well-defined stages as follows:

Stage 1: The first stage, the discovery of potential new drug products, is neither covered by a regulatory standard, nor are studies demonstrating proof of concept. This area may well require some international standards or guidance documents in the future.

Stage 2: The position of GLP studies within the drug development process is specific to the second stage. These studies are termed "nonclinical" as they are not performed in human. Their primary purpose is safety testing. Toxicology and safety pharmacology studies, with a potential extension to pharmacokinetics and bioavailability, are those studies where the compliance with GLP is required, which is the rather restricted scope of GLP.

Stage 3: The third stage, following on from safety studies, encompasses the clinical studies in human. Here, GCP is the basis for quality standards, ethical conduct, and regulatory compliance. GCP must be instituted in all clinical trials from Phase I (to demonstrate tolerance of the test drug and to define human pharmacokinetics), through Phase II (where the dose–effect relationship is confirmed), to Phase III (full-scale, often multicenter, clinical efficacy trials in hundreds and thousands of patients).

Stage 4: The fourth stage is postapproval. Here the drug is registered and available on the market. However, even after marketing, the use of the drug is monitored through formalized pharmacovigilance procedures. Any subsequent clinical trials (Phase IV) must also comply with GCP.

A brief summary of different stages is shown in Table 1.1.

TABLE 1.1 Stages Defined Within Discovery and Development Programs

Stage I	Stage II	Stage III	Stage IV
Establish discovery assessment of compounds with *in vitro* and/or *in vivo* data (not regulated under GxP)	Demonstrate efficacy, identify side effects including Tox and assessment of pharmacokinetics (GLP and GCP)	Gain more data on safety and effectiveness in multicenters with thousands of patients (GLP, GCP, cGMP)	Monitor claims or demonstrate new indications; examine special drug–drug interactions; assess pharmacokinetics (GLP, GCP, cGMP)

1.1.2 Bioanalytical Laboratories—Bioanalysis

Bioanalytical laboratories have increasingly become center of excellence and critically important in data generation for discovery, preclinical and clinical development in life science industries. Bioanalysis is a broad term that is derived from analytical applications to biological materials (matrices) such as human and/or animal biological fluids and materials (blood, plasma, serum, urine, feces, tissues, etc.), biopharmaceutical (peptides, protein, etc.), and biochemistry (DNA, RNA, organonucleotides, etc.). The main focus of this book is within the aspects of liquid chromatography–tandem mass spectrometry (LC–MS/MS) and to certain extent of immunochemistry assays—enyzme-linked immunosorbent assays (ELISA) or ligand-binding assays (LBAs). Bioanalysis is mainly referred to the quantitative determination of drugs and their metabolites, and other life science products in various sample matrices. However, it should also apply to qualitative analysis (identification and elucidations) of drug degradants, metabolites, impurities, and other analytes of interests. The techniques (chromatographic-based and ligand-binding-based assays) are used very early in the drug discovery and development process to provide support to product discovery programs on metabolite fate and pharmacokinetics of chemicals in living cells and animals. They are referred by FDA Guidance for Industry Bioanalytical Method Validation for chromatographic-based and ligand-binding-based assays [5]. Their uses continue throughout the nonclinical and clinical product development phases into postmarketing support and may sometimes extend into clinical therapeutic monitoring. Recent developments and industry trends for rapid sample throughput and data generation are introduced and discussed in following chapters, together with examples of how these high throughput needs are met in bioanalysis.

1.1.2.1 ***High-Throughput Bioanalytical Sample Preparation*** Methods and automation strategies are authoritative reference on the current state-of-the-art in sample preparation techniques for bioanalysis. The following related chapters focus on high-throughput (rapid productivity) techniques and describe exactly how to perform and automate these methodologies, including useful strategies for method development and optimization. A thorough review of the literature is included describing high-throughput sample preparation techniques: protein removal by precipitation; equilibrium dialysis and ultrafiltration; liquid–liquid extraction; solid phase extraction; and various online techniques. A schematic diagram of analytical/bioanalytical techniques used in automation is shown in Figure 1.1.

Among the sample preparation scheme, protein precipitation (PPT) is the most commonly used approach for a simple, fast, and unique process of removing unwanted materials from analyte(s) of interest for analysis or in some case for further cleanup. High selectivity and sensitivity are also imperative for bioanalytical laboratories to deal with sample analyses with great demand in method limits of quantitation (LOQ), wide dynamic range (linearity and range), free of interferences (specificity and selectivity), and other highly challenging requirements such as multiple compounds (analytes—parent drugs, prodrugs, and their degradants/

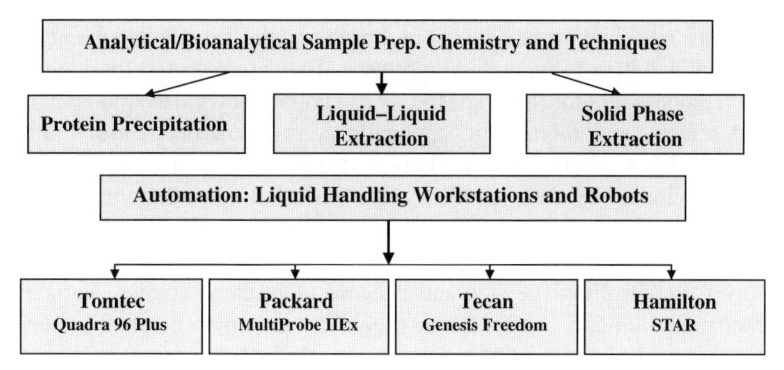

FIGURE 1.1 General schematic of analytical/bioanalytical techniques used in laboratory operations/automation.

metabolites), various sample types (matrices), and different analytical techniques including LC–MS/MS, GC–MS/MS, LC–NMR, ICP–MS, and other advanced hybrid techniques. In addition to above analytical techniques, immunoassays (ELISA and/or ligand-binding assays—LBAs or alike) are also widely used within bioanalytical laboratories, especially in biopharmaceutical and biotechnology industries where relatively large molecules are dealt such as peptides and proteins as part of drug development compounds and applying to different therapeutic areas. Rapid advances in chromatographic as well as ligand-binding assay technologies have been observed to meet the needs in product research and development processes. More details of description are elaborated on above analytical and bioanalytical techniques as powerful methodologies in trace level qualitative and quantitative analyses.

There have been varieties of separation and detection techniques involved in analytical and bioanalytical methodologies as indicated in Figure 1.2. More recent years, biomarker analysis in various therapeutic areas has become incredibly significant in drug/product development and monitoring programs. Without any doubt, this has increasingly become part of bioanalytical capabilities. Biomarker

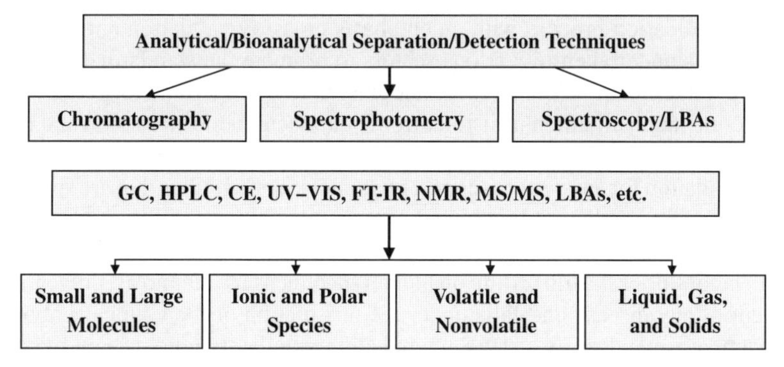

FIGURE 1.2 Commonly used techniques in analytical/bioanalytical separation and detection.

measurements now support key decisions throughout the drug development process, from lead optimization to regulatory approvals. They are essential for documenting exposure–response relationships, specificity and potency toward the molecular target, untoward effects, and therapeutic applications. In a broader sense, biomarkers constitute the basis of clinical pathology and laboratory medicine. The utility of biomarkers is limited by their specificity and sensitivity toward the drug or disease process and by their overall variability. Understanding and controlling sources of variability is not only imperative for delivering high-quality assay results, but ultimately for controlling the size and expense of research studies. Variability in biomarker measurements is affected by biological and environmental factors (e.g., gender, age, posture, diet, and biorhythms), sample collection factors (e.g., preservatives, transport and storage conditions, and collection technique), and analytical factors (e.g., purity of reference material, pipetting precision, and antibody specificity). The quality standards for biomarker assays used in support of nonclinical safety studies fall under GLP (FDA) regulations, whereas, those assays used to support human diagnostics and healthcare are established by Clinical Laboratory Improvement Amendments (CLIAs) and Centers for Medicare & Medicaid Services (CMSs) regulations and accrediting organizations such as the College of American Pathologists (CAPs). While most research applications of biomarkers are not regulated, biomarker laboratories in all settings are adopting similar laboratory practices in order to deliver high-quality data. Because of the escalation in demand for biomarker measurements, the highly parallel (multiplexed) assay platforms that have fueled the rise of genomics will likely evolve into the analytical engines that drive the biomarker laboratories of tomorrow. The role of biomarkers in drug discovery and development has gained precedence over the years. As biomarkers become integrated into drug development and clinical trials, quality assurance and, in particular, assay validation become essential with the need to establish standardized guidelines for bioanalytical methods used in biomarker measurements. New biomarkers can revolutionize both the development and use of therapeutics but are contingent on the establishment of a concrete validation process that addresses technology integration and method validation as well as regulatory pathways for efficient biomarker development. Perspective focuses on the general principles of the biomarker validation process with an emphasis on assay validation and the collaborative efforts undertaken by various sectors to promote the standardization of this procedure for efficient biomarker development. It is important to point out that biomarker method validation is distinct from pharmacokinetic validation and routine laboratory validation. The FDA has issued guidance for industry [5] on bioanalytical method validation for assays that support pharmacokinetic studies that are specific for chromatographic and ligand-binding assays, and that are not directly related to the qualification or validation of biomarker assays. Whereas routine laboratory validation refers to laboratories that do testing on human specimens for diagnosis, prevention, or treatment of any disease and falls under the jurisdiction of the Clinical Laboratory Improvement Amendments of 1988, there is little regulatory guidance on biomarker assay validation. Hence, a “fit-for-purpose” approach for biomarker method development and validation is derived with the idea that assay qualification or validation should be tailored to meet the intended purpose of the biomarker study. Numerous applications using bioanalytical techniques have generated enormous interests and some case reveal ultimate solutions in drug efficacies and other indications that are critical to the success in drug/product development and approval processes.

1.1.3 Good Laboratory Practices Versus Bioanalytical Labs/Bioanalysis

Recently, more and more debate and discussion around the connection between GLP and Bioanalysis are surfaced. It is noted that there is no direct reference from GLP regulations to bioanalysis. However, it has become common terminology and acceptance when people refer to GLP–Bioanalysis. In a regulatory term, it may be referred as regulated bioanalysis to support programs or studies under GLP compliance. There is a misconception in some quarters that GLP is required for the conduct of clinical studies. This is not correct. The introduction to the OECD Principles of GLP (and the introduction to the USFDA GLPs in 21 CFR part 58) makes clear that they apply only to the portions of nonclinical (preclinical) studies. The relevant documents for clinical studies are the various codes of GC(R)P (e.g., ICH; TGA). The USFDA and other registration authorities do require a demonstration of the quality of test data from clinical studies. In the United States, this may well be by means of conformance with Clinical Laboratories Improvement Act (CLIA) [6]. In Australia, this is best demonstrated by the testing laboratory's NATA accreditation (in Medical Testing, Chemical Testing, etc.)[1]. Nevertheless, bioanalytical laboratories generate data in support of clinical studies and ultimately as part of data submissions to regulatory agencies. More detailed discussions on above techniques and guidelines are available in respective chapters of this book. The regulatory environment in which clinical trials are conducted continues to evolve. The changes are generally focused on requiring more rigorous control within the organizations performing clinical trials in order to ensure patient safety and the reliability of data produced. The global acceptance of the ICH Guideline for GCP and the implementation of the European Union Clinical Trials Directive (2001/20/EC) are two clear examples of such change. For some years, it has been internationally recognized that clinical laboratories processing specimens from clinical trials require an appropriate set of standards to guide good practices. With that aim in mind, the *Good Clinical Laboratory Practice Guidelines* [7] were drafted and published in 2003 by a working party of the Clinical Committee of the British Association of Research Quality Assurance (BARQA). This guidance identifies systems required and procedures to be followed within an organization conducting analysis of samples from clinical trials in compliance with the requirements of GCP. It thus provides sponsors, laboratory management, project managers, clinical research associates (CRAs), and quality assurance personnel with the framework for a quality system in analysis of clinical trial samples, ensuring GCP compliance overall of processes and results.

[1] The National Association of Testing Authorities (NATA)—Australia's national laboratory accreditation authority. NATA accreditation recognizes and promotes facilities competent in specific types of testing, measurement, inspection, and calibration.

1.2 OBJECTIVES AND KEY REQUIREMENTS FOR GLP REGULATIONS

The ability to provide timely, accurate, and reliable data is essential to the role of analytical and bioanalytical chemists and is especially true in the discovery, development, and manufacture of pharmaceuticals and life science products. Analytical and bioanalytical data are used to screen potential drug candidates, aid in the development of drug syntheses, support formulation studies, animal PK/Tox, clinical safety and efficacy programs, monitor the stability of bulk pharmaceuticals and formulated products, and test final products for release. The quality of analytical and bioanalytical data is a key factor in the success of a drug or product development program. The process of method development and validation has a direct impact on the quality of these data.

Although a thorough validation cannot rule out some potential problems, the process of method development and validation should address the most common ones. Examples of typical problems that can be minimized or avoided are synthesis impurities that coelute with the analyte peak in an HPLC assay; a particular type of column that no longer produces the separation needed because the supplier of the column has changed the manufacturing process; an assay method that is transferred to a second laboratory where they are unable to achieve the same detection limit; and a quality assurance audit of a validation report that finds no documentation on how the method was performed during the validation.

Problems increase as additional people, laboratories, and equipment are used to perform the method. When the method is used in the developer's laboratory, a small adjustment can usually be made to make the method work, but the flexibility to change it is lost once the method is transferred to other laboratories or used for official product testing. This is especially true in the pharmaceutical and life science industries, where methods are submitted to regulatory agencies and changes may require formal approval before they can be implemented for official testing/intended use. The best way to minimize method problems is to perform adequate validation experiments during development and establishment. Analysis of chemicals/drugs in the complex environments/matrices in which they occur are carried out by a vast range of institutions for a variety of purposes, from pharmaceutical and agrochemical companies to hospital biochemistry labs and industry laboratories, from environmental monitoring to safety and toxicity testing of new drugs/products. The range of compounds for analysis is enormous, from naturally occurring compounds such as vitamins to man-made chemicals from the pharmaceutical and agrochemical industries. The following chapters offer an integrated, readable reference text describing the full range of analytical techniques and regulatory requirements available for such small molecules (mostly) and large molecules in an up-to-date manner and should be useful and appeal to all involved in the rapidly growing field of bioanalytical sciences.

- Responsibilities should be defined for the sponsor management, for the study management, and for the quality assurance unit.

- All routine work should follow written standard operating procedures (SOPs).
- Facilities such as laboratories should be large enough and have the right construction/facility to ensure the integrity of a study, for example, to avoid cross contamination during implementation and processes.
- Test and control articles should have the right quality and instruments should be calibrated and well maintained.
- People should be trained or otherwise qualified for the job.
- Raw data and other data should be acquired, processed, and archived to ensure integrity of data.

The main objective is clearly stated within GLP regulations and guidelines—embodies a set of principles that provides a framework for *A quality system concerned with the organizational process and the conditions under which* laboratory studies are planned, performed, monitored, recorded, reported, and archived. These studies are undertaken to generate data by which the *hazards and risks* to users, consumers, and third parties, including the environment, can be assessed for pharmaceuticals, agrochemicals, veterinary medicines, industrial chemicals, cosmetics, food and feed additives, and biocides. GLP helps assure regulatory authorities that the data submitted are a true reflection of the results obtained during the study and can therefore *be relied upon when making risk/safety assessments*. GLP regulations were established by the regulatory bodies to ensure that research submitted to them is not only properly executed but also documented thoroughly enough so that any scientist/organization skilled or qualified can follow the documentation and replicate the results. The level of detail required to achieve this level of documentation is substantial to ensure the integrity, quality, and accuracy of data for product approval.

Unfortunately most laboratories are in situations where they have had to interpret the regulations. Procedures have been developed on an ad hoc basis, in isolation, in response to inspections by both their company's Quality Assurance Unit (QAU) and regulatory bodies. Under such duress, many scientists in industry have developed procedures to validate their instrumentation even though the same approach will already have been applied at the instrument manufacturer's sites. SOPs written to accompany such validation efforts often duplicate extracts from operation manuals—why don't the manufacturers provide the SOPs directly? When it comes to validating the instrument's application software, the person responsible has to take the manufacturer's word for it that the software has been validated and hope that supporting documents, such as test results and source code are available to regulatory agencies upon request, as part of basic requirements for GLP quality system and implementation.

- To assign responsibility for sponsor management, study management, and quality assurance.
- Standard Operating Procedures must be followed.
- Calibration and maintenance of instruments.
- Right construction of laboratories to maintain integrity of the study.

- Raw data should be processed and achieved.
- Employee should be well qualified and trained as per the job assigned.

1.3 FUNDAMENTAL UNDERSTANDING OF GLP REGULATIONS AND PRINCIPLES

Scientific measurements (whether they pertain to monitoring contaminants and active ingredients in pharmaceutical products, clinical determinations of diversified functional elements, characterization of forensic evidence, or testing materials for intermediates and/or final products) are generally recognized as affecting decisions literally concerned with life and death issues. As personal acknowledgement of their responsibility, scientists have traditionally adopted sound laboratory practices directed at assuring the quality of their data. However, until recently these practices were not consistently adopted, enforced, or audited. Because of some notorious historic examples where erroneous data have lead to tragic consequences, national and international agencies have developed guidelines directed at various industries (food, agriculture, pharmaceutical, clinical, environmental, etc.), which fall in the general category of GLPs.

Good laboratory practice regulations became part of the regulatory landscape in the latter part of the 1970s in response to malpractice in R&D activities of pharmaceutical companies and contract facilities used by them. The malpractice included some cases of fraud, but by far the most important aspect of poor practice was the lack of proper management and organization of studies used to complete regulatory dossiers. The investigations of the US Food and Drug Administration (FDA) in the toxicology laboratories in the United States demonstrated a lack of organization and poor management which, it was decided, could only be dealt with by imposing regulations. These regulations are the GLP regulations. First the US FDA, then the US Environmental Protection Agency (EPA), instituted GLP regulations, and eventually many nations of the world followed suit. In 1981, the OECD also published GLP Principles and these have now dominated the international scene—so far 30 countries (the member states of the OECD) have signed agreements that make the OECD GLP Principles binding on them. This effectively makes the OECD Principles an international text. The intent of GLP was to regulate the practices of scientists working on the safety testing of prospective drugs. With the obvious potential impact on consumers and patients recruited for clinical trials, the safety of drugs became a key issue and GLP was seen as a means of ensuring that scientists did not invent or manipulate safety data and a means of ensuring that GLP compliant studies are properly managed and conducted. Hence GLP became the champion of the consumer, the regulatory safeguard, and the guarantee that the safety data were being honestly reported to the registration or receiving authorities as the basis of a decision whether or not to allow a new drug onto the market. GLP was imposed on the industry by regulatory authorities, in the same way as GMP had been before, and GCP was to be afterwards.

Within the United States, federal agencies such as FDA and EPA have produced documents defining laboratory operational requirements, which must be met so that technical data from laboratory studies may be acceptable by those agencies for any legal or contractual purposes. Laboratories doing business with and/or for these agencies must therefore comply with the specified GLP regulations. Not only the issue of GLP is obviously so crucial to modern laboratory operations, but also most importantly because good laboratory practice is an essential ingredient for any professional scientist, this chapter will incorporate many of the principles that are part of GLP in contemporary laboratories. A brief summary of GLP principles is described and presented below.

1.3.1 Elements of Good Laboratory Practices

In general, basic elements of GLP may be defined as follows (but not limited to):

- Qualification of test facility management and personnel
- Standard operating procedures
- Quality assurance program
- Qualification of facilities (e.g., bioanalytical/analytical testing facilities)
- Qualification and validation of apparatus (equipment, computers, or computerized systems), materials, and reagents
- Test systems, and test and reference items
- Performance of the study and reporting of study results
- Storage and retention of records and materials
- Documentation and maintenance of records

1.3.1.1 Qualification of Test Facility Management and Personnel The test facility (TF) management means the person(s) who has (have) the authority and formal responsibility for the organization and functioning of the TF according to the GLP regulations. This requires the identification of management and the need of a job description, qualification background, and training records (CVs or resumes). The organization has to describe in an ad hoc document the way the TF is structured. The TF management must ensure the availability of a master schedule, appropriate facilities, equipment, and materials for the timely and proper conduct of the study. A statement has to be in place that identifies the individual(s) within the TF by whom the responsibilities of management are fulfilled.

1.3.1.2 Standard Operating Procedures SOPs provide standard working tools that can be used to document routine quality system management and technical activities. The development and use of SOPs are an integral part of a successful quality system as it provides individuals with the information to perform a job or complete a project properly, and facilitates consistency in the quality and integrity of a product or end-result. The term "SOP" may not always be appropriate and terms such as protocols, instructions, worksheets, and laboratory operating procedures may also be used. SOPs detail the regularly recurring work processes that are to be conducted or followed within an organization. They document the way activities are to be performed to facilitate consistent conformance to technical and quality system requirements and to support data quality. They may describe, for example, fundamental programmatic actions and technical actions such as analytical processes, and processes for maintaining, calibrating, and using equipment. SOPs are intended to be specific to the organization or facility whose activities are described and assist that organization to maintain their quality control and quality assurance processes and ensure compliance with governmental regulations.

1.3.1.3 Quality Assurance The primary products of any laboratory concerned with qualitative and quantitative analysis are the analytical data reported for specimens examined by that laboratory. QA for such a laboratory includes all of the activities associated with insuring that chemical and physical measurements are made properly, interpreted correctly, and reported with appropriate estimates of error and confidence levels. QA activities also include those maintaining appropriate records of specimen/sample origins and history (sample tracking), as well as procedures, raw data, and results associated with each specimen/sample. The various elements of Quality Assurance are itemized here: (1) SOPs; (2) instrumentation validation; (3) reagent/materials certification; (4) analyst qualification/certification; (5) lab facilities qualification/certification; and (6) specimen/sample tracking.

Many volumes could be written regarding each of the QA elements itemized above. However, a brief discussion is presented here. SOPs are what the name implies ... procedures which have been tested and approved for conducting a particular determination. Often, these procedures will have been evaluated and published by the regulatory agency involved (e.g., EPA or FDA); these agencies may not accept analytical data obtained by other procedures for particular analytes. Within the context of laboratory work, the experimental procedures provided in Laboratory Manual correspond to the SOPs. Within any commercial laboratory, SOPs should be either available or developed to acceptable standards, so that any analytical data collected and reported can be tied according to a documented procedure. Presumably, this implies that a given determination can be repeated at any later time, for an identical specimen, using the SOP specified.

1.3.1.4 Qualification/Certification of Laboratory Facilities These are normally done by some external agency. For example, an analytical and bioanalytical laboratory might be audited by representatives of a federal agency with which they have a contract. An independent laboratory might file documentation with a responsible state or federal agency. The evaluation is concerned with such issues as space (amount, quality, and relevance), ventilation, equipment, storage, and hygiene. Routine chemistry laboratories are generally evaluated by the American Chemical Society, as part of the process of granting approval for the overall chemistry program presented by the college or university. This latter approval process is not as detailed regarding analytical facilities as the certification processes pursued by agencies, concerned specifically with quality assurance.

1.3.1.5 Instrumentation/Apparatus Qualification and Validation It is a process inherently necessary for any analytical and bioanalytical laboratory. Data produced by "faulty" instruments may give the appearance of valid data. These events are particularly difficult to detect with modern computer-controlled systems, which remove the analyst from the data collection/instrument control functions. Thus, it is essential that some objective procedures be implemented for continuously assessing the validity of instrumental data. These procedures, when executed on a regular basis, will establish the continuing acceptable operation of laboratory instruments within prescribed specifications.

1.3.1.6 Reagent/Materials Certification It is an obvious element of quality assurance. However, GLP guidelines emphasize that certification must follow accepted procedures, and must be adequately documented. Moreover, some guidelines will specify that each container for laboratory reagents/materials must be labeled with information related to its certification value, date, and expiration time. This policy is meant to assure that reagents used are as specified in the SOPs.

1.3.1.7 Qualification/Certification of Analysts (Quality Personnel) This is a required part of QA. Some acceptable proof of satisfactory training and/or competence with specific laboratory procedures must be established for each analyst. Because the American Chemical Society does not currently have a policy regarding "certification" of chemists or analysts, the requirements for "certification" vary, and are usually prescribed by the laboratory in question. These standards would have to be accepted by any agency or client obtaining results from that laboratory. For routine laboratory, the requirement for certification as an analyst is satisfactory completion of the predefined assignments (specified by relevant SOPs). Execution of these basic procedures will be repeated, if necessary, until satisfactory results are obtained (evaluated based on analytical accuracy and precision).

1.3.1.8 Specimen/Sample Tracking This is an aspect of quality assurance that has received a great deal of attention with the advent of computer-based Laboratory Information Management Systems (LIMSs). However, whether done by hand with paper files, or by computer with modern bar-coding techniques, sample tracking is a crucial part of quality assurance. The terms "specimen" and "sample" are often used interchangeably. However, "specimen" usually refers to an item to be characterized chemically; whereas "sample" usually refers to a finite portion of the specimen, which is taken for analysis. When the specimen is homogeneous (such as a stable solution), the sample represents the overall composition of the specimen. However, if specimens are heterogeneous (e.g., metal alloys, rock, soil, textiles, foods, polymer composites, vitamin capsules, etc.), then the samples may not represent the overall compositions. Maintaining the distinction in records of analytical results can be crucial to the interpretation of data.

Procedures for assuring adequate specimen/sample tracking will vary among laboratories. The bottom line, however, is that these procedures must maintain the unmistakable connection between a set of analytical data and the specimen and/or samples from which they were obtained. In addition, the original source of the specimen/sample(s) must be recorded and likewise unmistakably connected with the set of analytical data. Finally, in many cases the "chain-of-custody" must be specified and validated. This is particularly true for forensic samples (related to criminal prosecution), but can also be essential for many other situations as well. For example, a pharmaceutical company developing a new product may be called upon at some time to defend their interpretation of clinical trial tests. Such defense may require the company to establish that specimens collected during these trials could not have been deliberately tampered. That is, they may have to establish an unbroken chain-of-custody, which would remove all doubt regarding the integrity of specimens submitted to sample analysis.

1.3.1.9 Performance of the Study and Reporting of Study Results A critical process to the success and quality of a study and data generated within a study. For each study, a written protocol/plan should exist prior to the initiation of the study. Acceptance criteria should be defined and followed. Deviations (e.g., amendments) from the study plan and criteria should be justified, described, acknowledged, approved (if necessary/applicable), and documented during the process/execution. Upon completion of a study, results should be reported in a timely manner. Study reports should be prepared in a format that is compliant with GLP requirements (including GLP compliance statement where applicable) and signed/approved by Study Director, which is a formal record to confirm that the study was/is conducted and reported in accordance with GLP, clearly identifying, as appropriate, where the study deviated from GLP. The GLP compliance statement should not be confused with the QA statement, also presented in the study report, which is a distinct and separate record of QA study monitoring.

1.3.1.10 Documentation and Maintenance of Records A central feature of GLP guidelines is documentation along with the maintenance of records of specimen/sample origins, chain-of-custody, raw analytical data, processed analytical data, SOPs, instrument validation results, reagent certification results, analyst certification documents, etc. Maintenance of instrument and reagent certification records provides for postevaluation of results, even after the passage of several years. Maintenance of all records specified provides documentation, which may be required in the event of legal challenges due to repercussions of decisions based on the original analytical results.

So important is this record-keeping feature of GLP that many vendors are now providing many of these capabilities as part of computer packages for operating modern instruments. For example, many modern computer-based instruments will provide for the indefinite storage of raw analytical data for specific samples in a protected (tamper proof) environment. They also provide for maintenance of historical records of control chart data establishing the operational quality of instruments in any period during which analytical data have been acquired by that instrument.

The length of time over which laboratory records should be maintained will vary with the situation. However, the general guidelines followed in regulated laboratories are to maintain records for at least 5 years. In practice, these records are being maintained much longer. The development of higher density storage devices for digitized data is making this kind of record-keeping possible. The increasing frequency of litigation regarding chemistry-related commercial products is making this kind of record-keeping essential. Moreover, establishing the integrity of the stored data is becoming a high-level security issue for companies concerned about future litigation.

1.3.1.11 Accountability GLP procedures inherently establish accountability for laboratory results. Analysts, instruments, reagents, and analytical methods cannot (and should not) maintain the anonymity that might be associated with a lack of GLP policy. Responsibility for all aspects of the laboratory processes leading to technical results and conclusions is clearly defined and documented. This situation should place appropriate pressure on analysts to conduct studies with adequate care and concern. Moreover, it allows the possibility of identifying more quickly and succinctly the source(s) of error(s) and taking corrective action to maintain acceptable quality of laboratory data.

The OCED GLP Principles simply state that the fundamental points of GLP help the research to perform his/her work in compliance with reestablished plan and standardized procedures worldwide. The regulations/principles do not concern the scientific or technical content of the research programs. All GLP texts, whatever their origins or the industry targeted, stress the importance of the following points:

(1) *Resources*: Organization, personnel, facilities, and equipment.
(2) *Rules*: Protocols and written procedures.
(3) *Characterization*: Test items and test systems.
(4) *Documentation*: Raw data, final report, and archives.
(5) Quality assurance unit.

The training program of the WHO takes each of these five fundamental points in turn and explains the rules of GLP in each case. The major points are summarized here.

(1) *Resources (Organization and Personnel):* GLP regulations require that the structure of the research organization and the responsibilities of the research personnel be clearly defined. GLP also stresses that staffing levels must be sufficient to perform the tasks required. The qualifications and the training of staff must also be defined and documented. Facilities and Equipment— The regulations emphasize the need for sufficient facilities and equipment in order to perform the studies. All equipment must be in working order. A strict program of qualification, calibration, and maintenance attains this.
(2) *Rules (Protocols and Written Procedures):* The main steps of research studies are described in the study plan or protocol. However, the protocol

does not contain all the technical details necessary to exactly repeat the study. Since being able to repeat studies and obtain similar results is a *sine qua non* of mutual acceptance of data (and, indeed, a central tenet in the scientific method), the routine procedures are described in written SOPs. Laboratories may also need to standardize certain techniques to facilitate comparison of results; here again written SOPs are an invaluable tool.

(3) *Characterization:* In order to perform a study correctly, it is essential to know as much as possible about the materials used during the study. For studies to evaluate the properties of pharmaceutical compounds during the preclinical phase, it is a prerequisite to have details about the test item and about the test system (often an animal or plant) to which it is administered.

(4) *Documentation (Raw Data):* All studies generate raw data. These are the fruits of research and represent the basis for establishing results and arriving at conclusions. The raw data must also reflect the procedures and conditions of the study. Final Report—The study report, just like all other aspects of the study, is the responsibility of the study director. He/she must ensure that the contents of the report describe the study accurately. The study director is also responsible for the scientific interpretation of the results. Archives—Storage of records must ensure safekeeping for many years, coupled with logical and prompt retrieval.

(5) *Quality Assurance:* QA as defined by GLP is a team of persons charged with assuring management that GLP compliance has been attained within the laboratory. They are organized independently of the operational and study program, and function as witnesses to the whole preclinical research process.

1.4 KEY ELEMENTS OF BIOANALYTICAL METHODS VALIDATION

It is apparent that the quality of bioanalytical data is critical to supporting regulatory filing and approval process. BMV employed for the quantitative determination of drugs and their metabolites in biological fluids plays a significant role in the evaluation and interpretation of BA, bioequivalence (BE), PK, and toxicokinetic (TK) study data. The quality of these studies is directly related to the quality and integrity of the underlying bioanalytical data. It is therefore important that guiding principles for the validation of these analytical methods be established and disseminated to the pharmaceutical and life sciences communities.

FDA Bioanalytical Method Validation Guidance for Industry (May 2001) [5] provides assistance to sponsors of investigational new drug (INDs) applications, new drug applications (NDAs), abbreviated new drug applications (ANDAs), and supplements in developing bioanalytical method validation information used in human clinical pharmacology, BA, and BE studies requiring PK evaluation. The guidance also applies to bioanalytical methods used for nonhuman pharmacology/toxicology studies and nonclinical studies. For studies related to the veterinary drug approval process, the guidance applies only to blood and urine in BA, BE, and PK studies.

The information in the guidance generally applies to bioanalytical procedures such as gas chromatography (GC), high-performance liquid chromatography (HPLC), combined GC and LC mass spectrometric (MS) procedures such as LC–MS, LC–MS/MS, GC–MS, and GC–MS/MS performed for the quantitative determination of drugs and/or metabolites in biological matrices such as blood, serum, plasma, or urine. The guidance also applies to other bioanalytical methods, such as immunological and microbiological procedures, and to other biological matrices, such as tissue, skin, feces, and other samples/specimens. The guidance provides general recommendations for bioanalytical method validation. The recommendations may be adjusted or modified depending on the specific type of analytical method for intended use. The guidance should be an excellent reference to other similar method validation in life science industries.

Selective and sensitive analytical methods for the quantitative evaluation of drugs and their metabolites (analytes of interest) are critical for the successful conduct of nonclinical and/or biopharmaceutics and clinical pharmacology studies. Bioanalytical method validation includes all of the procedures that demonstrate that a particular method used for quantitative measurement of analytes in a given biological matrix, such as blood, plasma, serum, or urine, is reliable and reproducible for the intended purpose, scope, and use. The fundamental parameters for validation include: (1) accuracy, (2) precision, (3) selectivity, (4) sensitivity, (5) reproducibility, (6) stability, and (7) ruggedness and robustness (not clearly addressed within the Guidance). Validation involves documenting, through the use of specific laboratory investigations, that the performance characteristics of the method are suitable and reliable for the intended analytical applications. The acceptability of analytical data corresponds directly to the criteria used to validate the method.

Published methods of analysis are often modified to suit or meet the requirements of the laboratory performing the assay. These modifications should be validated to ensure suitable performance of the analytical method. When changes are made to a previously validated method, the analyst should exercise judgment as to how much additional validation is needed. During the course of a typical drug development program, a defined bioanalytical method undergoes many modifications. The evolutionary changes to support specific studies and different levels of validation demonstrate the validity of an assay's performance. Different types and levels of validation are defined and characterized as follows:

Full Validation: It is important when developing and implementing a bioanalytical method for the first time or is considered to be method official establishment? Full validation is important for a new drug entity. A full validation of the revised assay is important if metabolites are added to an existing assay for quantification.

Partial Validations: These are modifications of already validated bioanalytical methods. Partial validation can range from as little as one intra-assay accuracy and precision determination to a nearly full validation. Typical bioanalytical method changes that fall into this category include, but are not limited to

- bioanalytical method transfers between laboratories or analysts;
- change in analytical methodology (e.g., change in detection systems);
- change in anticoagulant in harvesting biological fluid;
- change in matrix within species (e.g., human plasma to human urine);
- change in sample processing procedures;
- change in species within matrix (e.g., rat plasma to mouse plasma);
- change in relevant concentration range;
- changes in instruments and/or software platforms;
- limited sample volume (e.g., pediatric study);
- rare matrices;
- selectivity demonstration of an analyte in the presence of concomitant medications; and
- selectivity demonstration of an analyte in the presence of specific metabolites.

Typical recommendation of Method Partial and Full Validation is given in Table 1.2.

Cross-Validation: It is a comparison of validation parameters when two or more bioanalytical methods are used to generate data within the same study or across different studies. An example of cross-validation would be a situation where an original validated bioanalytical method serves as the *reference* and the revised bioanalytical method is the *comparator*. The comparisons should be done both ways.

TABLE 1.2 Typical Recommendation of Method Partial and Full Validation

Items	Full Validation	Partial Validation
Different matrices	+	
Extend dynamic range		+
Add metabolite(s)	+	
Reduce matrix volume		+
Check for comed		+
Different anticoagulant		+
LLOQ		+
Change analysts		+
Change instruments		+
Change extraction (mechanism)	+	
Change detection system		+
Change chromatography		+

Please note that some of the partial validation may be up to a full validation (e.g., anticoagulant with different types and chromatographic conditions).

When sample analyses within a single study are conducted at more than one site or more than one laboratory, cross-validation with spiked matrix standards and subject samples should be conducted at each site or laboratory to establish interlaboratory reliability. Cross-validation should also be considered when data generated using different analytical techniques (e.g., LC–MS/MS versus LBAs) in different studies are included in a regulatory submission.

All modifications should be assessed to determine the recommended degree of validation. The analytical laboratory conducting pharmacology/toxicology and other nonclinical studies for regulatory submissions should adhere to FDA's GLPs (21 CFR part 58) and to sound principles of quality assurance throughout the testing process. The bioanalytical method for human BA, BE, PK, and drug interaction studies must meet the criteria in 21 CFR 320.29. The analytical laboratory should have a written set of SOPs to ensure a complete system of quality control and assurance. The SOPs should cover all aspects of analysis from the time the sample is collected and reaches the laboratory until the results of the analysis are generated and reported. The SOPs also should include record keeping, security and chain of sample custody (accountability systems that ensure integrity of test articles), sample preparation, and analytical tools such as methods, reagents, equipment, instrumentation, and procedures for quality control and verification of results. Here are typical recommendations of what need to be performed as for a full validation or a partial validation.

The process by which a specific bioanalytical method is developed, validated, and used in routine sample analysis can be divided into (1) reference standard preparation, (2) bioanalytical method development and establishment of assay procedure, and (3) application of validated bioanalytical method to routine sample analysis and acceptance criteria for the analytical run and/or batch. These three processes are described below.

1.4.1 Reference Standards

Analysis of drugs and their metabolites in a biological matrix is carried out using samples spiked with calibration (reference) standards and using QC samples. The purity of the reference standard used to prepare spiked samples can affect study data. For this reason, an authenticated analytical reference standard of known identity and purity should be used to prepare solutions of known concentrations. If possible, the reference standard should be identical to the analyte. When this is not possible, an established chemical form (free base or acid, salt, or ester) of known purity can be used. Three types of reference standards are usually used (1) certified reference standards (e.g., USP compendial standards); (2) commercially supplied reference standards obtained from a reputable commercial source; and/or (3) other materials of documented purity custom synthesized by an analytical laboratory or other noncommercial establishment. The source and lot number, expiration date, certificates of analyses when available, and/or internally or externally generated evidence of identity and purity should be furnished for each reference standard.

1.4.2 Method Development—Chemical/Chromatographic Assay

The method development and establishment phase defines the chemical assay. The fundamental parameters for a bioanalytical method validation are accuracy, precision, selectivity, sensitivity, reproducibility, and stability. Measurements for each analyte in the biological matrix should be validated. In addition, the stability of the analyte in spiked samples should be determined. Typical method development and establishment for a bioanalytical method include determination of (1) selectivity; (2) accuracy, precision, recovery; (3) calibration curve; and (4) stability of analyte in spiked samples.

Selectivity/specificity is the ability of an analytical method to differentiate and quantify the analyte in the presence of other components in the sample. For selectivity, analyses of blank samples of the appropriate biological matrix (plasma, urine, or other matrix) should be obtained from at least six sources. Each blank sample should be tested for interference, and selectivity should be ensured at the lower limit of quantification (LLOQ).

Potential interfering substances in a biological matrix include endogenous matrix components, metabolites, decomposition products, and in the actual study, concomitant medication and other exogenous xenobiotics. If the method is intended to quantify more than one analyte, each analyte should be tested to ensure that there is no interference.

Accuracy of an analytical method describes the closeness of mean test results obtained by the method to the true value (concentration) of the analyte. Accuracy is determined by replicate analysis of samples containing known amounts of the analyte. Accuracy should be measured using a minimum of five determinations per concentration. A minimum of three concentrations in the range of expected concentrations is recommended. The mean value should be within 15% of the actual value except at LLOQ, where it should not deviate by more than 20%. The deviation of the mean from the true value serves as the measure of accuracy.

Precision of an analytical method describes the closeness of individual measures of an analyte when the procedure is applied repeatedly to multiple aliquots of a single homogeneous volume of biological matrix. Precision should be measured using a minimum of five determinations per concentration. A minimum of three concentrations in the range of expected concentrations is recommended. The precision determined at each concentration level should not exceed 15% of the coefficient of variation (CV) except for the LLOQ, where it should not exceed 20% of the CV. Precision is further subdivided into within-run, intrabatch precision, or repeatability, which assesses precision during a single analytical run, and between-run, interbatch precision, or repeatability, which measures precision with time, and may involve different analysts, equipment, reagents, and laboratories.

Recovery of an analyte in an assay is the detector response obtained from an amount of the analyte added to and extracted from the biological matrix, compared to the detector response obtained for the true concentration of the pure authentic standard. Recovery pertains to the extraction efficiency of an analytical method within the limits of variability. Recovery of the analyte need not be 100%, but the extent of recovery of an analyte and of the internal standard should be consistent, precise, and reproducible. Recovery experiments should be performed by comparing the analytical results for extracted samples at three concentrations (low, medium, and high) with unextracted standards that represent 100% recovery.

1.4.3 Calibration/Standard Curve

A calibration (standard) curve is the relationship between instrument response and known concentrations of the analyte. A calibration curve should be generated for each analyte in the sample. A sufficient number of standards should be used to adequately define the relationship between concentration and response. A calibration curve should be prepared in the same biological matrix as the samples in the intended study by spiking the matrix with known concentrations of the analyte. The number of standards used in constructing a calibration curve will be a function of the anticipated range of analytical values and the nature of the analyte–response relationship. Concentrations of standards should be chosen on the basis of the concentration range expected in a particular study. A calibration curve should consist of a blank sample (matrix sample processed without internal standard), a zero sample (matrix sample processed with internal standard), and six to eight nonzero samples covering the expected range, including LLOQ.

(1) *Lower Limit of Quantification:* The lowest standard on the calibration curve should be accepted as the limit of quantification if the following conditions are met: The analyte response at the LLOQ should be at least five times the response compared to blank response. Analyte peak (response) should be identifiable, discrete, and reproducible with a precision of 20% and accuracy of 80–120%.

(2) *Calibration Curve/Standard Curve/Concentration–Response:* The simplest model that adequately describes the concentration–response relationship should be used. Selection of weighting and use of a complex regression equation should be justified. The following conditions should be met in developing a calibration curve: 20% deviation of the LLOQ from nominal concentration; 15% deviation of standards other than LLOQ from nominal concentration. At least four out of six nonzero standards should meet the above criteria, including the LLOQ and the calibration standard at the highest concentration. Excluding the standards should not change the model used.

1.4.4 Stability

Drug stability in a biological fluid is a function of the storage conditions, the chemical properties of the drug, the matrix, and the container system. The stability of an analyte in a particular matrix and container system is relevant only to that matrix and container system and should not be extrapolated to other matrices and container systems. Stability procedures should evaluate the stability of the analytes during sample collection and handling, after long-term (frozen at the intended storage temperature) and short-term (bench top, room temperature) storage, and after going through freeze and thaw cycles and the analytical process. Conditions used in stability experiments should reflect situations likely to be encountered during actual sample handling and analysis. The procedure should also include an evaluation of analyte stability in stock solution. All stability determinations should use a set of samples prepared from a freshly made stock solution of the analyte in the appropriate analyte-free, interference-free biological matrix. Stock solutions of the analyte for stability evaluation should be prepared in an appropriate solvent at known concentrations.

Freeze and Thaw Stability: Analyte stability should be determined after three freeze and thaw cycles. At least three aliquots at each of the low and high concentrations should be stored at the intended storage temperature for 24 h and thawed unassisted at room temperature. When completely thawed, the samples should be refrozen for 12–24 h under the same conditions. The freeze–thaw cycle should be repeated two more times, and then analyzed on the third cycle. If an analyte is unstable at the intended storage temperature, the stability sample should be frozen at -70°C during the three freeze and thaw cycles.

Short-Term Temperature Stability: Three aliquots of each of the low and high concentrations should be thawed at room temperature and kept at this temperature from 4 to 24 h (based on the expected duration that samples will be maintained at room temperature in the intended study) and analyzed.

Long-Term Stability: The storage time in a long-term stability evaluation should exceed the time between the date of first sample collection and the date of last sample analysis. Long-term stability should be determined by storing at least three aliquots of each of the low and high concentrations under the same conditions as the study samples. The volume of samples should be sufficient for analysis on three or more separate occasions. The concentrations of all the stability samples should be compared to the mean of back-calculated values for the standards at the appropriate concentrations from the first day of long-term stability testing.

Stock Solution Stability: The stability of stock solutions of drug and the internal standard should be evaluated at room temperature for at least 6 h. If the stock solutions are refrigerated or frozen for the relevant period, the stability should be documented. After completion of the desired storage time, the stability should be tested by comparing the instrument response with that of freshly prepared solutions.

Postpreparative Stability: The stability of processed samples, including the resident time in the auto-sampler, should be determined. The stability of the drug and the internal standard should be assessed over the anticipated run time for the batch size in validation samples by determining concentrations on the basis of original calibration standards. Although the traditional approach of comparing analytical results for stored samples with those for freshly prepared samples has been referred to in this guidance, other statistical approaches based on confidence limits for evaluation of an analyte's stability in a biological matrix can be used. SOPs should clearly describe the statistical method and rules used. Additional validation may include investigation of samples from dosed subjects.

1.4.5 Reproducibility

It is part of evaluation for the closeness of the agreement between the results of successive measurements of the same analyte in identical material made by the same method under different conditions, for example, different operators and different laboratories and considerably separated in time. Incurred sample reanalysis (ISR) is highly recommended for this investigation. Results should be expressed in terms of the reproducibility standard deviation, the reproducibility coefficient of variation or the confidence interval of the mean value.

1.4.6 Robustness or Ruggedness

It measures the capacity of a test to remain unaffected by small variations in the procedures. It is measured by deliberately introducing small changes to the method and examining the consequences, which is not mentioned within BMV Guidelines, but it is important to evaluate and establish rugged/robust methods, especially for pivotal clinical trial sample analyses.

1.5 BASIC PRINCIPLES OF BIOANALYTICAL METHOD VALIDATION AND ESTABLISHMENT

The fundamental parameters to ensure the acceptability of the performance of a bioanalytical method validation are accuracy, precision, selectivity, sensitivity, reproducibility, and stability. A specific, detailed description of the bioanalytical method (test method) should be written. This can be in the form of a protocol, study plan, report, and/or SOP. Each step in the method should be investigated to determine the extent to which environmental, matrix, material, or procedural variables can affect the estimation of analyte in the matrix from the time of collection of the material up to and including the time of analysis.

It may be important to consider the variability of the matrix due to the physiological nature of the sample. In the case of LC–MS/MS-based procedures, appropriate steps should be taken to ensure the lack of matrix effects throughout the application of the method, especially if the nature of the matrix changes from the matrix used during method validation. A bioanalytical method should be validated for the intended use or application. All experiments used to make claims or draw conclusions about the validity of the method should be presented in a report (method validation report). Whenever possible, the same biological matrix as the matrix in the intended samples should be used for validation purposes (for tissues of limited availability, such as bone marrow, physiologically appropriate proxy matrices can be substituted). The stability of the analyte (drug and/or metabolite) in the matrix during the collection process and the sample storage period should be assessed, preferably prior to sample analysis. For compounds with potentially labile metabolites, the stability of analyte in matrix from dosed subjects (or species) should be evaluated and established.

The evaluation of accuracy, precision, reproducibility, response function, and selectivity of the method for endogenous substances, metabolites, and known degradation products should be performed and established for the biological matrix. For selectivity, there should be evidence that the substance being quantified is the intended analyte. The concentration range over which the analyte will be determined should be defined in the bioanalytical method, based on evaluation of actual standard samples over the range, including their statistical variation. This defines the *standard curve.* A sufficient number of standards should be used to adequately define the relationship between concentration and response. The relationship between response and concentration should be demonstrated to be continuous and reproducible. The number of standards used should be a function of the dynamic range and nature of the concentration–response relationship. In many cases, six to eight concentrations (excluding blank values) can define the standard curve. More standard concentrations may be recommended for nonlinear than for linear relationships. The ability to dilute samples originally above the upper limit of the standard curve should be demonstrated by accuracy and precision parameters in the validation.

In consideration of high-throughput analyses, including but not limited to multiplexing, multicolumn, and parallel systems, sufficient QC samples should be used to ensure control of the assay. The number of QC samples to ensure proper control of the assay should be determined based on the run/batch size. The placement of QC samples should be judiciously considered in the run/batch sequences. For a bioanalytical method to be considered valid, specific acceptance criteria should be set in advance and achieved for accuracy and precision for the validation of QC samples over the range of the standards.

1.5.1 Specific Recommendations for Method Validation

The matrix-based standard curve should consist of a minimum of six standard points, excluding blanks, using single or replicate samples. The standard curve should cover the entire range of expected concentrations in support of various studies. Please note that different dynamic ranges, possibly including dilution evaluation may be necessary to cover various regulated programs/studies including nonclinical toxicology studies. Standard curve fitting is determined by applying the simplest model that adequately describes the concentration–response relationship using appropriate weighting and statistical tests for *goodness of fit.* LLOQ is the lowest concentration of the standard curve that can be measured with acceptable accuracy and precision. The LLOQ should be established using at least five samples independent of standards and determining the coefficient of variation and/or appropriate confidence interval. The LLOQ should serve as the lowest concentration on the standard curve and should not be confused with the limit of detection and/or the low QC sample. The highest standard will define the upper limit of quantification (ULOQ) of an analytical method.

For validation of the bioanalytical method, accuracy and precision should be determined using a minimum of five determinations per concentration level (excluding blank samples). The mean value should be within 15% of the theoretical value, except at LLOQ, where it should not deviate by more than 20%. The precision around the mean value should not exceed 15% of the CV, except for LLOQ, where it should not exceed 20% of the CV. Other methods of assessing accuracy and precision that meet these limits may be equally acceptable. The accuracy and precision with which known concentrations of analyte in biological matrix can be determined should be demonstrated. This can be accomplished by analysis of replicate sets of analyte samples of known concentrations QC samples from an equivalent biological matrix. At a minimum, three concentrations representing the entire range of the standard curve should be studied: one within 3× the LLOQ (low QC sample), one near the center (middle QC), and one near the upper boundary of the standard curve (high QC—being approximately 80% of ULOQ).

Reported method validation data and the determination of accuracy and precision should include all outliers; however, calculations of accuracy and precision excluding values that are statistically determined as outliers can also be reported. The stability of the analyte in biological matrix at intended storage temperatures should be established. The influence of freeze–thaw cycles (a minimum of three cycles at two concentrations in triplicate) should be studied. The stability of the analyte in matrix at ambient temperature should be evaluated over a time period equal to or even greater than the typical sample preparation, sample handling, and analytical run times. Reinjection reproducibility should be evaluated to determine if an analytical run could be reanalyzed in the case of instrument failure.

The specificity of the assay methodology should be established using a minimum of six independent sources of the same matrix. For hyphenated mass spectrometry-based methods, however, testing six independent matrices for interference may not be absolutely important. In the case of LC–MS and LC–MS/MS-based procedures, matrix effects should be investigated to ensure that precision, selectivity, and sensitivity will not be compromised. Method selectivity should be evaluated during method development and throughout method validation and can continue throughout application of the method to actual study samples. Acceptance/rejection criteria for spiked, matrix-based calibration standards, and validation QC samples should be based on the nominal (theoretical) concentration of analytes. Specific criteria can be set up in advance and achieved for accuracy and precision over the range of the standards, if so desired.

1.5.1.1 Method Development for Microbiological and Ligand-Binding Assays

Many of the bioanalytical validation parameters and principles discussed above are also applicable to microbiological and ligand-binding assays. However, these assays possess some unique characteristics that should be considered during method validation.

1.5.1.2 Selectivity Issues As with chromatographic methods, microbiological and ligand-binding assays should be shown to be selective for the analyte. The following recommendations for dealing with two selectivity issues should be considered:

Interference from Substances Physiochemically Similar to the Analyte: Cross-reactivity of metabolites, concomitant medications, or endogenous compounds should be evaluated individually and in combination with the analyte of interest. When possible, the immunoassay should be compared with a validated reference method (such as LC–MS) using incurred samples and predetermined criteria for agreement of accuracy of immunoassay and reference method. The dilution linearity to the reference standard should be assessed using study (incurred) samples. Selectivity may be improved for some analytes by incorporation of separation steps prior to immunoassay.

Matrix Effects Unrelated to the Analyte: The standard curve in biological fluids should be compared with standard in buffer to detect matrix effects. Parallelism of diluted study samples should be evaluated with diluted standards to detect matrix effects. Nonspecific binding should be determined where applicable.

1.5.1.3 Quantification Issues Microbiological and immunoassay standard curves are inherently nonlinear and, in general, more concentration points may be recommended to define the fit over the standard curve range than for chemical assays. In addition to their nonlinear characteristics, the response–error relationship for immunoassay standard curves is a nonconstant function of the mean response (heteroscadisticity). For these reasons, a minimum of six nonzero calibrator concentrations, run in duplicate, are recommended. The concentration–response relationship is most often fitted to a 4- or 5-parameter logistic model, although others may be used with suitable validation. The use of *anchoring points* in the asymptotic high- and low-concentration ends of the standard curve may improve the overall curve fit. Generally, these anchoring points will be at concentrations that are below the established LLOQ and above the established ULOQ. Whenever possible, calibrators should be prepared in the same matrix as the study samples or in an alternate matrix of equivalent performance. Both ULOQ and LLOQ should be defined by acceptable accuracy, precision, or confidence interval criteria based on the study requirements. For all assays, the key factor is the accuracy of the *reported results.* This accuracy can be improved by the use of replicate samples. In the case where replicate samples should be measured during the validation to improve accuracy, the same procedure should be followed as for unknown samples. The following recommendations apply to quantification issues.

If separation is used prior to assay for study samples but not for standards, it is important to establish recovery and use it in determining results. Possible approaches to assess efficiency and reproducibility of recovery are (1) the use of radiolabeled tracer analyte (quantity too small to affect the assay); (2) the advance establishment of reproducible recovery; and (3) the use of an internal standard that is not recognized by the antibody but can be measured by another technique. Key reagents, such as antibody, tracer, reference standard, and matrix should be characterized appropriately and stored under defined conditions. Assessments of analyte stability should be conducted in true study matrix (e.g., should not use a matrix stripped to remove endogenous interferences). Acceptance criteria: At least 67% (four out of six) of QC samples should be within 15% of their respective nominal value, 33% of the QC samples (not all replicates at the same concentration) may be outside 15% of nominal value. In certain situations, wider acceptance criteria may be justified.

Assay reoptimization or validation may be important when there are changes in key reagents, as follows (1) labeled analyte (tracer); (2) binding should be reoptimized; (3) performance should be verified with standard curve and QCs; and (4) antibody. Key cross-reactivities should be checked and evaluated. Tracer experiments above should be repeated. Matrix–tracer experiments above should be repeated. Method development experiments should include a minimum of six runs conducted over several days, with at least four concentrations (LLOQ, low, medium, and high) analyzed in duplicate in each run.

1.5.1.4 Application of Validated Method to Routine Sample Analysis Assays of all samples of an analyte in a biological matrix should be completed within the time period for which stability data are available. In general, biological samples can be analyzed with a single determination without duplicate or replicate analysis if the assay method has acceptable variability as defined by validation data. This is true for procedures where precision and accuracy variabilities routinely fall within acceptable tolerance limits. For a difficult procedure with a labile analyte where high precision and accuracy specifications may be difficult to achieve, duplicate or even triplicate analyses can be performed for a better estimate of analyte concentration. A typical flowchart of method development and validation is shown in Figure 1.3.

A calibration curve should be generated for each analyte to assay samples in each analytical run and should be used to calculate the concentration of the analyte in the unknown samples in the run. The spiked samples can contain more than one analyte. An analytical run can consist of QC samples, calibration standards, and either (1) all the processed samples to be analyzed as one batch or (2) a batch composed of processed unknown samples of one or more volunteers in a study. The calibration (standard) curve should cover the expected unknown sample concentration range in addition to a calibrator sample at LLOQ. Estimation of concentration in unknown samples by extrapolation of standard curves below LLOQ or above the highest standard is not recommended. Instead, the standard curve should be redefined or samples with higher concentration should be diluted and reassayed. It is preferable to analyze all study samples from a subject in a single run.

Once the analytical method has been validated for intended use, its accuracy and precision should be monitored regularly to ensure that the method continues to perform satisfactorily. To achieve this objective, a number of QC samples prepared separately should be analyzed with processed test samples at intervals based on the total number of samples. The QC samples in duplicate at three concentrations (one near the LLOQ (i.e., $3\times$ LLOQ), one in midrange, and one close to the high end of the

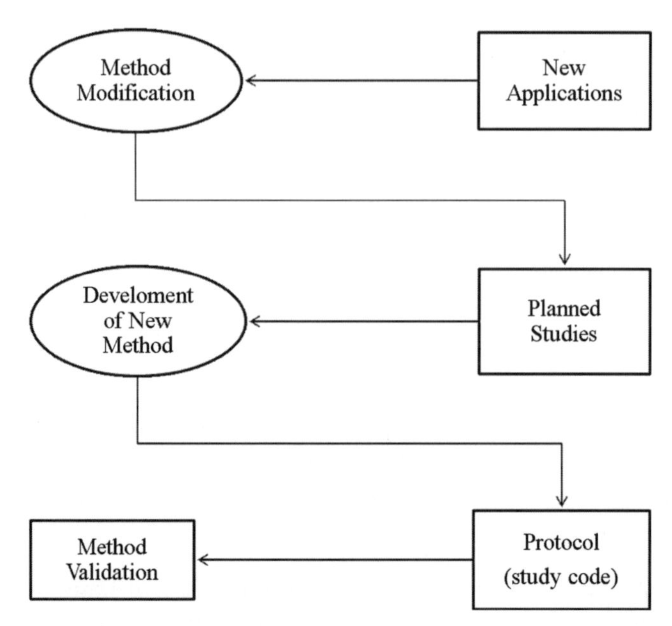

FIGURE 1.3 Typical flowchart of method development and validation.

range) should be incorporated in each assay run. The number of QC samples (in multiples of three) will depend on the total number of samples in the run/batch. The results of the QC samples provide the basis of accepting or rejecting the run. At least four of every six QC samples should be within ±15% of their respective nominal value. Two of the six QC samples may be outside the ±15% of their respective nominal value, but not both at the same concentration.

The following recommendations should be noted in applying a bioanalytical method to routine sample analysis: A matrix-based standard curve should consist of a minimum of six standard points, excluding blanks (either single or replicate), covering the entire range.

Response Function: Typically, the same curve fitting, weighting, and goodness of fit determined during prestudy validation should be used for the standard curve within the study. Response function is determined by appropriate statistical tests based on the actual standard points during each run in the validation. Changes in the response–function relationship between prestudy validation and routine run validation indicate potential problems. The QC samples should be used to accept or reject the run/batch. These QC samples are matrix spiked with analyte.

System Suitability: Based on the analyte and technique, a specific SOP (or sample) should be identified to ensure optimum operation of the system used. Any required sample dilutions should use like/similar matrix (e.g., human to human) obviating the need to incorporate actual within-study dilution matrix QC samples.

Repeat Analysis: It is important to establish an SOP or guideline for repeat analysis and acceptance criteria. This SOP or guideline should explain the reasons for repeating sample analysis. Reasons for repeat analyses could include repeat analysis of clinical or preclinical samples for regulatory purposes, inconsistent replicate analysis, samples outside of the assay range, sample processing errors, equipment failure, poor chromatography, and inconsistent pharmacokinetic data. Reassays should be done in triplicate if sample volume allows. The rationale for the repeat analysis and the reporting of the repeat analysis should be clearly defined or justified and documented.

Sample Data Reintegration: An SOP or guideline for sample data reintegration should be established. This SOP or guideline should explain the reasons for reintegration and how the reintegration is to be performed. The rationale for the reintegration should be clearly described and documented. Reintegration data along with original result(s) should be reported.

1.5.1.5 Documentation The validity of an analytical method should be established and verified by laboratory studies, and documentation of successful completion of such studies should be provided in the assay validation report. General and specific SOPs and good record keeping are an essential part of a validated analytical method. The data generated for bioanalytical method establishment and the QCs should be documented and available for data audit and inspection. Documentation for submission to the Agency should include (1) summary information, (2) method development and establishment, (3) bioanalytical reports of the application of any methods to routine sample analysis, and (4) other information applicable to method development and establishment and/or to routine sample analysis.

1.5.2 Acceptance Criteria for Analytical Run

It is vitally important to establish analytical run (batch) acceptance criteria to ensure the quality of data for regulated studies. An in-house SOP specifying batch pass/fail criteria should be written in accordance with FDA BMV Guidelines. The following acceptance criteria should be considered for accepting an analytical run.

Standards and QC samples can be prepared from the same spiking stock solution, provided the solution stability and accuracy have been verified. A single source of matrix (pooled one) may also be used, provided selectivity has been verified. Standard curve samples, blanks, QCs, and study samples can be arranged as considered appropriate within the run. Placement of standards and QC samples within a run should be designed to detect assay drift over the run. Matrix-based standard calibration samples: 75% or a minimum of six standards, when back-calculated (including ULOQ) should fall within 15%, except for LLOQ, when it should be within 20% of the nominal value. Values falling outside these limits can be discarded, provided they do not change the established model (e.g., regression/weighing model). Acceptance criteria for accuracy and precision should be provided for both the intraday and intrarun experiments.

Quality Control Samples: Quality control samples replicated (at least once) at a minimum of three concentrations (one within 3× of the LLOQ (low QC), one in the midrange (middle QC), and one approaching the high end of the range (high QC) should be incorporated into each run. The results of the QC samples provide the basis of accepting or rejecting the run. At least 67% (four out of six) of the QC samples should be within 15% of their respective nominal (theoretical) values; 33% of the QC samples (not all replicates at the same concentration) can be outside the 15% of the nominal value. A confidence interval approach yielding comparable accuracy and precision is an appropriate alternative. The minimum number of samples (in multiples of three) should be at least 5% of the number of unknown samples or six total QCs, whichever is greater. Samples involving multiple analytes should not be rejected based on the data from one analyte failing the acceptance criteria. The data from rejected runs need not be documented, but the fact that a run was rejected and the reason for failure should be explained and recorded.

1.5.2.1 Reasons for Repeat Analysis

Documentation for Repeat Analyses: Documentation should include the initial and repeat analysis results, the reported result, assay run identification, the reason for the repeat analysis, the requestor of the repeat analysis, and the manager authorizing reanalysis. Repeat analysis of a clinical or preclinical sample should be performed only under a predefined SOP.

Documentation for Reintegrated Data: Documentation should include the initial and repeat integration results, the method used for reintegration, the reported result, assay run identification, the reason for the reintegration, the requestor of the reintegration, and the manager authorizing reintegration. Reintegration of a clinical or preclinical sample should be performed only under a predefined SOP. Deviations from the analysis protocol or SOP, with reasons and justifications should be documented for the deviations.

Summary Information Summary information should include the following: summary tables of validation reports include (but not limited to) analytical method validation, partial revalidation, and cross-validation reports. The table should be in chronological sequence, and include assay method identification code, type of assay, and the reason for the new method or additional validation (e.g., to lower the limit of quantitation). Summary should also include a list, by protocol, of assay methods used and so on. The protocol number, protocol title, assay type, assay method identification code, and bioanalytical report code should be provided. A summary table allowing cross-referencing of multiple identification codes should be provided (e.g., when an assay has different codes for the assay method, validation reports, and bioanalytical reports, especially when the sponsor and a contract laboratory assign different codes).

Documentation for Method Establishment Documentation for method development and establishment should include the following:

(1) An operational description of the analytical method (e.g., lab or test method)
(2) Evidence of purity and identity of drug standards, metabolite standards, and internal standards used in validation experiments
(3) A description of stability studies and supporting data
(4) A description of experiments conducted to determine accuracy, precision, recovery, selectivity, limit of quantification, calibration curve (equations and weighting functions used, if any), and relevant data obtained from these studies
(5) Documentation of intra- and interassay precision and accuracy
(6) In NDA submissions, information about cross-validation study data, if applicable
(7) Legible annotated chromatograms or mass spectrograms, if applicable
(8) Any deviations from SOPs, protocols, or GLPs (if applicable), and justifications for deviations.

Application to Sample Analysis Documentation of the application of validated bioanalytical methods to routine drug/sample analysis should include: (1) evidence of purity and identity of drug standards, metabolite standards, and internal standards used during routine analyses; (2) summary tables should contain information on sample processing and storage; (3) tables should also include sample identification, collection dates, storage prior to shipment, information on shipment batch, and storage prior to analysis; and (4) information should include dates, times, sample condition, and any deviation from protocols.

For summary tables of analytical runs of clinical or preclinical samples, information should include: (1) assay run identification, date and time of analysis, assay method, analysts, start and stop times, duration, significant equipment and material changes, and any potential issues or deviation from the established method; (2) equations used for back calculation of results; (3) tables of calibration curve data used in analyzing samples and calibration curve summary data; (4) summary information on intra- and interassay values of QC samples and data on intra- and interassay accuracy and precision from calibration curves and QC samples used for accepting the analytical run; (5) QC graphs and trend analyses in addition to raw data and summary statistics are encouraged; and (6) data tables from analytical runs of clinical or preclinical samples. Tables should include (1) assay run identification, sample identification, raw data and back-calculated results, integration codes, and/or other reporting codes; and (2) complete serial chromatograms from 5% to 20% of subjects, with standards and QC samples from those analytical runs. For pivotal bioequivalence studies for marketing, chromatograms from 20% of serially selected subjects should be included. In other studies, chromatograms from 5% of randomly selected subjects in each study should be included. Subjects whose chromatograms are to be submitted should be defined prior to the analysis of any clinical samples.

1.5.2.2 Other Information Other information applicable to both method development and establishment and/or to routine sample analysis could include: lists of abbreviations and any additional codes used, including sample condition codes, integration codes, and reporting codes.

(1) Reference lists and legible copies of any references
(2) SOPs or protocols covering the following areas:
   - Calibration standard acceptance or rejection criteria
   - Calibration curve acceptance or rejection criteria
   - Quality control sample and assay run acceptance or rejection criteria

Acceptance criteria should be defined for reported values when all unknown samples are assayed in duplicate. Sample code designations, including clinical or preclinical sample codes and bioassay sample code should be documented. Information such as sample collection, processing, storage, and repeat analyses of samples; reintegration of samples, and so on should also be included.

Bioanalysis and the production of pharmacokinetic, toxicokinetic, and metabolic data play a fundamental role in pharmaceutical life sciences research and development; therefore, the data must be produced to acceptable scientific standards and regulatory (GLP) compliance. For this reason and the need to satisfy regulatory authority requirements, all bioanalytical methods should be properly validated for intended purposes and uses. It is hoped that these validation guidelines not only have taken into account the statistical arguments described in the literature but also have regard to the practicalities of performing bioanalytical method validations for the pharmaceutical industry in this highly competitive era and that they aid further standardization in this field.

It is very important to note that the validation of standard or collaboratively tested methods should not be taken for granted, no matter how impeccable the method's pedigree—the laboratory should satisfy itself that the degree of validation of a particular method is adequate for the required or intended purpose, and that the laboratory is itself able to match any stated performance data. There are two important requirements in this excerpt:

(1) The standard's method validation data are adequate and sufficient to meet the laboratory's method requirements.
(2) The laboratory must be able to match the performance data as described in the standard.

The main objectives of GLP regulations/principles and Bioanalytical Method Validation Guidelines are to help scientists obtain results that are: (1) reliable, (2) repeatable, (3) auditable, and (4) recognized by scientists worldwide. These may also enhance the opportunity of *L*imiting waste of resources, *E*nsuring high quality of data, *A*cquiring comparability of results, and *D*eriving to mutual recognition of scientific findings worldwide, and ultimately *S*ecuring the health and well-being of our societies, as being *LEADS* concept per author's perspectives. The fundamental requirements for GLP are to define conditions under which studies are planned, performed, recorded, monitored, reported, and archived based on study's designs.

REFERENCES

1. Code of Federal Regulation. Food Drug and Cosmetic Act, 21 CFR, Part 58—Good laboratory practice for nonclinical laboratory studies, 1978.
2. International Conference on Harmonization (ICH)/World Health Organization (WHO). Topic E 6 (R1) Guideline for Good Clinical Practice, 1996.
3. Code of Federal Regulation. Food Drug and Cosmetic Act 21 CFR Part 210 and 211 Current Good Manufacturing Practices, 1995–2007.
4. Organization for Economic Co-operation and Development. Principles on Good Laboratory Practice (as revised in 1997) OECD, Paris, 1998 (Series on principles of GLP and compliance monitoring, No. 1, ENV/MC/CHEM(98)17).
5. USFDA Guidance for Industry: *Bioanalytical Method Validation*, 2001.
6. Code of Federal Regulation. Department of Health and Human Services, Centers for Disease Control and Prevention 21 CFR, Part 493—Clinical Laboratory Improvement Act, 2004.
7. World Health Organization (WHO). *Good Clinical Laboratory Practices (GCLP)*, 2009.

2

HISTORIC PERSPECTIVES OF GLP REGULATIONS, APPLICABILITY, AND RELATION TO OTHER REGULATIONS

2.1 HISTORIC PERSPECTIVES OF GLP REGULATIONS

GLP has its roots in a history of mistakes by food producers and drug manufacturers that resulted in harm to the general public. Because of these mistakes, the federal government, under the auspices of the Food and Drug Administration (FDA), stepped in to regulate the food and drug industry. The FDA, the federal agency whose mission is to protect consumers from harmful foods, drugs, and cosmetics, was created with the enactment of the Food and Drug Act of 1906. This law gave the FDA the power to ban the manufacture and interstate sale of misbranded or impure food and drugs. Food and drug alteration, the mixing of harmful substances with food and drugs, was defined under the inaugural act. The powers given to the FDA increased with the Food and Drug Act of 1938, which was passed in response to the 1937 incident when the incorrectly labeled elixir sulfanilamide containing diethylene glycol killed 100 people in 2 months. This new act authorized injunction and seizure, set specific standards, required ingredient listing, established acceptable pesticide residue levels in foods, required drug screening by the FDA, regulated cosmetics and medical devices and increased FDA's enforcement power. In response to the European drug market approval of thalidomide, a sedative that caused severe birth defects, the New Drugs Amendments of 1962 was approved. This act gave the FDA more time to review new drug applications, required more safety and efficacy testing, and authorized records inspections. Between 1962 and 1976, the FDA acquired even more power, including the requirements for the manufacturer to prove drug

Regulated Bioanalytical Laboratories: Technical and Regulatory Aspects from Global Perspectives, By Michael Zhou

effectiveness and to present to the FDA postapproval reporting. Also, the FDA was charged with the responsibility of determining if the benefits from a drug outweigh its risks.

In the 1950s, 1960s, and 1970s, Industrial Bio-Test Laboratories (IBT) performed about 35–40% of all US toxicology testing. Of the 867 audits of IBT performed by the FDA under the 1962 law, 618 were found to be invalid because of numerous discrepancies between the study conducted and data. FDA flexed its muscles and found four IBT managers guilty of fraud. As a result of the IBT incident, the FDA decided to regulate laboratory testing. Prior to 1976, the FDA conducted inspections of nonclinical laboratory studies and study sites on a "for-cause" basis to ensure the quality of the nonclinical study data submitted to the FDA. The for-cause inspections were usually triggered when data submitted to the FDA contained inconsistencies or had questionable integrity, the purpose or design of the study was questionable, or there was simultaneous performance of an unusually large number of complex studies. These for-cause inspections identified serious deficiencies, which presented a concern to the FDA that the quality and integrity of the studies and study data were much worse than they had previously assumed.

To address this problem, the FDA established the bioresearch monitoring (BIMO) program to address domestic and international inspections of nonclinical testing laboratories under GLPs, clinical investigators under good clinical practice regulations, and sponsor/monitors and institutional review boards (IRBs). The GLP regulations were proposed on November 19, 1976 [1]; promulgated on December 22, 1978; and effective June 20, 1979. Prior to the effective date of the GLP regulations, the FDA held a management briefing on the regulations. The questions posted during that briefing session were answered by the FDA and published in a postconference report in August 1979. After the effective date of the regulations, additional questions were presented to the FDA regarding the new regulations. The FDA compiled the questions and their responses and published the Good Laboratory Practice Questions and Answers in June 1981. Minor formatting and editorial changes were made to this document in December 1999 and September 2000, and the FDA reissued it as Guidance for Industry [2].

A brief history of GLP until the early 1970s, many private and public laboratories applied GLP-type principles in one form or another. Repeat-dose safety studies involving a variety of techniques concerned with animal handling, dosing, and observation were regularly conducted. In 1972, New Zealand formally introduced GLP as the Testing Laboratory Registration Act, which covered staff records, procedures, equipment, and facilities. The act prompted the establishment of a related council "to promote the development and maintenance of GLP in testing." In the same year, Denmark also introduced a law to promote GLP.

Significant events took place in 1975 when Senator Edward Kennedy (Dem-MA) and members of FDA made allegations against research laboratories in the United States (Searle and Hazelton) related to preclinical research studies. Both sites were subsequently investigated, which revealed serious problems with the conduct of safety studies submitted to the agency. Violations included poor record keeping and data storage, inadequate personnel training, poor test facility management, and even fraud.

In December 1978, FDA published final GLP regulations and made compliance with them the law in the United States in June 1979. United States Environmental Protection Agency (EPA) adopted its own set of GLP regulations shortly thereafter, governing the research surrounding pesticides and toxic chemicals [3]. These regulations were collected in Title 21: "Food and Drugs" of the *Code of Federal Regulations* (CFR) as Part 58: "Good Laboratory Practice for Nonclinical Laboratory Studies," and they applied to all nonclinical safety studies intended to support research permits or marketing authorizations of products regulated by FDA. Subsequently, FDA's Office of Regulatory Affairs (ORAs) released two *Guidance for Industry* documents to ensure the proper and consistent interpretation of the directives by industry and by FDA's field investigators. Further changes to the GLP rules were proposed in 1984, and in September 1987, FDA published its "Final Rule"—*Compliance Program Bioresearch Monitoring: Good Laboratory Practices*, which was expanded to incorporate the following:

- The requirement for a QA department
- The requirement for protocol preparation (study or work plan)
- The characterization of test and control materials
- The requirement to retain specimens and samples.

Since then, the requirement for laboratories to apply and comply with GLP principles has extended from pharmaceutical companies to many other types of research and testing establishments throughout the developed world. In Europe, adherence to the principles of GLP is governed by European Union (EU) law and, in compliance with EU Directives, an inspection program confirms that "toxicological studies for the regulatory assessment of industrial chemicals, medicines, veterinary medicines, food and animal feed additives, cosmetics, and pesticides must be conducted in accordance with GLP."

By January 1986, scientists at G.D. Searle (acquired by Pharmacia and currently part of Pfizer) had developed a document, *Good Laboratory Practice*, which was designed to be used as guidance to evaluate research activities, and submitted it to both FDA and the Pharmaceutical Research and Manufacturers Association of America (PhRMA). In August of the same year, FDA released a draft GLP document based on the Searle paper and published GLP regulations in the *Federal Register*. At the same time, the agency was creating 606 new positions to monitor biological research, and it also began a pilot inspection program to establish baseline levels of competence. Items listed as "major failings" included failure to have a quality assurance (QA) department, failure to test every batch of manufactured product, and failure to maintain standard operating procedures (SOPs).

Within 14 years, therefore, GLP moved from an ad hoc concept to legally enforceable code, designed to control and regulate the quality of laboratory-based operations. Since that time, the principles have remained fairly consistent and have not changed much; they have simply expanded to include other scientific activities and devices (such as computers). In the Federal Register of October 29, 1984 (49 FR

43530), FDA published a proposal to amend the agency's regulations in 21 CFR Part 58, which prescribe good laboratory practice for conducting nonclinical laboratory studies (the GLP regulations). The proposal was the result of an evaluation of the GLP regulations and of the data obtained by the agency's inspection program to assess laboratory compliance with the regulations. The evaluation led the agency to conclude that some of the provisions of the regulations could be revised to permit nonclinical testing laboratories greater flexibility in conducting nonclinical laboratory studies without compromising public protection. FDA invited comments on all aspects of the proposal and provided 60 days for interested persons to submit comments, views, data, and information on the need to revise any other provisions of Part 58. FDA received 33 comments: 19 from manufacturers of articles regulated by FDA, 4 from associations, 8 from foreign or domestic testing or consulting laboratories, and 2 from individuals within FDA. The majority of these comments endorsed the proposed changes. Many of the comments suggested additional revisions to the GLP regulations or modifications to the proposed changes. On October 29, 1984, the FDA published a proposed amendment to the GLP regulations to allow more flexibility for conducting nonclinical laboratory studies. This final amendment was issued on September 4, 1987, and became effective on October 5, 1987.

The FDA had revised the regulations that specify good laboratory practice for nonclinical laboratory studies since its inception in 1970s. The revisions are based on an agency determination that several provisions of the regulations should be clarified, amended, or deleted to reduce regulatory burdens on testing facilities. Major changes are proposed in the provisions on quality assurance, protocol preparation, test and control article characterization, and retention of specimens and samples. The changes proposed will not compromise the regulations' objective, which is to ensure the quality and integrity of the safety data submitted in support of the approval of regulated products. The action is intended to reduce the burden of compliance with the regulations. FDA discussed the need for regulations on good laboratory practice (GLP) for nonclinical laboratory studies in the preamble to the proposed regulations (41 FR 51206; November 19, 1976). Agency inspections of toxicology laboratories had revealed problems in the conduct of some nonclinical laboratory studies so severe that in many instances the studies could not be relied on for regulatory decision making. The importance of proper safety testing to the product approval process prompted the agency to begin its Toxicology Laboratory Monitoring Program. To ensure the quality and integrity of the safety data submitted to it in support of the approval of any application for a research or marketing permit, FDA issued regulations specifying standards for adequate safety testing, prepared an inventory of domestic and foreign toxicology laboratories engaged in safety testing, conducted training sessions for agency investigators to develop proficiency in evaluating testing facilities, and instituted a compliance program that provided for periodic inspections of the testing facilities. These steps have been completed. In the Federal Register of December 22, 1978 (43 FR 60013), FDA issued final GLP regulations. The regulations, which were codified as 21 CFR Part 58, became effective on June 20, 1979. In the Federal Register of April 11, 1980 (45 FR 24865), FDA amended § 58.113(b) to delete the requirement that reserve samples of test and control article-carrier mixtures be retained.

FDA has conducted numerous training courses at its National Center for Toxicological Research in Jefferson, AR, to provide training in good laboratory practice and the associated laboratory inspection techniques as well as "hands-on" exercises in toxicology experimentation. Finally, in 1976 the agency began an inspection program to assess laboratory compliance with the GLP regulations. Program features include biennial surveillance inspections to assess compliance with the procedural provisions of the GLP regulations and data audit inspections to assess the accuracy of the information contained in final study reports as required by §58.185. By the end of the 1982 fiscal year, the agency had evaluated the reports of 710 laboratory inspections made under the Toxicology Laboratory Monitoring Program.

FDA believes that the proposed changes will reduce record keeping for a large number of nonclinical laboratory studies. Reserve samples of feed and water administered to the control groups of the test system would not have to be retained; characterization and stability assessment would not have to be done; and strict accountability records of receipt, use, and disposition would not need to be generated and maintained for such articles. FDA proposes to modify the definition of "nonclinical laboratory study" to allow the conduct of several experiments using the same test article under a single, comprehensive protocol. For example, a battery of several studies of one test article conduct in several animal species to determine the safety of the test article could be conducted under one protocol. Similarly, where several test articles are to be studied concurrently using a single common procedure, for example, mutagenicity testing, a single protocol could be developed and followed. FDA believes that this approach will reduce the amount of required paper work with no loss in the quality or accuracy of test article data developed by a testing facility.

2.1.1 Economic Assessment

FDA had examined the economic consequences of the proposed changes in accordance with Executive Order 12291 and the Regulatory Flexibility Act (Pub. L. 96-354). The agency tentatively concludes that the revisions would have favorable economic impacts on product sponsors and testing facilities without compromising the quality and integrity of the safety data submitted to support product approval. Although agency estimates were imprecise, cost savings were expected to accrue from changes affecting test and control articles ($5.6 million per year) from revisions in protocol requirements ($6.2 million per year), and from changes in quality assurance procedures ($12.9 million per year), and possibly others. Accordingly, the agency concludes that the proposed revisions do not constitute a major rule as defined in Executive Order 12291 and that no regulatory flexibility analysis is required. The agency also certifies that the revisions will not have a significant impact on a substantial number of small entities. The vast majority of laboratories are not considered small businesses under the Regulatory Flexibility Act and the agency estimates that the impact on those laboratories that are small is not significant because toxicology testing is but a small portion of the work performed by many laboratories. The savings described above will accrue to all sponsors and testing facilities regardless of size.

2.1.2 Environmental Impact

The agency had determined pursuant to 21 CFR 25.24(d)(14) (proposed December 11, 1979; 44 FR 71742) that the changes are of a type that does not individually or cumulatively have a significant impact on the human environment. Therefore, neither an environmental assessment nor an environmental impact statement is required.

In order to avoid foreseeable trade restrictions, the Organization for Economic Cooperation and Development (OECD) became—a union of the 26 mostly important western restaurant nations then—arranges itself to concern with the topic GLP. In the year 1979 the chemical department of the OECD appointed a team of experts, which advanced both the international adjustment of the testing methods and internationally principles of good laboratory practice recognized first compiled, which were officially published in May 1981 as resolution. The OECD left it not with the principles, but was concerned intensively with the possibility, national control of the adherence to these GLP principles for reasons, in order to give each State of security that in each case different the country these principles are used correctly and sufficiently. Guidelines and manual were recognized 1989 of the OECD states as binding and form nowadays the basis for the national monitoring. Starting from this time the OECD was driving Kraft for all sues of good laboratory practice. In the year 1990 the OECD created the GLP panel, a forum of GLP responsible person of the OECD member states, who should advance the harmonization of GLP by various activities worldwide. Annual meetings GLP panel, of it used working groups as well as representatives from monitoring and industry, adopted a set from documents to the interpretation and conversion of the GLP principles in different ranges of application.

The European Union, whose member states are also at the same time OECD members, reacted in each case to the OECD and fixed the adherence to the principles of the GLP for certain investigations and uniform control for all European Union states. The EU commission is a quite slow-acting thing and limps in the range GLP temporally strongly behind the recommendations of the innovative OECD ago, in order to take over these temporally unchanged, and are evenly so slow-acting despite binding European Union guidelines, some of their member nations with the complete implementation of GLP. This inertia under the European Union member nations led to substantial criticism on the part of the official contacts (the United States, Japan, Israel, etc.) with the discussions across agreements and even to a provisional suspension of these discussions. It was denounced by US side that some European Union nations are still far distant from a recognize-worth GLP. For the recognition of the existing differences in the practical conversion of the GLP guidelines in the individual states and for harmonization the European Union commission 1995 initiated the introduction of mutual attendance in each case three supervisors with observer status. Within 2 years each member state should be visited once. The attendance was locked in each case by a common report of the observers. The results were presented in a working group and the harmonization need resulting from it was discussed. The reports confirmed the opinion of the official contacts and the European Union will hopefully arrange to give some member states helping instruction within GLP. Despite the obvious lack, the resolution of the European

Union states had further existed that within the European Union under adherence to produced test data mutually are to be recognized. However, as mentioned above in the report it was pointed out that inspection results from certain member nations should be particularly examined by the evaluation or monitoring authorities.

In the years 1983–1989, with few exceptions, only testing facilities were examined, which accomplished toxicological examinations in connection with the permission of medicaments. Starting point for the desire of the industry for national inspections was the international acknowledgment of German GLP examinations. Due to increasing pressure on the part of the United States, with threatening permission delays, the chemical industry in Germany began to already state in 1978, GLP standard in their laboratories without obligatory regulations on national side. With the novella of the chemical law in August 1990 the appropriate guidelines of the European Union community were converted into German right and thus the adherence to the principles of the GLP was prescribed obligatorily on nonclinical experimental examinations by chemical materials or preparing, whose results are to make the official evaluation possible of potential dangers for humans and environment.

It was specified extremely clearly that without proof of the adherence to the GLP principles an evaluation of the examination results may not take place via the appropriate authority. The examinations prescribed legally for the risk evaluation must be accomplished in one nationally GLP recognizing testing facility under adherence to the GLP principles. Demonstrating compliance with GLP has become a prerequisite for clinical testing and drug registration in developing countries, and certainly if drug products are to be exported to countries other than the country of origin. It is a general concern that it is essential to avoid the coexistence of two or more international regulatory GLP standards for nonclinical safety testing. However, more clear guidance is needed for the implementation of GLP at worldwide level including OECD principles of GLP [4] (see Figure 2.1).

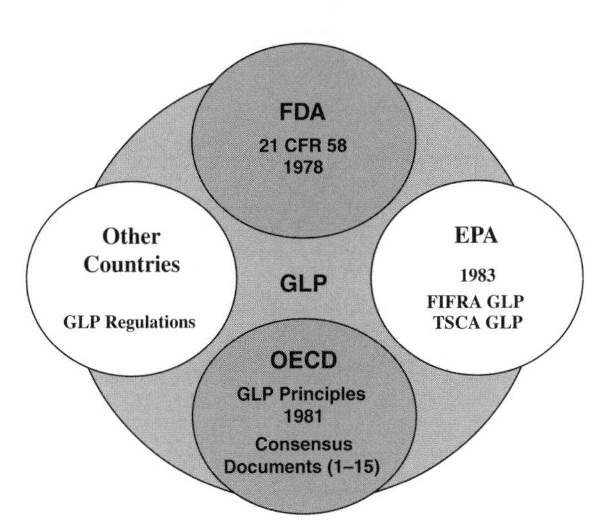

FIGURE 2.1 GLP versus OECD applicable to GLP regulated laboratories.

2.2 APPLICABILITY AND RELATIONS TO OTHER REGULATIONS/ PRINCIPLES

As known, there are several applicable good practices within our industries as referred to GxPs, which are related to various stages/processes during research and development programs. Up to date, it is very important for laboratories, facilities, and organizations to think about GxPs from global and cross-functional perspectives to develop and market drugs or medicines regardless of its origin.

2.2.1 GLP, GCP, GMP, and Part 11

Figure 2.2 illustrates how the so-called Good Practices regulations correlate to the life of a drug, starting from basic discovery on the left side through preclinical development in the middle and clinical trials and manufacturing at the right.

Typically research and drug discovery are not regulated at all. However, regulatory agencies have increasingly issued more and more guidance documents to facilitate and ensure proper and adequate conduct of studies for data submission and approval processes. GLP starts with certain preclinical development, for example, "GLP" toxicology studies. Clinical trials are regulated by good clinical practice regulations [5] (part of laboratory operation applicable to GLP), and manufacturing through GMPs [6] for active pharmaceutical ingredients (APIs) and drug products. In fact, during clinical trials, scale-up or manufacturing process (materials/batch suppliers) is also involved and regulated by cGMP. There is a frequent misunderstanding that all laboratory operations are regulated by GLP. This is not true. For example, Quality Control laboratories in manufacturing are regulated by cGMPs and not by GLPs. Also Good Laboratory Practice regulations are frequently mixed up with good analytical or bioanalytical practice ("GAP or GBP"). Applying good analytical or bioanalytical practices is important but not sufficient, as no clear rules/acceptance criteria are established. When small quantities of active ingredients are prepared in a research or development laboratory for use in samples for clinical trials or finished drugs, that activity has been covered by cGMP and not by GLP. 21 CFR Part 11 is FDA's

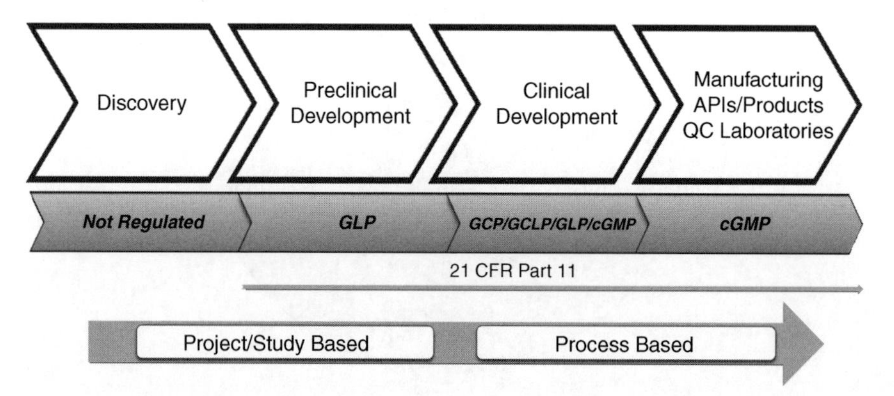

FIGURE 2.2 Relationships among GxPs including 21 CFR Part 11.

regulation on electronic records and signatures and applies for electronic records or to computerized systems in all FDA regulated areas [7]. For example, it applies for computers that are used in GLP studies. Characteristic for GLPs is that they are study based where as cGMPs are process based. GLPs regulate all nonclinical safety studies that support or are intended to support applications for research or marketing permits for products regulated by the FDA, or by other similar national agencies. This includes not only drugs for human and animal use but also aroma and color additives in food, biological products, and medical devices. The duration and location of the study is of no importance.

With existing regulations at international level, to adopt the revised Organization for Economic Cooperation and Development Principles of Good Laboratory Practices as the officially endorsed WHO/TDR regulation in the area of nonclinical safety testing was seen as the most rational way forward. On the international level, the OECD, in order to avoid nontariff barriers to trade in chemicals, to promote mutual acceptance of nonclinical safety test data, and to eliminate unnecessary duplication of experiments, followed suit by assembling an expert group who formulated the first OECD principles of GLP. Their proposals were subsequently adopted by the OECD Council in 1981 through its "Decision Concerning the Mutual Acceptance of Data in the Assessment of Chemicals" C(81)30(Final), in which they were included as Annex II. In this document, the Council decided that data generated in the testing of chemicals in an OECD member country in accordance with the applicable OECD Test Guidelines and with the OECD Principles of Good Laboratory Practice shall be accepted in other member countries for the purposes of assessment and other uses relating to the protection of man and the environment. It was soon recognized that these principles needed explanation and interpretation, as well as further development, and a number of consensus workshops dealt with various issues in subsequent years. The outcome of these workshops was then published by OECD in the form of consensus documents. After some 15 years of successful application, the OECD principles were revised by an international group of experts and were adopted by the OECD Council on November 26, 1997 C(97)186/Final by a formal amendment of Annex II of the 1981 Council Decision. A number of OECD member countries have adopted these principles in their national legislation, notably the amendment of the European Union in Commission Directive 1999/11/EC of March 8, 1999 to the Council Directive 87/18/EEC of December 18, 1986, where GLP had first been introduced formally into the European legislation.

Internationally, the observance of GLP has thus been defined as a prerequisite for the mutual acceptance of data (MAD) [8], which means that different countries or regulatory authorities accept laboratory studies from other countries as long as they follow the internationally accepted GLP principles. This mutual acceptance of safety test data will also prevent the unnecessary repetition of studies carried out in order to comply with any requirements of Chemicals' C(81)30(Final) and C(89)87(Final). C(97)114/Final, wherein interested nonmember countries are given the possibility of voluntarily adhering to the standards set by the different OECD Council Acts and thus, after satisfactory implementation, to join the corresponding part of the OECD Chemicals Program. Mutual acceptance of conformity of test facilities and studies with respect to their adherence to GLP, on the other hand, necessitated the establishment of national procedures for monitoring compliance. According to the OECD Council "Decision-Recommendation on Compliance with Principles of Good Laboratory Practice" of October 2, 1989 C(89)87(Final), these procedures should be based on nationally performed laboratory inspections and study audits. The respective national compliance monitoring authorities should not only exchange information on the compliance of test facilities inspected but also provide relevant information concerning the countries' procedures for monitoring compliance. Although devoid of such officially recognized national compliance monitoring authorities existed, some developing countries do have an important pharmaceutical industry, where preclinical safety data are already developed under GLP. In these cases, individual studies are—whenever necessary—audited by foreign GLP inspectors (e.g., of FDA, the Netherlands or Germany).

During the global implementation of GLP principles since 1980s, OECD has proactively drafted and revised documents to comprehensively cover as many aspects of regulated soon-to-be-regulated programs. Please note that these documents were prepared based on current documents of OECD Series on Principles of Good Laboratory Practice and Compliance Monitoring. These documents are regularly reviewed; therefore the user of these should also refer to the OECD for updated version. The OECD series would be considered as final. The following documents are available (Table 2.1) and more are coming.

Good Laboratory Practice is defined in the OECD principles as "... a quality system concerned with the organizational process and the conditions under which nonclinical health and environmental safety studies are planned, performed, monitored, recorded, archived, and reported." The purpose of these principles of Good Laboratory Practice is thus to promote the development of quality test data and to provide a managerial tool to ensure a sound approach to the management, including conduct, reporting, and archiving, of laboratory studies. The principles may be considered as a set of criteria to be satisfied as a basis for ensuring the quality, reliability, and integrity of studies, the reporting of verifiable conclusions, and the traceability of data. Consequently the principles require institutions to allocate roles and responsibilities in order to improve the operational management of each study, and to focus on those aspects of study execution (planning, monitoring, recording, reporting, archiving), which are of special importance for the reconstructability of the whole study. Since all these aspects are of equal importance for compliance with the principles of GLP, there cannot be any possibility of using only a choice of requirements and still claiming GLP compliance. No test facility may thus rightfully claim GLP compliance if it has not implemented, and if it does not comply with, the full array of GLP rules. The GLP principles in their strict, regulatory sense apply to such studies on pharmaceuticals which

- are nonclinical, for example, are mostly conducted in animals or *in vitro*, and include analytical aspects;
- are conceived to obtain data on the properties and/or safety with respect to human health and/or the environment of the tested substances;

TABLE 2.1 A Series of Guidance Documents from OECD GLP Principles

• **Document Number 1**	*The OECD Principles of Good Laboratory Practice* (ENV/MC/CHEM(98)17) 1998
• **Document Number 2**	*Revised Guides for Compliance Monitoring Procedures for Good Laboratory Practice* (Environment Monograph No. 110) 1995
• **Document Number 3**	*Revised Guidance for the Conduct of Laboratory Inspections and Study Audits* (Environment Monograph No. 111) 1995
• **Document Number 4**	*Quality Assurance and GLP* (Environment Monograph No. 48) 1999
• **Document Number 5**	*Compliance of Laboratory Suppliers with GLP Principles* (Environment Monograph No. 49) 2000
• **Document Number 6**	*The Application of the GLP Principles to Field Studies* (Environment Monograph No. 50) 1999
• **Document Number 7**	*The Application of the GLP Principles to Short-Term Studies* (Environment Monograph No. 73) 1999
• **Document Number 8**	*The Role and Responsibilities of the Study Director in GLP Studies* (Environment Monograph No. 74) 1999
• **Document Number 9**	*Guidance for the Preparation of GLP Inspection Reports* (Environment Monograph No. 115) 1995
• **Document Number 10**	*The Application of the Principles of GLP to Computerized Systems* (Environment Monograph No. 116) 1995
• **Document Number 11**	*The Role and Responsibilities of the Sponsor in the Application of the Principles of GLP* (ENV/MC/CHEM (98)16) 1998
• **Document Number 12**	*Requesting and Carrying Out Inspections and Study Audits in Another Country* 2000
• **Document Number 13**	*The Application of the OECD Principles of GLP to the Organization and Management of Multi-Site Studies* 2002
• **Document Number 14**	*The Application of the Principles of GLP to In Vitro Studies* 2004
• **Document Number 15**	*Advisory Document of the Working Group on GLP: Establishment and Control of Archives that Operate in Compliance with the Principles of GLP* 2007

- are intended to be submitted to a national registration authority for the purposes of registering or licensing the tested substance or any product derived from it.

In general, and depending on national legal requirements, the GLP requirements for nonclinical laboratory studies conducted for safety evaluation in the field of drug safety testing cover the following classes of studies:

- Single dose toxicity.
- Repeated dose toxicity (subacute and chronic).
- Reproductive toxicity (fertility, embryo–fetal toxicity, and teratogenicity, peri-/postnatal toxicity).

- Mutagenic potential.
- Carcinogenic potential.
- Toxicokinetics (pharmacokinetic studies that provide systemic exposure data for the above studies).
- Pharmacodynamic studies designed to test the potential for adverse effects (safety pharmacology).
- Local tolerance studies, including phototoxicity, irritation and sensitization studies, and testing for suspected addictivity and/or withdrawal effects of drugs.

Irrespective of the place of study conduct, the GLP principles generally apply to the relevant studies planned and conducted in a manufacturer's laboratories, at a contract or subcontract facility or in a university or governmental laboratory. GLP is not directly concerned with the scientific design of a study. The scientific design of a study (e.g., with regard to the test system used and the scientific state-of-the-art of its conduct) is governed by the applicable testing guidelines and its scientific rationale. It has to be reiterated that, in introducing GLP in a test facility, and in the training for its application, it is important to clearly differentiate between the formal, regulatory use of the term "good laboratory practice," as opposed to the general application of "good practices" in scientific investigations. Since the term "good laboratory practice" is not a trademark protected term, any laboratory that may consider itself to be following good practices in its daily work might be tempted to describe its adherence to these (possibly even self-defined) quality standards by this terminology. It has to be clearly stated, however, that only adherence to, and compliance with, all the requirements set forth in the OECD principles will constitute real compliance with GLP, and that therefore any use of similar terminology for the description of quality practices outside of the scope of GLP should be emphatically discouraged.

While the regulations fix the rules for good laboratory practices in a managerial, administrative way, they also help the researcher to perform his/her work in compliance with his/her own preestablished plan, and they standardize planning, recording, reporting, and archiving procedures. The regulations are not concerned with the scientific or technical content of the studies conducted in compliance with them, and they do not aim to evaluate the scientific value of the studies, as this latter task is reserved for the registration authorities. However, the requirements for proper planning, for the faithful recording of all observations and for the complete archiving of all raw data obtained, will serve to eliminate many sources of error in the studies conducted in compliance with these regulations. Good laboratory practice, applied in whatever the industry targeted, stresses the importance of the following main points:

- *Resources:* Organization, personnel, facilities, equipment.
- *Rules:* Protocols, standard operating procedures, concept of the Study Director as the pivotal point of study control.
- *Characterization:* Test items, test systems.
- *Documentation:* Raw data, final report, archives.
- *Quality Assurance:* Independence from study conduct.

2.2.2 General Terminologies and Definitions of GxPs (GLP, GCP, and cGMP)

Obviously, GxPs may have applied to different stages and areas of drug development processes to ensure the integrity, quality, and safety of the data and products for intended uses and purposes. There have been numerous presentations and review articles defining the relationships, roles, and responsibilities during the implementation as well as operational infrastructure and functionality. Table 2.2 illustrates and summaries some basic definitions, concepts, and requirements to demonstrate their similarity and differences.

2.3 COMPARISON OF FDA GLP, EPA GLP REGULATIONS, AND OECD GLP PRINCIPLES

Since the inception of the FDA good laboratory practice (GLP) in 1979, the Organization for Economic Cooperation and Development principles of GLP in 1981 and the finalization of the EPA GLP in 1983 there have been recognizable differences among the compliance programs [1–4]. All have been revised since their initial publication. Nonetheless, there remain differences in verbiage, and in some cases content, among the FDA, EPA, and OECD GLP principles. The end result for each is the assurance that the experimental information generated under each program is of sufficient quality and integrity to support the reports for the various studies. These differences, while not affecting the data quality, can result in issues when submitting studies globally. The US GLP programs, in general, tend to be more prescriptive than the OECD principles. In some cases they are more stringent, for example; as regards animal care requirements, or requiring the reporting of all circumstances that may have affected the quality or integrity of the data be included in the final report. Another notable difference is in the US test facilities (TF) are not required to submit an application to the monitoring authorities (MA) and receive GLP qualification or certification prior to generating or submitting GLP data to Regulatory Authorities (RA). Further, in the United States, laboratories do not pay a fee to the MA for the performance of a compliance audit. The original OECD GLP principles were based on the 1979 FDA GLP. The principles were written and approved with the input and approval of the OECD membership. The OECD principles provide a framework for countries to implement their own national programs. The principles are written to allow for variations among national programs such as archival storage periods or the approval sequence of the study plan by the TF management (TFM) and sponsor. Additionally, the OECD GLP principles have numerous consensus documents published (15 in total) that give additional definition and clarity to areas of the GLP principles such as (i) the quality assurance unit; (ii) application of the GLP principles to field studies; (iii) application of the GLP principles to short-term studies; (iv) role and responsibilities of the Study Director (SD) and sponsor; (v) application of GLP to computerized systems; (vi) organization and management of multisite studies; (vii) application to *in vitro* studies; and (viii) establishment and control of archives. Many of the consensus documents are used in the US GLP community as important

TABLE 2.2 Basic Definitions, Concepts, and Requirements for GLP, GCP, and cGMP (Similarity and Differences)

	GLP	GCP	cGMP
General definition	GLP is a quality system concerned with the organizational process and the conditions under which nonclinical health and environmental safety studies are planned, performed, monitored, recorded, archived, and reported	GCP is an international clinical and scientific quality standard for designing, conducting, recording, and reporting trials that involve the participation of human subjects. Compliance with this standard provides public assurance that the rights, safety, and well being of trial subjects are protected, consistent with the principles that have their origin in the Declaration of Helsinki, and that the clinical trial data are credible	cGMP is that part of Quality Assurance which ensures that products are consistently produced and controlled to the quality standards appropriate to their intended use and as required by the Marketing Authorization of product specification. cGMP is concerned with both production and quality control
Scope of Roles/ Responsibilities			
Ownership	Organization/facility management	Sponsor	Manufacturer/holder of the authorization
Main responsibility for the activity	Study Director	Medical Director/Principal Investigator	Qualified person
Responsibility for specific "program or production"	Principal Investigator	Pharmacist	Head of production
Quality	QAU	Monitor quality assurance	Head of quality control
Archive	Archivist	Archivist	Not specified yet
The quality unit and organization	QAU is fully independent from the activities being assured. No clear definition has been defined yet for Quality Control	QAU	Quality Assurance, as organization, is not described in the regulations. Most of the duties are under Quality Control functions

Document management of the SOP system	Study personnel and management approve SOPs	The Sponsor has the responsibility for the SOP system	The historical files and original copies are filed by QAU. The whole system is generally handled and under QAU. Generally QAU helps department in drafting the SOPs
	QAU review and verify GLP compliance. QAU does not approve the SOPs (not required) The SOP management is not under the umbrella of the QAU but under the responsibility of the Site management		
Documentation			
Approval	The Study Director is responsible for the scientific contents of any protocols, amendments, memos, final study report	Not applicable or clearly defined yet	Quality Control and management issue the Certificates for authorization for use of the batch manufactured and QAU has the duty to review all the documentation
	QAU is responsible to inspect these documents to confirm that the methods, procedures, and observations are accurately and completely reflect the raw data of the studies		
Activities (material management)	Records including test articles and reference articles characterization, date of receipt, expiry date, quantities received, and used in studies should be maintained securely	Traceability, accountability, and reconciliation of the material is mandatory	Very strictly controls on the materials (from arrival to the facility, warehouse through the shipping of the APIs and final product)—some quarantine process may be applied

(*continued*)

TABLE 2.2 (*Continued*)

	GLP	GCP	cGMP
Facilities and equipment	Basic criteria: prevention of cross contamination and mix-ups; separation of the different activities to assure the proper conduct of the study	Not applicable or clearly defined yet	Premises and equipment must be located, designed, constructed, adapted, and maintained to suit the operation to be carried out. Their layout and design must aim to minimize the risk of errors and permit effective cleaning and maintenance in order to avoid cross contamination, build up of dust or dirt and, in general, any adverse effect on the quality of APIs and products
Facilities and equipment qualification	Maintenance is required to ensure the reliability and integrity of data generation and processing. Some basic qualification (IQ and OQ) is highly recommended	Not applicable or clearly defined yet	Facilities, equipment, and utilities should be qualified to assure the adequate performance
Facilities, equipment, and laboratory instruments: calibration	Apparatus used in a study should be periodically inspected, cleaned, maintained, and calibrated in accordance with related SOPs. Records of these activities should be kept securely. Calibration should, where appropriate, be traceable to national or international standards of measurement	Not applicable or clearly defined yet	Calibration for critical instrumentation should be described in appropriate SOPs and a master calibration plan is available

Computerized system validation	Requested and implemented at present industry wide. The OECD and FDA guidelines are not precise yet about the validation steps and requirements	Requested; there is a need for a full life cycle validation (validation plan, IQ/OQ/PQ along with validation report and associated SOPs)	Requested; there is a need for a full life cycle validation (validation plan, IQ/OQ/PQ along with validation report and associated SOPs)
Laboratory controls: method validation	FDA Bioanalytical Method Validation Guidance for Industry May 2001	Not applicable or clearly defined yet	Validation of the method is essential for the cGMP and scientific values of the data generated during the analytical work
Laboratory control: reserve samples	A sample for analytical purposes from each batch of test item should be retained for all studies except for short-term studies	Not applicable	It is mandatory to keep appropriate amount of reference samples of starting materials, packaging materials or finished products, and retention samples of finished products
Quality Management			
Audits	Generally three types of audits - Study-based audit - Facility-based audit - Process-based audit	The purpose of a sponsors' audit, which is independent of and separate from routine monitoring or quality control functions, should be to evaluate trial conduct and compliance with the protocol, SOPs, GCP, and the applicable regulatory requirements	Personnel matters, premises, equipment, documentation, productions, quality control, distribution of the medicinal products, arrangements for dealing with complaints and recalls, and self inspection, should be examined at intervals following a prearranged program in order to verify their conformity with the principles of Quality Assurance

(*continued*)

TABLE 2.2 (*Continued*)

	GLP	GCP	cGMP
	A requirement of GLP is to audit a "live" analysis or "live" sample preparation—as known as in-process audit, in addition to the historical type		
Deviations, OOS, change control, etc.	Deviations from SOPs, protocols related to the study should be documented and acknowledged by the Study Director and the Principal Investigator(s) as applicable. Corrective action should be taken and root cause should be defined and documented. However, there is no clear requirements for initiating with investigations	Not applicable or clearly defined yet	Need formal investigation with specific SOPs addressing the way to perform the investigation

reference sources. This is especially true of the OECD Consensus Document No. 13 dealing with multisite studies [9].

2.3.1 US and OECD GLP Similarity and Differences

General: There are numerous similarities between the United States (FDA and EPA) and the OECD GLP principles. All require (i) a study plan that is approved by the sponsor and signed by the Study Director; (ii) a QA Unit; (iii) a substance/article/item that is appropriately characterized; (iv) trained, qualified personnel to conduct the study; (v) raw data collection and change procedures; (vi) a final report that reflects the data generated; and (vii) the final report and all associated raw data and records be archived. Not all aspects of the US and OECD GLP principles are the same. There exist many terms in the published documents that differ, although the intent is essentially the same, as set forth. These differences, at times, can cause confusion because of the common usage of these words and phrases in different countries. For instance, the word "should" in the OECD GLP principles is intended to be interpreted as required, but in the US GLP principles should imply that it is recommended, but not absolutely required. The main differences between the US and OECD GLP programs can be traced back to the following key aspects: (i) responsibilities and compliance; (ii) statement of compliance; (iii) approvals; (iv) laboratory certification; (v) authority inspections; and (vi) archiving requirements. These differences are illustrated in detail hereafter.

Responsibilities and Compliance: In the United States, the SD is responsible to assure that all applicable GLP regulations are followed. If there are any deviations from the EPA or FDA GLP principles this must be noted in the final report and in the case of the EPA GLP regulations noted in the GLP statement of compliance. In the OECD context, on the other hand, TFM is mainly responsible to ensure GLP compliance. This is not to say that the SD is not responsible, but only that the emphasis for compliance is weighted to the TFM. The OECD requires that the TFM issues a declaration, where applicable, that a study was carried out in accordance with the GLP principles. Moreover, under the OECD GLP program, the final report should be signed and dated by the SD to indicate responsibility for data and should indicate compliance with the GLP principles. The EPA requires that each study include a true and correct statement, signed by the applicant, the sponsor, and the SD indicating the level of compliance with 40 CFR Part 160 (GLP statement of compliance) and is required to be page three of the final report [9]. It should be noted that the FDA has no similar requirement.

Statement of Compliance in the Final Report: The EPA requires that each study submitted be accompanied by a GLP compliance statement specifying one of the following: (i) the study was conducted under the EPA GLP regulations; (ii) the study was not conducted under EPA GLP regulations with description of the ways it differs; and (iii) the submitter is not the study sponsor, did not conduct the study and does not know whether the study was conducted under the EPA GLP regulations. Again, unlike the EPA, the FDA does not require such a statement to accompany the final report. Even though it is not a requirement of many countries to supply a GLP compliance statement as such, they do create a statement for US submissions. They can have different appearance and wording, but covey what is required by the EPA. Because many of these studies are intended for global regulatory submissions the GLP compliance statement has been modified to meet multiple regulatory agency acceptability requirements. Some TF generate a compliance statement that claims compliance with several GLP programs including programs that are outside the scope of their countries' MAs. One question that might be asked is whether it is helpful to claim compliance with GLP regulations other than those of the national program. Is there value in making these claims given that the additional claims are not monitored by any MA?

Protocol Approval: The US GLP regulations require sponsor approval prior to SD signature on protocol/study plan. On the other hand, the OECD allows the individual countries to determine when sponsor and TFM need to sign the study plan.

Assignment of Study Director: In the United States, GLP program there is a single SD for a given study. With the OECD it is allowed by some RA that the TFM assigns deputy SD for defined durations during a study (only one SD at any given time).

Laboratory Certification: Laboratory certificates, for example, official certification documents from country MA are not a specific requirement of the OECD GLP program. In the United States, authorities do not certify TF. They perform what is commonly referred to as neutral scheme compliance inspections on a 2–4 year cycle. They can perform audits more frequently if there is evidence to indicate it is warranted. At the completion of the audit neither the EPA nor the FDA issue a certificate of compliance. The majority authorities following the OECD GLP principles certify TF for 2–4 year intervals. At the completion of the TF inspection a GLP certificate is issued. In lieu of certification in some countries RA have requested documentation from the US GLP authorities, or have requested a more specific document from the EPA Laboratory Data Integrity Branch.

Authority Inspections: Within the OECD, authority inspections are normally requested by the TF. If the facility is new they must request an inspection and obtain a GLP certificate prior to claiming GLP compliance for the studies that are conducted. Certificates generally are valid for 2–4 years. Recertification is requested by the TF. There is generally a fairly long lead time between the notice of inspection and the actual inspection event. In the United States, selection of the TF to be inspected is by the RA. New laboratories are not required to apply for certification from the monitoring authority prior to conducting or claiming GLP compliance. In the case of EPA there is a 10 day notice (every 2–4 years), whereas the TF receives no advance notice of the GLP inspection by the FDA.

Archiving Requirements: The length of time that the original raw data and final reports are required to be maintained differs between regulations and MA. The

OECD allows the individual RA to determine the appropriate storage interval. In the United States, on the other hand, there are different rules between the FDA and the EPA. Some examples of the OECD approach are, for instance, Germany requires archival for 15 years and Switzerland requires 10 years. The FDA has defined an interval of 5 years after results are submitted for research or marketing permit and 2 years after termination if not submitted to the agency. The EPA indirectly defers to the FIFRA books and records requirements (e.g., retained as long as the registration is valid and the producer is in business) or 2 years after study termination if not submitted to the agency [3]. The EPA and the FDA require government notification if materials are transferred to sponsor archives when a contract facility goes out of business. Notification is not a specific OECD requirement.

More and more companies are performing or contracting work across the globe. Studies conducted are not used only in the country of origin, but are also submitted to Regulatory Authority in many countries. Global efforts at harmonization within technical arenas of the OECD are currently being developed. One area is the development of endocrine disruptor endpoints. In addition, there is currently an OECD group working to harmonize GLP field residue studies. This will allow for the acceptance of up to 50% field studies from one country to be accepted by other national authorities. When the residue harmonization effort is realized one question that must be asked is whether RA that have a certification program accept data from field facilities in the United States that have not yet been inspected by the US GLP authorities even if they are claiming compliance. FDA and EPA have both begun the evaluations of their GLP programs. Among other things, both of these programs are looking at opportunities for greater global harmonization. With regards to the OECD mutual acceptance of data program, it should be noted that all of the new applicants' programs are based on the OECD GLP principles, although they have their own unique requirements as allowed by the OECD GLP principles. Moreover, all current and applying members are or have been audited by other member countries to assure the appropriate level of data quality. Upon careful analysis one can see that no country can simply apply the OECD GLP principles as written, but must make individual decisions on such aspects as length of archival storage or the timing of sponsor approval. For any MA GLP regulations that have gone through the MAD evaluation and approval process has demonstrated itself to be at a sufficient level as to assure the quality and integrity of the data regardless the differences that might exist between it and other MAD countries. These differences should not become obstacles to global acceptance of any GLP studies by any country that is a MAD member.

2.4 APPLICATIONS OF GLP TO MULTIPLE SITE STUDIES

General Considerations: The planning, performance, monitoring, recording, reporting, and archiving of a multisite study present a number of potential problems that should be addressed to ensure that the GLP compliance of the study is not compromised. The fact that different study activities are being conducted at different sites means that the planning, communication, and control of the study are of vital importance. Although a multisite study will consist of work being conducted at more than one site (which includes the TF and all test sites), it is still a single study that should be conducted in accordance with the OECD principles of GLP. This means that there should be a single study plan, a single Study Director, and ultimately, a single final report. It is therefore essential that, when the study is first planned, personnel and management at the contributing sites are made aware that the work they will perform is part of a study under the control of the Study Director and is not to be carried out as a separate study [10].

It is imperative that the work to be carried out by the various sites is clearly identified at an early stage of planning, so that the necessary control measures can be agreed upon by the parties concerned before the study plan is finalized. Many of the problems associated with the conduct of multisite studies can be prevented by clear allocation of responsibilities and effective communication among all parties involved in the conduct of the study. This will include the sponsor, the Study Director, and the management, the Principal Investigator(s), Quality Assurance, and study personnel at each site. All of these parties should be aware that when a multisite study is conducted in more than one country there might be additional issues due to differences in national culture, language, and GLP compliance monitoring programs. In these situations, it may be necessary to seek the advice of the national GLP compliance monitoring authority where the site is located. These proposed guidances apply to all types of nonclinical health and environmental safety studies.

Management and Control of Multisite Studies: A multisite study means any study that has phases conducted at more than one site. Multisite studies become necessary if there is a need to use sites that are geographically remote, organizationally distinct, or otherwise separated. This could include a department of an organization acting as a test site when another department of the same organization acts as the test facility. A phase is a defined activity or set of activities in the conduct of a study. The decision to conduct a multisite study should be carefully considered by the sponsor in consultation with test facility management assigned by the sponsor before study initiation. The use of multiple test sites increases the complexity of study design and management tasks, resulting in additional risks to study integrity. It is therefore important that all of the potential threats to study integrity presented by a multisite configuration are evaluated, that responsibilities are clear and that risks are minimized. Full consideration should be given to the technical/scientific expertise, GLP compliance status, resources and commercial viability of all of the test sites that may be used.

Communication: For a multisite study to be conducted successfully, it is imperative that all parties involved are aware of their responsibilities. In order to discharge these responsibilities, and to deal with any events that may need to be addressed during the conduct of the study, the flow of information and effective communication among the sponsor, management at sites, the Study Director, Principal Investigator(s), Quality Assurance, and study personnel is of paramount importance. The mechanism for communication of study-related information among these parties should be agreed in advance and documented. The Study Director should be kept informed of the progress of the study at all sites.

Study Management: The sponsor will assign a study to a test facility. Test facility management will appoint the Study Director who need not necessarily be located at the site where the majority of the experimental work is done. The decision to conduct study activities at other sites will usually be made by test facility management in consultation with the Study Director and the sponsor, where necessary. When the Study Director is unable to perform his/her duties at a test site because of geographical or organizational separation, the need to appoint a Principal Investigator(s) at a test site(s) arises. The performance of duties may be impracticable, for example, because of travel time, time zones, or delays in language interpretation. Geographical separation may relate to distance or to the need for simultaneous attention at more than one location. Test facility management should facilitate good working relationships with test site management to ensure study integrity. The preferences of the different groups involved, or commercial and confidentiality agreements, should not preclude the exchange of information necessary to ensure proper study conduct.

2.4.1 Roles and Responsibilities

Sponsor: The decision to conduct a multisite study should be carefully considered by the sponsor in consultation with test facility management before study initiation. The sponsor should specify whether compliance with the OECD principles of GLP and applicable national legislation is required. The sponsor should understand that a multisite study must result in one final report. The sponsor should be aware that, if its site acts as a test site undertaking a phase(s) of a multisite study, its operations and staff involved in the study are subject to control of the Study Director. According to the specific situation, this may include visits from test facility management, the Study Director, and/or inspections by the lead Quality Assurance. The Study Director has to indicate the extent to which the study complies with GLP, including any work conducted by the sponsor.

Test Facility Management: Test facility management should approve the selection of test sites. Issues to consider will include, but are not limited to, practicality of communication, adequacy of Quality Assurance arrangements, and the availability of appropriate equipment and expertise. Test facility management should designate a lead Quality Assurance that has the overall responsibility for quality assurance of the entire study. Test facility management should inform all test site quality assurance units of the location of the lead

Quality Assurance. If it is necessary to use a test site that is not included in a national GLP compliance monitoring program, the rationale for selection of this test site should be documented. Test facility management should make test site management aware that it may be subject to inspection by the national GLP compliance monitoring authority of the country in which the test site is located. If there is no national GLP compliance monitoring authority in that country, the test site may be subject to inspection by the GLP compliance monitoring authority from the country to which the study has been submitted.

Test Site Management: Test site management is responsible for the provision of adequate site resources and for selection of appropriately skilled Principal Investigator(s). If it becomes necessary to replace a Principal Investigator, test site management will appoint a replacement Principal Investigator in consultation with the sponsor, the Study Director and test facility management where necessary. Details should be provided to the Study Director in a timely manner so that a study plan amendment can be issued. The replacement Principal Investigator should assess the GLP compliance status of the work conducted up to the time of replacement.

Study Director: The Study Director should ensure that the test sites selected are acceptable. This may involve visits to test sites and meetings with test site personnel. If the Study Director considers that the work to be done at one of the test sites can be adequately controlled directly by him/herself without the need for a Principal Investigator to be appointed, he/she should advise test facility management of this possibility. Test facility management should ensure that appropriate quality assurance monitoring of that site is arranged. This could be by the test site's own Quality Assurance or by the lead Quality Assurance. The Study Director is responsible for the approval of the study plan, including the incorporation of contributions from Principal Investigators. The Study Director will approve and issue amendments to and acknowledge deviations from the study plan, including those relating to work undertaken at sites. The Study Director is responsible for ensuring that all staffs are clearly aware of the requirements of the study and should ensure that the study plan and amendments are available to all relevant personnel. The Study Director should set up, test, and maintain appropriate communication systems between him/herself and each Principal Investigator. For example, it is prudent to verify telephone numbers and electronic mail addresses by test transmissions, to consider signal strength at rural field stations, etc. Differences in time zones may need to be taken into account. The Study Director should liaise directly with each Principal Investigator and not via an intermediary except where this is unavoidable (e.g., the need for language interpreters).

Throughout the conduct of the study, the Study Director should be readily available to the Principal Investigators. The Study Director should facilitate the coordination and timing of events and movement of samples, specimens, or data between sites, and ensure that Principal Investigators understand chain of custody procedures. The Study Director should liaise with Principal Investigators about test site quality assurance findings as necessary. All communication between the Study Director and Principal Investigators or test site quality assurance in relation to these findings should be documented. The Study Director should ensure that the final report is prepared, incorporating any contributions from Principal Investigators. The Study Director should ensure that the final report is submitted to the lead Quality Assurance for inspection. The Study Director will sign and date the final report to indicate the acceptance of responsibility for the validity of the data and to indicate the extent to which the study complies with the OECD principles of Good Laboratory Practice. This may be based partly on written assurances provided by the Principal Investigator(s). At sites where no Principal Investigator has been appointed, the Study Director should liaise directly with the personnel conducting the work at those sites. These personnel should be identified in the study plan.

Principal Investigator: The Principal Investigator acts on behalf of the Study Director for the delegated phase and is responsible for ensuring compliance with the principles of GLP for that phase. A fully cooperative, open working relationship between the Principal Investigator and the Study Director is essential. There should be documented agreement that the Principal Investigator will conduct the delegated phase in accordance with the study plan and the principles of GLP. Signature of the study plan by the Principal Investigator would constitute acceptable documentation. Deviations from the study plan or Standard Operating Procedures related to the study should be documented at the test site, be acknowledged by the Principal Investigator and reported to and acknowledged by the Study Director in a timely manner. The Principal Investigator should provide the Study Director with contributions that enable the preparation of the final report. These contributions should include written assurance from the Principal Investigator confirming the GLP compliance of the work for which he/she is responsible. The Principal Investigator should ensure that all data and specimens for which he/she is responsible are transferred to the Study Director or archived as described in the study plan. If these are not transferred to the Study Director, the Principal Investigator should notify the Study Director when and where they have been archived. During the study, the Principal Investigator should not dispose of any specimens without the prior written permission from the Study Director.

Study Personnel: The GLP principles require that all professional and technical personnel involved in the conduct of a study have a job description and a record of the training, qualifications, and experience, which support their ability to undertake the tasks assigned to them. Where study personnel are required to follow approved SOPs from another test site, any additional training required should be documented. There may be some sites where temporarily employed personnel carry out aspects of study conduct. Where these persons have generated or entered raw data, or have performed activities relevant to the conduct of the study, records of their qualifications, training, and experience should be maintained. Where these individuals have carried out routine operations such as livestock handling subject to supervision by more highly qualified staff, no such personnel records need be maintained.

2.4.1.1 Quality Assurance The quality assurance of multisite studies needs to be carefully planned and organized to ensure that the overall GLP compliance of the study is assured. Because there is more than one site, issues may arise with multiple management organizations and Quality Assurance programs.

Responsibilities of Lead Quality Assurance: The lead Quality Assurance should liaise with test site quality assurance to ensure adequate quality assurance inspection coverage throughout the study. Particular attention should be paid to the operation and documentation relating to communication among sites. Responsibilities for quality assurance activities at the various sites should be established before experimental work commences at those sites. The lead Quality Assurance will ensure that the study plan is verified and that the final report is inspected for compliance with the principles of GLP. Quality assurance inspections of the final report should include verification that the Principal Investigator contributions (including evidence of quality assurance at the test site) have been properly incorporated. The lead Quality Assurance will ensure that a Quality Assurance Statement is prepared relating to the work undertaken by the test facility including or referencing quality assurance statements from all test sites.

Responsibilities of Test Site Quality Assurance: Each test site management is usually responsible for ensuring that there is appropriate quality assurance for the part of the study conducted at their site. Quality assurance at each test site should review sections of the study plan relating to operations to be conducted at their site. They should maintain a copy of the approved study plan and study plan amendments. Quality assurance at the test site should inspect study-related work at their site according to their own SOPs, unless required to do otherwise by the lead Quality Assurance, reporting any inspection results promptly in writing to the Principal Investigator, test site management, Study Director, test facility management, and lead Quality Assurance. Quality assurance at the test site should inspect the Principal Investigator's contribution to the study according to their own test site SOPs and provide a statement relating to the quality assurance activities at the test site.

2.4.1.2 Master Schedules A multisite study in which one or more Principal Investigators have been appointed should feature on the master schedule of all sites concerned. It is the responsibility of test facility management and test site management to ensure that this is done. The unique identification of the study must appear on the master schedule in each site, cross-referenced as necessary to test site identifiers. The Study Director should be identified on the master schedule(s), and the relevant Principal Investigator shown on each site master schedule. At all sites, the start and completion dates of the study phase(s) for which they are responsible should appear on their master schedule.

2.4.1.3 ***Study Plan*** For each multisite study, a single study plan should be issued. The study plan should clearly identify the names and addresses of all sites involved. The study plan should include the name and address of any Principal Investigators and the phase of the study delegated to them. It is recommended that sufficient information be included to permit direct contact by the Study Director, for example, telephone number. The study plan should identify how data generated at sites will be provided to the Study Director for inclusion in the final report. It is useful, if known, to describe in the study plan the location(s) at which the data, samples of test and reference items and specimens generated at the different sites are to be retained. It is recommended that the draft study plan be made available to Principal Investigators for consideration and acknowledgement of their capability to undertake the work assigned to them, and to enable them to make any specialized technical contribution to the study plan if required. The study plan is normally written in a single language, usually that of the Study Director. For multinational studies, it may be necessary for the study plan to be issued in more than one language; this intention should be indicated in the original study plan, the translated study plan(s) and the original language(s) should be identified in all versions. There will need to be a mechanism to verify the accuracy and completeness of the translated study plan. The responsibility for the accuracy of the translation can be delegated by the Study Director to a language expert and should be documented.

2.4.2 Performance of the Studies

This section repeats the most important requirements from the principles of GLP and recommendations from the *Consensus Document on the Application of the GLP Principles to Field Studies* in order to provide useful guidance for organization of multisite studies. These documents should be consulted for further details.

Facilities: Sites may not have a full-time staff presence during the working day. In this situation, it may be necessary to take additional measures to maintain the physical security of the test item, specimens, and data. When it is necessary to transfer data or any materials among sites, mechanisms to maintain their integrity need to be established. Special care needs to be taken when transferring data electronically (email, internet, etc.).

Equipment: Equipment being used in a study should be fit for its intended purpose. This is also applicable to large mechanical vehicles or highly specialized equipment that may be used at some sites. There should be maintenance and calibration records for such equipment that serve to indicate their "fitness for intended purpose" at the time of use. Some apparatus (e.g., leased or rented equipment such as large animal scales and analytical equipment) may not have records of periodic inspection, cleaning, maintenance, and calibration. In such cases, information should be recorded in the study-specific raw data to demonstrate "fitness for intended purpose" of the equipment.

Control and Accountability of Study Materials: Procedures should be in place that will ensure timely delivery of study related materials to sites. Maintaining integrity/stability during transport is essential, so the use of reliable means of transportation and chain of custody documentation is critical. Clearly defined procedures for transportation, and responsibilities for who does what, are essential. Adequate documentation should accompany each shipment of study material to satisfy any applicable legal requirements, for example, customs, health, and safety legislation. This documentation should also provide relevant information sufficient to ensure that it is suitable for its intended purpose on arrival at any site. These aspects should be resolved prior to shipment. When study materials are transported between sites in the same consignment it is essential that there is adequate separation and identification to avoid mix-ups or cross contamination. This is of particular importance if materials from more than one study are transported together. If the materials being transported might be adversely affected by environmental conditions encountered during transportation, procedures should be established to preserve their integrity. It may be appropriate for monitoring to be carried out to confirm that required conditions were maintained. Attention should be given to the storage, return, or disposal of excess test and reference items being used at sites.

2.4.2.1 Reporting of Study Results A single final report should be issued for each multisite study. The final report should include data from all phases of the study. It may be useful for the Principal Investigators to produce a signed and dated report of the phase delegated to them, for incorporation into the final report. If prepared, such reports should include evidence that appropriate quality assurance monitoring was performed at that test site and contain sufficient commentary to enable the Study Director to write a valid final report covering the whole study. Alternatively, raw data may be transferred from the Principal Investigator to the Study Director, who should ensure that the data are presented in the final report. The final report produced in this way should identify the Principal Investigator(s) and the phase(s) for which they were responsible. The Principal Investigators should indicate the extent to which the work for which they were responsible complies with the GLP principles, and provide evidence of the quality assurance inspections performed at that test site. This may be incorporated directly into the final report, or the required details may be extracted and included in the Study Director's compliance claim and Quality Assurance statement in the final report. When details have been extracted the source should be referenced and retained. The Study Director must sign and date the final report to indicate acceptance of responsibility for the validity of all the data.

The extent of compliance with the GLP principles should be indicated with specific reference to the OECD principles of GLP and Regulations with which compliance is being claimed. This claim of compliance will cover all phases of the study and should be consistent with the information presented in the Principal Investigator claims. Any sites not compliant with the OECD principles of GLP should be indicated in the final report. The final report should identify the storage location(s) of the study plan, samples of test and reference items, specimens, raw data, and the final report. Reports produced by Principal Investigators should provide information concerning the retention of materials for which they were responsible. Amendments to the final report may only be produced by the Study Director. Where the necessary amendment relates to a phase conducted at any test site the Study Director should contact the Principal Investigator(s) to agree appropriate corrective actions. These corrective actions must be fully documented. If a Principal Investigator prepares a report that report should, where appropriate, comply with the same requirements that apply to the final report.

2.4.2.2 Standard Operating Procedures The GLP principles require that appropriate and technically valid SOPs are established and followed. The following examples are procedures specific to multisite studies:

- selection and monitoring of test sites;
- appointment and replacement of Principal Investigators;
- transfer of data, specimens, and samples between sites;
- verification or approval of foreign language translations of study plans or SOPs; and
- storage, return or disposal of test and reference items being used at remote test sites.

The principles of GLP require that SOPs should be immediately available to study personnel when they are conducting activities, regardless of where they are carrying out the work. It is recommended that test site personnel should follow test site SOPs. When they are required to follow other procedures specified by the Study Director, for example, SOPs provided by the test facility management, this requirement should be identified in the study plan. The Principal Investigator is responsible for ensuring that test site personnel are aware of the procedures to be followed and have access to the appropriate documentation. If personnel at a test site are required to follow SOPs provided by the test facility management, it is necessary for test site management to give written acceptance. When SOPs from a test facility have been issued for use at a test site, test facility management should ensure that any subsequent SOP revisions produced during the course of the study are also sent to the test site and the superseded versions are removed from use. The Principal Investigator should ensure that all test site personnel are aware of the revision and only have access to the current version.

When SOPs from a test facility are to be followed at test sites, it may be necessary for the SOPs to be translated into other languages. In this situation, it is essential that any translations be thoroughly checked to ensure that the instructions and meaning of the different language versions remain identical. The original language should be defined in the translated SOPs.

2.4.2.3 Storage and Retention of Records and Materials During the conduct of multisite studies, attention should be given to the temporary storage of materials. Such storage facilities should be secure and protect the integrity of their contents. When data are stored away from the test facility, assurance will be needed of the site's ability to readily retrieve data, which may be needed for review. Records and materials need to be stored in a manner that complies with GLP principles. When test site storage facilities are not adequate to satisfy GLP requirements, records and materials should be transferred to a GLP compliant archive. Test site management should ensure that adequate records are available to demonstrate test site involvement in the study.

2.4.3 Applications of GLP to *In Vitro* Studies for Regulatory Submissions

The early versions of the GLPs were written with long-term, animal-based toxicology studies in mind. Up to date, many of these animal-based studies are being replaced by *in vitro* studies, which are being used by many companies to make safety decisions without accompanying animal data. In order to assure that the data from these validated *in vitro* methods are universally accepted as being of high quality, the studies must be carried out in compliance with Good Laboratory Practices. Confusion in the interpretation of the regulations arises when the traditional GLPs are applied to short-term, nonanimal studies. Differences in test system handling and basic assay design require that the GLPs be supplemented with guidelines that fit the unique structure of these *in vitro* assays. Workshops were sponsored by such organizations as the European Center for the Validation of Alternative Methods (ECVAM) and Institute for *In Vitro* Sciences, Inc (IIVS) where interested individuals (such as cell culture technologists, toxicologists and quality assurance personnel from industry, academia, and government) were able to discuss the possible design of GLPs for these methods. Following these consensus building workshops, the OECD published "*The Application of the GLP Principles to Short Term Studies*" (1993, Rev. 1999), Number 7 [11] in their series on the Principles of GLP and Compliance Monitoring. These documents supplement the GLPs to effectively assure that there are GLP guidelines to cover all critical aspects of *in vitro* assay design.

The principles of GLP were originally written to address animal-based toxicology. However, there has been a growing appreciation that certain additions/modifications are required to meet the current state-of-the-art for *in vitro* studies and to assure the quality of the data they provide. To this end, ECVAM and the Institute for *In Vitro* Sciences convened the workshop on GLP. It should be emphasized that this workshop was not an effort to diminish the principles of GLP, but to enhance them. It drew on successful approaches used by industry, quality assurance professionals, and regulatory agencies with *in vitro* bioassays. The primary goals were

(A) to recommend additions/modifications to the OECD principles of GLP needed to address the specific needs of *in vitro* bioassays (e.g., test system characterization, facilities);

(B) to formalize standards of practice to ensure quality data from *in vitro* studies (e.g., use of controls and performance standards);

(C) to provide assistance in the implementation of the principles of GLP; and

(D) to address the specific needs of prevalidation and validation with respect to *in vitro* toxicology multistudy trials.

Characterization of the *in vitro* test system is of the utmost importance. The handling of *in vitro* systems is perhaps more critical than that of *in vivo* systems. The following points should be considered:

(1) Proper conditions should be established and maintained for the storage, handling, and care of test systems, in order to ensure the quality of the data. High-quality cell and tissue culture practices and good aseptic techniques, which are an essential part of *in vitro* work, should be enforced.
(2) Characterization of the test system is necessary, to ensure that it is what it claims to be, and any significant changes in the test system should be evaluated.
(3) Newly received test systems should be controlled for their purity, lack of contamination, suitability, and identity. Until the test system has met these predefined criteria, it should not be used in GLP-compliant studies and should be appropriately treated or destroyed. Any future contamination or defects that could affect the quality of the data should be investigated, and the test system returned to quarantine until it is "cleared." When beginning the experimental work in a study, the test system should be free of any contamination or defects, or any conditions that might interfere with the purpose or conduct of the study. Test systems that do not conform (e.g., those with contamination, defects) during the course of a study should be appropriately treated, if possible, or destroyed, if necessary, to maintain the integrity of the study. Any diagnosis and treatment of nonconformity before, or during, a study should be recorded.
(4) Records of origin (e.g., species, tissue, etc.), maintenance requirements, identity, source, date of arrival, and arrival condition of *in vitro* systems should be maintained.
(5) Test systems should be acclimatized (if necessary) to the test environment for an adequate period before the first administration or application of the test, control, or reference item. The propagation of the *in vitro* test system after receipt by the test facility should be consistent with the study plan and/or SOPs.
(6) All information needed to properly identify the *in vitro* test system should be adequately recorded throughout the course of the study. Individual test materials (e.g., flasks, plates) that are prepared and used during the course of the study should bear appropriate identification, wherever possible. Individual test system containers (e.g., transwells) too small to be individual labeled should be contained in a properly labeled outer container.
(7) Any material that comes into contact with the test system should be free of contaminants. Sterile equipment and good culture techniques should be used where appropriate.

Applying the OECD guidelines in a practical way to day-to-day laboratory work can be challenging. Three areas requiring unique interpretation for *in vitro* assays are the use of positive and negative controls and benchmarks;the care and maintenance of test systems; andquality assurance monitoring. Although each laboratory will need to apply the principles of GLP in a way that will complement their facility structure and infrastructure, there are some common concepts that could be applied to *in vitro* work performed in any laboratory.

2.5 21 CFR PART 11 IN RELATION TO GLP PROGRAMS

In March 1997, FDA issued final Part 11 regulations [7] that provided criteria for acceptance by FDA, under certain circumstances, of electronic records, electronic signatures, and handwritten signatures executed to electronic records as equivalent to paper records and handwritten signatures executed on paper. These regulations, which apply to all FDA program areas, were intended to permit the widest possible use of electronic technology, compatible with FDA's responsibility to protect the public health. After Part 11 became effective in August 1997, significant discussions ensued between industry, contractors, and the agency concerning the interpretation and implementation of the rule.

In April 2002, the US Food and Drug Administration published an update to the Final Rule for Part 11 "*Electronic Records; Electronic Signatures*" of Title 21 of the Code of Federal Regulations (21 CFR Part 11). This rule states the conditions under which the FDA considers electronic signatures and electronic records to be trustworthy, reliable, and equivalent to traditional handwritten signatures on paper. It defines the conditions under which a manufacturer or supplier must operate to meet its record keeping and record submission requirements when it uses electronic records and signatures rather than paper records and handwritten signatures. The FDA has issued formal written guidance to help clarify its expectations with regards to 21 CFR Part 11 compliance. From Guidance for Industry Part 11, Electronic Records; Electronic Signatures—Scope and Application, August 2003: The new mantra at the FDA concerning 21 CFR Part 11 is that "the agency will use its enforcement discretion" during the audit process for 21 CFR Part 11—compliant systems. The latest regulatory guidance for 21 CFR Part 11 represents the agency's attempt to make Part 11 more flexible and easier to implement. The agency acknowledges some of the challenges faced by industry as a result of Part 11, which restricted the use of some electronic systems, discouraged innovation in some cases, and resulted in significant costs.

The agency makes very clear that Part 11 is still in effect. When reviewing the "enforcement discretion" statement, one must be very careful. Enforcement discretion does not mean that the agency will not enforce! Upon careful inspection of the guidance the FDA states. We intend to enforce all other provisions of Part 11 including the following:

- Limited system access to authorized individuals
- Use of operational system checks

- Use of authority checks
- Use of device checks
- Determination that persons who develop, maintain, or use electronic systems have the education, training, and experience to perform their assigned tasks
- Establishment of and adherence to written policies that hold individuals accountable for actions initiated under their electronic signatures
- Appropriate controls over systems documentation
- Controls for open systems corresponding to controls for closed systems bulleted above (§ 11.30)
- Requirements related to electronic signatures (e.g., §§11.50, 11.70, 11.100, 11.200, and 11.300).

2.5.1 A New Risk-Based Approach

According to leading industry analysts, the costs associated with 21 CFR Part 11 compliance are staggering and could vary from $5 million to $400 million per organization, depending on the size and state of compliance of the company. The new guidance suggests a risk-based approach to Part 11 compliance. When reviewing Part 11, one must first understand the legal, regulatory, and practical implications of electronic records. The basic qualities of good electronic records are as follows:

- Authenticity
- Reliability
- Trustworthiness
- Integrity
- Accessibility as needed

From a legal perspective, the integrity of electronic records is a key element. The system and supporting processes all must be of the highest quality. Therefore, it is clear why the agency stipulated its continued enforcement of the above-mentioned principles of Part 11. In other words, the agency will maintain enforcement for all aspects of Part 11 that ensure record integrity.

2.5.2 Understanding Predicate Rule Requirements

Predicate rule requirements provide governance for most regulatory activities within a life sciences organization. Predicate rules are preexisting regulatory requirements such as GLP, GMP, and GCP guidelines. These requirements are essential to Part 11 in that they provide the ground rules for management of electronic records produced in accordance with Part 11 guidelines. It is important to understand the risk associated with documents required under current predicate rules, and incorporate this risk assessment in the development of Part 11—compliant systems.

2.5.3 21 CFR Part 11 Best Practices

Compliance with 21 CFR Part 11 cannot be achieved with technology alone. It may only be achieved through technology coupled with policies and procedures to ensure compliance. The following best practices are organized according to technology and policy procedures. These best practices were adopted based on practical experience in developing Part 11—compliant systems since August 20, 1997.

Technology-Best Practices may consist of the following:

(1) *Audit Trail must be Independently Generated:* All changes to records within a Part 11—compliant system should include time and date-stamped audit trails. Failure to do so violates 21 CFR Part 11.10(e), which clearly stipulates an "independently-recorded" audit trail. If an audit trail is independently recorded that means it cannot and should not be turned on or off on demand. Further, as a control for closed systems, 21 CFR Part 11.10(j) states its intent is to "deter record and signature falsification." Notice that the requirement sites both record and signature falsification.

(2) *Ensure that System Maintains an "Irrefutable Link" Between Documents, Metadata, and the Electronic Signature:* When electronic records are signed within Part 11—compliant systems, there must be an irrefutable link between the electronic record (document and/or associated metadata) and the electronic signature. Current best practice is to design systems such that whenever the electronic record is displayed, printed, or otherwise accessed the signature manifestation is always displayed. Further, it is best practice to maintain this irrefutable link when archiving electronic records in order to maintain integrity and authenticity of the records.

(3) *Establish Clear Electronic Signature Manifestations for all Electronic Records:* This is the most clearly defined part of the Final Rule; yet, it is the most misunderstood within the vendor community and the industry in general. If any record is signed in accordance with 21 CFR Part 11, it must include the following components, according to §11.50(a)(1),(2),(3):
   - Printed name of the signer (e.g., John R. Smith).
   - Date and time of signature execution (it is typically best practice to include the time on the server versus the time of the user desktop).
   - Meaning of signature (such as review, approval, responsibility, or authorship).

   It is current best practice to design systems in which these signature elements appear on all signed electronic records.

(4) *Validate the System:* 21 CFR Part 11.10(a) stipulates that "validation of systems to ensure accuracy, reliability, consistent intended performance, and the ability to discern invalid or altered records." Although the agency relaxed some Part 11 requirements, it is still the best practice to validate any Part 11—compliant system.

(5) *Establish Role-Based Access and Control:* The system must provide adequate security controls to prevent unauthorized access. Current best practice is

to establish security based on the role of the user. In most systems, users play multiple roles. For instance, any given user may be both an author and a reviewer; security controls should thus support these rules and allow authorized system administrators to configure security access accordingly.

(6) *Establish Password and Identification Controls:* 21 CFR Part 11.300 requires that password controls be unique and protected. Current best practice is to require password and ID control changes every 60–90 days. When employees are terminated or otherwise no longer require system access, they should be immediately disabled from the system in accordance with Part 11 guidelines.

(7) *Series of Signings:* Best practice for execution of a series of signings in accordance with Part 11.200(1)(i) is to require BOTH signature components for each signing. Although Part 11.200(1)(i) allows, after the first signing "subsequent signings shall be executed using at least one electronic signature component."

(8) *Avoid Hybrid Systems, Where Practical:* A hybrid system is defined as an automated process that combines electronic records and manual paper records. Current best practice is to migrate to a fully automated electronic records/electronic signature system. Although the FDA does not prohibit hybrid systems, they have expressed concern over the years about their acceptability and ability to achieve sustained compliance. Compliance process control systems allow organizations to avoid hybrid systems and processes and move to higher levels of compliance through automated compliance process control systems.

(9) *Do not Over Customize Technology Solutions:* Over customization has resulted in accelerated costs for compliance with Part 11. Be practical. All pharmaceutical, medical device, and biotechnology organizations strive for the same goal when it comes to Part 11. Current best practice is to leverage off-the-shelf technology where applicable. There are many Part 11—compliant systems on the market apply the principle of Caveat Emptor: "let the buyer beware;"examine your vendors' compliance with Part 11 requirements;conduct supplier audits to ensure a quality-oriented software development process.

Policy and Procedure Best Practices may include (but not limited to):

(1) *Establish Corporate Internal Policies and Guidelines for Part 11:* Written procedures codify management's intent and criteria for operational execution and excellence. 21 CFR Part 11 Subpart B 11.10(j) requires "the establishment of, and adherence to, written policies that hold individuals accountable and responsible for actions initiated under their electronic signatures, in order to deter record and signature falsification." It is clear that the intent is to avoid fraudulent activity. Part 11 stipulates that such policies be established by each organization to ensure compliance. As a best practice, it is recommended that,

at a minimum, the following policies and procedures be established in association with any system developed in compliance with Part 11:

- Validation policy and procedures (21 CFR Part 11.10(a))
- Disaster recovery
- Revision and change control procedures (21 CFR 11.10(k)(2))
- System access and security procedures (21 CFR Part 11.10(c),(d),(g))
- Training procedures (21 CFR Part 11.10(i))
- Document control procedures (21 CFR Part 11.10(k)(1), (2))

(2) *Develop a Clear, Comprehensive Migration Strategy:* As Part 11 systems mature, electronic records captured within the system must be migrated to near line or far line storage. Migration requires the transfer of the audit trail, electronic signatures, and their associated records. It should be clear in the above technology best practices section that an irrefutable link be maintained until ultimate destruction of the electronic record. In most cases, migration is often an afterthought. Current best practice for Part 11 is to consider a comprehensive migration strategy up front. This involves review of predicate rule requirements and migrating complete electronic records to ensure integrity and authenticity.

(3) *Understand the Impact of Part 11 "Open" or "Closed" System Definitions:* An open system environment according to 21 CFR Part 11.3(9) means "an environment in which system access is not controlled by persons who are responsible for the content of electronic records that are on the system." A closed-system environment according to Part 11.3(4) means "an environment in which system access is controlled by persons who are responsible for the content of electronic records that are on the system." There are differences in the requirements depending upon the system type, so it is important to ensure that the company understands what type of system it has and if changes in that system will change its type. If so, they must ensure that the system still meets the Part 11 requirements.

(4) *Establish Retention Policies Based on Current Predicate Rule Requirements:* Part 11 systems are designed to support the electronic management (signature/records) of records required by current predicate rules. Part 11 is not a mandate. It is a guideline for those organizations that choose to use electronic records and signatures. Prior to the development and deployment of any electronic records/signature system, current best practice is to establish retention policies based on current predicate rule requirements.

When determining the guidelines for Part 11 compliance, company should review the above current best practices. Best practices for Part 11 provide a baseline for acceptable systems implementation. Best practices by definition can be legally derived or based on acceptable industry standards. Part 11 is considered a legal best practice in and of itself. It is strongly recommended adopting a top-down approach to compliance. The new guidelines recently issued by the FDA are good and acceptable business practices, even in the absence of a strict regulatory requirement. Compliance with Part 11 makes good business sense—determine how Part 11 can be best applied to your organization and use your best judgment as to applied technology.

2.5.4 Use of Electronic Signatures

Electronic signatures are used to ensure that actions/records are attributable in a legally binding manner. The following are excerpts from the 21 CFR Part 11 regulations. The regulation covers the controls and procedures that should be in place to ensure that the electronic signature can be considered binding and verifiable. Three important sections define *system controls*, *signature controls*, and *password controls*.

System Controls include controls in workflows, procedures, and system design to ensure that attribution and electronic signatures can be verified. Most important are

- System validation (discussed above)
- Audit trails (also discussed above)
- Operational system checks—enforcing permitted sequencing of events (workflow management)
- Authority checks—ensuring that only authorized individuals can use the system
- Device (e.g., terminal) checks to determine, as appropriate, the validity of the source of data input or operational instruction (IP Address/MAC Address logging)
- Documentation of adequate user education, training, and experience
- Policies for individual accountability for actions initiated under their electronic signatures
- Controls over distribution of, access to, and use of system documentation
- Change control procedures documenting time-sequenced modification of system documentation

Signature Controls (for signatures based on passwords) include the following:

- Use of at least two distinct ID components such as an user ID and password
- Signature verification

When an individual executes a series of signings during a single, continuous period of controlled system access, the first signing shall be executed using all electronic signature components; subsequent signings shall be executed using at least one electronic signature component that is only executable by, and designed to be used only by, the individual.

When an individual executes one or more signings not performed during a single, continuous period of controlled system access, each signing shall be executed using all of the electronic signature components.

Password Controls include the following:

- Password expiration
- Rigid and rigorous password deactivation and temporary generation protocols
- Encryption and transaction safeguards to prevent sniffing (SSL, JavaScript MD5)

To ensure the integrity and quality of raw data collected from regulated studies, the Bioanalytical Laboratories abide by United States Food & Drug Administration (US FDA) Regulations for Good Laboratory Practices should consider both 21 CFR Part 58 Good Laboratory Practice for Nonclinical Laboratory Studies and 21 CFR Part 11 Electronic Records and Electronic Signatures.

At present, the use of electronic records as well as their submission to the FDA is voluntary. However, where an organization does decide to use electronic records and electronic signatures, the requirements of the rule must be met in full for all relevant electronic records. This applies to the manufacturer, as well as the vendors supplying the manufacturer, and should a vendor not be in compliance, both vendor and manufacturer face penalties. Compliance is required of anyone involved in the development, manufacturing and marketing of life sciences products, including drugs, diagnostics, and medical devices. It is also required of all companies operating under GMP, GLP, or GCP guidelines. Within the pharmaceutical industry, such regulatory requirements are nothing new. The FDA has long enforced strict guidelines for pharmaceutical manufacturers regarding the keeping of detailed records that define processes and procedures—from the designing stages through packaging, sales, and distribution. 21 CFR Part 11 was seen as a much-needed means to bring these record keeping procedures into the digital information age while still protecting public health.

Electronic record keeping provides a much less cumbersome process than traditional paper trails, and saves companies both time and money. It reduces errors, ensuring the reliability and authenticity of data, providing for increased process efficiencies, data management, and company profitability. In fact, for this very reason many pharmaceutical companies have adopted company-wide, CFR-compliant data-handling policies, and are purchasing CFR-compliant equipment even for areas where it is not required. The elements of compliance the FDA is looking for under 21 CFR Part 11 are the ability to prevent unauthorized changes to electronic records, the monitoring of all electronic signatures, the retention of electronic records, and the authentication of users with access rights to data. In addition, all changes to process and device configurations must be archived and be traceable via an electronic signature auditing trail. In short, 21 CFR Part 11 requires the creation of a detailed electronic trail of who did what, where, when, and for what reason.

The application of regulatory guidances/guidelines, which generally elaborate specific and documentary requirements, should be differentiated from the "compliance" perspectives of GxPs (e.g., GLP, GCP, and cGMP). Similarly, the FDA Guidance for Industry: Bioanalytical Method Validation, May 2001 can be implemented in a GLP or non-GLP program; implementation of this Guidance is not synonymous with GLP.

Indeed, in the context of the FDA, bioanalysis is an integral part of a number of non-GLP-based guidances (e.g., 21 CFR 320—Bioavailability and Bioequivalence Requirements (Drug for Human Use) [12]. A similar EU document, Committee on Proprietary Medicinal Products (CPMP) Note for Guidance—has been published and has a similar requirement for bioanalysis. On the other hand, the ICH-53A: Toxicokinetics Guidance on the assessment of systemic exposure in Toxicity Studies Topics 3A [13], as the ICH appellation (International Conference on Harmonization) implies, in addition to being contained within the Federal Register, is incorporated into the national legislation of those countries committed to the concept of the ICH. This document definitely requires the application of GLP to both the toxicological aspect and to the related bioanalysis.

The FDA requirements state "the analytical laboratory conducting pharmacology/toxicology and other preclinical studies for regulatory submission should adhere to the FDA's GLP and to sound principles of quality assurance throughout the testing process," this strongly suggests that the bioanalytical method validation for preclinical studies be carried out to GLP standards. On the other hand, bioanalytical methods for human bioavailability (BA), bioequivalence (BE), pharmacokinetic (PK) studies, and drug–drug interaction (DDI) studies must meet the criteria of 21 CFR 320.29 and are essentially not covered by GLP. The FDA Bioanalytical Method Validation Guidance for Industry [14] provides some insight, *viz.* "The analytical laboratory conducting BA and BE studies should closely adhere to FDA's Good Laboratory Practices and to sound principles of quality assurance throughout the testing process." The FDA compliance manual for inspecting BA and BE studies, which states "The analytical laboratory should have a written set of standard operating procedures to ensure a complete system of quality control and assurance." The SOPs should cover all aspects of analysis from the time the sample is collected and reaches the laboratory until the results of the analysis are generated and reported. The SOPs should include record keeping, security, and chain of sample custody (and integrity). Thus, at the laboratory bench level the differences between GLP or regulated based studies and clinical studies are minimal and tend to be restricted to the use of terminology where terms such as Study Director and protocols, have rigid legal meanings in the GLP context but are more loosely applied in clinical studies.

In addition to the regulatory guidances, which govern the specific aspects of the drug development programs (e.g., bioequivalence, toxicology, etc.), there are a number of requirements that overlay all aspects of laboratory work. One of these requirements is designed to ensure that the quality of the instrumentation used for both GLP and cGMP studies is fit for purpose. However, there are differences in the wording of GLP and cGMP requirements as to what is to be expected. Thus EU GMP regulations require equipment, which is critical for the quality of the products (not quality of data), to be subjected to appropriate qualification for human use. On the other hand, the FDA GLP regulations require that equipment which is used for generation, measurement, or assessment of data adequately tested, calibrated, and/or standardized. The qualification process is generally involved in a series of steps including Design Qualification (DQ), Installation Qualification (IQ), Operation Qualification (OQ), and Performance Qualification (PQ). Qualification is the responsibility of the end user and it is designed to ensure that the instrument is both fit for purpose and continues to function as expected, over the "lifetime" and/or during the course of a study. Responsibility cannot be abrogated to the vendor/supplier. However, this does not prevent suppliers from providing assistance in the form of protocols and targeted services. Instrument qualification and calibration will be elaborated in respective chapters within this book.

2.6 GLP, cGMP, AND ISO APPLICABILITIES, SIMILARITY, AND DIFFERENCES

2.6.1 GLPs, cGMPs, ISO 17025:2005: How Do They Differ?

Increasingly, a number of laboratories must meet both Good Laboratory Practices compliance and International Organization for Standardization/International Electrotechnical Commission (ISO/IEC) 17025:1999 general requirements for the competence of testing and calibration laboratories [15] and 2005 accreditation. Furthermore, GLPs are sometimes confused and/or connected with current Good Manufacturing Practices (cGMPs) requirements when active pharmaceutical ingredients need to be certified for use in clinical trials. How do these requirements differ and what should laboratories that must comply with both GLPs and ISO accreditation do?

GLP is more than simply good analytical/bioanalytical practice. Although applying good analytical/bioanalytical practices is important, it is not sufficient for GLP compliance, which is study based. GLPs are used to ensure the quality and integrity of safety data submitted in support of the approval of regulated products. For example, GLPs are used specifically in nonclinical studies by the US Food and Drug Administration. In the hands of the US Environmental Protection Agency, they ensure the quality and integrity of data submitted to support applications for research or marketing permits for regulated pesticides products. For the Organization for Economic Cooperation and Development, GLPs have been adopted on an international scale for preclinical safety studies.

2.6.2 GLPs Versus GMPs

On the other hand, GMPs are process based and are used to regulate quality control laboratories in manufacturing. GMP regulations require that manufacturers, processors, and packagers of drugs, medical devices, some food, and blood take proactive steps to ensure that their products are safe, pure as specified, and effective, thus minimizing or eliminating instances of contamination, errors, or mix-ups. They protect the consumer from purchasing a product that is not effective or may be dangerous. GMP regulations address issues including record keeping, personnel qualification, sanitation, cleanliness, equipment verification, process validation, and complaint handling. Quite often, GMP is referred to as "cGMP" (current Good Manufacturing Practices). A distinct difference is between laboratory research and development (GLP) and manufacturing and production of APIs and/or drug products (GMP).

2.6.3 GLPs Versus ISO/IEC 17025:2005

An Accreditation Update column on GLP versus ISO/IEC 17025 predates ISO/IEC 17025:2005. Inquiries have been received as to whether this revision of ISO/IEC 17025 has resulted in significant changes for laboratories that must meet both GLP and ISO/IEC 17025:2005 requirements. Since the column, GLP compliance has been more strictly enforced by inspections. Previously, laboratories frequently simply considered themselves GLP compliant and were not evaluated by a third-party independent body. Now, more and more, accrediting bodies with checklists verify compliance. The European Parliament and Council Directive 2004/10/EC specify the harmonization of laws, regulations, and administrative provisions relating to the principles of GLP. Directive 2004/9/EC provides a harmonized system for study audit and inspection of laboratories to ensure that they are working under GLP conditions. The provisions for inspections and verification of GLP compliance is found in Annex I of the OECD Mutual Acceptance of Data, a multinational agreement that requires that results in nonclinical safety data must be accepted by OECD and adhering countries for the purposes of assessment and other uses related to the protection of human health and the environment. Participation in the MAD system begins by provisional adherence, during which time nonmembers work with OECD countries to make their GLP monitoring program acceptable to all members. After a team of OECD experts from three OECD countries have evaluated the nonmember GLP compliance monitoring program onsite, and based on the outcome of this evaluation, the OECD council can invite the provisional adherent to become a full adherent to the Council Acts, with the same rights and obligations as OECD counties. India, Singapore, Brazil, Argentina, and Malaysia are provisional adherents. Full adherence to the OECD system means that nonmembers will accept data from OECD countries under MAD conditions.

In today's world, it is more likely to specifically require a GLP inspection so a laboratory might be able to coordinate GLP inspections and ISO/IEC 17025:2005 assessments. For instance, if a test facility in Belgium seeks accreditation to both ISO 17025 and GLP, it can submit separate requests to both organizations and use certain findings for both. Some aspects, such as the separation of activities, identification, storage and handling of test and reference materials, maintenance, and calibration, can be inspected in common. ISO 17025:2005 requirements and GLP regulations are both concerned with traceability and integrity of data. Both sets of requirements continue to be designed to ensure data quality and acceptance of data. However, their means to this end are still markedly different. ISO 17025 underscores all of the requirements that testing and calibration laboratories must meet to demonstrate that they operate a management system, are technically competent, and are able to generate technically valid results. ISO 17025:2005 especially emphasizes meeting the needs of the customer and continuing improvement. To meet customer needs, a laboratory that is accredited to ISO/IEC 17025:2005 must maintain records of feedback and complaints and provide service to the customer.

GLPs have a unique requirement for a study director who is responsible for the technical conduct of safety studies, as well as the interpretation, analysis, documentation, and reporting of results. While both GLPs and ISO/IEC 17025:2005 have important roles for a Quality Assurance Unit (QAU) or a quality manager, the functions and responsibilities differ somewhat depending on the requirement. Only the QAU is required to maintain a master schedule of all studies. For GLP studies, the required information that must be recorded is much more clearly specified. GLP regulations go beyond requirements for laboratory analysis and include those for animal facilities.

2.6.4 ISO Versus GLPs

More and more organizations/producers submit their company laboratories to an assessment in order to acquire accreditation or abandons running an own laboratory and uses the services or an external laboratory, which posses an accreditation certificate. Laboratory accreditation is an acknowledgement of the laboratory's competences to perform certain actions, by the certifying agency. Accreditation is given upon application of laboratories, after their evaluation and confirmation that they meet the specified requirements and conditions. The basis for meeting the requirements by the laboratory is the PN-EN ISO/IEC 17025:2001 standard "General requirements regarding the competences of research and calibrating laboratories." Compliance of laboratory's operations with the requirements of the international PN-EN ISO/IEC 17025:2001 standard certifies about its competences. The requirements included in the standard, which regard the laboratory's management and the technical requirements, regard each laboratory, regardless of its type, size or structure, and adopted methods. Meeting the requirements regarding the laboratory's management is equal to meeting the quality management system's requirements included in the ISO 9001 standard, but is not sufficient to confirm the laboratory's competences to perform specific tests or calibrations.

Thus, the second group of requirements of the PN-EN ISO/IEC 17025:2001 standard is related with the laboratory's technical competences and concerns: the laboratory's equipment, measurement consistency, the methods of testing and calibrating as well as their validation, personnel, premises and environmental conditions of taking samples, handling of the testing and calibration objects; ensuring the quality of results and presenting the results. Confirming the competences to perform certain tests through accreditation, conducted according to worldwide adopted criteria is to ensure that the results are reliable, unbiased, and credible and that they can be recognized not only at the state/national level, but also at the international level. If conducting tests independently from the supplier and the recipient is not required, the producer may submit the products to testing at the company laboratory. Such a laboratory conducts the tests for the organization within the frames of internal production control. The company laboratory, which wishes to achieve reliability, credibility, and precision of testing results, should adopt the rules of the Good Laboratory Practice. Good Laboratory Practice specifies the requirements for laboratories regarding personnel, equipments, testing, and recording methods. General GLP requirements—Good Laboratory Practice for a testing laboratory cover the following:

(1) Properly trained personnel
(2) Standard analytical methods
(3) Schedule of frequency of taking test samples, the way samples are taken
(4) Proper equipment (controlled and calibrated on a regular basis)
(5) Storage of the tests' results
(6) System of recordings

A laboratory without accreditation performs the same tasks as an accredited company and/or laboratory. However, the producer may trust that the results acquired during the tests are reliable and repeatable. On the other hand, independent accredited laboratories are competitive in the services' market as they possess a confirmation that they utilize the quality management system, are technically competent and are capable of producing reliable results.

Some laboratories must comply with both GLP regulations and ISO 17025:2005 requirements. If so, it is recommended that these laboratories generate a quality manual and Standard Operating Procedures that meet the general requirements of ISO 17025:2005 and include additional specifications for GLP studies. The laboratory must meet the needs of customers and achieve continuing improvement to satisfy the requirements of ISO 17025:2005. If possible, a laboratory seeking compliance to both sets of requirements should coordinate the inspection and assessment to cover common aspects. In some instances, both GLP compliance and ISO 17025:2005 accreditation can be achieved through one accreditation body, so a laboratory might consider such an option. Finally, because GLP requirements go beyond the laboratory to animal facilities, it is not sufficient to have only the laboratory to meet the requirements. An organization must ensure that requirements for animal facilities are met as well.

ISO/IEC 17025:2005 specifies the general requirements for the competence to carry out tests and/or calibrations, including sampling. It covers testing and calibration performed using standard methods, nonstandard methods, and laboratory-developed methods. It is applicable to all organizations performing tests and/or calibrations. These include, for example, first-, second-, and third-party laboratories, and laboratories where testing and/or calibration forms part of inspection and product certification. ISO/IEC 17025:2005 is applicable to all laboratories regardless of the number of personnel or the extent of the scope of testing and/or calibration activities. When a laboratory does not undertake one or more of the activities covered by ISO/IEC 17025:2005, such as sampling and the design/development of new methods, the requirements of those clauses do not apply. ISO/IEC 17025:2005 is for use by laboratories in developing their management system for quality, administrative, and technical operations. Laboratory customers, regulatory authorities, and accreditation bodies may also use it in confirming or recognizing the competence of laboratories. ISO/IEC 17025:2005 is not intended to be used as the basis for certification of laboratories. Compliance with regulatory and safety requirements on the operation of laboratories is not covered by ISO/IEC 17025:2005.

ISO 9001:2000, on the other hands, specifies requirements for a quality management system where an organization

(1) needs to demonstrate its ability to consistently provide product that meets customer and applicable regulatory requirements, and
(2) aims to enhance customer satisfaction through the effective application of the system, including processes for continual improvement of the system and the assurance of conformity to customer and applicable regulatory requirements.

All requirements of this International Standard are generic and are intended to be applicable to all organizations, regardless of type, size, and product provided: (A) where any requirement(s) of this International Standard cannot be applied due to the nature of an organization and its product, this can be considered for exclusion. (B) where exclusions are made, claims of conformity to this International Standard are not acceptable unless these exclusions are limited to requirements within clause 7, and such exclusions do not affect the organization's ability, or responsibility, to provide product that meets customer and applicable regulatory requirements.

2.7 GOOD CLINICAL PRACTICES AND GOOD CLINICAL LABORATORY PRACTICES

The regulatory environment in which clinical trials are conducted continues to evolve. The changes are generally focused on requiring more rigorous control within the organizations performing clinical trials in order to ensure patient safety and the reliability of data produced. The global acceptance of the ICH Guideline for Good Clinical Practice (GCP) and the implementation of the European Union Clinical Trials Directive (2001/20/EC) are two clear examples of such change. While the EU Clinical Trials Directive and ICH GCP Guideline clearly specify roles such as that of the Ethics Committee, the sponsor and the investigator to name just a few, they only vaguely define the standards to be applied in the analysis of samples from a clinical trial. The EU Clinical Trial Directive states that guidance documents may be issued to define the requirements for various aspects of trials, but it is not clear at this time whether these will include the analyses of trial samples. The most applicable reference within ICH that indicate the standards required for the analysis of samples are in Section 2.13 "Systems with procedures that assure the quality of every aspect of the trial should be implemented," and in Section "Essential Documents" Parts 8.2.12 and 8.3.7.

For some years, it has been internationally recognized that clinical laboratories processing specimens from clinical trials require an appropriate set of standards to guide good practices. With that aim in mind, the *Good Clinical Laboratory Practice Guidelines* were drafted and published in 2003 by a working party of the Clinical Committee of the British Association of Research Quality Assurance (BARQA). This guidance identifies systems required and procedures to be followed within an organization conducting analysis of samples from clinical trials in compliance with the requirements of Good Clinical Practice. It thus provides sponsors, laboratory management, project managers, clinical research associates (CRAs), and quality assurance personnel with the framework for a quality system in analysis of clinical trial samples, ensuring GCP compliance overall of processes and results.

In 2006, WHO/TDR convened a meeting of organizations engaged in clinical trials in disease endemic countries to discuss the applicability of GCLP guidelines to their work. It was agreed that GCLP would be a valuable tool for improving quality laboratory practice. In line with that agreement, TDR/WHO recently acquired copyright to GCLP guidelines [16] that were originally published in 2003 by a working party of the Clinical Committee of the BARQA, with the aim of disseminating them widely in developing countries and developing related training materials. Compliance with them will allow clinical laboratories to ensure that safety and efficacy data is repeatable, reliable, auditable, and easily reconstructed in a research setting. GCLP guidelines set a standard for compliance by laboratories involved in the analysis of samples from TDR-supported clinical trials.

GCLP is considered to be a critical need as a hybrid between GLP and GCP for medical testing laboratories conducting clinical trials. This guideline is intended to provide a framework for the analysis of samples from clinical trials on the facilities, systems, and procedures that should be present to assure the reliability, quality, and integrity of the work and results generated by their contribution to a clinical trial. The principles are intended to be applied equally to the analysis of a blood sample for routine safety screening of volunteers (hematology/biochemistry) as to pharmacokinetics or even the process for the analysis of ECG traces. The types of facilities undertaking analyses of clinical samples may include pharmaceutical company laboratories, contract research organizations (CROs), central laboratories, pharmacogenetic laboratories, hospital laboratories, clinics, investigator sites, and specialized analytical/bioanalytical services.

Good Clinical Laboratory Practice Guidelines was published by WHO/TDR under the terms of an agreement between WHO and BARQA. Meanwhile, GCLP training materials specifically addressing the conduct of clinical trials in tropical countries also are under development by WHO/TDR and its partners (double check if available by now). Here are quick recaps for GCP, GLP, and GCLP as follows.

Good Clinical Practice is an international ethical and scientific quality standard for designing, conducting, recording, and reporting trials that involve the participation of human subjects. Compliance with this standard provides public assurance that the rights, safety, and well being of trial subjects are protected, consistent with the principles that have their origin in the Declaration of Helsinki (ICH GCP Guideline).

Good Laboratory Practice is intended to promote the quality and validity of test data. It is a managerial concept covering the organizational process and the conditions under which laboratory studies are planned, performed, monitored, recorded, reported, and archived (OECD GLP principles).

Good Clinical Laboratory Practice (*GCLP*) applies those principles established under GLP for data generation used in regulatory submissions relevant to the analysis of samples from a clinical trial. At the same time it ensures that the objectives of the GCP principles are carried out. This ensures the reliability and integrity of data generated by analytical laboratories. It is recognized that a number of countries are already applying the GCLP principles to the analysis of clinical trial samples. Indeed it is possible within some of these countries for clinical laboratories to be accredited by the National Monitoring Authority. Some organizations and indeed countries operate proficiency testing schemes to which laboratories subscribe. While these ensure the integrity of the analytical process they may not assure compliance with GCP. The GCLP is intended to provide a unified framework for sample analysis to lend credibility to the data generated and facilitate the acceptance of clinical data by regulatory authorities from around the world. It is important to recognize that the framework outlined in this document will be applied across a diverse set of disciplines involved in the analysis of samples from clinical trials. It is therefore important to understand that this framework should be interpreted and applied to the work of those organizations that undertake such analyses with the objective of assuring the quality of every aspect of the work that they perform.

2.8 GAP AND CURRENT INITIATIVES ON REGULATING LABORATORY ANALYSIS IN SUPPORT OF CLINICAL TRIALS

It is widely recognized within industries that neither the US FDA or EMEA regulation nor ICH harmonized nor WHO guidelines address laboratory analysis and bioanalysis in support of clinical trials.

The FDA and OECD good laboratory practices for nonclinical laboratory studies regulations clearly address the quality principles for safety studies prior to clinical trials. The FDA regulations for Bioavailability and Bioequivalence Requirements do provide that methods must be demonstrated to be accurate, sufficiently sensitive and have appropriate precision, but do not address any current thinking regarding quality principles beyond requiring that the drug products under consideration are manufactured according to appropriate levels of GMP. The collection of regulations that fall within the scope of FDA good clinical practices do not address quality systems or guidelines for laboratory studies, with the exception of 21 CFR 314—applications for FDA approval to market a new drug, which requires that the nonclinical data are generated according to GLP regulations and the drugs are manufactured according to GMP regulations. The EMEA guidance on the investigation of bioavailability and bioequivalence, 140198en states that the studies "should be conducted according to the applicable principles of GLP, however, the studies fall outside the formal scope of GLP." The FDA guidance on Bioanalytical Method Validation (May 2001) [14], and the subsequent AAPS/FDA Bioanalytical workshops, resulting in the whitepaper "Quantitative Bioanalytical Methods Validation and Implementation: Best Practices for Chromatographic and Ligand-Binding Assays" [17] represents the most proactive effort to address the existing regulatory gap for Bioanalysis supporting clinical trials.

Even with the recent efforts of the joint AAPA/FDA Bioanalytical workshops and publications, adequate guidance for the industry is still absent. The current methodology for Bioanalysis is not uniformly applied in labs, examples include the implementation of incurred sample reanalysis, repeat analysis, and "event resolution" (investigation of unexpected or unforeseen circumstances, and the resultant corrective and preventative actions). The roles of sponsor, Study Director and contributing scientists, as well as the role of the quality assurance and quality control unit (QAU/QCU) is not addressed in this publication, leaving the application of the "best practices" to the individual labs. As a result of the inconsistencies in the application of the various regulatory standards and guidance documents, regulatory agencies have had to apply a higher level of scrutiny to the bioanalytical results submitted in support of NDA and ANDA submissions. Further, the lack of uniform application of the regulations and absence of clear guidance has resulted in inconsistent regulatory enforcement applied to studies and the organizations conducting the Bioanalysis for clinical studies. Until the industry has developed a coordinated approach to identifying and adopting the best practices for quality management systems applicable to the laboratory execution of clinical studies, the regulatory agencies will necessarily have to establish standards through the issuance of nonconformance reports and warning letters.

Because of the high level of outsourcing by Generic Pharmaceutical companies for execution of the Bioavailability and Bioequivalence studies required for FDA marketing approval, Generic Pharmaceutical companies are most vulnerable to the inconsistent practices and applications of regulatory principles during the execution of the studies. An expert working group launched an initiative with the goal to develop an industry wide acceptable standard comprising the "best practices" and the principles of the regulations and guidelines. The Bioanalytical Quality Standard Initiative under development is based on the quality management system that is fully, and effectively, implemented by increasing number of companies since its inception in 2009. The Standard completely integrates the best practices from the core regulatory requirements of the FDA, GMP for drug products and medical devices, GLP and Bioavailability and Bioequivalence, as well as the ICH guidelines for quality, including recently adopted Q9 (Quality Risk Management) and Q10 (Pharmaceutical Quality Management). The quality standard also addresses the major issues noted in FDA warning letters and 483's issued over the last couple of years; the top nonconformance relating to inadequate or failure to perform or complete event resolution (investigation along with appropriate corrective and preventative actions in response to unexpected for unforeseen circumstances). Other significant areas addressed in the Bioanalytical Quality Standard include the responsibility of Executive Management in the quality planning process and ongoing assessment and evaluation of the effectiveness of the quality system, through both internal auditing and annual review of key metrics and system indicators.

The format of the standard follows that provided by the ICH Q7A standard, providing guidance appropriate for a Bioanalytical Laboratory for each of the 19 sections. For example, each clinical trial is unique, and as such, the specific laboratory activities are best provided in a protocol that is specific to the study, reviewed and approved by a study director or project manager, the sponsor, and consistent with the principles of GMP, the quality unit. On the other hand, a Bioanalytical Laboratory should perform studies consistently, and standard operating procedures for development of the protocol, coordinating workflow through the laboratory, communication with the sponsor, resolving events and reporting results should be developed. In keeping with the principles of GMP and ICH, the quality assurance unit has greater responsibility for the development of the quality management system and for ensuring overall compliance with the system and the principles of the regulatory requirements; however, everyone has responsibility for quality and compliance to the quality management system. The quality unit is responsible for ensuring that events are resolved, and the study director is responsible for the investigation and the subsequent corrective and preventative actions that are implemented as a result of the event. The primary objective was to develop a broad standard that would be adopted by the laboratories performing the analysis for the clinical trials, resulting in consistent approach and application of the best practices and principles of the regulations. Sponsor companies could expect comparable execution, data and results obtained from different laboratories. When required, sponsor companies could assess compliance against the standard in advance of contracting with the laboratory and potentially reduce exposure to risk. If appropriate, third party certification to the standard could be considered. An appropriate, comprehensive, industry accepted and adopted standard would contribute to more consistent enforcement by the regulatory agencies, moving the industry once again in the direction of self regulating to the highest quality standards and current "best practices."

Ultimately, the goal of any increased control or standardization of the processes is to reduce risk, to both the company and the consumer. The clear guidance provided by this critical initiative moves in the direction of reducing risk while providing greater benefits. Studies executed under the guidelines established by this standard are expected to withstand the increased scrutiny by the agencies, provide process transparency and highest levels of accuracy and integrity in the results. This could, in fact, lead to reduction in the registration approval cycle due to fewer issues or concerns resulting from the bioanalytical results in the clinical trials.

In the United States, laboratories testing products regulated by the Food and Drug Administration or the Environmental Protection Agency must operate within the guidelines published by these agencies and codified as law in the *Code of Federal Regulations* and *Federal Register* (*FR*). These guidelines are known as GLP, cGMP, and GCP. cGMP guidelines are codified under 21 CFR Parts 210 and 211 for finished pharmaceuticals, and 21 CFR Parts 225 and 226 for Type A medicated articles and finished feeds regulated by FDA. GLP guidelines are codified under 21 CFR Part 58 for nonclinical laboratory studies under FDA. The EPA regulates GLP through 40 CFR Part 160 for products/laboratory analysis regulated by the *Federal Insecticide, Fungicide, and Rodenticide Act of 1996* (*FIFRA*) and 40 CFR Part 792 160 for products/laboratory analysis regulated by the *Toxic Substances Control Act of 1976* (*TSCA*). Global acceptance for GCPs written by the International Conference on Harmonization of Technical Requirements for Registration of Pharmaceuticals for Human Use (ICH) have been published, and thus are codified under 62 FR 25692 for products/laboratory analysis regulated by FDA. For clinical laboratories, GCPs are codified under 21 CFR Part 493, the

Clinical Laboratories Improvement Act of 2003 (CLIA). GCPs can often be interpreted for the needs of each business concern. More recently GCLP was established to bridge regulated laboratory work (clinical, analytical, bioanalytical laboratories, etc.) and clinical trials for ensuring the reliability, quality, and integrity of data to support regulatory submission for approval. If work is being done in these areas, it is recommended that the quality assurance department be consulted to determine which GxP regulations apply.

GLP, cGMP, GCP, and GCLP interpretations for analytical laboratories contain many common elements and have now become collectively known in the pharmaceutical industry as "GxP." Grouping these guidelines as GxP compliance has created a harmonized norm and helped many companies address the varied interpretations that exist within the industry. In terms of instrumentation, it is arguable whether cGMP or GLP is stricter in their requirements, simply because their interpretation is left up to the business concern and each company interprets the guidelines to suit their own business practices. Setting up policies to comply with both cGMP and GLP as GxP guidelines eliminates this issue within many companies.

Whether automated or not, GxP guidelines require analytical methods to be scientifically valid prior to implementation. The fact that automation can run more

TABLE 2.3 Historic Progress on GLP and Related GxP Regulations/Guidelines

1963	GMP Regulation by US FDA
1976	21 CFR 58 GLP Regulation by US FDA
1978	OECD establishes expert group
1978	GCP established by US FDA
1981	OECD Council Decision on Mutual Acceptance of Data
1983	FIFRA GLP Regulation by US EPA
1987/88	EU adopts GLP Directives
1989	OECD Council Decision
1993	GCP Guidelines by WHO
1995	Guidance for GLP Monitoring Authorities
1997	Adoption of Revised GLP Principles
1997	Guidelines on GCP Operational by ICH
1999	Amendment of EU GLP Directives
2000	21 CFR 320 Bioequivalence/Bioavailability (BE/BA) by US FDA Revised
2000	Guidance for Industry: Analytical Method Validation by US FDA
2001	21 CFR 310 New Drug Application (NDA) by US FDA Revised
2001	Guidance for Industry: Bioanalytical Method Validation by US FDA
2003	21 CFR 11 Regulation by US FDA
2003	GCLP by BARQA originated
2005	Handbook for GCP Guidelines for Implementation by WHO
2006	21 CFR 210&211 cGMP by US FDA (Drug) Revised
2006	21 CFR 110 cGMP by US FDA (Medical Device) Revised
2006	21 CFR 820 cGMP by US FDA (Food) Revised
2009	GCLP by WHO/TDR
2009	EMEA Draft Guideline on Validation of Bioanalytical Methods (BMV)

precisely or more accurately than manual methods is irrelevant. Documentation must exist for regulatory inspection that clearly demonstrates that the automated or manual method is appropriate for the phase of development and is sufficiently accurate, precise, and robust. Even though the compendial methods published by United States Pharmacopoeia (USP), American Society for Testing and Materials (ASTMs), and Association of Analytical Communities (AOAC) [18] have been validated and fully tested prior to publication, it is still up to the user to demonstrate and document that the method is performing properly on their equipment. A summary table with different regulations, principles and guidance documents is given in Table 2.3 with historic perspectives.

REFERENCES

1. FDA Good Laboratory Practices (GLP) for Non-Clinical Laboratory Studies 21 CFR Part 58 Supporting Statement, 2000.
2. Code of Federal Regulation. Food Drug and Cosmetic Act, 21 CFR Part 58—Good Laboratory Practice for Nonclinical Laboratory Studies, 1978.
3. EPA Code of Federal Regulation. Federal Insecticide, Fungicide, and Rodenticide Act, 40 CFR Part 160—Good Laboratory Practice Standards.
4. Kathrin E, Preu M. *Ann Ist Super Sanità* 2008; 44(4):390–394.
5. International Conference on Harmonization (ICH)/World Health Organization (WHO). Topic E 6 (R1) Guideline for Good Clinical Practice, 1996.
6. International Conference on Harmonization (ICH). Harmonized Tripartite Guideline for Good Manufacturing Practices for Active Pharmaceutical Ingredients Q7, 2000.
7. Code of Federal Regulation: 21 CFR Part 11—Electronic Records and Signature Standards, 2002.
8. Mutual Acceptance of Data (MAD) Under the Organization for Economic Cooperation and Development (OECD) Council Act, 1981.
9. Ulrey AK, Curren RD, Raabe HA. AATEX 14, Special Issue. *Proceedings of 6th World Congress on Alternatives and Animal Use in the Life Sciences,* Tokyo, Japan; August 21–25 2007. pp. 595–600.
10. OECD Document Number 13. *The Application of the OECD Principles of GLP to the Organisation and Management of Multi-Site Studies*, 2002.
11. OECD Document Number 7. *The Application of the GLP Principles to Short-Term Studies* (Environment Monograph No. 73), 1999.
12. Food and Drug Administration. *21 CFR 320—Bioavailability and Bioequivalence Requirements (Drugs for Human Use).* US Department of Health and Human Services, FDA, CDER, Rockville, MD, 2002.
13. ICH-53A. Note for Guidance on Toxicokinetics: A Guidance for Assessing Systemic Exposure in Toxicology Studies (CPMP/ICH/384/95), 1995.
14. Guidance for Industry: Bioanalytical Method Validation, US Department of Health and Human Services, FDA, CDER, CVM, May 2001.
15. ISO/IEC 17025. General requirements for the competence of testing and calibration laboratories—Accreditation by International Standardization Organization (ISO), 2005.

16. World Health Organization Special Programs for Research and Training in Tropical Diseases (TDR)—Good Clinical Laboratory Practices (GCLP), 2009.
17. Viswanathan CT, Bansal S, Booth B, et al. Quantitative bioanalytical methods validation and implementation: best practices for chromatographic and ligand binding assays. *J AAPS* 2007; 9 (Article 4): E30–E42.
18. Fox A. *Inside Laboratory Management*—AOAC International January/February, 2009.

3

GLP QUALITY SYSTEM AND IMPLEMENTATION

3.1 GLP QUALITY SYSTEM

Since the inception of the FDA good laboratory practice (GLP) regulations in 1979, the Organization for Economic Cooperation and Development (OECD) principles of GLP in 1981 and the finalization of the EPA GLP program in 1983 there have been recognizable differences among the three compliance programs. All have been revised since their initial publication, but still there remain differences in verbiage, and in some cases content, among the FDA, EPA, and OECD GLP principles, but the end result for each is the assurance that the experimental information generated under each program is of sufficient quality and integrity to support the reports for the various studies. These differences, while not affecting the data quality, can result in issues when submitting studies globally. Please note that some of the differences that exist between the US and OECD GLP principles and the challenges global companies face when making regulatory submissions.

All regulations promulgated by the FDA are written in very general terms to ensure suitable application to the variety of products regulated by the FDA and to promote flexibility to employ the latest technology and innovation to achieve compliance. The general regulations specify "what to do" but allow latitude in how to fulfill the regulation requirements. A GLP-regulated facility must interpret the general GLP regulations to apply them effectively to the product and studies that they are performing and to ensure compliance with all other FDA requirements and expectations. The most effective method for interpreting FDA regulations is to consult the

Regulated Bioanalytical Laboratories: Technical and Regulatory Aspects from Global Perspectives, By Michael Zhou

preamble published with the final regulations, the FDA guidelines and guidance that are issued to industry, and the internal FDA documents, such as compliance programs and compliance policy guides.

Most compliance initiatives usually start as projects as companies race to meet the timelines and/or deadlines to comply with a specific regulation. However, compliance is not a onetime event. As a result, organizations are redesigning and improving their compliance programs to meet ever scrutinizing (tighten) regulations and thus make them repeatable process that can be sustained cost-effectively as part of competitive advantages.

Quality systems, as whole, usually cover product life cycle that is applicable to all types of life science industries. A general product life cycle spans from initial concept to prototype (discovery), nonclinical development, clinical programs, manufacture, marketing, commercial, and product withdrawal or renewal (Figure 3.1). GLP quality system certainly plays a significant role in the entire process. GLP is about recognition of a quality system that has organizational processes and conditions that in accordance with OECD criteria. ISO/IEC 17025 accreditation is a formal recognition for technical competence to undertake specific tests or calibrations. In general GLP is used for nonclinical health and safety studies that are often called for in regulations

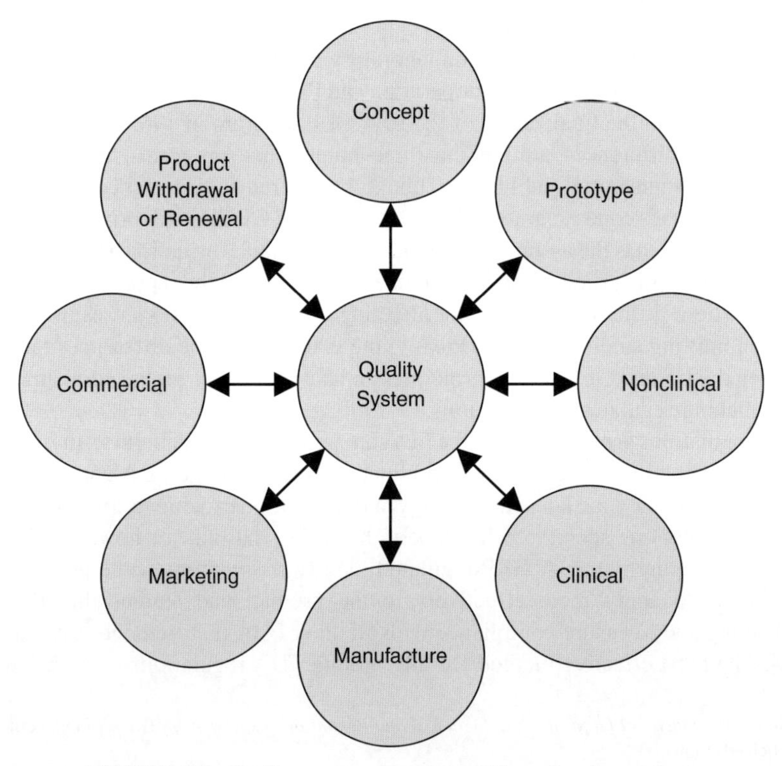

FIGURE 3.1 Quality system covering a life cycle of a product.

while ISO/IEC 17025 accreditation is used normally voluntary. Laboratory accreditation to ISO/IEC 17025 is for routine testing and measurement while GLP covers often nonroutine work or a series of experiments. Examples of nonroutine work may include dosing animals or spraying fields. Only GLP recognition can provide an effective quality system that gives regulators the necessary confidence in relation to how nonclinical health and safety studies are planned, performed, monitored, recorded, reported, and archived. There is a misconception that GLP is not part of the conduct of clinical studies required for analytical and bioanalytical sample analyses. The relevant documents for clinical studies are the various codes such as International Conference on Harmonization (ICH). The US FDA and other registration authorities do require a demonstration of the quality of test data from clinical studies. In the United States, this may be by means of conformance with CLIA (Clinical Laboratories Improvement Act) and lately with GCLP (Good Clinical Laboratory Practices—WHO/TDR).

It is widely recognized within industries that neither the US FDA or EMEA regulation nor ICH-harmonized guidelines address laboratory and analysis and bioanalysis in support of clinical trials. The FDA and OECD good laboratory practices for nonclinical laboratory studies regulations clearly address the quality principles for safety studies prior to clinical trials. The FDA regulations for bioavailability and bioequivalence requirements do provide that methods must be demonstrated to be accurate, sufficiently sensitive and have appropriate precision, but do not address any current thinking regarding quality principles beyond requiring that the drug products under consideration are manufactured according to appropriate levels of GMP. The collection of regulations that fall within the scope of FDA good clinical practices do not address quality systems or guidelines for laboratory studies, with the exception of 21 CFR 314—applications for FDA approval to market a new drug, which requires that the nonclinical data are generated according to GLP regulations and the drugs are manufactured according to GMP regulations. The EMEA guidance on the investigation of bioavailability and bioequivalence, 140198en states that the studies "should be conducted according to the applicable principles of GLP, however, the studies fall outside the formal scope of GLP."

More recently the Good Clinical Laboratory Practices concepts were developed with the objective of providing a single, unified document that encompasses IND sponsor requirements to guide the conduct of laboratory testing for human clinical trials. The intent of GCLP guidance [1] is that when laboratories adhere to this process, it ensures the quality and integrity of data, allows accurate reconstruction of experiments, monitors data quality, and allows comparison of test results regardless of performance location. The GCLP core elements include the following: organization and personnel; laboratory equipment; testing facility operations; quality control program; verification of performance specifications; records and reports; physical facilities; specimen transport and management; personnel safety; laboratory information systems and quality management. A quality system usually covers life cycle of a product as indicated in Figure 3.1.

An important aspect of a quality system is to work according to unambiguous Standard Operating Procedures (SOPs). In fact the whole process from sampling to

TABLE 3.1 Typical SOPs for a Regulated Laboratory/Facility

-Fundamental SOPs give instructions how to make SOPs of the other categories
-Methodology SOPs describe a complete testing system or method of investigation
-SOPs for safety precautions
-SOPs for operating, maintaining and calibrating instruments, apparatus, and other equipment
-SOPs for analytical methods
-SOPs for the preparation of reagents
-SOPs for receiving and registration of samples
-SOPs for quality assurance
-SOPs for generating, analyzing, and archiving data

the filing of the analytical result should be described by a continuous series of SOPs. A SOP (see a generic template of SOP: Chapter 10 Appendices) is a compulsory instruction. If deviations from this instruction are allowed, the conditions for these should be documented including who can give permission for this and what exactly the complete procedure will be. The original should rest at a secure place while working copies should be authenticated with stamps and/or signatures of authorized persons. A number of important SOP types should be (but not limited to) as given in Table 3.1.

GLP quality system consists of quality management (QM), quality assurance (QA), and quality control (QC—might not be absolutely required within GLP quality system) for implementation and execution, and continued improvement. Quality management is the assembly and management of all activities aimed at the production of quality by organizations of various kinds. In the present case, this implies the introduction and proper running of a "quality system" in laboratories. A statement of objectives and policy to produce quality should be made for the organization or department concerned (by the institute's directorate). This statement also identifies the internal organization and responsibilities for the effective operation of the quality system.

Quality management can be considered a somewhat wider interpretation of the concept of "Good Laboratory Practice." Therefore, inevitably the basics of the present guidelines largely coincide with those of GLP. An even wider concept of quality management is presently coming into vogue "Total Quality Management" (TQM). This concept includes additional aspects such as leadership style, ethics of the work, social aspects, and relation to society. Proper quality management implies consequent implementation of the next level quality assurance. The general definition of QA reads "the assembly of all planned and systematic actions necessary to provide adequate confidence that a product, process, or service will satisfy given quality requirements." The result of these actions, aimed at the production of quality, should ideally be checked by someone independent of the work the Quality Assurance Officer. If no QA officer is available, then usually the Head of Laboratory performs this job as part of his quality management task. In case of special projects, customers may require special quality assurance measures or a quality plan. A major part of the quality assurance is the quality control defined by referring as "the operational techniques and activities that are used to satisfy quality requirements." An important part of the quality control is the quality assessment the system of activities to verify if the quality control activities are effective, in other words an evaluation of the products themselves.

Quality management in the present context can be considered a modern version of the hitherto much used concept "Good Laboratory Practice" with a somewhat wider interpretation. The OECD Document defines GLP as follows: "*Good Laboratory Practice(GLP) is concerned with the organizational process and the conditions under which laboratory studies are planned, performed, monitored, recorded, and reported.*" Thus, GLP prescribes a laboratory to work according to a system of procedures and protocols. This implies the organization of the activities and the conditions under which these take place are controlled, reported, and filed. GLP is a policy for all aspects of the laboratory that influence the quality of the analytical work. When properly applied, GLP should then

- allow better laboratory management (including quality management);
- improve efficiency (thus reducing costs);
- minimize errors;
- allow quality control (including tracking of errors and their cause);
- stimulate and motivate all personnel;
- improve safety;
- enhance communication possibilities, both internally and externally.

The result of GLP is that the performance of a laboratory is improved and its working is effectively controlled. An important aspect is also that the standards of quality are documented and can be demonstrated to authorities and clients. This results in an improved reputation for the laboratory (and for the institute as a whole). In short, the message is—say what you do; do what you say; do it better and ultimately—be able to show what you have done. The basic rule is that all relevant plans, activities, conditions, and situations are recorded and that these records are safely filed and can be produced or retrieved when necessary. These aspects differ strongly in character and need to be attended to individually. As an assembly, the involved documents constitute a so-called *Quality Manual.* This comprises then all relevant information on organization and personnel; facilities; equipment and working materials; analytical or testing systems, quality control; reporting and filing of results.

Since institutions having a laboratory are of divergent natures, there is no standard format and each has to make its own Quality Manual. The present guidelines contain examples of forms, protocols, procedures, and artificial situations. They need at least to be adapted and many new ones will have to be made according to the specific needs, but all have to fulfill the basic requirement of usefulness and verifiability. As already indicated, the guidelines for quality management described here are mainly based on the principles of Good Laboratory Practice as they are laid down in various relevant documents such as ISO and ISO/IEC guides, ISO 9000 series, OECD, and CEN (EN 45000 series) documents, as well as a number of text books. This concerns mainly the so-called Standard Operating Procedures and Protocols. Sometimes these documents are hard to acquire, as they are classified information for reasons of competitiveness.

The organizational setup of an institute can conveniently be represented in a diagram, the *organizational chart* (also called OrgChart). An OrgChart should be drawn by the department of Personnel and Organization (P&O) (or equivalent) on behalf of the Directorate. Since the organization of an institute is usually quite dynamic, frequent updating of this document might be necessary. For the laboratory, an important aspect of the OrgChart is the hierarchical line of roles and responsibilities. Not all details of these responsibilities can be given in the main OrgChart. Such details are to be documented in suborgcharts, the various job descriptions as well as in regulations and statutes of the institute as a whole. The way work is organized in the laboratory should be described in a SOP. This includes the kind and frequency of consultations and meetings, how jobs are assigned to laboratory personnel, how instructions are given, and how results are reported. The statement that personnel are protected from improper pressure of work can also be made in this SOP.

Quality assurance in the laboratory requires that all work is done by staff, which is qualified for the job. Thus, to ensure a job is done by the right person, it is essential for the management to have records of all personal skills and qualifications of staff as well as of the required qualifications for the various jobs. To maintain or improve the quality of the work, it is essential that staff members follow training or refresher courses from time to time. These may concern new developments in analytical techniques or approaches, data handling, the use of computers, laboratory management (such as quality management and LIMS) or training in the use of newly acquired instruments. Such training can be given within the institute, by outside specialists, or centrally conducted courses can be attended, if necessary abroad. In certain cases, it may be worthwhile sending someone to another laboratory for a certain period to get in-service training and experience in a different laboratory culture. Ideally, after training or attending a course, the staff member should report and convey his experience or knowledge to colleagues and make proposals for any change of existing procedures or adoption of new practices to improve the performance of the laboratory. When a new employee is appointed in the laboratory, he or she should be properly introduced to the other staff, to the rules of the laboratory in general and in particular to details of his/her new job, roles, and responsibilities. In order to ensure that this is properly done, it is useful to draw up a SOP with a checklist of all aspects involved. A program of training and monitoring the settling into the job has to be established.

If an institute or organization establishes a laboratory and expects (or demands) quality analytical data then the directorate should provide the necessary means to achieve this goal. The most important requirements that should be fulfilled, in addition to skilled staff, are the supply of adequate equipment and working materials, the presence of suitable housing, and the enforcement of proper safety measures. Often, the laboratory has to be housed in an existing building or sometimes in a few rooms or a shed. On the other hand, even when a laboratory is planned in a new prospective building not all wishes or requirements can be fulfilled. Whatever the case, conditions should be made optimal so that the desired quality can be assured.

Quality analytical work can only be performed if all materials used are suitable for the projects/studies, properly organized, and well cared for. This means that the tools are adequate and in good conditions (well maintained and calibrated), and that sample material receives attention with respect to proper handling, storing, and disposal. The tools used for analysis may be subdivided into four categories:

(1) Primary measuring equipment (pipettes, diluters, burettes, balances, thermometers, flow meters, etc.)
(2) Analytical apparatus or instruments (HPLC, GC, LC–MS/MS, etc.)
(3) Miscellaneous equipment and materials (ovens, furnaces, fridges, stills, glassware, etc.)
(4) Reagents and chemicals

The saying that a chain is as strong as its weakest link applies particularly to these items. An analyst may have gone out of his/her way (as he/she should) to prepare extracts, if the cuvette of the spectrophotometer is dirty, or if the wavelength dial does not indicate the correct wavelength, mass calibration is off, the measurements are in jeopardy. Both the blank and the control sample (and a possible "blind" sample or spike) most likely will reveal that something is wrong, but the harm is already done: the problem has to be found and resolved, and the batch might have to be repeated. This is a costly affair and has to be minimized (it is an illusion to think that it can be totally prevented) by proper handling and maintenance of the equipment.

Also the quality and condition of a number of other working materials have to be watched closely. The calibration of thermometers, burettes, and pipettes, detection system (UV or MS) particularly the adjustable types, may exceed the acceptable tolerance (and be put out of use). A similar reasoning applies to reagents. One of the most prominent sources of the errors made in a laboratory is the use of wrongly prepared or old reagents. Therefore, reagents have to be prepared very carefully and exactly following the prescriptions and/or descriptions, they have to be well labeled and expiry dates have to be observed closely. Filtering a pH buffer solution in which fungi are flourishing may save time and reagent but is penny-wise and pound-foolish. Of equal importance for the quality of the work is the proper handling of the sample material. Not only the technical aspects such as sample preparation, but particularly the safeguarding of identity and integrity of the samples as well as the final storage or disposal *(chain of custody)* are all vitally important for successful projects and programs.

The most important part of laboratory work is of course the analytical procedures meticulously performed according to the corresponding SOPs. Relevant aspects include calibration, use of blanks, performance characteristics of the procedure, and reporting of results. All activities associated with these aspects are aimed at one target: the production of reliable and accurate data with a minimum of errors. In addition, it must be ensured that reliable data are produced consistently. To achieve this, an appropriate program of quality control must be implemented. Quality control is the term used to describe the practical steps undertaken to ensure that errors in the analytical data are of a magnitude appropriate for the use to which the data will be put. This implies that the errors (which are unavoidably made) have to be quantified to enable a decision whether they are of an acceptable magnitude, and that unacceptable errors are discovered so that corrective action can be taken. Clearly, quality control must detect both random and systematic errors. The procedures for QC primarily monitor the accuracy of the work by checking the bias of data with the help of (certified) reference samples and control samples and the precision by means of replicate analyses of test samples as well as of reference and/or control samples.

Method validation is the process of determining the performance characteristics of a method/procedure or process. It is a prerequisite for judgment of the suitability of produced analytical data for the intended use. This implies that a method may be valid in one situation and invalid in another. Consequently, the requirements for data may, or rather must, decide which method is to be used. When this is ill considered, the analysis can be unnecessarily accurate (and expensive), inadequate if the method is less accurate than required, or useless if the accuracy is unknown. It should be noted that the trueness of an analytical result may be sensitive to varying conditions (level of analyte, matrix, extraction, temperature, etc.). If a method is applied to a wide range of materials, for proper validation different samples at different levels of analyte should be performed to demonstrate the ruggedness and robustness of a method. The various activities in a laboratory produce a large number of data streams, which have to be recorded, processed, and archived. Some of the main streams are (1) sample registration; (2) desired analytical program; (3) work planning and progress monitoring; (4) calibration; (5) raw data storage; (6) data processing; (7) data quality control; (8) reporting; and (9) data retention and archiving.

Each of these aspects requires its own typical paperwork most of which is done with the help of computers. As discussed in previous chapters, it is the responsibility of the laboratory manager to keep track of all aspects and tie them up for proper functioning of the laboratory as a whole. To assist him/her in this task, the manager will have to develop a working system of records and journals. In laboratories of any appreciable size, but even with more than two analysts, this can be a tedious and error-sensitive job. Consequently, from about 1980, computer programs appeared on the market that could take over much of this work. Subsequently, the capability of Laboratory Information Management Systems (LIMSs) has been further developed and widely utilized within the industries.

The main benefit of a LIMS is a drastic reduction of the paperwork and improved data recording, leading to higher efficiency and increased quality of reported analytical results. Thus, a LIMS can be a very important tool in quality management. As discussed above, when the desired quality level of the output of the laboratory is reached, it must be maintained and, where necessary, improved. To achieve this, the Quality Manual should contain a plan for regular checking of all quality assurance measures as they have been discussed so far. Such a plan would include a regular reporting to the management of the institute or company. This is usually done by the head of laboratory and/or, if applicable, by the quality assurance officer. In addition to such a continuous internal inspection, particularly for larger laboratories it is very useful to have the quality system reviewed by an independent external auditor. For accreditation this is even an inherent part of the process. An external audit can assist the organization to recognize bottlenecks and flaws. Such shortcomings often result from insufficient and inefficient measures and activities that remain unnoticed or are ignored. An audit can be requested by the laboratory itself or by the management of the institute and involves basically the inspection of the Quality Manual, that is, all the protocols, procedures, registration forms, logbooks, control charts, and other documents related to the laboratory work. Attention is given not only to the contents of the documents but also to the practical implementation ("say what you do, do what you say, and be able to show what you have done"). Laboratory staff members sometimes see these audits as a sign of suspicion about their performance, and sometimes audits may be (mis)used to get things organized or changed under the guise of quality. Yet, the auditor should not be seen as a policeman but as someone who was asked to help. Therefore, good cooperation with the auditor is essential for the effectiveness of the audit. Conversely, the auditor should be selected carefully for the same reason.

In large laboratories it may be advisable to have the audit done by more than one person, for instance an organization specialist and an analytical expert. The audit should result in a report of findings and recommendations to improve possible shortcomings. Subsequently, the management which will have to decide to what extent the report will remain confidential, and if and what actions will have to be taken.

3.1.1 Regulatory Inspection for GLP Quality System

Regulatory inspection from FDA or related international agencies is part of continued monitoring processes for implementation and improvement. The FDA typically preannounces the routine GLP inspections, but the agency is not required to preannounce inspections of commercial laboratories or inspections for special compliance or investigations [2]. The refusal to permit FDA inspection or the review of records is a serious violation of the GLP regulations, as evidenced by an FDA Warning Letter issued to a nonclinical laboratory on April 27, 2001. The study laboratory refused to provide the master schedule sheet, protocol, husbandry and surgery records, and minutes of study reviews. The Warning Letter concluded that the refusal, along with other GLP deficiencies, constituted serious deviations from the GLP regulations, and the FDA scheduled the facility for reinspection. The Warning Letter also directed the study laboratory to notify each potential study sponsor that their studies were not conducted in accordance with the GLP regulations [3]. The specific areas to inspect and the related inspection techniques are outlined in Part III of Compliance Program CP7348.808. One area that is likely to receive special inspection coverage is computerized systems, as outlined in Attachment A of the Compliance Program [4].

The QAU is required by 21 CFR §58.35 to inspect each nonclinical laboratory study and to submit the inspection report to management and the study director. The FDA adopted a policy of not routinely requesting the review of these internal QAU inspection reports to encourage the study laboratories to perform meaningful self-inspection. As an alternative, the FDA investigator may request that company management provide a written certification that the required QAU inspections had been performed and documented, and that corrective actions have been implemented. The FDA will continue to request and review these internal inspection reports for directed or for-cause inspections, in litigations, and for inspections based on an inspection warrant or judicial search warrant [5]. Quality system implementation comes with two essential elements: internal periodical audits (study-based or projected-based audits) and external inspections (study-based, process-based, and system/facility-based inspections). The main purpose of these audits/inspections is to continue improvement of GxP quality systems in place, and identify any flaws, gaps, and refinements in these quality systems.

A search of the FDA website found the posting of nine Warning Letters, issued between January 2000 and December 2004, for GLP deficiencies. Five Warning Letters were issued to academic institutions, one was issued to a pharmaceutical research sponsor, and the remaining three were issued to testing laboratories. The review of these Warning Letters found that the most prevalent GLP violations involved the responsibilities and qualifications of the study director and the responsibilities of the QAU. Some of the other violations cited involved areas concerning the reporting of nonclinical results, testing facility management, standard operating procedures, study protocols, instrument calibration, and the storage and retrieval of records, and data [6–13]. The deficiencies associated with the qualification and responsibilities of the study director include the failure to follow the basic requirements specified in §58.33. The responsibilities of the study director are clearly stated in the GLP regulations such that there are very few interpretations of study director responsibilities in the 1978 GLP preamble, the 1979 Postconference Report, and the 1981 GLP Questions and Answers guidance. It is interesting to note that study director deficiencies were cited in all five Warning Letters issued to academic research facilities and were cited in only one of the three Warning Letters to testing laboratories.

This observation indicates that either the academic institution management or their study directors did not correctly interpret the GLP regulations or did not realize or were not informed by the study sponsors that the studies they were performing were covered by the GLP regulations. Another explanation might be the misunderstanding that these studies were performed under an academic research environment and therefore were not subject to the GLP regulations. The GLP compliance problems with studies performed by academic institutions have been an ongoing issue dating to the promulgation of the GLP regulations. The preamble (comment 15) to the 1978 GLP regulations stated: "Insofar as academic institutions are concerned, the Commissioner notes that such institutions conduct significant amounts of commercial testing pursuant to contracts. He also notes that significant levels of noncompliance with GLP requirements have been found in such institutions."

The Warning Letter citations concerning the QAU also represented the failure to follow the basic requirements specified in the GLP regulations. The citations included those for the failure to establish a QAU; the failure of the QAU to maintain independence from the personnel who conducted the studies; the failure to maintain records required by the GLPs such as a master schedule; and the inadequate inspection, review, and detection of errors by the QAU.

As in the case discussed for study directors, the QAU deficiencies were listed in each Warning Letter issued to an academic institution. Another likely reason, in addition to those discussed for the study director deficiencies, might be the limited resources at an academic institution to establish and maintain an independent QAU as required by the GLP regulations. Careful review and interpretation of the GLP regulations, however, indicate that independence does not mean a separate department or dedicated personnel. For example, a study director for a particular study may serve as the QAU for a different study (1978 GLP preamble, comment 75). Preamble comment 12 in the 1987 GLP revision confirmed this earlier FDA policy with the following statement: "FDA was aware that many small laboratories could not afford the operation of a permanently staffed QAU. For this reason, the agency noted the separation of functions on a study-by-study basis as permitted in the existing, and revised regulations would provide effective quality assurance" [14]. In one Warning Letter, the FDA expressed concerns over the use of a part-time individual for performing QAU inspections. The FDA found that the part-time individual was performing only a limited inspection, monitoring only one phase of a study, and the inspection report lacked details on the noted deviations. Although the FDA relaxed the GLP regulations in 1987 to remove the requirement to inspect each phase of a nonclinical study, the FDA would not accept a limited review. As indicated in the 1987 GLP revision, preamble comment 18, the FDA expects the QAU inspection to be thorough and in depth and to include the review of all operations required to accomplish a particular study phase (e.g., the examination of personnel, facilities, equipment, SOP, raw data, and data collection procedures).

A similar concern was expressed in another Warning Letter; the contract laboratory claimed that because their sponsor performed audits every 1–2 months, the sponsor was performing the QAU inspections as required by the GLP regulations. The FDA rejected this response and indicated that general audits performed by the sponsor did not fulfill the requirements for an in-process QAU inspection of each GLP study. This policy was made clear in the 1978 preamble, comment 79, in which the FDA commissioner stated that the QAU of the nonclinical testing facility is solely responsible for fulfilling the quality assurance functions for the studies. The only exception when the contractor might be allowed to act as the QAU is for the portion of the study contracted to a site that is not involved in the performance of nonclinical studies (e.g., a testing laboratory). The word inspect was specifically incorporated in the GMP in preference over the word audit; therefore, an audit of the nonclinical study would not fulfill the requirement for a QAU inspection of the study. The 1978 preamble, comment 82, stated: "'Inspect' more accurately conveys the intent that the QAU actually examine and observe the facilities and operations for a given study, while the study is in progress, whereas 'audit' could be interpreted to mean simply a detailed review of the records of a study."

What is the scope of the QAU inspection? What is the QAU reviewer inspecting? Does the QAU representative need extensive education and scientific experience to perform this assigned job function? These questions were answered by the FDA in the 1978 preamble (comments 75, 77, and 90); the GLP Postconference Report of August 1979 (Subpart B, questions 5 and 11); and the June 1981 GLP Questions and Answers guidance (QAU, question 1). The FDA indicated in these references that the function of the QAU is administrative rather than scientific; therefore, the QAU would not have to evaluate the scientific merits of the final report or "second guess" the scientific procedures that were used. The FDA also did not expect the QAU to verify and trace all raw data to the study and to check the accuracy of all calculations. The FDA recommended a random sampling of raw data for a more in-depth review. The FDA stated the following objectives for the QAU inspection:

- The study was conducted in accordance with the requirements of the GLP regulations, the study laboratory's SOPs, and the study protocol.
- No changes or deviations from the SOPs and study protocols were made without proper authorization.
- The study report contains the information and data required by the study protocol.
- The summary data accurately support the conclusion of the study.
- The identification of significant problems that may adversely affect the integrity of the study.

The FDA also expects that the QAU inspections have sufficient depth to reveal inadequacies in the final report. This expectation is reflected in the FDA's Compliance Program CP7348.008 (Part III, pages 6 and 7), which directs the FDA investigator to determine whether the QAU representative is spending sufficient time to detect problems in critical study phases and whether there are adequate personnel to perform these inspections. The preamble to the regulations addresses all questions and comments submitted to the FDA based on the initial proposal of the regulations. The responses to these questions and comments represent the most authoritative interpretations of the FDA regulations because each response is supported by the highest level person at the FDA: the FDA commissioner. The guidance documents issued by the FDA (e.g., GLP Questions and Answers) represent the current thinking and policy of the FDA. Unlike the regulations, these guidance documents are not binding on industry; that is, the failure to follow the guidance does not constitute a violation of the regulations. The firm must, however, show evidence that the procedures and controls actually in use are equivalent or superior to those specified in the guidance document. The FDA internal documents (e.g., compliance programs and compliance policy guides) provide valuable information on how the FDA interprets the regulations and their position on the enforcement of those regulations (e.g., whether flexibility is allowed or alternate procedures are considered acceptable).

To illustrate the importance of the GLP preamble and FDA guidance documents for the interpretation of the regulations, consider the general statement in the scope of the regulations (21 CFR §58.1), which states that the regulations apply to "nonclinical laboratory studies that support or are intended to support applications for research or marketing permits for products regulated by the Food and Drug Administration." Fifteen examples of studies and tests that were excluded from GLP requirements were listed in the preamble to the 1978 GLP regulations [5], and many other examples of specific studies that were covered or excluded from the GLP regulations were discussed in eight pages of the 1979 Postconference Briefing Report [6]. One interesting interpretation of the application of the GLP regulations is that although the regulations were promulgated for nonclinical laboratory studies, they are applicable to chemical procedures used to analyze clinical specimens (e.g., clinical chemistry and urinarysis).

3.1.2 Good Laboratory Practice Inspections

The FDA conducts a GLP inspection according to their Compliance Program 7348.808 to verify the quality and integrity of data submitted to the agency and to evaluate the level of GLP compliance. The FDA conducts four types of inspections: routine GLP inspection, data audit, directed inspections, and follow-up inspections. The routine GLP inspection is scheduled on a biennial basis (except for overseas laboratories, which are inspected based on the submission of significant studies of important products) and covers ongoing and completed studies for compliance with the GLP regulations. The data audit focuses on the raw data to support the accuracy of information submitted to the FDA. A directed inspection is performed in response to specific reasons, such as questionable or suspect data submitted to the FDA or tips from informers. A follow-up inspection is performed following a previous violative inspection and focuses on the corrective actions taken to bring the facilities into GLP compliance [4].

An audit or an inspection is a methodical evaluation that should be performed in cooperation with the people concerned. The internal audit is not an inquisition or a punitive exercise. In addition to the QAU review of planning activities, the QAU performs three types of audits/inspections (1) study-based inspections/audits, (2) facility/systems-based inspections/audits, and (3) process-based inspections/audits. QA may also inspect contractors and suppliers.

3.1.2.1 Study-Based Inspections/Audits Study-based inspections target specific "critical" phases of the study. Inspections are performed as planned, with additional or follow-up inspections, if necessary. There are numerous useful guides on inspection and audit techniques. Some general points are as follows:

- SOPs for inspections and for audit reports should ideally be prepared in dialogue with the operational staff.
- The inspector should prepare for the inspection. Usually this means reviewing the protocol, applicable SOPs, and past inspections beforehand.
- The inspector/auditor must follow all rules of access, safety, and hygiene and must not disrupt the work.
- The inspector/auditor must allow sufficient time for the inspection.
- Checklists may or may not be used, as considered necessary. Adherence to a checklist is no guarantee of completeness but is useful for training and as a memory aide. Also checklists enable management to approve QAU methods and

coverage and provide technical staff with a means of auto-control. Checklists are usually established formally and are updated as needed. However, the checklist may engender the risk that an unexpected finding be missed.

- Logically, and out of consideration for study staff, at the close of the inspection, or at least before a report is generated, the inspector should discuss all problems with the persons inspected. Any error (e.g., dosing error, animal ID) should, obviously, be pointed out immediately.
- Comments should be clear and specific.
- Comments should be constructive. The best means to ensure this is to propose a solution to each problem reported in the inspection report.
- The report circulated to management (with or without a separate summary) should include comments and responses with or without a separate report in summary form to management. Rules for the writing, approval, distribution, and archiving of inspection/audit reports as well as arbitration procedures should be included in the SOPs.
- As a general rule, internal QAU inspections and audits target events and organization, not people. The more problems uncovered and resolved the better the level of quality.

3.1.2.2 System- or Facility-Based Inspections/Audits These are performed independently of studies. Frequency should be justifiable in terms of efficiency versus costs. The results of a system/facility inspection are reported to the appropriate manager of the test facility rather than to a study director. The follow-up procedure will, however, be exactly the same as for a study specific inspection. System-/facility-based inspections typically cover such areas as: personnel records, archives, animal receipt, cleaning, computer operations and security, access and security, SOP management, utilities supply (water, electricity), and metrology.

3.1.2.3 Process-Based Inspections/Audits These are also performed independently of specific studies. They are conducted to monitor procedures or processes of a repetitive nature. Frequency is justified by efficiency and costs. These process-based inspections are performed because it is considered inefficient or inappropriate to conduct study-based inspections on repetitive phases. It is worth noting that the OECD at least recognizes "that the performance of process-based inspections covering phases which occur with a very high frequency may result in some studies not being inspected on an individual basis during their experimental phases." Other useful process-based inspections are those that focus on cross-organizational processes—for example, the transfer of test samples from the animal facilities to the bioanalysis laboratory.

3.1.2.4 Final Report/Raw Data Audits The QAU should audit all reports from GLP studies with reference to the protocol, SOPs and raw data. A full audit does not mean a 100% check of all data contained in the report. Enough data should be audited to convince QA that the report gives a faithful account of the way in which the study was performed and provides an accurate representation of the data. The QAU is also looking for evidence of authenticity and GLP compliance in the data, for example, signatures, dates, handling of corrections and deviations, and consistency.

Typically, QA may cover the following during the report audit: contents, data completeness, protocol compliance, animal environment records, test item QC/accountability, dose preparation/dosing/QC records, individual tables versus raw data (sample basis), summary tables, appendices, and conclusions. The QAU statement that is placed in the report provides the dates on which the study was inspected and findings reported to the Study Director and management. QAU also reports the study phases inspected, along with the dates, as recommended by OECD. The QAU statement is not a GLP compliance statement. The Study Director provides this statement—that the study had been conducted in compliance with the applicable principles of GLP. However, recommendations of the OECD with regard to the QAU statement should be noted: "It is recommended that the QA statement only be completed if the Study Director's claim to GLP compliance can be supported. The QA statement should indicate that the study report accurately reflects the study data. It remains the Study Director's responsibility to ensure that any areas of noncompliance with the GLP Principles are identified in the final report."

In this way, the signed QAU statement becomes a sort of "Release" document that assures that (1) the study report is complete and accurately reflects the conduct and data of the study; (2) the study was performed to GLP; and (3) all audit comments have been satisfactorily resolved. Most QAU organizations also inspect/audit suppliers of major materials (animals, feed, etc.). In the same manner, QAU may also inspect contract facilities before contracting out work. This applies whether the work concerned is a whole study or part of a study (e.g., analytical work). For pivotal studies, QAU may program periodic visits to the contract facility to ensure that the contractor is in compliance throughout the duration of the study and/or audits the final report independently.

3.1.2.5 The Distribution and Archiving of QAU Files and Reports The QAU has a dual role as an internal control and as the public guarantee that preclinical studies are performed in a way intended to provide valid data. QAU reports are distributed to the Study Director and to management, and are absolutely to be regarded as internal working documents. They are particularly valuable if important findings are picked up during the QAU activities, reported accurately, discussed, and acted on. Therefore, the provision that the QAU audit reports are not normally available to regulatory authorities will encourage the QAU to report findings honestly, without tactical fears that the facility will be damaged in the eyes of the outside world. It follows that the QAU reports are not for general distribution, and should be handled with discretion. It is best to archive reports separately from the study files so that regulatory authorities or external auditors do not access to them by mistake during inspections.

The implementation of the OECD principles of Good Laboratory Practice presents challenges to novice organizations not because it necessarily requires a large financial investment, but because it represents and requires an organizational attention and commitment. TDR requisitioned the compilation of this document as part of an initiative to encourage laboratories working in the field of product development to comply with GLP. The aim of the document is to provide a sequential framework for implementation. Although there are many ways to achieve GLP compliance, the steps recommended here are based upon the practical experience of scientists who have already implemented GLP. The implementation of GLP must be a collaborative effort, preferably enthusiastically supported by top management, and involving personnel practicing a multitude of disciplines such as research, quality assurance, maintenance, metrology, human resources, documentation, and archives. Implementation is best organized in the form of a project, with a team of persons deriving their authority directly from upper management. It is important to remember that the whole effort depends on management setting the scope of GLP implementation by giving the project team its mandate. Management support is essential if the project is to move forward at the agreed pace. Top management should not become directly involved in the day-to-day business of the implementation process so as to retain some measure of impartiality for conflict resolution. The project team members will need to have the explicit support of their immediate superiors, since they will need relief from their usual duties, and they will need a mandate from their superiors when attending team meetings. Finally, the project team leader must have immediate access to all levels of management in all departments concerned. The project team should draw up the list of steps to be achieved within an agreed timeframe. In order not to disturb the regular work of the organization, it is unwise to be too ambitious when setting the overall time allotted to implement GLP. Experience shows that allowing 20–24 months for implementation is fairly reasonable. However, it is possible to do some tasks in parallel and "overlapping." These would reduce the implementation time to 18 months or so. This schedule will allow staff to continue their other work, albeit at a slightly reduced pace, and yet requires that a momentum be maintained in order to reach the goal of implementation.

One can maintain the momentum by setting up the main, high-level steps for the project, and identifying individual tasks within each step. Each task has a responsible person and a finishing date. In addition to the person responsible for the task in question, it is advisable to appoint a second person (not necessarily senior to the first) who will critically review the work of the first person. This process of verification during the life of the project assures timely completion of each task, and helps coordinate implementation. The project team should meet regularly (monthly) and review progress. Someone who is well versed in GLP should manage the project team. This person should be appointed by, and report directly to, upper management. The project manager must, in addition to a scientific profile, have excellent management and communication skills. If the necessary skills do not appear to be available from within the organization, it would be appropriate to request aid from external resources.

General steps for effective implementations should be as follows:

- Upper management appoints a project team leader.
- Project team leader draws up a formal document to inform personnel of the missions and objectives of the team.
- Upper management circulates the formal document to all staff and holds a project launch meeting to explain the importance of the project.
- Project objectives include an overall time plan for completion of the project.
- The project team leader appoints a multidisciplinary team.
- Calendar dates are set for the project team meetings.
- Establish table of tasks to be achieved during the life of the project.
- An expert (internal or external to the organization) should evaluate the shortfall in meeting GLP compliance (gap analysis). The gap analysis is based on an audit of the organization over a 4–5 day period.
- The project team will then jointly agree on the priorities and details of the tasks necessary to achieve GLP compliance. These will be assembled in the project task table. To achieve this, the first project meetings will be held more frequently than during the rest of the project life span.

3.1.3 GLP Quality System Objectives

GLP quality system may be aimed as follows:

- GLP quality system guarantees the control of every parameter, which might influence results generated during regulated studies and from the laboratory operation.
- GLP Quality Manual and procedures constitute a set of rules and guidelines in which all GLP personnel are trained and with which everyone must comply. These documents, their revision, and distribution are managed by QAU department.
- Strict follow-up of deviations and out-of-specification (OOS) and out-of-trend (OOT) results is also an essential part of the system.

3.1.3.1 Personnel and Responsibilities

- Each member of staff has a formal job description and a corresponding training program.
- Continuous training is given in GxP, the good practices of the pharmaceutical sector, and in the procedures of quality assurance.
- Training and evaluation of all personnel is carried out on a regular basis and is systematically recorded.

3.1.3.2 Audits

- By means of both internal audits and those carried out by regulatory bodies and by our customers (3 or 4 a month), quality system should be under regular surveillance and are regularly made aware of any improvements necessary.

3.1.3.3 Equipment and Premises

- Computerized management of access and maintenance schedules
- Qualification: URS, DQ, IQ, OQ, and PQ for each piece of equipment
- Personnel dedicated to maintenance 24/7
- Qualified environmental chambers/cabinets and freezers
- Continuous recording (with alarms) of sample storage premises and equipment
- Validation of computer systems according to GAMP5
- Biosafety laboratories (BSL2 and BSL2 +) and clean rooms dedicated to sterility testing are validated, as are the pressure cascades, and are continuously monitored (with alarms)

3.1.3.4 Environmental and Health and Safety Policy Principal objectives (1) to guarantee a safe working environment for laboratory personnel and (2) to respect regulatory requirements concerning the environment. The following should have therefore been set up: (1) waste disposal management in compliance with current legislations, (2) drawing up of procedures and a code regulating work practices, (3) continuous training of personnel, and (4) regular evaluation of work practices.

Quality system is a management/process system to the highly regulated and normalized pharmaceutical and life science industries. This should include (but not limited to) the following:

(1) Guarantee the *competence* of the personnel by
  - recruiting appropriately qualified people;
  - organizing continuous training;
  - periodically evaluating competence;
  - updating the personnel on advances in the sector, thanks to scientific, technological and regulatory input from a particular committee.

(2) Guarantee the *quality* of data generated by
  - respecting rules set out in the Quality Manual;
  - respecting organizational and technical procedures, written and revised periodically;

- respecting the quality references of the pharmaceutical sector: *GLP*, *cGMP*, *GCP*, and *GCLP*;
- keeping up to date with national and international regulations applicable to the development, registration, and manufacture of *medicinal* products.

(3) Respect the *timelines* for the work by

- reliably evaluating the work required and the overall workload at all times;
- using suitable tools to allow transparent planning for all people involved in the work.

(4) Establish efficient *communication* by

- organizing transfer of information and communications within entire organization;
- exchanging contact details of the members of the commercial, technical, operations, and quality departments involved in each project;
- ensuring the management, by a project coordinator, of regular contacts between the personnel in charge of the work and executors, especially in case of deviations from approved protocols, schedule modifications, and unexpected results;
- computerized traceability of all contacts, internal and external, commercial and technical, for each project.

(5) Ensure the *safety* of the personnel and all items entrusted by

- giving information on, and training in, risks specific to our activity via committees dealing with *biosafety* and the protection of company personnel;
- ensuring the personnel has thorough knowledge of safety data sheets for the products handled;
- limiting access to customer documents, samples, and products to authorized people;
- protecting electronic data via a dedicated restricted access server room, daily backups, and a secured network.

(6) Guarantee the *confidentiality* of commercial relationships by

- giving customers a formal commitment to respect the confidentiality of all projects;
- making the personnel aware of their responsibilities regarding access to confidential data related to the development of medicinal products.

(7) Conservation of the *environment* by

- giving appropriate information on, and training in, the environmental risks linked to sector of activity;
- using natural resources and energy efficiently and rationally;
- managing disposal of solid and liquid waste appropriately.

A quality system for GLP operations may be summarized in Figure 3.2.

FIGURE 3.2 Quality system in GLP operations.

3.2 GLOBAL GLP REGULATIONS AND PRINCIPLES

3.2.1 General

There are numerous similarities between the US (FDA and EPA) and the OECD GLP principles. All require (i) a study plan that is approved by the sponsor and signed by the SD; (ii) a QA unit; (iii) a substance/article/item that is appropriately characterized; (iv) trained, qualified personnel to conduct the study; (v) raw data collection and change procedures; (vi) a final report that reflects the data generated; and (vii) the final report and all associated raw data and records be archived. Not all aspects of the US and OECD GLP principles are the same. There exist many terms in the published documents that differ, although the intent is essentially the same, as set forth in Table 3.2. These differences, at times, can cause confusion because of the common usage of these words and phrases in different countries. For instance, the word "should" in the OECD GLP principles is intended to be interpreted as required, but in the US GLP principles should imply that it is recommended, but not absolutely

TABLE 3.2 Comparison of Terms Used in the US GLP Regulations and OECD GLP Principles with the Same Meaning

US GLP Regulations	OECD GLP Principles
Shall	Should
Substance/article	Item
Equipment	Apparatus
QA unit	QA Program
Experimental termination	Experimental completion

required. The main differences between the US and OECD GLP programs can be traced back to the following key aspects: (i) responsibilities and compliance; (ii) statement of compliance; (iii) approvals; (iv) laboratory certification; (v) authority inspections; and (vi) archiving requirements. These differences are illustrated in detail hereafter. More detailed comparisons of USEPA, FDA, and OECD Principles of GLP may be located within "Comparison Charts for GLP:FDA_EPA_OECD" in Chapter 10.

3.2.2 Responsibilities and Compliance

In the United States, the SD is responsible to assure that all applicable GLP regulations are followed. If there are any deviations from the EPA or FDA GLP principles this must be noted in the final report and in the case of the EPA GLP regulations noted in the GLP statement of compliance. In the OECD context, on the other hand, Test Facility Manager (TFM) is mainly responsible to ensure GLP compliance. This is not to say that the SD is not responsible, but only that the emphasis for compliance is weighted to the TFM. The OECD requires that the TFM issues a declaration, where applicable, that a study was carried out in accordance with the GLP principles. Moreover, under the OECD GLP program, the final report should be signed and dated by the SD to indicate responsibility for data and should indicate compliance with the GLP principles. The EPA requires that each study include a true and correct statement, signed by the applicant, the sponsor, and the SD indicating the level of compliance with 40 CFR, part 160 (GLP statement of compliance) and is required to be page three of the final report [5]. It should be noted that the FDA has no similar requirement.

3.2.3 Statement of Compliance in the Final Report

The EPA requires that each study submitted be accompanied by a GLP compliance statement specifying one of the following: (i) the study was conducted under the EPA GLP regulations; (ii) the study was not conducted under EPA GLP regulations with description of the ways it differs; and (iii) the submitter is not the study sponsor, did not conduct the study and does not know whether the study was conducted under the EPA GLP regulations. Again, unlike the EPA, the FDA does not require such a statement to accompany the final report. Even though it is not a requirement of many countries to supply a GLP compliance statement as such, they do create a statement for US submissions. They can have different appearance and wording, but convey what is required by the EPA. Because many of these studies are intended for global regulatory submissions the GLP compliance statement has been modified to meet multiple regulatory agency acceptability requirements. Some TFs generate a compliance statement that claims compliance with several GLP programs including programs that are outside the scope of their countries' Monitoring Authorities (MAs). One question that might be asked is whether it is helpful to claim compliance with GLP regulations other than those of the national program. Is there value in making these claims given that the additional claims are not monitored by any MA?

3.2.4 Protocol Approval

The US GLP regulations require sponsor approval prior to SD signature on protocol/study plan. On the other hand, the OECD allows the individual countries to determine when sponsor and TFM need to sign the study plan.

3.2.5 Assignment of Study Director

In the US GLP program, there is a single SD for a given study. With the OECD it is allowed by some Regulatory Authority (RA) that the TFM assigns deputy SD for defined durations during a study (only one SD at any given time).

3.2.6 Laboratory Qualification/Certification

Laboratory qualification/certificates, for example, official certification documents from country MA are not a specific requirement of the OECD GLP program. In the United State, authorities do not certify TF. They perform what is commonly referred to as neutral scheme compliance inspections on a 2–4 year cycle. They can perform audits more frequently if there is evidence to indicate it is warranted. At the completion of the audit neither the EPA nor FDA issue a certificate of compliance. The majority authorities following the OECD GLP principles certify TF for 2–4 year intervals. At the completion of the TF inspection a GLP certificate is issued. In lieu of certification in some countries, RAs have requested documentation from the US GLP authorities, or have requested a more specific document from the EPA Laboratory Data Integrity Branch.

3.2.7 Authority Inspections

Within the OECD, authority inspections are normally requested by the TF. If the facility is new they must request an inspection and obtain a GLP certificate prior to claiming GLP compliance for the studies that are conducted. Certificates generally are valid for 2–4 years. Recertification is requested by the TF. There is generally a fairly long lead time between the notice of inspection and the actual inspection event. In the United States, selection of the TF to be inspected is by the RA. New laboratories are not required to apply for certification from the monitoring authority prior to conducting or claiming GLP compliance. In the case of EPA there is a 10 day notice (every 2–4 years), whereas the TF receives no advanced notice of the GLP inspection by the FDA.

3.2.8 Archiving Requirements

The length of time that the original raw data and final reports are required to be maintained differs between regulations and MA. The OECD allows the individual RA to determine the appropriate storage interval. In the United States, on the other hand, there are different rules between the FDA and EPA. Some examples of the OECD approach are, for instance, Germany requires archival for 15 years and Switzerland requires 10 years. The FDA has defined an interval of 5 years after results are submitted for research or marketing permit and 2 years after termination if not submitted to the agency. The EPA indirectly defers to the FIFRA books and records requirements (e.g., retained as long as the registration is valid and the producer is in business) or 2 years after study termination if not submitted to the agency. The EPA and FDA require government notification if materials are transferred to sponsor archives when a contract facility goes out of business. Notification is not a specific OECD requirement.

3.3 IMPLEMENTATION OF GLP REGULATIONS AND OECD PRINCIPLES

The following sections will elaborate different areas, functions, roles and responsibilities, resources, processes, and requirements that are associated with applicability and implementations of GLPs and OECDs regulations and principles.

Organization and Personnel: GLP regulations require the structure of the research organization and the responsibilities of the research personnel to be clearly defined. This means that the organizational chart should reflect the reality and be kept up-to-date. The organizational chart and job descriptions give an immediate idea of the way in which the laboratory functions, and of the relationships between the different posts. GLP stresses that the number of personnel must be sufficient to perform the tasks required in a timely and GLP-compliant way. The responsibilities of all personnel should be defined and recorded in job descriptions and their qualifications and competence defined in training and education records. GLP attaches considerable importance to the qualifications of staff, and to both internal and external training given to personnel, in order to maintain levels of excellence.

A point of major importance in GLP is the position of the Study Director, who is the pivotal point of control for the whole study. This individual will have to assume full responsibility for the GLP compliance of all activities within the study, and she/he will have to assert this at the end of the study in a dated and signed compliance statement. She/he has therefore to be aware of all occurrences that may influence the quality and integrity of the study, to judge their impact and to institute corrective actions, as necessary.

Facilities and Equipment: The GLP Principles emphasize the adequacy of facilities and equipment, which have to be sufficient to perform the studies. The facilities should be spacious enough to avoid problems such as overcrowding, cross contamination, confusion between projects and cramped working conditions. Utilities (water, electricity, etc.) must be adequate and stable. All equipment must be in working order. A strict program of validation, qualification, calibration, and maintenance attains this. Keeping records of use and maintenance is essential in order to know, at any point in time, the precise state of the equipment.

Protocols: The principal steps of studies conducted in compliance with GLP have to be described in the study protocol. Thus, the Study Plan or Protocol serves to outline the conduct of the study demonstrating at the same time that adequate planning is provided. The Protocol has therefore to be adopted by the Study Director through dated signature before the study starts, and alterations to the study design cannot be made unless by formal amendment procedures. All this will ensure the reconstructability of the study at a later point in time.

Written Procedures: It will not be possible to describe in the protocol all the technical details of the study. Since the possibility for exact reconstruction of studies is a *sine qua non* for the mutual acceptance of data, routine procedures are described in written Standard Operating Procedures. Laboratories may also need to standardize certain techniques and approaches to facilitate comparison of results; here again written SOPs are an invaluable tool. However, procedures cannot be fixed for all time, since this would lead to the use of outdated methods and procedures; they have to be adapted to developments in knowledge and technical progress. They must, therefore, be reviewed regularly, and modified, if necessary, so that they reflect the actual "state-of-the-art." Finally, it is important that, for ease of consultation, SOPs are available directly at the workplace and in the current version only.

Study Director: This is the single most important individual in a GLP study, as he/she represents the pivotal point of study control. The Study Director is the person fully responsible for GLP compliance in a study: is responsible for the adequacy of the protocol, and the GLP compliant conduct of the study. The Study Director has to formally accept responsibility for GLP compliance by signing the compliance statement.

Characterization: In order to perform a study correctly, it is essential to know as much as possible about the materials used during the study. For studies intended to evaluate the safety-related properties of pharmaceutical compounds during the preclinical phase, it is a prerequisition for test articles. Characteristics such as identity, purity, composition, stability, impurity profile, should be known for the test item, for the vehicle, and for any reference material. If the test system is an animal (which is very often the case), it is essential to know such details as its strain, health status, and normal biological values.

Documentation—Raw Data: All studies generate raw data which are, on the one hand, the results of the investigations and represent the basis of the conclusions, but which, on the other hand, also document the procedures and circumstances under which the study was conducted. Some of the study results will be treated statistically, while others may be used directly. Whatever the case, the results and their interpretation provided by the scientist in the study report must be a true and accurate reflection of the raw data.

Study Report: The study report, just like all other aspects of the study, is the responsibility of the Study Director. He/she must ensure that the contents of the report describe the study accurately. The study director is also responsible for the scientific interpretation of the results.

Archives: Since a study may have to be reconstructed after many years, and the records must be stored for long periods of time but be available for prompt retrieval, the safekeeping of all records must be ensured. Archiving of the raw data and other essential documents must be such that data are kept in an integer state and can neither be lost nor altered; to achieve this goal, it is usual practice to restrict access to archives to a limited number of people and to maintain records of log-in and log-out for both documents and people.

Quality Assurance: Quality assurance, as defined by GLP, is a team of persons charged with assuring the management that GLP compliance has been attained in a test facility as a whole as well as within each study. QA has to be independent of the operational conduct of the studies, and it functions as witness to the whole preclinical research process.

Resources—Facilities: Buildings and Equipment: Testing facilities should be of suitable size, construction, and location to meet the requirements of the study and to minimize disturbances that would interfere with the validity of the study. They should be designed so as to provide an adequate degree of separation of the various aspects of the study. The purpose of these requirements is to ensure that the study is not compromised because of inadequate facilities. It is important to remember that fulfilling the requirements of the study does not necessarily mean providing state-of-the-art construction, but does mean careful consideration of the objectives and how to achieve them. Separation ensures that different functions or activities do not interfere with one another or affect the study, and minimizes disturbances. This can be done by

- Physical separation, for example, by walls, doors, or filters. In new buildings or those under conversion, separation will be part of the design. Otherwise separation can be achieved by the use of isolators, for example.
- Separation by organization, for example, carrying out different activities in the same area at different times, allowing for cleaning and preparation between operations, maintaining separation of staff, or establishing defined work areas within a laboratory. As an illustration of the principles involved we shall consider:
    - Areas concerned with test material control and mixing with vehicles (although the same considerations would apply to other areas such as analytical or histopathology laboratories).
    - Animal facilities.

The pharmacy and dose mixing area is a laboratory dealing with test item work flow: receipt, storage, dispensing, weighing, mixing, dispatch to the animal house, and waste disposal.

Size: The laboratory should be big enough to accommodate the number of staff working in it and allow them to carry on their work without risk of getting in each other's way or of mixing up different materials. Each operator should have a workstation sufficiently large to enable him/her to carry out the operation efficiently. There should also be a degree of physical separation between the workstations to reduce the chance of mix-up of materials or cross contamination. The pharmacy is a sensitive area, and to such facilities access should be restricted so as to limit the possible contamination of one study compound by another.

Construction: The laboratory has to be built of materials that allow easy cleaning but do not allow test materials to accumulate and cross contaminate others. There should be a ventilation system that provides airflow away from the operator through filters which both protect personnel and prevent cross contamination. Most modern dose mixing areas are now designed in a "box" fashion, each box having an independent air system.

Arrangement: There should be separate areas for (1) storage of test items under different conditions; (2) storage of control items; (3) handling of volatile materials; (4) weighing; (5) mixing of different dose formulations, for example, in the diet or as solutions or suspensions; (6) storage of prepared doses; (7) cleaning equipment; (8) offices and refreshment rooms; and (9) changing rooms.

Animal Facilities: To minimize the effects of environmental variables on the animal, the facility should be designed and operated so as to prevent the animals coming into contact with disease, or with a test item other than the one under investigation. Requirements will differ depending upon the nature and duration of the studies being performed in the facilities. The risks of contamination can be reduced by a "barrier" system, where all supplies, staff, and services cross the barrier in a controlled way, as well as by providing "clean" and "dirty" corridors for the movement of new and used supplies. A typical animal house would therefore have this required separation maintained by provision of areas for (a) species; (b) studies; (c) quarantine; (d) changing rooms; (e) receipt of materials; (f) storage (bedding and diet, test doses, cages, etc.); (g) cleaning equipment; (h) necropsy; (i) laboratory procedures; (j) utilities; and (k) waste disposal.

The building and its rooms should provide space for sufficient animals and studies, allowing the operators to work efficiently. The environment at control system should maintain the temperature, humidity, and airflow constantly at the defined levels for the species concerned. The surfaces of walls, doors, floors, and ceilings should be easy to clean completely, with no gaps or ledges where dirt and dust can build up, or no uneven floors where water can build up. Whatever the capabilities or needs of your laboratory are, sensible working procedures reduce the potential danger to the study from outside influences and maintain a degree of separation. This can be achieved by (1) minimizing the number of staff allowed to enter the building; (2) restricting entry into animal rooms; (3) organizing work flow so that clean and dirty materials are moved around the facility at different times of day, if the construction of the facility does not permit other solutions, with corridors being cleaned between these times; (4) requiring staff to put on different clothing for different zones within the animal facility; and (5) ensuring that rooms are cleaned and sanitized, if necessary, between studies.

Equipment: Appropriate equipment of adequate capacity should be available for the proper conduct of the study. All equipment should be suitable for its intended use, and it should be properly calibrated and maintained to ensure accurate performance. Records of repairs and routine maintenance, and of any nonroutine work, should be retained. The purpose of these GLP requirements is to ensure the reliability of data generated and to ensure that data are not lost as a result of inaccurate, inadequate, or faulty equipment.

Suitability: This can only be assessed by consideration of the tasks that the equipment is expected to perform. Just as there is no need to have a balance capable of weighing to decimals of a milligram to obtain the weekly weight of a rat, a balance of this precision may well be required in the analytical laboratory. Adequate capacity is also needed to perform the tasks in a timely manner.

Calibration: Equipment that is performing to specification, whether it is used to generate data (e.g., analytical equipment or balances) or to maintain standard conditions (e.g., refrigerators or air conditioning equipment), should have some proof that the specification is being achieved. This will generally be furnished by periodic checking. In the case of measuring equipment, this is likely to involve the use of standards. For example, a balance will be calibrated by the use of known standard weights. In the case of a piece of analytical equipment, a sample of known concentration will be used to ensure that the equipment is functioning as expected and provides a basis from which to calculate the final result. Other equipment, such as air conditioning systems for animal housing or constant temperature storage rooms, will be checked periodically, at a frequency that allows action to be taken in time to prevent any adverse effect on the study should the equipment be demonstrated to be operating out of specification limits.

Maintenance: The GLP requirement that equipment should be maintained, that ensures that equipment performs constantly to specification and reduces the likelihood of unexpected breakdown and consequent loss of data. Maintenance may be carried out in two quite distinct ways:

- Planned, when a regular check is made, irrespective of the performance of the equipment, and reparative work is undertaken when the calibration or regular checking suggests that the machine is not functioning according to specification. Planned maintenance may be a useful precaution for large items of equipment or items that do not possess suitable backup or alternatives. Regular maintenance therefore reduces the risk of breakdown.
- For equipment, such as modern computer-driven analyzers or electronic balances, that does not easily lend itself to routine maintenance of this sort, a better approach may be to check it regularly and ensure that suitable contingencies are available should a problem occur. These contingencies may include having equipment duplicated or having immediate access to an engineer. Backup for vital equipment should be available whenever possible as well as backup in the event of service failures, such as power cuts. A laboratory should have the ability to continue with essential services to prevent

animals or data being lost, and studies irretrievably affected. A laboratory carrying out animal studies, for example, may, as a minimum, need a stand-by generator capable of maintaining the animal room environment even if it does not allow all the laboratory functions to continue, because the loss of the animals would irretrievably affect the study whereas samples may be stored for a period until power is returned. Early warning that equipment is malfunctioning is important. The checking interval should be assigned to assure this, but the use of alarms will often assist in this, particularly, if the problem occurs at a time when the staff is not present in the laboratory.

Documentation: The planning of routine maintenance, as mentioned above, should be documented in such a way that users of the equipment can be assured that it is adequately maintained and is not outside its service interval. A "sticker" attached to equipment, or provision of a clear service plan, may ensure this. The records should also demonstrate that the required action was taken as a result of the checks that had been made. Records should show that all relevant staff knew about, and took, appropriate action when parameters exceeded acceptable limits.

Personnel: Although laboratory management and organizational requirements occupy about 15% of GLP texts, unfortunately they are still seen by regulators and QA as one of the principal sources of noncompliance with the letter, if not the spirit, of GLP. Indeed, without full management commitment and the formal involvement of all personnel, GLP systems lack credibility and will not function as they should. Management and organizational systems therefore are a critical element of setting up GLP in a laboratory. The management of a test facility, of course, has overall responsibility for the implementation of both good science and good organization, including compliance with GLP.

Good Sciences: It should carry (1) careful definition of experimental design and parameters based on known scientific procedures; (2) control and documentation of experimental and environmental variables; (3) careful, complete evaluation, and reporting of results; and (4) results become part of accepted scientific knowledge.

Good Organization: It should establish (1) provision of adequate physical facilities and qualified staff; (2) planning of studies and allocation of resources; (3) definition of staff responsibilities and training of staff; (4) good record keeping and organized archives; (5) implementation of a process for the verification of results; and (6) compliance with GLP.

3.3.1 Planning (Master Schedule)

The requirement for a master planning system seems obvious but how many laboratories suffer from "Monday morning syndrome" when their project activities have been modified but without adequate allowance made for needed resources or the impact on existing work? It is a management responsibility to ensure that sufficient personnel resources are allocated to specific studies and support areas. The planning/resource allocation system required by GLP is called the master schedule or plan. These systems may take many forms but each system must ensure that (1) all studies (contracted and in-house) are included; (2) change control reflects shifts in dates and workload; (3) time-consuming activities such as protocol review and report preparation are allocated sufficient time; (4) the system is the "official" one (i.e., there are no competing systems for the same purpose); (5) the system is described in an approved SOP; (6) responsibility for maintenance and updating of the master schedule are defined; (7) the various versions of the master schedule are approved and maintained in the archive as raw data; and (8) distribution is adequate and key responsibilities are identified.

In most laboratories, the system includes these elements. Once the protocol is signed and distributed, the study is entered into the master schedule. This may or may not be a QA function in a laboratory. Often it is a project management function and is computerized for efficiency and ease of cross-indexing. The master schedule system is described in an SOP. Typically, QA has "Read-only" and "Print" access to this data file. Signed hard copies are usually archived regularly as raw data. In contract facilities, sponsor and product names are usually coded to provide confidentiality. The QA inspection plan will be described later.

3.3.2 Personnel Organization

Management has the responsibility for overall organization of the test facility. With respect to personnel, this organization is usually reflected in the organization chart. This is often the first document requested by national monitoring authorities to obtain an idea of how the facility functions. GLP requires personnel to have the competence (education, experience, training, etc.) necessary to perform their functions. Personnel competence is reflected in job descriptions, curricula vitae (CVs), and training records. These documents should be defined in SOPs, regularly updated, and verified by QA audits. Definition of tasks and responsibilities/job descriptions and quality system is based on making people responsible for their actions. This responsibility may be described in the two sentences given below (1) do not do something where you do not understand the reason, the context and the consequences; and (2) each person signs her/his work and feels completely responsible for its correct completion.

There must be a clear definition of tasks and responsibilities. The contents of job descriptions should correspond to the qualifications as described in the CV. In addition, they should be (1) updated at a minimum required interval (fixed by an SOP) and (2) signed by the person occupying the post and by at least one appropriate member of management supervising the post.

Rules of delegation should be defined at the test facility. Tasks can be delegated, but the final responsibility remains with the person who delegates the task. A review of all job descriptions annually, or in the event of any reorganization, helps a facility's management to ensure that their organization is coherent.

3.3.3 Curriculum Vitae

A procedure should ensure that CVs (1) exist for all personnel in a standard approved format; (2) are kept up-to-date; (3) exist in required languages (local and sometimes

English for regulatory submissions); and (4) are carefully archived to ensure historical reconstruction. All staff should have a CV. Even if some staff do not have extensive qualifications, they will have professional experience which should be listed in their CV. It is usual to include in a CV: Name, Sex/gender of the person (optional), Education, including diplomas and qualifications awarded by recognized institutions, Professional experience both within the institution and before joining it, Publications, Membership of associations, Languages spoken, and so on.

Training: Finally, training records complement CVs and job descriptions: job competence depends largely on internal and external specialized training. GLP explicitly requires that all personnel understand GLP, its importance, and the position of their own job within GLP activities. Training must be formally planned and documented. New objectives and new activities always involve some training. Training systems are usually SOP based. A new SOP therefore requires new certification of the involved personnel. Some companies have advanced training schemes linking training to motivation, professional advancement, and reward. The training system will have elements common to all GLP management systems, for example, it will be formal, approved, documented to a standard format, described in a Standard Operating Procedure, and with historical reconstruction possible through the archive.

3.3.4 Rules of the Conducts of Studies

General Aspects: The laboratory should have different types of document, which direct the scientific conduct of the studies. The purpose of these is to (1) state general policies, decisions, and principles; (2) inform staff as to the correct performance of operations; and (3) provide retrospective documentation of what was planned.

The Study Plan or Protocol: The protocol is the central document whereby the Study Director communicates the planned organization and development of the study to the staff, and by which she/he informs third parties involved, such as the quality assurance unit (QAU) or, if the study is contracted to a contract research organization, the sponsor. The protocol may also function as the basis for a contract in the latter situation. The protocol contains the overall plan of the study, and describes the methods and materials that will be used, thus demonstrating the adequacy of advance planning. It is very important to remember that, since the protocol is the principal means of instruction of staff during conduct of the study, the contents, style and layout should be suitable for that purpose.

3.3.5 Content of Study Protocol

The content of the Protocol is designed to meet the scientific requirements of the study and to comply with GLP.

Identification: The study number provides a means of uniquely identifying all records of the laboratory connected to a particular study and of confirming the identity of all data generated. There are no set rules for the numbering system.

Title and Statement of Purpose: The title should be both informative and short. It should state the name of the compound, the type of study, and the test system as a minimum. It is particularly important to define why a study is being done. A study must be planned and designed in advance. This cannot be done adequately unless the designer has a clear understanding of the purpose of the work. Stating this purpose in the protocol ensures that the results of the study are only utilized for purposes for which they are suited. The purpose of the study may include both scientific and regulatory reasons.

Identification of Test (and Control) Items: This includes not only the chemical name and/or code number of the test item but also its specifications or characterization and its stability, or details about how these will be determined. The protocol must also detail any active control materials, which are to be used in addition to the vehicle.

Name of Sponsor and Address of Test Facility: The sponsor and the test facility may or may not be the same organization. The protocol should indicate the location where the test is to be carried out and also the address of any consultants planned to be used. The name of the sponsor should also be included.

Name of Study Director and Other Responsible Personnel: The name of the Study Director must appear in the protocol. It is also a requirement to identify any other responsible scientists who are going to contribute significantly to the study. As a rule of thumb, most laboratories include the names of scientists who will be responsible for the interpretation of data generated under their responsibility (e. g., pathologists, clinical pathologists). For contract studies it is usual to include the name of the monitor or sponsor contact person. The proposed dates for the study are the start and finish dates (corresponding to the date when the protocol is signed and the date when the final report is signed by the Study Director) and the experimental dates. The latter correspond to the dates when the first and last experimental data are collected. To help study personnel performing the work, the protocol may include a more detailed time plan or this may be produced separately. Dates are notorious for slipping. Rules for changing dates, either by making protocol amendments or by updating an independent project planning system, should be defined in the SOP for protocol administration.

Justification for Selection of the Test System: In the case of experiments using animals, the species and possibly the strain may have been defined in test guidelines. However, it is still important that the protocol contains a reason why the test system has been chosen. Often this is based on the test facility's background (historical) data with the strain concerned, but there may be special scientific or regulatory reasons.

For animal experiments, this will include the proposed species, strain, age, weight, and source of animals and how they are to be identified. It will also contain details of the animal husbandry including environmental conditions, diet, and its source.

Experimental design includes the following:

- *Dosing Details*: Dose levels; dosing route; frequency of dosing; vehicles used; method of preparation/administration; storage conditions of formulation; and quality control.
- Assignment to groups or randomization of animals.
- Parameters to be measured and examined. These identify the measurements to be made and the frequency of measurement. They will also detail any additions to, or planned deviations from, the SOPs and give complete details of nonstandard procedures or references to them. Analytical methods are not included in detail in most protocols but will be available as SOPs or "Methods" documents which are held in the analytical laboratory with the study data.
- Statistical methods and power.
- Data to be retained after the study.
- *Quality Assurance*. Frequently, the Protocol outlines the proposed QA program but this is not mandatory.

3.3.6 Approval of Study Protocol

A GLP study cannot be started before the Protocol/Plan is approved. This is done by dated signature of the Study Director. The draft Protocol should also be controlled by QA in order to assess its compliance with GLP requirements. It is a good practice also that the sponsor should have agreed to the design of the study before it begins. Protocol approval should be early enough before the study starts to ensure that all staff know their scheduled duties. Also, QA should receive a copy of the approved Study Protocol/Plan in order to allow them to plan their audit/inspection activities.

Allowance of insufficient time between producing the Protocol and starting the study may lead to serious problems later in the study. Sufficient time must therefore be allowed to (1) produce the Protocol; (2) discuss its implications with staff concerned; (3) circulate the Protocol for QA review; (4) circulate the Protocol for approval; and (5) circulate the approved version to all staff involved in the study.

3.3.7 Distribution of Study Protocol

All involved staff should receive a copy of the Protocol/Plan. In order to ensure that everybody does get a copy, a distribution list should be prepared, and it is often worthwhile to obtain a signature from each person and to hold meetings before the study begins, to ensure that everybody is aware of their roles in the study.

3.3.8 Protocol Amendment

Although the Protocol is the document that directs the conduct of the study, it should never be thought of as being immutable, or "cast in tablets of stone." It is a document that can be amended to allow the Study Director to react to results or to other factors during the course of the study. However, any change to the study design must be justified. A protocol amendment has to be issued to document a prospective change in the study design or conduct. If a change in a procedure needs to be instituted before a formal protocol amendment can be generated, this needs to be recorded, and a protocol amendment is issued as soon as possible afterwards.

It is not an acceptable practice to use the amendment to retrospectively legalize omissions or errors that occurred during the study. Such unplanned occurrences should be documented in study file noted as "deviations," and be attached to the relevant raw data. The important elements of a protocol amendment are (1) the study being amended is clearly identified; (2) the amendment is uniquely numbered; (3) the reason for the amendment is clear and complete; (4) the section of the original protocol being amended is clearly identified; (5) the new instruction is clear; and (6) the circulation is the same as that of the original protocol.

As with the original protocol, the Study Director is the person who approves and is responsible for issuing the document. He/she is also responsible for ensuring that the new instruction is performed correctly. It is as essential to review amendments as the main protocol per GLP compliance. This is a QA function. Because amendments are by their very nature required extremely urgently by study staff, this review is often performed retrospectively. The original signed protocol and all amendments must be lodged in the archives as part of the study file. It would be a good idea to archive the protocol at the beginning of the study, and work from authorized photocopies.

3.3.9 Standard Operating Procedures

A collection of good SOPs is a prerequisite for successful GLP compliance. Setting up an SOP system is often seen as the most important and time-consuming compliance task. Even without GLP regulations, classical quality assurance techniques, indeed good management, require standardized, approved, written working procedures. Remember the Deming quote: "Use standards e.g., SOPs as the liberator that relegates the problems that have already been solved to the field of routine, and leaves the creative faculties free for the problems that are still unsolved" (W. Edwards Deming).

The successful implementation of SOPs requires (1) sustained and enthusiastic support from all levels of management with commitment to establishing SOPs as an essential element in the organization and culture of the laboratory; (2) SOP-based education and training of personnel so that the procedures are performed in the same way by all personnel; and (3) a sound SOP management system to ensure that current SOPs are available in the right place.

3.3.10 SOP System Overview

The system should include the following characteristics: (1) total integration into the laboratory's system of master documentation (e.g., not a separate system in potential conflict with memos or other means of conveying directives to laboratory personnel) and (2) comprehensive coverage of (a) all critical phases of study design, management, conduct, and reporting; (b) scientific' administrative policy and procedures (e.g., formats, safety and hygiene, security, personnel management systems, etc.); and (c) standard scientific techniques, equipment, and so on.

Readability: The SOPs should follow a standard layout (standards and guidelines exist for this). The procedures should be written (or translated) into the local language of the operational personnel and expressed in an appropriate vocabulary. All personnel should be encouraged to improve the SOPs. Ideally, the people who do the work should also write the SOPs, thus promoting their sense of responsibility for the work they perform.

Usability and Traceability: For reasons of traceability and ease of use, a two-tier system of SOPs is often the preferred approach. For example, one tier reflects general policies and procedures (e.g., protocol writing, review, approval, distribution and modification, SOPs, general rules for equipment use and maintenance, archives, etc.), the second represents technical methods (e.g., methods of staining in histology, analytical methods, specific procedures for use and maintenance of equipment). It is advisable to present the SOPs (SOP manuals) as a binder with an up-to-date table of contents, logical chapter divisions and selective distribution, to avoid a mushrooming packet of dust-gathering paper that often gets misplaced. In some laboratories, SOPs are available directly from a screen, but in this case you will need to implement special rules about printing out the SOPs (expiry dates, etc.) and rules about signatures. All revisions to SOPs have to be made through formal revision process; notes and changes as handwritten margin comments are not admissible.

Responsibility: Somebody should be responsible for each SOP (author or person responsible), to handle queries and keep each procedure updated. It is a good idea to impose a minimal requirement for periodic review. For example, SOPs should be reviewed and revised, if applicable, every 2 years.

Change Control: A formal system should be in place that ensures historical reconstruction. An SOP system, if working properly, tends to seem perpetually incomplete because of additions, deletions, and modifications reflecting the normal rate of improvements or changes. Indeed, changes and amendments are good evidence that the laboratory uses the SOPs. Therefore updating should be easy and rapid, and authorization should not involve too many signatures.

Centralized Organization: This concerns issues of reporting such as formatting, numbering, issuance, modification and withdrawal, incoherence, delays, lack of traceability, and incomplete distribution. Centralized organization avoids duplication of effort.

Availability: SOPs should be made immediately and readily available to the person doing the work.

Archiving: All withdrawn SOPs, whether no longer used or superseded by a revised version, must be archived carefully in order to make a complete historical record of the test facility's procedures. Properly designed SOPs will bring the following benefits to the laboratory (1) standardized, consistent procedures (person-to-person, test-to-test variability minimized); (2) an opportunity to optimize processes; (3) technical and administrative improvements; (4) demonstration of management commitment to quality as part of the SOP approval process; (5) ease of documenting complicated techniques in study protocols and reports (a simple reference to the procedure should often suffice); (6) continuity in case of personnel turnover; (7) use as training manual; (8) a means of study reconstruction after the event, also after a lapse of years; and (9) a means of communication in case of audit, visits, technology transfer, and so on.

In summary, most laboratories incorporate the necessary characteristics into the following approach: (1) a two-tier system; (2) a defined format; (3) thorough review, including QA review; (4) formal approval by at least two people (a designated author and an appropriate member of test facility management); and (5) a formal change control system, coordinated by a designated person/group. During the course of the study, a general SOP (tier 1) requires all deliberate deviations to operational SOPs to be approved in advance by the Study Director. If this is impossible, he/she should be informed in writing. This record, along with the technical person's and/or the Study Director's assessment of the deviation (e.g., no impact on the study nor, if so, the extent of impact on the study), is maintained as raw data in the study file for audit and consideration when writing the final report.

3.3.11 Characterization

The test item—The identity, activity, stability, and bioavailability of the test item are central to the validity of the study. It must be demonstrable that the test system has received the correct amount of material. This is assured by proper control of the test item at all stages of its use, and by preparation of detailed records which document every stage of the test item's disposition. A GLP quality assurance program should systematically attempt to minimize the possibility that the test item is affected by any quality problems.

3.3.12 Test Item/Article Control before Formulation

Receipt: The test article/item will be delivered from the manufacturer. This may be a section within the same organization as the test facility or a separate organization altogether. The responsible person should be aware in advance of test item arrival so as to ensure correct storage conditions and necessary handling requirements. In the case of a study conducted by a contract laboratory (CRO), the sponsor should provide this information, as well as other details, which may help in the preparation of the dose formulation, to the CRO. A standard form on which to provide this information is helpful. During development of the protocol, the sponsor fills it out to give the testing facility the essential information necessary for safe and adequate handling of the test item. The sponsor will either supply, or indicate that he has obtained, the necessary data on chemical characterization and stability of the test material. The manufacturer, meanwhile, will archive and store batch records.

The test item container should be robust enough to withstand transfer between facilities, and should ideally be suitable for further use. Packaging of the test item is very important. The sponsor should consider the method of transport used and the duration of the journey. This is particularly true when the material is packed in fragile containers, for example, glass bottles, or needs to be transported long distances using public transport under special conditions, for example, kept frozen. Consideration should always be given to the unexpected, such as airport delays, strikes, or bad weather. The test item should be accompanied by a delivery form detailing: (1) manufacturer's name or sponsor's name; (2) date of dispatch; (3) number of containers or items, type, amount of contents; (4) identity of test item; (5) batch number(s); (6) identity of person responsible for dispatch; and (7) name of carrier.

Each test material container should be clearly labeled with sufficient information to identify it and allow the testing facility to confirm its contents. Ideally, labels should contain the following information: (1) test item name; (2) batch number; (3) expiry date; (4) storage conditions; (5) container number; (6) tare weight; and (7) initial gross weight. On arrival of the test item, the testing facility should have a procedure for handling and documentation of receipt. It is most important that the compound is logged in immediately to ensure a complete audit trail and to demonstrate that it has not been held under conditions, which might compromise its chemical activity. The receipt procedure should include instructions for handling if the designated person is absent or if the container is damaged on receipt. The Study Director should be informed of the arrival of the test item.

The test facility's documentation, on arrival of the test item, normally includes the following information: (1) compound name; (2) batch number (s); (3) description of the test item that is completed on its arrival at the laboratory and compared with the description supplied by the sponsor. This ensures that any concern about the identity of the material can be sorted out at an early stage; (4) container number, to allow identification of the container in use; (5) container type; (6) net weight of the contents and container tare weight; (7) storage conditions and location of the container; (8) initials of the person receiving the container; (9) date of arrival of the container at the laboratory; and (10) condition of goods on arrival.

Storage of the Test Article/Item: Test articles/items must be stored under closely controlled conditions, particularly with respect to access and environment. The construction of the store should ensure that only designated staff has access to the material. The stores are kept locked when not in use. Separate areas should be available for storage at ambient temperature, +4°C and −20°C. The storage of the test item is arranged to minimize the risk of any cross contamination between compounds and containers.

On arrival at the test facility, a sample of the batch of test item is taken and stored in a separate container. This "reserve sample" is ideally held in a separate compound archive under the same conditions as the main bulk of the test material. It carries the following information on its label: (1) test material identification (name or code number); (2) batch number; (3) storage conditions; (4) net weight; and (5) date on which sample was taken. This information will be retained by the test facility in the compound archive for the same duration as are the study raw data and specimens. Normally this sample will not be used unless some test item is required, for example, for confirmatory analysis.

Test Article/Item Use: Documenting each use of test article/item on a record form allows a running check to be kept. Not only does this provide a complete trail of all test item used, but it also provides a means of monitoring actual use against expected use. The type of information includes (1) *date of use*; (2) *study number*—This is important if the same batch of test item is being used for more than one study (some laboratories split the material into separate containers for each study); (3) *gross weight before use*—The container and contents are weighed prior to each use. The initials of the person carrying out the weighing are recorded; (4) *gross weight after use*—The container and contents are weighed after use; (5) *weight of material used*—This is the amount of material disappearing from the container on each occasion; (6) *weight from dose preparation records*—This is the amount of material recorded as used in the preparation of the dose form. Comparison between this record and the amount that has been removed from the container provides a useful double check on the amount weighed out; (7) *discrepancy*—This allows explanation of any discrepancy (e.g., spillage); and (8) *stock remaining*—This provides a running total of the quantity of material in the container, and gives a warning of the need to order additional material.

Disposal: Following the completion of a study, surplus amounts of the test item should be disposed of in an environmentally acceptable way. This final event must again be documented so as to account for the total amount of test item.

3.3.13 Preparation of the Dose Formulation

If the test system receives an incorrect dose, or if there is doubt about the dose, the rest of the experiment is almost certainly compromised. The following well-specified procedures and the documentation of every stage of the process are necessary.

3.3.13.1 Initial Preparation and Planning Before the study begins, a number of factors must be considered and communicated to staff by the Study Director. Some of these may be considered before the protocol is finally signed:

- *Dose levels, number of animals, and dose volume.* This information in the protocol allows the Study Director to estimate how much test item is required and ensure that sufficient is available throughout the course of the study. As a part of this consideration, he/she also checks on the purity of the test item. In most studies, the test item is assumed to be 100% active ingredient, but, if significantly less, it will be necessary to adjust the amounts to be weighed out (and to investigate what impact the impurities may have on the validity of the study).

- *Concentration of the dose, amount, or volume required.* The volume required will vary throughout the study with the animals' weight, and the Study Director will keep this under review. To ensure that this is done regularly, the Study Director is often required to produce a request form every 2 weeks.
- SOPs must exist for each procedure in the preparation of the formulation, analysis of the documentation and data required, and operation of all equipment.
- The method of preparation of the dose form should be tested prior to starting the study. This entails a trial preparation of at least the highest dose level, to confirm that the various standard procedures detailed in the SOPs produce an acceptable dose of the right concentration and homogeneity.
- This trial preparation may indicate the need for further development of the method, for example, experimentation with other vehicles or different mixing techniques.
- The stability of the dose form must also be assessed in the vehicle used. Following the trial preparation, the SOP for the formulation may need amending.

Formulating the Test Item In many test facilities, an independent group formulates the test item. This situation emphasizes the importance of clearly recording what is planned and what is actually done. Even if the Study Director carries out the whole process, the formulation plan is an important part of the final record.

Before the container of material is opened, the persons carrying out the procedure will have ensured that (1) there is a dedicated workstation of adequate size for the procedure; (2) the preparation surface is clean. This is often best achieved by covering it with a clean sheet of paper or plastic, which is disposed of after each test item preparation; (3) there are adequate clean containers, spatulas, and other small equipment at hand; (4) labels have been made out and are available; and (5) no other compound is being handled at the same time. This minimizes the possibility of confusion or cross contamination.

The test item is obtained from the store. The identity is checked against the protocol instructions or order. Following these instructions the correct amount is weighed out. The control mixes are usually done first. Then the test item is mixed with the vehicle exactly following, without deviation, the method determined during the trial preparation before the start of the study. In most cases, this involves making up each concentration from a separately weighed out amount of test item, mixing it first with a small volume of vehicle, and gradually increasing the amount of vehicle to achieve the required total volume. In some cases, where the test item is dissolved in the vehicle or where the diet is the vehicle, it may be preferable to make up the highest concentration and dilute samples of this stock concentration to obtain the lower dose levels.

Following preparation, the dosing material is placed in suitable containers before being passed to the animal room for dosing. The suitability of the containers should be considered quite carefully in order to preserve the integrity of the dose form including: (1) composition—the container must not react with either test item or vehicle and (2) size—if the formulation needs to be mixed using a magnetic stirrer in the animal house to keep it in homogeneous suspension, the container must be big enough to develop a vortex, but not so big, in relation to the volume made up, as to prevent the mixer working. The final container (and any intermediate containers) should be labeled to allow identification. The container sent to the animal house should carry at least the following information: (1) study number; (2) group number (and if relevant, sex); (3) weight of container and contents; (4) date formulated; and (5) storage conditions. It may be useful to color code the label for each dose, with the same colors as those on the cage labels.

3.3.14 Sampling and Quality Control of Dose Formulation

Analysis of the formulation is required by the protocol to fulfill GLP requirements and to ensure that concentration, stability, and homogeneity of test item/vehicle mixtures is assessed. This information may be generated after the start of the study. It is an advantage, however, to conduct some of these analyses before the study starts, to prevent waste of time and resources as well as unnecessary dosing of test system by using a dose form that is subsequently shown to be unsuitable for the experiment.

As indicated above, the measurement of stability and homogeneity of the test material/vehicle formulation should have been done on a trial preparation. Samples of this preparation are taken under conditions as closely identical to the dosing situation as possible. The dose is left for the same period of time as will be the case between preparation and administration in the real situation. Then samples are taken from different positions in the dosing vessel. For long-term studies where a stock solution is made for generating dose formulation throughout the study, aliquots will also be taken and analyzed periodically to assess the "shelf-life" of the formulation.

The samples taken as indicated above give a good estimate of the effectiveness of the dose preparation process. However, periodic checks are also required to confirm that the process is being carried out correctly throughout the study even if doses are made up fresh each time. Only the chemist who takes the samples (but not the persons making up the mixture or performing the dosing) knows the day they will be taken. It is preferable to take the sample in the animal room from the residue following dosing, as this gives not only information on the concentration dosed to the animals but also some further confirmation of homogeneity and stability of the test article in real use. The following records are made of the formulation process: (1) date; (2) confirmation of test item identity; (3) identity of formulation instruction (request); (4) weight of empty container; (5) weight of container + test item; (6) weight of added vehicle; (7) final weight of mixture; and (8) signature/initials of all staff carrying out procedures.

3.3.14.1 ***Dosing*** The purpose of dosing is to deliver the required amount of test material to the animal accurately and consistently. Therefore, the procedure must be very conscientiously carried out and the records must confirm that all the animals have been dosed with the correct volume and concentration. Detailed records with built in cross-references document the fact that the dosing has been correctly carried out. The staff must be well trained, both to ensure that the amount is accurately delivered and also to assure the well being of the animals. In many countries, the staff who dose animals must be licensed or formally qualified in some way under animal welfare laws.

On arrival in the animal area, the dose should be checked for identity and that the amount is the same as the amount issued from the formulation department. Staff should ensure that the container is still intact. The containers are then kept under appropriate conditions (e.g., placed on a magnetic stirrer, on ice, etc.) until dosing starts. The dosing procedure is done in a fixed order, taking into account the need to minimize the possibility of cross contamination and confusion between animals, dose groups, and different formulations. Consequently, the following precautions are typical of those that most laboratories take when dosing animals orally by gavage:

- The animals are dosed group by group, working in ascending dose levels.
- Only one dose container is open at any onetime, and each dose level has its own catheter and syringe.
- All cages from one group are identified before the group is dosed, using the group number and label color code as a confirmatory check.
- A new catheter and syringe are used for each dose level.
- The container, catheter, and syringe are removed from the dosing station before the new group is dosed.
- The outside of the catheter is wiped with a clean tissue before each animal is dosed. This prevents the possibility of test material being drawn into the lung.
- Only one cage of animals is opened at a time. If the study animals are individually housed, each is returned to its cage following dosing. If multiply housed, the animals should be placed in a second container until all animals from the cage have been dosed and then returned to their cage.
- Each animal is positively identified (e.g., from its tattoo), not merely from the cage number.

The dose volume is calculated from the body weight, using a list giving the required volume for each weight to avoid the risk of calculation error during dosing.

Records should identify the following: (1) the staff involved in dosing; (2) the dose given to each animal. These acts both as a confirmation of dosing of each individual and as a record that can be checked against the expected weight; (3) the date and time dosing took place; and (4) the weight of each dose level container before and after dosing. This allows some check to be made of the expected use against actual use of formulation.

3.4 INITIATIVES AND IMPLEMENTATION OF BIOANALYTICAL METHOD VALIDATION (GUIDANCE FOR INDUSTRY BMV—MAY 2001)

In 1990 leaders in the field of bioanalytical sciences met at Crystal City Conference Center in Arlington, VA to discuss important issues in the field of bioanalysis. A consensus paper was published [15] from this meeting, which set the industry standard for bioanalysis and provided the foundation for the FDA guidelines. Following the first workshop report and the experience gained at the FDA, the draft Guidance on Bioanalytical Methods Validation was issued by the FDA in January 1999. This draft guidance provided stimulus and opportunity for further discussion at the 2nd AAPS/FDA Bioanalytical Workshop in January 2000. In addition, newer technology, such as chromatography coupled to tandem mass spectrometry (LC–MS/MS), was discussed along with an update on ligand-binding assays. This workshop resulted in a report "Bioanalytical Method Validation—A Revisit with a Decade of Progress" [16] and formed the basis for the FDA Guidance on Bioanalytical Methods Validation in May 2001 [17].

The evolution of divergent analytical technologies for conventional small molecules and macromolecules, and the growth in marketing interest in macromolecular therapies, led to the workshop held in 2000 to specifically discuss bioanalytical methods validation for macromolecules. Because of the complexity of the issues, the workshop failed to achieve a consensus. To address the need for guiding principles for the validation of bioanalytical methods for macromolecules, the AAPS Ligand-Binding Assay Bioanalytical Focus Group developed and published recommendations for the development and validation of ligand-binding assays in 2003 [18].

A third meeting occurred in May 2006 and the working group from this 3rd AAPS/FDA Bioanalytical Workshop has published a consensus report [19] which is likely to be as standard-setting as the original report was 15 years ago. Sponsors need to ensure that the laboratories where they have bioanalytical testing performed are aware of these recommendations and are taking steps to come into compliance with them. The FDA was represented at the conference and it is reasonable to expect that the FDA will start requiring compliance with some or all of these recommendations, even if it is slow to publish its own follow-up guidelines.

There are many individual recommendations published in the paper for both ligand-binding assays and chromatographic assays. Some of the recommendations on implements are elaborated in following chapters.

3.4.1 Summary

More and more companies are performing or contracting work across the globe. Studies conducted are not only used in the country of origin but also submitted to RA in many countries. Global efforts at harmonization within technical arenas of the OECD are currently being developed. One area is the development of endocrine disruptor endpoints. In addition, there is currently an OECD group working to harmonize GLP field residue studies. This will allow for the acceptance of up to 50% field studies from one country to be accepted by other national authorities. When the residue harmonization effort is realized one question that must be asked is whether RA that have a certification program accept data from field facilities in the United States that have not yet been inspected by the US GLP authorities even if they are claiming compliance. FDA and EPA have both begun evaluations of their GLP programs. Among other things, both of these programs are looking at opportunities for greater global harmonization. As regards the OECD mutual acceptance of data

(MAD) program, it should be noted that all of the new applicants programs are based on the OECD GLP principles, although they have their own unique requirements as allowed by the OECD GLP principles. Moreover, all current and applying members are or have been audited by other member countries to assure the appropriate level of data quality. Upon careful analysis one can see that no country can simply apply the OECD GLP principles as written, but must make individual decisions on such aspects as length of archival storage or the timing of sponsor approval. For any MA GLP regulations that have gone through the MAD evaluation and approval process has demonstrated itself to be at a sufficient level as to assure the quality and integrity of the data regardless the differences that might exist between it and other MAD countries. These differences should not become obstacles to global acceptance of any GLP studies by any country that is a MAD member.

The objectives of the GxP quality guidelines are to ensure a product is safe and meets its intended use. GxP guides quality manufacture in regulated industries including food, drugs, medical devices, and cosmetics. The most central aspects of GxP are (1) *Traceability*: the ability to reconstruct the development history of a drug or medical device and (2) *Accountability*: the ability to resolve who has contributed what to the development and when. For a drug to be produced in a GxP compliant manner, some specific information technology practices must be followed. Computer and/or computerized systems involved in the development, manufacture, and sale of regulated product must meet certain requirements.

The pharmaceutical and life science industries therefore must heed various things that are somewhat neglected in other industries. The business case for any overhead cost of technical measures in this field is frequently justified by considering the costs associated with the potential losses associated with litigation. Investment in information technology security infrastructure and procedures are weighed against the risks and costs associated with litigation. System developments in the pharmaceutical and life sciences industries are associated with stringent record-keeping requirements. Traceability is of central importance. That is, it is necessary to create a chain of decisions that lead from user needs and business goals down to the system design decisions, and the verification of proper system installation and operation.

REFERENCES

1. World Health Organization Special Programs for Research & Training in Tropical Diseases (TDR)—*Good Clinical Laboratory Practices* (GCLP), 2009.
2. Food and Drug Administration, Bioresearch Monitoring Staff (HFC-30), Guidance for Industry, Good Laboratory Practices Questions and Answers, Rockville, MD, June 1981, revised December 1999 and September 2000; pp. 2–10.
3. FDA Warning Letter. Center for Devices and Radiological Health, Office of Compliance, Rockville, MD, April 27, 2001.
4. Food and Drug Administration. Office of Regulatory Affairs, Compliance Program Guidance Manual, Program 7348.808, Good laboratory practice (nonclinical laboratories), Appendix A, Rockville, MD, February 21, 2001.

5. Food and Drug Administration. Office of Regulatory Affairs, Compliance Policy Guide, FDA access to results of quality assurance program audits and inspections (CPG 7151.02), Rockville, MD, Revised January 3, 1996, pp. 1–2.
6. FDA Warning Letter. Reference no. CBER-00-026, Rockville, MD, July 3, 2000.
7. FDA Warning Letter. Reference no. 01HFD-45-1101, Rockville, MD, November 29, 2001.
8. FDA Warning Letter. Reference no. 01-HFD-45-0402, Rockville, MD, October 4, 2001.
9. FDA Warning Letter. Center for Veterinary Medicine, Division of Compliance, Rockville, MD, September 12, 2002.
10. FDA Warning Letter. Center for Devices and Radiological Health, Office of Compliance, Rockville, MD, July 2, 2004.
11. FDA Warning Letter. Center for Devices and Radiological Health, Office of Compliance, Rockville, MD, June 15, 2004.
12. FDA Warning Letter. Center for Devices and Radiological Health, Office of Compliance, Rockville, MD, April 16, 2004.
13. FDA Warning Letter. Center for Devices and Radiological Health, Office of Compliance, Rockville, MD, December 21, 2004.
14. Food and Drug Administration. 21 CFR part 58 Good Laboratory Practice Regulations; final rule. 52 Federal Register, September 4, 1987, pp. 33,768–33,782.
15. Shah VP, et al. Analytical methods validation: bioavailability, bioequivalence and pharmacokinetic studies. *Pharm Res* 1992;9:588–592.
16. Shah VP, et al. Bioanalytical method validation—a revisit with a decade of progress. *Pharm Res* 2000;17(12):1551–1557.
17. Guidance for Industry—Bioanalytical Method Validation, 2001.
18. DeSilva B, Smith W, Weiner R, et al. Recommendations for the bioanalytical method validation of ligand-binding assays to support pharmacokinetic assessments of macromolecules. *Pharm Res* 2003;20:1885–1900.
19. Viswanathan CT, et al. Workshop/Conference Report—quantitative bioanalytical methods validation and implementation: best practices for chromatographic and ligand binding assays. *AAPS J* 2007; 9 (Article 4):E30–E42.

4

FUNDAMENTAL ELEMENTS AND STRUCTURES FOR REGULATED BIOANALYTICAL LABORATORIES

4.1 INTRODUCTION

Bioanalytical group/laboratory is essential to drug/product discovery and development processes. It provides competitive, high quality support/services from early discovery to product approval (NDA and/or ANDA) and registration. It offers a breadth of drug research and development expertise to ensure complete management, flexibility, fast sample turnaround, and focus on critical timelines. Bioanalytical laboratories are likely GLP accredited/regulated and can perform validations and data generation in accordance with FDA regulations and OECD principles/guidelines. From early drug discovery, to preclinical or clinical development stages, bioanalysis usually provide a wide range of supports as follows:

Discovery Bioanalysis: From high-throughput screening to lead optimization processes, bioanalytical group works closely with medicinal chemistry, biology, toxicology, drug metabolism, and pharmacokinetics (DMPK), and other multifunctional teams to generate timely data for decision-making purposes.

Preclinical Bioanalysis: Most bioanalytical laboratories have extensive toxicokinetic and pharmacology facilities. Bioanalytical group provides integral support for integrated packages for drug development, which provides supporting preclinical bioanalysis for study samples derived from pharmacokinetic and toxicokinetic studies.

Regulated Bioanalytical Laboratories: Technical and Regulatory Aspects from Global Perspectives, By Michael Zhou

Clinical Bioanalysis: Quite often bioanalytical group provides fast sample turnaround for clinical studies in all phases of the drug development process. Laboratories should have strong relationships with clinical Phase I units and can support clinical drug development programs through tailored bioanalytical methods.

General capabilities to support preclinical and clinical bioanalysis may include the following:

(1) High capacity qualitative and quantitative LC–MS/MS analysis on triple-quadrupole and linear ion trap mass spectrometers
(2) Immunoassay and immunometric assays for PK/PD therapeutic assessment
(3) Robotic automation for high-throughput sample analysis
(4) Fully integrated bioanalytical LIMS system
(5) 24/7 sample receipt management and extensive storage capabilities at -20 and -80°C

Bioanalytical scope usually includes the following:

(1) *Early ADME/Discovery Support*: Rapid generic "fit-for-purpose" method development and sample analysis support for discovery lead optimization PK/PD studies. Discovery support for unaudited feasibility studies and limited qualification or validation, GLP audited studies.
(2) *Pharmacology Support:* Rapid bioanalysis of plasma samples to provide useful PK/PD correlations to assist in lead candidate selection.
(3) Support of safety telemetry studies carried out to GLP using partially or fully validated methodologies to provide proof of exposure data and allow PK/PD correlation of results.
(4) Immunoassay and immunometric assays (aka ligand-binding assays) enable multiple configurations for high sensitivity analyses. Immunoassays are based on two broad formats: reagent excess (e.g., enzyme-linked immunosorbent assay ELISA) and reagent limiting.

    Reagent excess is a popular configuration that can be used as either photometric or chemiluminescent substrate endpoints. Dissociation enhanced lanthanide fluoroimmunoassay (DELFIA) is also a highly sensitive technique using time resolved fluorescence as the signal output.

    Reagent limiting: Competitive binding format (e.g., enzyme or radio immunoassay (EIA, RIA)). The reagent-limiting format is particularly useful if only one antibody specific for the analyte is available.

    Multiplex screening capabilities provide analysis of several analytes simultaneously to reduce sample volume requirements for multianalyte studies for use in preclinical cytokine release analyses. There include specific assays to analyze the following: peptide/protein therapeutics, protein biomarkers, and antitherapeutics/neutralizing antibodies.

It has become a center of discussion and topics around those establishing a new good laboratory practice s (GLPs) or regulated bioanalytical laboratory, reassessing an existing laboratory, or revamping a "spirit-of-GLP" laboratory to full GLP status. Bioanalytical laboratories have become increasingly important to all stage of drug/product research and development. Initially, liquid chromatography–mass spectrometers (LC–MS) and tandem mass spectrometers (LC–MS/MS) had been used largely in drug discovery and early development owing to their qualitative capabilities beyond rigorous regulation. This is clearly no longer the case. Mass spectrometers have been used increasingly as primary detectors in all facets of operations, which led to a column in 2004 exploring the changing nature of validation: "Taming the Regulatory Beast: Regulation Versus Functionalism" [1].

There have been a number of contract research organizations (CROs) within the areas of analytical and bioanalytical services intercepted during past decade. Tremendous demands in analytical and bioanalytical support of drug discovery, preclinical, and clinical development are evident. Many companies/organizations realize that performing GLP or regulated bioanalysis internally (as opposed to outsourcing them to CROs) saves time and money. This economic efficiency is particularly true for large-sample-number studies because pricing tends to be based upon a per-sample approach.

The principle rarely extends to contract laboratories performing small-sample-number studies, however, because smaller numbers of samples often lack profit potential. Consequently, those laboratories sometimes delay a sponsor's small-sample-number studies or assign them a lower priority. This frustrates pharmaceutical sponsors because early studies, no matter how small, can be time critical, as in the case of toxicology studies. Compared with a contract laboratory, internal resources are held accountable more easily for timeliness and compliance. A sponsor can always outsource, but a contract laboratory cannot always subcontract. Moreover, an internal resource that is active in performing regulated bioanalysis knows the latest trends in compliance and cost effectiveness. For that reason, such a resource is well suited to managing external resources, expecting them to implement those trends.

4.2 FUNDAMENTAL ELEMENTS FOR BIOANALYTICAL LABORATORIES

There are a few levels of laboratory functions and expertise within bioanalytical operations in support of discovery, non-GLP, and GLP programs. It is highly recommend treating a spirit-of-GLP laboratory as one would a non-GLP laboratory to convert it to full GLP status. To begin constructing a new GLP facility, it is suggested to first put into effect a master plan. The master plan should address these major categories of work frame or infrastructure, with each category assigned its own manager and allocated its own resources [2].

(a) Responsible management team along with sound quality system
(b) Qualified personnel selection, staffing, and training

(c) Adequate engineering and facility for laboratory operation
(d) Standard operating procedures
(e) Installation, operational, and performance qualification (IQ, OQ, and PQ, respectively) of facilities, instrumentation, and software
(f) Quality control (QC) procedures and staffing
(g) Quality assurance unit (QAU)
(h) Documentation and archival process
(i) Final gap analysis

Responsible Management Team: The responsibilities of the management of test facilities are as follows: Test facility management should ensure that the GLP Principles are complied with at the test facility and that qualified personnel, appropriate facilities, *equipment,* and *materials* are available. On the basis of these requirements, there is no case for suppliers of materials used in studies submitted to regulatory authorities to be included in national GLP compliance programs. As by definition in the GLP Principles, the responsibility for the quality and fitness for use of equipment and materials rests entirely with the management of the test facility. The acceptability of equipment and materials in GLP-compliant laboratories should therefore be guaranteed to any regulatory authority to which studies are submitted. It is advised that both test facility management and suppliers should understand to how they might meet GLP requirements through national accreditation schemes and/or working to formal national or international standards, or by adopting other measures, which may be appropriate to a particular product. National or international standards, which may be set by an accreditation organization, may be applied whenever they are acceptable to the facility's management. The management of facilities, individually or in cooperation with each other, should thus maintain close contacts with suppliers and with their accreditation organizations.

Qualification of Personnel: Like all regulations also GLPs have chapters on personnel. The assumption is that in order to conduct GLP studies with the right quality a couple of things are important: First of all, there should be sufficient people and second, the personnel should be qualified. The regulations are not specific at all what type of qualification or education people should have. Qualification can come from education, experience, or additional trainings, but it should be documented. This also requires a good documentation of the job descriptions, the tasks, and roles/responsibilities.

Facilities: All GLP regulations also have requirements for facilities, for example, animal care facilities are listed as well as animal supply facilities, facilities for handling test, and control articles, and laboratories and storage facilities. The main purpose of this is to ensure integrity of the study and of study data. Three main requirements for facilities are (1) limited access to buildings and rooms; (2) adequate size; and (3) adequate construction. For example, if a testing facility is too small to handle the specified volume of work there may be a risk to mix incompatible functions. Or if the air conditioning system is wrongly designed, there may be cross contamination between different areas.

Equipment and Computerized Systems: All GLP regulations also have requirements for equipment. They are related to design, calibration, maintenance, and validation. This includes analytical equipment such as chromatographs, spectrophotometers, and computerized equipment for instrument control and direct data capture, data evaluation, printing, archiving, and retrieval. Equipment used in generation, measurement, or assessment of data and equipment used for facility environmental control shall be of appropriate design and adequate capacity to function according to the protocol and shall be suitably located for operation, inspection, cleaning, calibration/qualification, and maintenance. The equipment should undergo a validation process to ensure that it will consistently function as intended. Examples are analytical equipment such as chromatographs, spectrophotometers, computerized equipment for direct data capture, and computers for statistical analysis of data.

Maintenance, Calibration, Testing, and Validation: Equipment shall be adequately inspected, cleaned, and maintained. Equipment used for generation, measurement, or assessment of data shall be adequately tested, calibrated, and/or standardized. These activities are frequently called qualification for equipment hardware and single modules and validation for software and complete systems. A laboratory shall establish schedules for such operations based on manufacturer's recommendations and laboratory experience and expectation along with testing specifications.

Quality Assurance: All GLP work requires a protocol, a study director, and a QAU. Without a QAU, a GLP group cannot exist. When drafting the master plan to build a GLP bioanalytical laboratory, the need to adequately staff the quality assurance unit is sometimes underestimated. The firm contemplating building or rebuilding the laboratory often possibly tries to combine this new work function with its existing in-life GLP or current good manufacturing practices (cGMPs) regulatory groups is not highly recommended. Regulated bioanalysis requires complex and highly cross-referenced documentation, and its regulatory environment is fast evolving. Form 483s and guidances are highly technical and require a focused effort to capture in an applied audit. If resources are limited, alternatives to hiring a new auditor exist, but it is not recommend avoiding the hire by appointing a cGMP or in-life GLP auditor in an "overtime" scenario.

It is evident that even for a relatively small facility, true 100% QC combined with metrics tracking, outsourcing management, SOP training, and SOP revisions justifies the services of a full-time employee. It seems to be effective in adopting a model in which a QC scientist, wherever possible, maintains a checklist system approved by a quality assurance unit and other procedures to maintain documentation, chromatograms, electronic archival resources, and all other QC mechanisms for the work. It is highly recommended that the same approach be processed in creating a dedicated resource for this function and staffing it with an experienced GLP bioanalyst whose sense of the urgency and necessity of performing unassailable bioanalytical work is fully developed. Management is responsible for ensuring adequate staffing for proper surveillance of the work, as stated in the preamble to the 1987 GLP regulations. If the QAU is able to deliver 100% QC service by retaining an experienced analyst whose time is fully dedicated to QC activities (as perhaps originally envisioned in the preamble and code), then doing so would suffice. Nevertheless, to maintain a connection to the process it is recommended the QC scientist performs some amount of bench work in addition to his or her QC responsibilities. Further recommendations for quality control measures are delineated in the following section.

Standard Operating Procedures: Standard operating procedures (SOPs) should preferably be written for different areas such as company general (administration, human resources, information technology, etc.), facilities and engineering, laboratory operation, instrument operation, maintenance and calibration, analytical methodologies, conduct of GLP studies, and other related GLP and/or non-GLP programs. SOPs should be readily accessible as reference. SOPs should not be written to explain how procedures are supposed to work, but how they work. This ensures that the information is adequate and that the document invites rather than discourages routine use.

4.2.1 Document Retention and Archiving

Electronic Records: It is recommended that one develops a strict file nomenclature and includes this task as a line item in a QC checklist to follow when reconstructing electronic data. Accepted and rejected runs should be indexed in the paper data, and any reinjections also should be indexed and explained. While many well-characterized chromatographic software systems have evolved to a highly "validatable" state, this is not always true of software developed to control mass spectrometers and other instruments in the laboratory that produce GLP-defined raw data. Electronic raw data also require specific archival procedures that are somewhat less established. For example, you might consider this approach (1) creating a new subdirectory for each run; (2) acquiring the data; (3) transferring the entire subdirectory into a validated, write-once, read-only, share drive; and (4) verifying and documenting that the data are retrievable and viewable.

With such a system, you can protect the data against hard-drive crashes and define a final version of the integrated data. Also, because the data are read-only, they are protected against corruption. It has been noted that there are numerous decisions to be made depending upon such things as archival needs (how long must data be maintained) and at what level of access.

Chromatogram Review: Acceptance of chromatographic quality remains a subjective assessment. The director of the facility and/or technical leader should approve a detailed guidance on chromatographic quality, and QC scientists should be prepared to implement this subjective guidance. Many companies and organizations have established relevant SOPs to define the process and acceptance criteria for chromatographic data review, which is highly recommended.

Integration Parameters: Current acquisition software makes it clear to an auditor whether different integration methods are applied during the course of a run. Whether it is preferable to improve baselines via individual integrations or objectively draw them using the same integration methods for all samples is a matter for legitimate debate. The latter approach (automated baselines only) should be adopted and recommend. It should be also recommend that chromatograms using manual or altered parameters be printed both ways and approved by the functional manager, who is usually the facility director.

Incurred Sample Reanalysis: The major problem with bioanalysis is that samples generated for method validation studies are prepared by spiking analytes into a biological matrix. Typically, the solution used to spike the drug concentrations is organic, for example, methanol or an organic-aqueous solution depending on the water solubility of the compounds being spiked. This is a source of a potential problem as the method validation samples and the QC samples used to monitor the performance of the assay when it is applied may not be characteristic of the samples taken from the animals and humans in the study. This problem of the difference between *in vivo* or incurred samples versus the *in vitro* or spiked samples has been debated since the first of the AAPS/FDA consensus conferences in 1990 [1]. There are many reasons for the differences between spiked and incurred samples such as protein binding, conversion between parent drug and metabolites, and matrix effects.

The discussion resurfaced in 2006 at the 3rd AAPS/FDA bioanalytical workshop held in 2006 (Crystal City Conference/Workshop) that indicated that there should be analysis of incurred samples. Incurred sample reanalysis is mainly important to pharmacokinetic studies to ensure that the conclusions drawn are based on robust bioanalytical data. Early in the study analysis, 20 samples are selected for reanalysis that are close to the highest and lowest concentrations measured from several subjects and 67% of the reanalyzed results should be within 20% of the original results. These results are reported in a separate table from the main study results.

Reintegration and Reanalysis: The criteria for data reintegration and repeat sample analysis should be documented in standard operating procedures. In the former case the SOP should define the criteria for a reintegration and how it will be performed along with documentation of the data and both the original and reintegrated results should be reported. In the latter case, the SOP should define the criteria for repeating sample analysis such as inconsistent replicate analysis, samples outside of the assay range, sample processing errors, equipment failure, poor chromatography, and inconsistent pharmacokinetic data.

[1] 21 CFR Part 58, and OECD Principals of Good Laboratory Practice.

However, reassays should be done in triplicate if there is sample volume available and the way of reporting the original, meaning or rejecting values must be defined and applied consistently.

Laboratory and Process Audits: In the 1987 GLP preamble, the FDA makes two important statements. The first is that the QAU auditor can be any scientist not involved in the study, and the second is that audits must be made in a timely fashion so that necessary correction can be made accordingly. This may be interpreted to indicate that the agency recognizes that some firms lack sufficient resources to recruit new employees for the QAU position. Compound/organization might therefore staff QAU positions with existing employees, provided those employees are not involved in a current study. This is a far cry from the level of separation often called for by QAU groups. As applied today, the industry trend calls for the QAU to report directly to the senior council rather than the department head. This minimizes conflicts of interest, insulates the auditor from the work, and provides each auditor with an enhanced career opportunity to audit multiple areas of work: cGMP, GLP in-life, cGMP dose formulations, Regulated Bioanalysis. However, it limits the specialization of the auditor in highly complex and increasingly specialized technical applications. This is not against centralized QAU groups or generalist models *per se*. Yet it is strongly recommended that the QAU be adequately staffed to enable satisfactory specialization and turnaround time to support a new lab, requirements that might supersede a few additional layers of independence or auditor career track.

Audit turnaround time, a critical issue for all organizations, is particularly critical for a new laboratory. An audit of several months' delay could catastrophically affect a new group. If the QAU group takes several months to return its audit of a single study while a new group continues to produce several studies' worth of data, the QAU would, in so doing, place the entire organization at risk of regulatory noncompliance. For lack of feedback, the resultant safety margin data for the several studies produced would all potentially contain the same flaws as the first study. Company's QAU committed from the beginning to achieve very short turnaround times for each component of the work it audited for the new group, including a run-to-run data review. It is highly recommended that speedy turnarounds be standard for contract research organizations and throughout the industry. Indeed, the 1987 preamble to the GLPs [3] states that audits "must be conducted in a timely fashion so that corrections can be made." As the accepted time for the making of corrections becomes shorter, the need for fast audits becomes more pressing. To fulfill this need when internal resources are unavailable, it is desirable to outsource the work to one of the many organizations that provide auditing services. Alternatively, as allowed for in the 1987 preamble, it is feasible and acceptable to adopt a peer audit model whose participants are not involved in the study.

Before rolling out a new group, it is recommended that one engage an external auditor to perform a full gap analysis. For this, one should choose an individual consultant or firm specializing only in quality assurance. Many CROs see internal laboratories as competitors or customers and so, either way, have a conflict of interest in performing the gap analysis.

4.3 BASIC REQUIREMENTS FOR GLP INFRASTRUCTURE AND OPERATIONS

Good Laboratory Practices covers the organizational process and conditions under which nonclinical and clinical studies (in conjunction with GCP) are conducted, monitored, recorded, reported, and archived. GLP is carried out to improve quality of data for its international acceptance. This assures the regulatory authorities that the submitted data are in accordance with the results and reliable while making decisions regarding safety assessments. Operational and functional processes for analytical laboratories are greatly influenced by the GLP regulations and guidelines. GLP functions as a regulation that deals with the specific organizational structure and documents related to laboratory work in order to maintain integrity and confidentiality of the data. The entire cost of GLP-based work is about 40% compared to non-GLP operations.

Essential Elements for GLP Infrastructure and Operation: This should include (1) assigning responsibility for sponsor management, study management, and quality assurance; (2) standard operating procedures must be followed; (3) calibration and maintenance of instruments; (4) right construction of laboratories to maintain integrity of the study; (5) raw data should be processed and achieved; and (6) employee should be well qualified and trained as per the job assigned.

(1) *Personnel:* There are two chapters of regulations for personnel. For GLP studies with right quality, things listed below are considered important:

- Sufficient number of employees
- Personnel should be qualified

Qualification can be relevant educational background, experience or training, but it must be documented. Apart from this, job profiles, tasks, and responsibilities must also be documented (job description).

(2) *Facilities:* This includes a number of responsibilities, which are listed below:

- Appointment of a study director
- Monitoring the entire study process. If there happens to be any discrepancy in the process, then the study director is replaced
- Ensuring that quality assurance unit is available or established
- Handling tests and control articles
- Availability of sufficient qualified personnel for study
- Animal care facility requirements where applicable
- Storage facilities

The three main requirements for facilities include limited access to the building rooms, adequate size and construction.

(3) *Study Director:* The study director is responsible for
- Technical conduct of safety and other regulated studies
- Interpretation of the results
- Analysis of the results
- Documentation of the results
- Reporting of the results

More details of descriptions for study director's responsibilities and specific duties are outlined in Table 4.1.

TABLE 4.1 Study Director's Responsibilities and Specific Duties

Main Responsibilities	Specific Duties
Approval of protocols and any subsequent changes/amendments	Review of protocols, amendments, and deviations
Ensuring that the current revision of protocol is followed	The determination of the appropriateness of the test system is a scientific decision made by management at the time of protocol approval. The study director needs only assure that protocol specifications are carefully followed
Ensuring correct documenting of experimental data	The study director is not required to observe daily activities, but should assure that data are generated and collected as specified by the study protocol. The study director should also review data periodically, or assure that such regular review occurs
Collating records of, and verifying, all experimental data, including observations and endorsement of unforeseen events	Circumstances that may affect the quality and integrity of the study must be noted, then corrective action taken and documented
Assure that all applicable GLP regulations are followed	Deviations from GLP requirements noted by QAU are required periodically to the management and to the study director. If those reports indicate that corrective action is needed for any deviation from regulatory requirements, it is the study director's responsibility to assure that corrective action occurs
Final statement on GLP compliance	A final statement is made in the study report that the study was conducted in compliance with GLP regulations
Assure timely archiving	All raw data, documentation, protocols, specimens, and final reports are transferred to the archives during or at the end of the study

(4) *Quality Assurance Unit:* It serves to control internal functions and processes within regulated programs. These include as examples
   - Ensures that facilities, devices, personnel, practice, protocol, records, SOPs, reports, and archives are in accordance to the GLP regulations.
   - Immediate reporting of problems along with actions recommended and taken, and scheduled date for inspection in a written format.
   - It is mandatory that the responsibilities applicable to QAU and records maintained by QAU with the indexing of these records are maintained in writing.
   - Master schedule sheet containing test system, nature and date of the study, identity of sponsor, and name of the study director are to be indexed by test articles.
   - Maintaining protocol of the study for which QAU is responsible.
   - Inspection of studies at regular intervals to assure integrity of the study and maintain properly signed records of each inspection.
   - Ensuring that the derivatives from protocols and SOPs were made with proper authorization and documentation.
   - Final review of the report to ensure that the results accurately reflect the raw data.
   - Prepare and sign a statement to be included in the final study report specifying dates of audits and reports to the management and study director.

(5) *Equipment:* GLP regulation also specifies requirements for equipment. These are pertaining to
   - *Design and Qualification:* Equipments used for generation, measurements, and assessments of study data shall be of appropriate design and capacity. They shall be suitably located for operation, inspection, and maintenance. Qualification/validation of equipment must be done to make sure that consistent intended functions are performed and expected.
   - *Maintenance:* Cleaning and maintenance of equipment is very critical. They must be adequately tested, calibrated, and standardized. The entire process is called maintenance and qualification for equipment. Time interval for calibration, re-validation, and testing of equipment depends on the equipment itself, laboratory experience, and extent of use.
   - *Records and Other Documents:* Written records shall be maintained for qualification of equipment and validation operations including date of operation, and specify within SOPs were followed for conducting maintenance operation. These equipment records can be maintained in the form of a logbook.

(6) *SOPs:* Standard operating procedures specify the manner in which protocol-specified activities are carried out. SOPs are written in the given series of steps:
   - Regular monitoring of the maintenance, testing, calibration, and standardization of equipment
   - Dealing with equipment failure (e.g., CAPA, etc.)

- Analytical operations and procedures
- Data handling, storage, and retrieval
- Safety precautions
- Recording, storing, and retrieval of data
- Data handling and use of computerized systems
- Facility and personnel qualification and continued improvement
- Report retention and archival

SOPs are preferably written and reviewed by instrument operators, laboratory personnel, technical specialists, and alike. The working of SOPs ensures that information is accurate and document is liable for routine use. FDA's warning letter contains three deviations for SOPs including unavailability, inadequacy, and non-consideration of SOPs. To begin with, a SOP is designed on writing SOPs, which defines who is responsible for initiating, authorizing, and approving SOPs.

(7) *Reagents:* The quality, purchasing, and testing of reagents and solutions used for GLP studies shall be handled by QAU.
- All reagents shall be labeled to specify titer, concentration, storage requirements, and expiry date.
- Reagents for GLP and non-GLP regulated studies must be labeled clearly and separately.
- Date opened must be included which can be critical for some chemicals.
- Reagents or solutions without any information about storage temperature do not require cool storage conditions.
- Expiration date can be mentioned based on literature or laboratory experience and test results.

(8) *Test and Control Articles:* Control articles are also referred to as reference substances by OECD. They are commonly used to calibrate the instruments. Test results depend on the accuracy of reference article. The requirements for control articles include:
- Determination and documentation of the features like identity, strength, purity, and other characteristics for each batch or lot.
- Method of synthesis, fabrication, and derivation of test and control article should be documented.
- Copies of this document must be available for FDA inspection upon request.
- Stability of test and control articles must be determined.
- Storage container for test and control articles must be labeled.
- Certified appropriate standards can be purchased from appropriate suppliers.

(9) *Retention and Retrieval of Records:* GLP regulation specifies how records are stored and retrieved.

Archived Data: This includes all the documents from raw data to final results along with protocols from meetings and decisions taken. GLP requires an archivist. Any document taken out of the archive must be documented and signed by the person who requested it.

Retention Period: In the United States, material supporting FDA submission should be retained for 2 years after FDA approval, and 5 years after FDA submission. According to GLP regulation original copy of the record can be retained. This exact copy can be a copy of an instable thermo paper or plain paper or paper records scanned into TIF or PDF files.

Raw Data: Raw data are defined as any laboratory worksheet, memorandum, or exact copies thereof that are the result of original observation of the study. Raw data may include hand-written notes, photographs, printouts, magnetic media, and recorded data from automated instruments.

Electronic Records: There is no specific requirement of electronic records as long as the printout includes every piece of information that is necessary to reconstruct the study. Refer to detailed discussion of 21CFR Part 11 (e-records and signature).

(10) *Inspection:* FDA is responsible for inspection of GLP studies before a drug or device or any other product is marketed in the United States. There are two types of inspection programs.

- *Routine Inspection:* This should be conducted every second year to ensure that laboratory's compliances are in accordance with the GLP regulations.
- *For-Cause Inspection:* They are conducted less frequently, only about 20% of all GLP inspections. FDA conducts such inspections only when it suspects non-compliance from/within new drug application (NDA) and/or abbreviated new drug applications (ANDAs) considerations.

If the laboratory refuses to accept FDA inspections, then FDA will not accept studies in support of NDA and/or ANDA.

(11) *Enforcement:* If the inspection team finds deviation, they mention these deviations in a particular form. These deviations are discussed in exit meeting. The lead inspector then prepares an establishment inspection report (EIR) and depending on the deviations, warning letter may be issued. This letter is issued to the company's management and corrective action plans are implemented by the company within 14 days.

Good Laboratory Practice regulations are applied to nonclinical and clinical safety of study items contained in pharmaceutical products, cosmetic products, veterinary drugs, devices as well as food additives. The purpose of testing these items is to obtain information on their safety with respect to human health and environment. GLP is also required for registration purpose and licensing of pharmaceuticals, pesticides, food additives, veterinary drug products, and some bio-products.

4.4 GxP QUALITY SYSTEMS

It is not quite sufficient to solely consider GLP quality system, as part of an island mentality from operational perspectives. The standard operating procedures are documents that describe how to perform various operations in a GLP based and GxP facility where applicable (cGMP in support of CMC, APIs, drug product suppliers to clinical trials, GCP in support of clinical research and clinical trials, etc.). They provide a general framework enabling the efficient implementation and performance of all the functions and activities, and contain step-by-step instructions that laboratory and production personnel consult on a daily basis to complete their tasks in a reliable and consistent manner. SOPs are, in essence, written commitments to the regulatory bodies that describe the performance of routine tasks; they are required to ensure the successful conduct of a study but can also serve as a valuable training tool.

General factors that must be covered by an SOP are as follows:

- The receipt, handling, and storage of test and control substances and related samples
- Test systems and methods
- The use, care, maintenance, and calibration of equipment
- Facility use (such as IT, QA, and in-life testing laboratories)
- Documentation and preparation of reports, data collection, handling, storage, and retrieval and archival
- The completion of documentation, such as *pro forma* result and data collection forms
- Any other routine procedure

SOPs must be comprehensive, covering all the necessary details to enable staff to accurately and correctly complete the procedure. Information that should be listed includes the preferred suppliers of chemicals, reagents, and equipment; the catalogue numbers of reagents and the model numbers of equipment; the storage conditions and stability of chemicals and test substances; and acceptance criteria for valid procedures. In some instances, it may also be necessary to specify such factors as centrifugation speeds listed as $\times g$ rather than rpm (unless the rotor radius is stated), incubation conditions with tolerances, volumes with tolerances (if relevant), and any other information that enables the accurate reproducibility of procedures.

SOPs: It should be written by personnel who are familiar with the procedure and then approved by the management team. They should be written in language that will be understood by the person using the SOP and checked for accuracy and quality. An SOP should be signed and dated by authorized personnel and, ideally, include the printed name of the signatory. All SOPs should be reviewed regularly, and the review date must appear on the documentation. They should be accessible to all and refer, when relevant, to product and process quality specifications. Above all, an SOP must be clearly presented and followed. Within the scope of GLP, several factors may not specifically be referred to in an SOP—those that should be addressed and considered under the general dogma of regulatory compliance. Examples of these items include equipment, experimental procedures, routine procedures, facilities, and laboratory reagents and test substances, which are described in further detail below.

Equipment: Types of equipment and apparatus that might be included are pipettes; balances; refrigerators and freezers; HVAC systems and laminar flow hoods; computers; pH meters, analytical instruments (ELISA, spectrophotometers, densitometers, and image analyzers); high performance liquid chromatography (HPLC), gas chromatography (GC), atomic absorption equipment; mass spectrometers (MS, MS/MS, and MS^n); centrifuges; and so on. The equipment should be for the intended purpose; kept in good, clean, working order; and used by trained personnel only—that is, according to SOPs describing use, care, calibration, and cleaning routines. Regular calibration and servicing are essential; full records of maintenance and equipment checks must be kept in an easily retrievable form. If maintenance is done by external contractors, such work is acceptable without an internal SOP. If equipment is serviced and calibrated by internal staff, however, SOPs must be in place to cover the work.

Experimental Procedures: It must be fully documented and contain all pertinent details. For example, lot numbers, catalog numbers, supplier details, and dates for when chemicals and reagents arrive and when they are opened must be listed, as well as temperatures and times, specific identification of equipment used and, perhaps most importantly, any deviations from the SOP. All the information must be recorded in lab notebooks with individual page numbers. Experimental details and results should be easily located; a log page at the front of a notebook can help track the recordings and observations. Any reference to computer files containing data should also be catalogued in the notebook. Data files should always be backed-up in case of computer failure, corruption, or deletion.

Routine Procedures: Daily or scheduled tasks must be fully described in written SOPs. Just as for other SOPs, SOPs on routine procedures must describe not only technical details but also a system for reporting results and the methods for cleaning and calibrating any equipment. Even routine procedures must be validated as appropriate for their intended use. Assays, for example, must be accurate in the conditions in which they will be used (for instance, cesium affects the detection of hepatitis B). The validation of an assay should include factors such as:

- *Specificity and Selectivity:* Evidence that cross reactivity by factors such as antibodies is detectable, or coelute for chromatographic assays
- *Accuracy:* As shown by comparison with other assays if possible
- *Precision:* With limited variability
- *Recovery (Spike Recovery):* Showing the assay detects known amounts of added analyte, for instance

Validation protocols and reports must be written and available for every procedure, and the most important criterion for any set of instructions is that the SOP is followed.

Facilities: It must be fit and suitable for the purpose of the work; that is, size, construction, and location should be appropriate, and the building should allow for the separation of activities. Animal handling facilities, in particular, should be large enough, allow for the isolation of incoming or diseased animals, allow for the separation of both species and studies, have storage space for feed and bedding, and allow for the adequate disposal of waste and refuse to minimize vermin, odor, disease, infestation, and environmental contamination.

SOPs should ensure that facilities are designed to permit separate areas for similar but unrelated work and sample preparation. In addition, buildings must be validated for the required functionality (e.g., air handling and sterilization) and monitored (for air pressure differences, flow patterns, or sterilization), with easily accessible results. The property must be regularly maintained and serviced, with full maintenance and modification records kept. All facilities, laboratory areas, benches, floors, corridors, and cupboards need to be neat, both during and after experimentation—which really only amounts to good housekeeping. Finally, the facility must be of sufficient size to accommodate data storage and archiving.

Reagents and Test Substances: Several GLP rules apply to all laboratory reagents and test substances. For instance, preparation of these ingredients should use a standardized and fully documented methodology that lists supplier information, lot numbers of component chemicals and reagents, dates of preparation, and the names of the staff involved in the preparation. Substances should be tested against reagents and standards of known reactivity before being released for use, and the results of these tests must be recorded, preferably using a Reagent Preparation Record. All prepared solutions should have unique lot numbers (a laboratory-based system is acceptable as long as the provided numbers are actually unique). This information log prevents duplication and should comprise lot numbers, dates, descriptions, initials, and signatures. Containers are often overlooked. For both test and control substances, the container should be inert to the stability of the substance and clearly labeled with the following details:

- Contents or identity
- Potency (e.g., titer, concentration, and activity)
- Storage temperature
- Preparation date
- Unique lot number
- Container number (in case there are multiple containers of the same lot)
- Date first opened
- Expiration date (determined by experimentation or reference to manufacturer's recommendation)
- Signature of the person who prepared the contents

Much of this information can be recorded separately with a unique identification number log. Test substances should also be retained for reference until a report is written, if tests last longer than 4 weeks.

Documentation: Although not actually referred to in many GLP guidelines, documentation quality is crucial for laboratory and manufacturing compliance. Some general documentation requirements are included in the UK Medicines Control Agency's *Orange Book*, ensuring all pro forma paper work, SOPs, results reports, product quality specification (PQS) sheets, and manufacturing instructions be

- typed, clearly and unambiguously written;
- written legibly with indelible ink;
- corrected, if necessary, with a single cross-out line, then initialed/signed and dated with a reason for the alteration, if relevant;
- officially authorized and signed;
- dated and also dated when reviewed;
- reviewed on time and retrieved when out of date;
- designed for ease of use and completion by appropriate staff;
- coded with document versions to ensure use of correct editions of similar forms;
- carefully controlled to ensure only correct documents are circulated.

Computer-generated documents should be closely controlled with password protection, and records should be kept of previous versions.

Recording Experimental Details: Experimental details must include lot numbers (internal and supplier's) of reagents used and the incubation temperatures, times, and conditions. These records should be stored in an unalterable, easily retrievable format, using laboratory notebooks that contain logs or indices and consecutively numbered pages. Cross-references to other data (e.g., charts, printouts, and computer files) should be straightforward and transparent. Above all, these records must be accurate and up-to-date.

Retaining Data: All results must be recorded and retained. Raw data (such as chromatograms and printouts) and processed results must be easily retrievable. Backups of computer-generated information should be maintained.

Reporting Results: All reports, whether final study reports or test request forms, must be checked for accuracy, signed, dated, and countersigned by authorized personnel. Ideally, the names of signatories should be printed because handwritten signatures can be difficult to identify. Reported results should be complete (containing all valid results) with a foolproof presentation—to avoid any ambiguity—according to the relevant SOP. Full study reports should also contain a QA statement that confirms work was performed according to GLPs, includes audit dates, and lists aspects of the study that did not conform to GLPs, if any.

Archiving and Retrieving: An effective archiving system must keep all related data (experimental, calibration details, slides, specimens, and reference material) together, under suitable storage conditions. Specimens and materials should be stored only as long as they are stable. The data should be indexed—with cross-references to any data or information stored at a different location—so that they are easy to locate and retrieve. Data require official authorization before disposal.

SOPs in Review: The compliance of all laboratories and analytical ones in particular is defined by their SOPs. FDA laboratory inspections focus on the adequacy of SOPs and on how those SOPs are followed. Therefore, properly written SOPs are critical. Poorly written SOPs or well-written SOPs that are not properly followed are a major source of preapproval inspection 483 observations by FDA. Yet SOPs should not necessarily be written for FDA; they should be written for those analysts or scientists who will use them. Although compiling SOPs can be somewhat daunting to the beginner, they are easily mastered and, with practice, can become almost second nature.

As described, two paramount rules must be followed. An SOP should be detailed enough to adequately define the task that it purports to describe, and an SOP should be general enough to prevent user confusion or limit his or her ability to work efficiently or impede decision-making processes. It is particularly important not to get restricted by semantics. An SOP for calibrating the temperature of a water bath could, for example, state that calibration frequency should be daily. This sounds harmless enough, but it would necessitate verifying the water bath every day, regardless of whether the instrument was used or not. Most laboratory water baths *are* used every day, but if the SOP states the frequency as "daily when in use," the bath would only need to be verified or calibrated on days when it was actually used. The difference in the text is minor, but it could save a lot of unnecessary work (and time).

Language Clarity and Accuracy: It is another crucial factor. Decision-making processes or scientific judgment can be limited by company used in an SOP. If a protocol is required, for example, a 0.1 M solution of trisodium citrate, the SOP might specify that 29.4 g of trisodium citrate dihydrate (MW 5294.1) should be weighed into a tared 1 L volumetric flask, dissolved in ultrapure water, and diluted to volume, even if only 300 mL of the solution is required for the experiment. If, on a given occasion, there was only enough trisodium citrate to make 0.5 L of the reagent (14.7 g in 500 mL of water), this could, according to FDA, constitute a breach of the SOP.

Common sense would suggest that the reagent should been prepared just as accurately, producing enough to complete the experiment, but the SOP had not, technically, been followed. A better option would be to write "prepare a sufficient quantity of the reagent at a concentration of 0.1 M," which allows the user some discretion concerning the choice of sample weight and final volume. Maintaining flexibility is an important factor in making an SOP useful and easy to follow without being unnecessarily deviated.

SOP Review and Approval: Every laboratory should have designated personnel who are authorized to have a copy of a manual containing the current version of documents, including analytical methods or SOPs. Only current versions of each document should be in use. Previous revisions of documents should be archived for reference purposes and for historical tracking of changes with time. Once a new document is approved, it should be issued by a central "project manager" and then included in the authorized manual. The previous version should be returned to the original author, and a receipt for its return obtained. This ensures that only current documents are used, and out-of-date versions are removed from common access. QA is responsible for auditing laboratories for compliance and for informing the study director of any observations or deviations. Everyone working in a laboratory is responsible for applying GLP principles and informing the study director (or the head of the department, the direct line manager, or the supervisor) if GLP is not being adhered to. Management is responsible for providing conditions that facilitate and enhance GLP adherence.

Consequences of noncompliance irrespective of what an SOP is for, how it is written, or who uses it, the cardinal rule is that the instructions are followed. It is much better to get an SOP right the first time than to have to go through several iterations or rounds of approval. FDA is less interested in the layout and style of the SOP but in whether it is being followed and a good level of compliance is being maintained. Noncompliance with GLPs is extremely serious and compromises the validity of results obtained in a study and, in extreme circumstances, could even render a study void. The consequences of noncompliance include:

- Rejection of study findings by regulatory authorities
- Requirement that a study be repeated (with attendant cost implications)
- Inability to complete the study
- Failure to achieve product registration and license
- Loss of client business and company credibility (with severe business consequences)
- Withdrawal of product license

All of these consequences are avoidable. Compliance with GLP/GxP is imperative, and the cornerstone of that compliance is a written, approved, and followed SOP. Perhaps as a final thought, a documentation writer should heed a quotation attributed to English poet, critic, and lexicographer Samuel Johnson: Language is the only instrument of science, and words are but the signs of ideas: I wish, however, that the instrument might be less apt to decay, and that signs might be more permanent, like the things which they denote.

4.4.1 Laboratory Instrument Qualification and Validation

The pharmaceutical industry relies on the precision and accuracy of analytical instruments to obtain valid data for research, development, manufacturing, and quality control. Indeed, advancements in the automation, precision, and accuracy of these instruments parallel those of the industries themselves. Through published regulations, regulatory agencies require pharmaceutical companies to establish procedures assuring that the users of analytical instruments are trained to perform their assigned tasks. The regulations also require the companies to establish procedures assuring that the instruments that generate data supporting regulated product testing are fit for intended use. The regulations, however, do not provide clear and authoritative guidance for validation/qualification of analytical instruments. Consequently, competing opinions abound regarding instrument validation procedures and the roles and responsibilities of the people who perform them. On the latter point, many believe that the users (analysts), who ultimately are responsible for the instrument operations and data quality, were not sufficiently involved when the various stakeholders attempted to establish criteria and procedures to determine the suitability of instruments for their intended use. Therefore, the American Association of Pharmaceutical Scientists sponsored a workshop entitled, "A Scientific Approach to Analytical Instrument Validation," which the International Pharmaceutical Federation (IPF) and International Society for Pharmaceutical Engineering (ISPE) cosponsored. Held in Arlington, VA, on March 3–5, 2003, the event drew a cross section of attendees: users, quality assurance specialists, regulatory scientists, validation experts, consultants, and representatives of instrument manufacturers. The conference's objectives were as follows:

- Review and propose an effective and efficient instrument validation process that focuses on outcomes, and not only on generating documentation.
- Propose a risk-based validation process founded on competent science.
- Define the roles and responsibilities of those associated with an instrument's validation.
- Determine whether differences exist between validations performed in laboratories that adopt Good Laboratory Practice regulations versus those that adopt Good Manufacturing Practice regulations.
- Establish the essential parameters for performing instrument validation.
- Establish common terminology.
- Publish a white paper on analytical instrument validation that may aid in the development of formal future guidelines, and submit it to regulatory agencies.

The various parties agreed that processes are "validated" and instruments are "qualified." It is author's intent, therefore, to use the phrase "Analytical Instrument Qualification (AIQ)," in lieu of "Analytical Instrument Validation." The term "validation" should henceforth be reserved for processes that include analytical procedures and software development.

4.4.2 Procedural Elements and Function that Maintain Bioanalytical Data Integrity for GLP Studies

Many aspects of bioanalytical method validation focus on the performance of a method as it is used in the analytical laboratory. However, the laboratory analysis is only one component controlling the overall quality of the data; several other procedural elements can also affect data integrity. These procedural or nonanalytical elements may be divided into three main categories. First, there are factors that could affect the actual, or apparent, measured amount of the analyte in the sample. These generally relate to the handling and storage of the samples prior to analysis and can result in either a change in the amount of analyte in a sample or a change in the ability of the analytical method to accurately measure the analyte (e.g., a matrix effect in both Liquid Chromatography–Mass Spectrometry/Mass Spectrometry LC–MS/MS analysis and ligand-binding assays—LBAs). Second, correct labeling and identification of the sample is critical; an ambiguously or incorrectly labeled sample will automatically result in an incorrect result. Finally, there are the processes, procedures, and documentation that support data security, data integrity, and the ability to reconstruct the analysis.

4.4.2.1 Factors that may Affect the Analytical Results

Sample Collection: Some type of processing is required for most biological samples immediately following their collection from an animal or human subject. Most commonly this involves collection of a venous blood sample followed by centrifugation to harvest plasma or serum to be frozen for later analysis. For most analytes, this should be a relatively simple procedure, and conditions such as temperature, centrifugation time and force, and maximum time from sampling to freezing the sample are specified in study documentation. While straightforward, these tasks will be performed a large number of times, and for a multisite study, often at many different locations. In contrast to the bioanalysis conducted in the laboratory, there are no calibrations or quality control samples being run alongside these sample collection processes to indicate if any problems occurred. The effect of variability in these processes will be highly dependent on the nature of the analyte and matrix. In some cases, sample collection conditions may be particularly important; for example, if an analyte is less stable in whole blood than in plasma, any delay in processing the sample or poor temperature control could result in analyte loss. Some analytes are not stable under standard sample collection conditions and may need stabilizers or other special sample handling conditions to be applied. There is also the potential for contamination of samples during collection, especially in toxicology studies where high doses of the drug are administered. This has resulted in significant discussion over the last few years, and a European guideline that addresses the issue of contamination in toxicology studies was published.

Because of the lack of quantifiable quality control procedures covering initial sample handling, it is essential that procedures be clearly defined in a protocol or study manual and be readily available to the staff processing the samples. Staff involved in this aspect of the work must also be fully trained and experienced in carrying out these activities. While sample collection and initial processing are often very simple technical procedures, if they are poorly performed, the quality of the samples will be compromised, nullifying any further activities performed in the analytical laboratory.

Sample Stability and Storage Conditions: The current US Food and Drug Administration (FDA) Guidance [4] and conference report [5] already describes in some detail how stability of analytes in the bio-matrices should be defined during validation and that during analysis of study samples, it is necessary to ensure that samples are stored under the same conditions and analyzed within the period of defined stability. Required stability experiments for the analyte in bio-matrices typically include short-term stability at room temperature, freeze/thaw stability, and long-term stability in frozen bio-matrices (typically at –20 or –70°C). Adequate documentation is also needed to track the location of the samples throughout their storage, from receipt until disposal, and to document the temperature in the storage freezers while the samples are stored. Adequate contingencies, including some backup capacity, should also be in place to protect the samples in the event of a failure of a freezer or the main power supply. Freezers, therefore, will require backup power, an alarm system to alert staff of temperature changes outside a prescribed range, and procedural arrangements to call-in staff outside of normal working hours to resolve a problem or transfer samples to another correctly functioning freezer comparable in characteristics to the defective freezer.

Sample Transport: Bio-matrix samples are usually shipped frozen in insulated containers with dry ice. The main concern is ensuring that the shipment is still frozen upon arrival. Data loggers can be included within shipment packages to monitor temperature; however, these are not generally followed, particularly as shipments are usually packed with sufficient dry ice to last for a significantly greater period than the anticipated shipment time. It is important that the sample condition be accurately recorded on receipt to document that samples were received still frozen and in good conditions. Additional attention may be warranted for shipments with particular risks, for example, where delays could occur in customs clearance or carrier schedule changes. The increasing sensitivity of modern analytical methods has resulted in sample volumes tending to be reduced. Often, samples are split into two aliquots at the collection site(s), for additional security; a set of reserve aliquots can then be safely stored until the first set is received for analysis.

Sample Identification and Labeling: The *FDA Compliance Program Guidance Manual*[2] refers to the possibility of sample mix-up on at least three separate occasions. Sample labeling and traceability are critical, because once a sample has been mislabeled; it will always provide an incorrect analytical result. Given the hundreds, often thousands, of samples collected in preclinical and clinical studies, it is almost inevitable that some labeling errors may occur, most likely when samples are transferred from one tube to another. These errors may be minimized by clear, simply described study designs, clear label design, and adequate workspace and procedures to minimize the risk of confusing samples during their processing. Other types of errors can occur when labels are

[2]FDA Compliance Program Guidance Manual (Chapter 48, Bioresearch Monitoring: Human Drugs).

ambiguous, do not match the protocol or case report forms, or have been altered or incorrectly completed. A key step when samples are received at the analytical laboratory is to reconcile their identity against the study protocol, sample accession list, and/or other study documentation. It is more likely that an error can be resolved if a problem is noted shortly after collection of the sample rather than many months later. Some laboratories relabeled tubes on receipt to provide their own unique number, especially when barcodes are to be used to monitor the audit trail; the additional labels should not obscure the original labels.

It is important to remember that sample labeling issues cannot be resolved by bioanalysis; repeat analysis can be used only to confirm whether an anomalous value may have been caused by an analytical error. There is some tendency to want to proceed with the analysis of ambiguously labeled samples to "see what you get." It is believed that analysis of poorly labeled samples should not be performed until or unless the labeling issue can be resolved before proceeding.

Documentation: Any analysis supporting a good laboratory practices or regulated clinical study needs to be fully documented, so that the study could be reconstructed. A variety of documents, from laboratory raw data, laboratory notebooks and worksheets, and facility and calibration records, to standard operating procedures and method documents, are typically required to fully document a study. At the end of a validation project, or on completion of sample analysis, a validation or analytical report is usually prepared and issued. Such reports not only provide details of study results and methods but also direct the reader to the location of all other records that support the study, including archived paper and electronic records, and references to methods and validations supporting the analysis. A separate validation report is always generated on completion of validation experiments; however, for analytical support studies, a separate report may not always be created, and reporting of the bioanalytical results may be integrated with the main study report, or with a subreport for a related part of the study (e.g., as a bioanalysis and toxicokinetic report).

Increasingly, a significant amount of raw data may exist only as electronic records. Special criteria apply to data that are generated and stored as electronic records by computerized systems. In these situations, laboratories must have policies and procedures in place to ensure that they meet the prevailing criteria for the acceptance of the electronic records/signatures as the equivalent of paper records/signatures by the regulatory authorities. FDA regulation 21 CFR Part 11 [6] allows for the use of electronic records and for electronic signatures when appropriate.

4.4.2.2 Data System Validation and Laboratory Information Management Systems Generation of data in a modern analytical laboratory is likely to involve the use of one or more computerized systems. Typically, the analytical instrument used to run the analysis (e.g., an LC–MS/MS system or a plate reader for an ELISA assay) will be controlled by a computer that will also capture data as it is being generated. These data will often subsequently be transferred to a laboratory information management system (LIMS) for further processing and evaluation, storage, and report generation. Further transfer of the data may then take place to allow statistical or pharmacokinetic analysis. While a LIMS is not an absolute requirement, it is unlikely that a modern bioanalytical laboratory could function efficiently without one. Computer validation is a major topic that is beyond the scope of this section. Computerized systems used to generate, manipulate, modify, or store electronic data should be validated, and key instrumentation should be appropriately qualified before use. When data are transferred between electronic systems, the link or transfer process should, ideally, be validated. If not, procedural and quality control processes will be needed to demonstrate that data integrity has been maintained. Some proposed validation requirements for computerized systems (close and open) are shown in Table 4.2.

Reporting: Reporting arrangements will depend on whether the analytical work undertaken formed part of a study or was a separate study. Similar principles apply to bioanalytical support for both preclinical GLP studies and regulated clinical studies, but for full GLP studies, the requirements for multisite studies should be followed where applicable. Both the current FDA Guidance and the conference report contain specific recommendations on the contents of bioanalytical reports to follow. Bioanalytical reports may be an integral part of preclinical or clinical reports or may be appendixes to such reports.

If the analytical work constitutes a complete study, there should be a final report containing the essential information required by the principles of GLP. Any data included in the report that were not generated following GLP principles, or were generated by a facility not claiming GLP compliance, should be fully identified (on the GLP compliance statement) by the study director. If the analytical work was the responsibility of a principal investigator, that person is responsible for producing a report detailing the work performed under his or her supervision and for sending the report to the study director. There should be a statement signed by the principal investigator certifying that the report accurately reflects all of the work performed and results obtained, and that the work was conducted in compliance with the principles of GLP. The principal investigator may present the original raw data as his or her report, accompanied by a statement of GLP compliance. If the work was conducted by a subcontractor laboratory, there should also be a quality assurance statement signed by that laboratory's quality assurance unit.

4.4.2.3 Laboratory Information Management System LIMS is probably best described as a program that watches everything that happens within the laboratory, and helps everyone who works there. It does this by storing important information within a database, and by performing everyday (and relatively tedious) operations such as calculations and reporting. If one thinks of the laboratory as a business, then LIMS is the Enterprise Resource Planning (ERP) system of the lab. At the heart of most modern laboratories is a LIMS that records, tracks, and organizes laboratory

TABLE 4.2 21 CFR Part 11 (Computerized System Validation)

Control Element for Closed System	Remark
Validation	System/software qualification/validation
Checks	• Operational system checks Enforces sequencing of steps/events • Authority checks on individuals • Device checks Determine validity of source of data input
Copying	• Ability to make copies of e-records that are Accurate and complete Both human readable and electronic Suitable for FDA inspection, review and copying
Archival	• Archival of e-records Ensure accurate and ready retrieval Maintain integrity of data records Throughout records retention period
Documentation controls	• Audit trail documentation Operator ID, time, date, and original entry Retain as long as electronic records Available for FDA review and copying • System documentation controls Distributions Access Revision and change control—audit trial
Written policies	• Relevant SOPs Data generation, handling, and maintenance IT server dedication and infrastructure frame work • Accountability Follow SOPs Hold personnel responsible for actions under their e-signature Purpose to deter falsification
Training	• Training for e-records/e-signature systems Persons who use, develop/generate, or maintain Education, training, and experience
Security/audit trails	• Security Limit system access (designated) • Authority checks • Audit trails Secure Computer generated Time and date-stamped Independent of operator
Control elements for open system	Quite similar to closed system, but added measures to ensure

(*continued*)

TABLE 4.2 (*Continued*)

Control Element for Closed System	Remark

Designated to ensure	Authenticity
	Integrity
	Confidentiality as appropriate
	Added measures such as
	  Digital signatures
	  Encryption

data. Laboratories themselves have evolved a lot in many years. It used to be the case that laboratories were almost a separate part of the business, an island within the organization. Today, the laboratory has become an integral part of the whole business. Laboratory managers tend to have business backgrounds (or at least, an understanding of the "bigger picture"), and one can even find labs that generate money by providing external services.

LIMS has evolved in tandem with the laboratories. In the days when the laboratory was an island, LIMS was also an island. LIMS would usually not communicate with other business systems (such as Material Requirement Planning and Process Control Systems) and the end users of LIMS were normally just the chemists and other laboratory personnel. Nowadays, a mainstream LIMS should offer a wide variety of interfaces, and LIMS is used by many people who are not working in the lab. LIMS has become part of the enterprise. Since LIMS came into widespread use in the mid-1980s, LIMS functionality, specialization, and applications have expanded to meet user demands. In larger organizations, the role of LIMS extends beyond the laboratory. The information that a LIMS can provide is valuable across an enterprise from manufacturing, to operations management, to business processes. Thus, interfacing LIMS with enterprise-wide computer systems has become a necessary step for many companies in order to make their operations more efficient, to comply with more stringent regulations and to save time and money. Such demand is making LIMS enterprise interfacing one of the fastest-growing and most dynamic sectors of the LIMS market. Industry needs, coupled with the latest product developments, changing software environments, and vendor strategies, have created a significant opportunity for LIMS makers. Among the LIMS companies focusing on the enterprise LIMS market are Thermo Electron, InnaPhase, and STARLIMS [7] formerly LIMS, USA). Watson LIMS is designed for bioanalytical work in the preclinical and clinical stages. The company's latest release, Newton, is for analytical development as well as quality control and stability during manufacturing.

One example of this change is that LIMS is one of the few business systems that are used from the very beginning to the very end of a product's lifecycle, the so-called "cradle to grave." For example, when a company is developing a new product, LIMS is used to test the prototype materials. When the product development is complete, LIMS will be involved in testing the scaling up of the manufacturing. As the product is manufactured, LIMS will manage the testing and release to market. Later on, LIMS can be used to test any complaint samples. In fact, in some industries, LIMS is used for testing samples for many years after the product was made and shipped, generally as a legal requirement. As corporate management realized that LIMS was holding lots of valuable information, it became critical to ensure that all other applicable business systems were interfaced to the LIMS. In many ways, LIMS is the hub of data activity within organizations.

These interfaces can be generally categorized as "downwards" and "upwards." Examples of downward interfaces are connections from laboratory instruments to LIMS. In some cases, an "instrument" is not just a piece of hardware such as an analytical balance, but an entire software application. One of the most common applications of this kind is Chromatography Data Systems (CDS), which are used in practically every industry. An emerging technology is Electronic Laboratory Notebooks (ELN), which promises to do for pen and paper what an asteroid allegedly did for the dinosaurs. The overall nature of these downwards interfaces is that LIMS will send a message to the other system, telling that system that there are some samples to test. When the instrument is run, the data is sent back electronically to LIMS for processing. This processing can include trend analysis and specification checking/monitoring.

An example of an upwards interface is the connection of LIMS to an ERP system. These systems are usually not specific to one department, but rather are owned by the entire business. Within this scenario, LIMS is typically viewed as a service. It is expected of upwards interfaces that LIMS data should be instantly available on a console screen, such as the ones supplied with process control systems. Laboratories have detailed workflows covering the processing of samples. These workflows include basic elements such as sample login and result entry; as well as more complex elements, such as sample sequencing and data review. In our modern age, companies are dependent on (and at the mercy of) the software applications they run. These programs hold large amounts of the company's knowledge and history. These same companies invest millions of dollars annually in ensuring that their personnel communications (weekly meetings, conference calls, etc.) are established and secured in database system. Isn't it time to do the same with their software? Interfacing system A to system B is much easier than most people realize, and the benefits are enormous. These benefits are particularly important when using a LIMS, as the majority of the organization's QA/QC data is stored within it. Creating a global village (rather then islands) of information will result in a truly enterprise LIMS solution [7].

This commercial-off-the-shelf LIMS has been developed over a decade to deliver precisely the functionality required for DMPK/bioanalytical work in drug development. Watson was developed in conjunction with many major pharmaceutical companies to accelerate bioanalysis and sample management in drug development. Watson is a comprehensive system for managing drug study information. It supports the entire drug study process, including study design, assay design, sample handling, bar-coding, shipment tracking, analytical run planning, instrument interfaces, regression of standards, analytical performance reports, GLP evaluation, table and graph summary of sample analysis results, PK calculations, and interfaces to word processing software. Within Watson, there are interfaces to handle the following types of instruments and their related assays: GC, HPLC, GC–MS, LC–MS, RIA, and

ELISA, as to quick to learn and easy to use. Watson also lets users conduct research and testing with full confidence of compliance with GLP regulations and 21 CFR Part 11 Guidance from the US FDA. The software offers maximum flexibility and configurability, while preserving data integrity, via the system's security and audit trail. In addition, Watson enables users to handle standard and complex study protocols, providing audit trails to track deviations and amendments to each study. Watson features a bidirectional interface with many of the analytical instruments in laboratories, it tracks shipments and samples through barcode labels that you design, it supports a wide range of PK/TK analyses, and it organizes study results in a unique document management system. Because it fully supports unit management, Watson allows users to consolidate data across studies and projects.

Current trends indicate that the need for consolidation of data systems is now a critical and strategic part of the global growth of any pharmaceutical company, no matter the stage of drug development that is the driving focus of the operation. This is especially true for those life sciences companies utilizing the services of contract research organizations or establishing their own proprietary CROs in developing parts of the world that are far removed from their US or European-based headquarters. Global deployments of integrated and web-based systems can now be relied upon to be more consistent and more rapid because purpose-built LIMS, like those offered by Thermo Fisher, can facilitate better data exchanges across organizations and bio-directional data flow from a CRO to its life sciences sponsor company. This freer exchange of data can mean improved business decisions that enable life science companies to deliver the right compounds faster, and save the time and expense of working on a compound that ultimately will prove ineffective. Purpose-built LIMS also proves valuable in simplifying routine system upgrades, dramatically minimizing project risks across sites, and sometimes enhancing compliance with the guidelines imposed by regulatory authorities. Manual capture, calculation, and verification of raw data result in a tremendous drain on human resources while also jeopardizing the integrity of the information. The administration of paper records is particularly inefficient and expensive and data cannot be easily integrated with other technologies employed by the organization. As a result, complying with the strict principles of GLP can prove a very time consuming and expensive process. There emerges a need to employ sophisticated, enterprise-wide Laboratory Information Management Systems capable of addressing the complexity of the regulations, ensuring compliance with current best practice, and satisfying the concerns and expectations of the regulators. Data generated from an instrument electronically and captured as a direct computer input can be identified at the time of the input by the individual(s) responsible for direct data entries. An efficient computerized system design always provides for the retention of full audit trails to show all changes to the data without obscuring the original data. It is also mandatory to associate all changes to data with the persons having made those changes by use of timed and dated electronic signatures. The justification of changes can be also recorded and saved with each entry. LIMS solutions can also generate the final report of the study automatically providing a comprehensive description of the methods and materials used and a presentation of the results, including calculations and determinations of statistical significance.

Laboratories generate enormous amounts of data from the hundreds of samples analyzed every day. The purpose of a Laboratory LIMS is to collect, process, store, and retrieve the data to ensure sample traceability and regulatory compliance, observance of industrial standards, as well as provide reporting, monitoring, and analysis capabilities. The latest release of Watson LIMS, the leading bioanalytical laboratory information management system, delivers improved efficiencies and reduced validation time, contributing to time and cost savings, and improved time to market. Of the many new features unique to Watson LIMS, key among them is the new functionality Watson 7.4 that simplifies daily workflow for bioanalytical laboratories by delivering enhanced functionality for incurred sample reanalysis (ISR). ISR has become an accepted way to assess the quality of bioanalytical assays and is widely recognized within the pharmaceutical industry and by regulatory agencies. The latest release of Watson LIMS allows researchers to perform incurred sample reanalysis on an individual sample basis, and eliminates the previously required mandatory repeat runs for bioanalytical samples. Watson's new ISR reporting features provide new comparison methods with built-in calculations and greater ease of use by enabling flexible, preconfigured reporting of ISR results using templates, for a truly out-of-the-box user experience.

Additional new functionality enhancements to Watson 7.4 include performance enhancements in the Design Summary Report, delivering a dramatic increase in reporting speed, enhancements to the Immune Response Module and the Watson Web Services Library, enabling users to more easily manage the status of samples, choose samples with a Positive Screen or a Positive Titer result for use in subsequent assay tests, and import samples and sample-related information from external applications. For small and medium sized laboratories the installation cost of a commercial LIMS is often prohibitive or not easy to justify. As a result, these enterprises frequently rely on solutions developed in-house, but the total cost of ownership of these solutions may actually be greater than anticipated. LIMS SaaS offers this group of users an alternative to costly commercial systems and the aggravation of maintaining a homegrown solution. SaaS alleviates the user's need to purchase servers, maintain software, and other ongoing support, enabling them to concentrate on core capabilities. LIMS solutions are to increase efficiency, facilitate standardization, reduce support costs, enable data sharing, and support other business requirements.

Compliance with the principles of GLP is of high importance when registering or licensing pharmaceuticals, pesticides, cosmetics, veterinary drugs, food and feed additives, and industrial chemicals. LIMS solutions have emerged as the most appropriate tool to assist toward compliance with the principles of GLP. Such systems can efficiently and safely record, report, store, and retrieve study plans, raw data, and final reports, thereby addressing the complexity of the regulations. In order to comply with the principles of GLP, LIMS solutions need to be fully validated. This may be a particularly expensive process but costs can be considerably minimized by following certain strategies; adhering to the principles of GLP, GCP, and GCLP, sourcing validation services from experienced vendors, employing a consistently trained validation team, standardizing on a single solution, and ensuring long-term maintenance of the validated state of the systems.

4.4.2.4 Needs of a LIMS System The main drivers for the implementation of a Laboratory Information Management System are twofold. First from the business perspective, increased efficiency, an increase in throughput, faster time to market for drug manufacture and increased speed of decision making make a LIMS system a valuable business tool. Companies, especially pharmaceuticals are expected to operate within a tighter regulatory framework. Second, a LIMS system can help an organization achieve this through the use of appropriately approved systems.

Types of System: There exist various flavors of LIMS systems that are typically used within the different stages of the drug discovery and development lifecycle. Many LIMS providers are now suggesting that customization should not be performed on a LIMS system as the cost of the customization may be relatively small compared to the cost of re-validation for GAMP compliance, for example. In this respect, it is often desirable to choose a LIMS for a specific purpose analogous to using a specific application for a particular need. Therefore, an organization may have different LIMSs for the various stages of the drug lifecycle, for example.

4.4.2.5 Drug Discovery and Early Development For drug discovery and early development purposes, many LIMS systems have the ability to design a workflow to suit the laboratory. Samples can then follow particular workflows, be linked to further processes within the laboratory, and contain functionality for plate handling and the associated pooling and replication of samples, sample aliquoting, and initial analysis. Typical systems are designed for medium to high-throughput laboratories and include functionality to design experiments, control consumables, reagents, and assays as well as subsequent analysis along with central database portal with an organization.

4.4.2.6 Preclinical and Clinical Development For nonclinical and clinical development purposes, LIMS systems have the ability to generate study protocols/ work plans, design a workflow across multifunctions to suit R&D laboratories for implementations. Most of these programs are subject to regulatory scrutiny (GLP, GCP, GCLP, and/or GxP compliance). Samples can then follow particular workflows (e.g., chain of custody of study samples), be linked to further processes within the laboratory, and contain functionality for qualitative and quantitative analyses. Typical systems are designed for medium to high-throughput laboratories and include functionality to design experiments, control consumables, reagents, and assays as well as subsequent analysis along with central database portal with an organization. Watson LIMS has a complete 21 CFR Part 11 compliant system, and full study audit trail with full bidirectional interface capability to analytical instruments. LIMS tracks shipments and samples through user-designed barcode labels, supports a wide range of analyses, and organizes study results in a unique document management system, which has enterprise-wide knowledge and collaboration. Robust, dynamic reporting and data analysis includes a regression calculation tool, and linear and nonlinear algorithms.

4.4.2.7 Quality Control/Quality Assurance For QC and QA purposes, LIMS are designed to ensure high productivity and quality for analytical testing laboratories. Typical systems allow for entering, tracking, receipt of samples, and automate the workflow of result entry, approval, and reporting through to generating certificate of analyses, for example. In order to meet the demands of the modern laboratory, modern LIMS systems are not one of many stand-alone islands of automation but are expected to integrate with Chromatography Data Systems and Enterprise Resource Planning systems such as SAP.

4.4.2.8 Functions of a LIMS System A typical LIMS has many functions. Some of the operations and abilities of LIMS are listed below:

- *Paperless Office:* Acquisition and manipulation of electronic data is faster than paper-based processes.
- *Automation of Regulatory Compliance:* Compliance to FDA 21 CFR Part 11 can be achieved via implementation of an appropriately approved system. A LIMS system will identify users as they access the system and will log specific actions that the user performs on entities within the system. All of this allows for proper auditing to be performed ensuring that it is known when, by whom, and why an entity was changed, for example.
- *Sample Tracking:* This includes the logging of samples in the LIMS, sample receipt (registering the physical presence of a sample), sample test result entry, and approval. A LIMS can then automatically generate a certificate of analysis or sample analysis summary and/or report, for example.
- *Integration with Patient Information Systems:* A LIMS system can be used to manage samples from patients within a hospital or forensic environment, for example. A hospital will use an PIS to manage patient core data that will then integrate with a LIMS in order to manage sample related data. Such interfaces are realized through the adoption of open-standards such as HL7 (see http://www.hl7.org.uk).
- *Integration with ERP Systems:* In order to manage resources within an organization in an automatic fashion, a LIMS will often need to integrate with ERP software such as SAP R/3. Such a system can alert a purchasing department if a raw material fails inspection or goods can be readily shipped if a manufactured lot passes QA checks. Many of the commercial LIMS providers work directly with SAP for instance so that an out-of-the-box LIMS is certified and compliant with the SAP Quality Management Inspection Data Interface (QM-IDI).
- *Integration with Instrumentation:* Many types of instruments and/or robots can be involved in the drug discovery/manufacture lifecycle. These instruments may be used for gene sequencing, high-throughput screening, or proteomics, for example. Data used by and generated by such instruments/robots will need to be maintained by a LIMS. In order to ensure compatibility with many systems, a LIMS will adopt standards-based mechanisms such as OLE for Process Control

(OPC) to provide such functions. Other standards are now emerging for the integration of software systems with laboratory instruments to streamline such processes and ensure future-proof systems. AnIML (see http://animl.sourceforge.net) is one such XML-based standard for the interchange of data between analytical instruments.

- *Business Intelligence Functions:* At a higher level than the laboratory, organizations will need to realize the costs of reagents, consumables, and staff costs. A LIMS will often be required to generate this information at an appropriate level.
- *Design of Analytical Methods:* A laboratory will have sets of analytical tests that it will perform that should be readily associated with samples. A LIMS allows for the design of complex and varied analytical methods that also allows for quality specifications to be assigned. Specifications may be associated with a method for the purposes of quality, and compliance.
- *Archiving:* A LIMS system used for manufacturing or drug discovery purposes will generate vast quantities of data. A LIMS system would be expected to archive data in order to leave the operational system running at peak performance.
- *Customization:* A global organization may require that localization of the system is required in order to present the system in the correct language, and date formats. Additionally, a LIMS is a very complex system and only a small subset of the system may be required. Therefore, a LIMS system may allow for the customization of screens in order to present a more simplified interface of commonly used functions. Many LIMS providers now offer the ability to customize their LIMS via an API. This can be more beneficial than "programming" customizations in that future upgrades of the software can be guaranteed to be compatible.
- *Extension of the LIMS:* LIMS have been developed over many years now and LIMS vendors have built systems to suit the modern laboratory. However, an organization may require that the system be tailored in order to meet specific business and/or regulatory requirements. An important aspect of a LIMS is the ability to extend and customize the system via an Application Programming Interface (API) without sacrificing the validity of the system.
- *Security of the Data:* A LIMS will be expected to have security implemented over who can view/modify data and also maintain roles so that users are restricted in the actions that they can perform within the LIMS.
- *Automatic Scheduling of Samples:* Often it is required that samples should be automatically logged into a LIMS system at frequent intervals. An environmental site, for example, may require that samples are taken on a daily basis from designated sampling sites.
- *Stability Studies:* Part of the drug manufacture lifecycle will involve the determination of the expiry date of a drug. These typically involve performing regular tests on many samples of a drug that are held in different storage

conditions of temperature, humidity, and packaging. Such studies will clearly last many years and involve thousands of samples of a drug.

- *Quality Assurance:* Quality assurance includes the testing of the raw materials of a drug, intermediate products through to the final manufactured product to ensure that the product meets its specification at each stage. A lot or batch is the common term within a LIMS used to describe a collection of samples associated with one or more steps of drug manufacture.
- *Radio-Frequency Identification Tags:* An emerging technology within the LIMS arena is the use of radio-frequency identification (RFID) tags. Laboratory applications such as LIMS are based upon tracking unique items (such as samples). RFID readers placed in different locations within a laboratory enable the automatic tracking of samples from one room to another, a task that was previously performed using time-consuming barcode scanners. A LIMS that supports such an event-driven framework can then log the complete chain-of-custody for a sample and who is responsible for it. Forensic and bioanalytical laboratories often require such stringent requirements on the chain-of-custody of a sample, for example.

Choosing a LIMS: One of the first issues to consider when deciding to implement a LIMS is whether a commercial system should be bought or a bespoke system to be developed. Every laboratory is different and while a bespoke solution may meet your needs exactly, most modern commercial LIMS sell to a variety of laboratories and hence flexibility is built into their products. Commercial LIMS represent many years of effort and experience in the field. A bespoke solution may require similar effort. Validation of a bespoke solution to satisfy regulatory requirements may take much longer than a commercial system. As mentioned above, most LIMSs incorporate an API in order that it can be tailored to meet the exact needs of a laboratory if the off-the-shelf solution does not meet them. Various Internet resources exist that provide information on selecting a LIMS as well as information on implementing a LIMS, regulatory compliance, and so on.

Other issues to consider when purchasing a LIMS are as follows:

- Compliance to regulatory requirements and the configuration of such parameters for an individual laboratory. A LIMS may need to fulfill its obligations for auditing and electronic signatures. However, this should be tailored to meet the demands of a laboratory since entering of electronic signatures will become burdensome if it is not required.
- The LIMS may have specific requirements for the infrastructure that is required in order for it to operate. This should be compatible with the organizations information technology (IT) infrastructure. As an example, a client-server LIMS architecture may be deployed via a Citrix metaframe or Terminal Services but only if it is appropriate to do so.
- The cost of a LIMS system extends to beyond the license fee. Costs of customizing the system and the subsequent re-validation of the system to meet specific regulatory and business requirements can be very expensive.

4.4.2.9 Process of Choosing a LIMS When choosing a LIMS, the following presents a structured process for the selection of a suitable solution.

- The existence of a valid business case for the new LIMS should exist as a prerequisite. This should include the expected benefits of the new LIMS and the expected timescales for return on investment. The return on investment should be expected to be no less than 3–5 years for such a system and budget support should be available over this time.
- Generation of requirements of LIMS—A system-neutral requirements gathering exercise should document the expected requirements of a LIMS. The requirements should be understandable both by users and by I.T. staff. They should be measurable in order that it is possible to determine whether a requirement has been met or not. This is far ranging and should include:
  - High-level management requirements of the system for reporting.
  - Analyst requirements for the processes of sample receipt and entering of analytical results.
  - Quality/Laboratory Management requirements covering sample approval, generation of certificates of analysis, audit reports.
  - Laboratory protocols that are in place or are expected to be in place. These are presumably in place because they are deemed to be good. Therefore, a LIMS should build on these strengths.
- Integration with other laboratory systems such as CDS, electronic laboratory notebooks, other instruments, and robots.
- Regulatory requirements covering the requirements to satisfy FDA regulations such as 21 CFR Part 11 (Electronic records and signatures) as well as GxP requirements.
- Training that will be required for all levels of staff.
- The existence of data predating the LIMS may mean that a migration strategy is required for such data. Regulatory compliance may impose constraints that this data are kept in a readily available fashion. The importing of such legacy data can be highly complex and is likely to require a great deal of effort.
- An analysis of the available commercial products should be made initially focusing on the type of LIMS that is required, for example, for manufacturing or drug discovery purposes. For this purpose, LIMS vendors should be able to arrange demonstrations of their products as well as provide further information. The expected outcome of this stage is a short-list of the most suitable systems.
- Request for Proposals (RFP). RFP should be sent to the short-listed vendors. This is based on the requirements document produced and it is expected that vendors should indicate how each of your requirements is or can be met by their system. The RFP should also request information about the financial health of the company and its future plans. The LIMS would be expected to be functioning for many years. During its lifetime, upgrades and patches would normally be expected in addition to consulting on modifications for specific organizational

needs. Another solution to this is to ensure that the LIMS source code is held in escrow to protect the organization against possible loss of support by the vendor.

- Evaluation of Proposals—The returned proposals should be evaluated using a scoring mechanism. A small number of the top respondents should be invited to provide detailed demonstrations of their systems. As most commercial LIMSs are designed for typical processes in the laboratory, most respondents will readily meet 80% of the requirements. The winner should be the one that best allows the organization to most easily achieve the remaining 20%.

Process of Implementing a LIMS: Following the selection of the most suitable system for your organization, there remains several steps that should be followed to ensure the correct implementation of the LIMS. These steps are detailed below

- *Pilot the System:* The system should be piloted initially for an identified laboratory within a business unit of the organization before rolling out to further laboratories and other business units if it is to be.
- *The System Installation Qualification:* This ensures that each step of the installation process is performed correctly and is accountable to an individual. The installation script will detail the step-by-step procedure for installing the system.
- *The System Operational Qualification:* An operational qualification (OQ) ensures that each component of the system performs as intended within intended or representative ranges. A qualification script will detail the step-by-step procedure for qualifying the system and should include security, screen flow, data validation, and data updates. A cross-reference between the operational qualification and the requirements should be maintained.
- *Preventive Maintenance:* Arrange on-going support and maintenance support with the vendor for "help-desk" support, and bug fixing.
- *Basic Training:* Users will require training in the new system and importantly will want to see the benefits to them as individuals in their day-to-day job.
- *Technical Support:* Full-time staff will need to be dedicated to administering the LIMS and providing technical support.
- *System Gate-Keeper:* An "executive" for the system will need to be appointed who has overall authority and responsibility for the system.
- *Relevant Activities:* Follow-on activities will need to be identified. A commercial LIMS will have met the majority of your organizations requirements. Now, the process of implementing specific enhancements for your organization will have to be planned.

Choosing a LIMS, whether you decide to build a bespoke system or purchase a commercial product is a large undertaking. A well-implemented LIMS can be an enormous benefit to an organization but a poor implementation may drag on for years and erode the confidence of users within the organization and customers. Once implemented, the LIMS needs continuing maintenance and assessment. Expert LIMS consultants can help users through this potential minefield.

4.4.2.10 Summary Bioanalysis is an important subdiscipline of analytical chemistry that covers the quantitative measurement of xenobiotics (drugs and their metabolites, biological molecules in unnatural locations or unnatural concentrations) and biotics (macromolecules, proteins, DNA, large molecule drugs, and metabolites), from biological systems. Many scientific endeavors are dependent upon accurate quantification of drugs and endogenous substances in biological samples; the focus of bioanalysis in the pharmaceutical industry is to provide a quantitative measure of the active drug and/or its metabolite(s) for the purpose of pharmacokinetics, toxicokinetics, bioequivalence, and exposure–response (pharmacokinetics/pharmacodynamics studies). Bioanalysis also applies to drugs used for illicit purposes, forensic investigations, antidoping testing in sports and environmental concerns.

Bioanalysis was traditionally thought of in terms of measuring small molecule drugs. However, the past 20 years has seen an increase in biopharmaceuticals (e.g., proteins and peptides), which have been developed to address many of the same diseases as small molecules. These larger biomolecules have presented their own unique challenges to quantification. Modern drugs are more potent, which has required more sensitive bioanalytical assays to accurately and reliably determine these drugs at lower concentrations. This has driven improvements in technology and analytical methods.

Bioanalytical functions have expanded to and cover the entire scope of drug discovery and development, its accuracy and reliability are essential to the success of drug/product research and development. It has been apparent that both capabilities and capacities are vitally imperative for any organization to generate reliable and speedy data during development processes and ultimately for regulatory approval.

REFERENCES

1. Balogh MP. *LCGC 22*. 2004;9:890–894.
2. Balogh MP. *LCGC 23*. 2005;6:576–584.
3. Shah VP, Midha KK, Dighe S, et al. Analytical methods validation: bioavailability, bioequivalence and pharmacokinetic studies. Conference Report. *Eur J Drug Metab Pharmacokinet* 1991;16:249–255.
4. Guidance for Industry: Bioanalytical Method Validation. U.S. Department of Health and Human Services, FDA, CDER, CVM, May 2001.
5. Shah VP, Midha KK, Findlay JW, et al. Bioanalytical method validation—a revisit with a decade of progress. Conference Report. *Pharm Res* 2000;17:1551–1557.
6. FDA Guidance for Industry Part 11, Electronic Records; Electronic Signatures—Scope and Application. U.S. Department of Health and Human Services, FDA, CDER, Rockville, MD, 2003.
7. Avery G, McGee C, Falk S. Product review: implementing LIMS: a "how-to" guide. A laboratory must devote the necessary time and resources to planning, selecting, and implementing a new system. *Anal Chem* 2000;72 (1):57A–62A.

5

TECHNICAL AND REGULATORY ASPECTS OF BIOANALYTICAL LABORATORIES

5.1 FUNDAMENTAL ROLES AND RESPONSIBILITIES OF BIOANALYTICAL LABORATORIES

Bioanalysis, employed for the qualitative and quantitative determination of drugs and their metabolites in biological matrices, plays a significant role in the evaluation and interpretation of bioequivalence (BE), bioavailability (BA), pharmacokinetic (PK), and toxicokinetic (TK) studies. The quality of these studies, which are often used to support regulatory filings, is directly related to the quality of the underlying bioanalytical data. It is therefore important that guiding principles for the validation of these analytical methods be established and disseminated to the pharmaceutical and life sciences communities. Obviously both technical and regulatory aspects are vitally important for bioanalytical laboratories to ensure the quality and integrity of data to be generated and delivered.

The first workshop organized by the AAPS and FDA (Crystal City Workshop 1990) raised awareness of the need for validated bioanalytical methods for the regulatory acceptance of bioequivalence and pharmacokinetic data. Although the workshop addressed bioanalysis in general, it also acknowledged the differences between chromatographic and ligand binding (nonchromatographic based) methods. The workshop identified the essential parameters for bioanalytical method validation, for example, accuracy, precision, selectivity, sensitivity, reproducibility, limit of detection (LOD)/limit of quantitation (LOQ), and stability. The outcome of the first

Regulated Bioanalytical Laboratories: Technical and Regulatory Aspects from Global Perspectives, By Michael Zhou

workshop and its report resulted in improved quality of data submissions to regulatory agencies.

Following the first workshop report [1] and the experience gained at the FDA, the draft Guidance on Bioanalytical Methods Validation was issued by the FDA in January 1999. This draft guidance provided stimulus and opportunity for further discussion at the 2nd AAPS/FDA Bioanalytical Workshop in January 2000. In addition, newer technology, such as chromatography coupled to tandem mass spectrometry (LC–MS/MS), was discussed along with an update on ligand-binding assays. This workshop resulted in a report "Bioanalytical Method Validation—A Revisit with a Decade of Progress" [2] and formed the basis for the FDA Guidance on Bioanalytical Methods Validation in May 2001 [3].

The evolution of divergent analytical technologies for conventional small molecules and macromolecules, and the growth in marketing interest in macromolecular therapies, led to the workshop held in 2000 to specifically discuss bioanalytical methods validation for macromolecules. Because of the complexity of the issues, the workshop failed to achieve a consensus. To address the need for guiding principles for the validation of bioanalytical methods for macromolecules, the AAPS Ligand-Binding Assay Bioanalytical Focus Group developed and published recommendations for the development and validation of ligand-binding assays in 2003 [4].

As bioanalytical tools and techniques have continued to evolve and significant scientific and regulatory experience has been gained, the bioanalytical community has continued its critical review of the scope, applicability, and success of the presently employed bioanalytical guiding principles. The purpose of the 3rd AAPS/FDA Bioanalytical Workshop was to identify, review, and evaluate the existing practices, white papers, and articles and clarify the FDA guideline documents. The focus of the workshop was primarily on quantitative bioanalytical methods' validation and their use in sample analysis, focusing on chromatographic and ligand-binding assays. A report of this workshop is detailed via theme issue of *The AAPS Journal* [5].

5.1.1 Technical Functions of Bioanalytical Laboratories

Bioanalysis plays a vital role in the qualitative (structural and metabolite identification) and quantitative (analyte concentration) determination of drugs and their metabolites in biological matrices. This technique is used very early in the drug development process to provide support to drug discovery programs on the metabolic fate and pharmacokinetics of chemicals in living cells and in animals. For some time, good bioanalytical practices (GBPs) had been introduced within the industries, even though these early development programs are not subject to regulatory scrutiny at this point, but rather to ensure that reliable data are going to be generated for decision making process. Its use continues throughout regulated (GLP) preclinical and clinical drug development phases, into postmarketing support and may sometimes extend into clinical therapeutic drug monitoring. The role of bioanalysis in pharmaceutical drug development is clearly noticeable, with focus on the particular activities that are performed within each stage of the development process and on the variety of sample processing/preparation matrices encountered. Recent developments and industry trends for rapid sample throughput and data generation are introduced, together with examples of how these high throughput needs are being met in bioanalysis (Figure 5.1).

5.1.1.1 Discovery and Early Development Pharmaceutical research process is looking at new and improved ways to develop medicines, in response to several important scientific advances that have recently evolved. These advances include the identification of new and more specific drug targets (as a result of establishment in genomics and proteomics); successes with tissue growth outside of the living organism; development of faster, more sensitive, and more selective analytical systems (mass spectrometry); higher throughput (as a result of robotics and laboratory automation); proliferation in synthesis techniques (combinatorial chemistry); and advances in computing and information systems (bioinformatics). In parallel with these scientific advances, business factors have changed with the consolidation of drug companies and the intense pressure to get drugs to market faster than ever before. The current focus of drug discovery research is on rapid data generation and analysis to identify promising candidates very early in the development cycle. An optimal lead candidate is selected for further evaluation. Combinatorial chemistry techniques allow the synthesis of compounds faster than ever before, and these greater numbers of compounds are quickly evaluated for potency and pharmacological activity using high-throughput screening (HTS) techniques. HTS involves performing various microplate-based immunoassays with synthesized compounds or compounds from

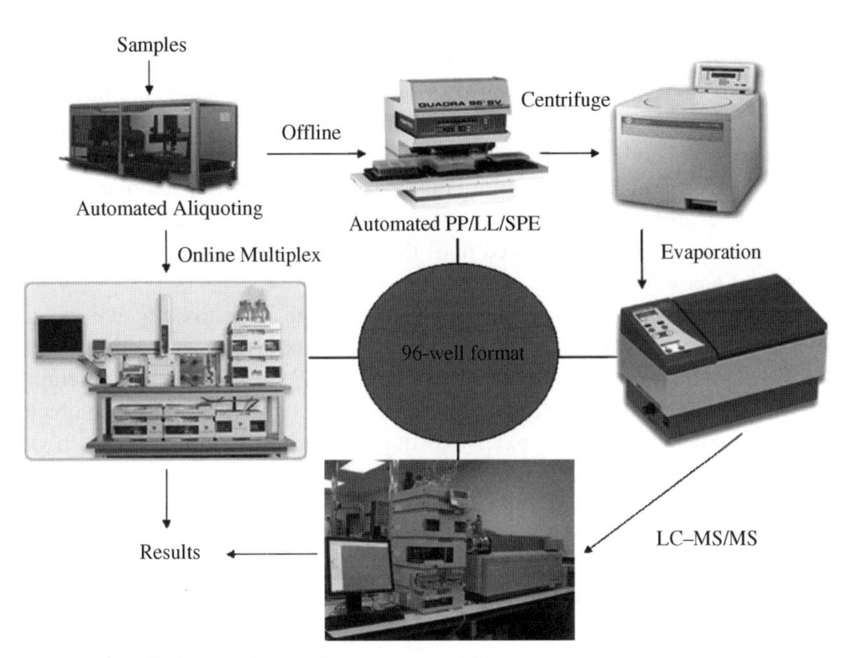

FIGURE 5.1 High-throughput of bioanalytical automation.

natural product isolation. Examples of assay types used in HTS are scintillation proximity assay (SPA), enzyme-linked immunosorbent assay (ELISA), fluorescent intensity, chemiluminescence, absorbance/colorimetry, and bioluminescence assays. These HTS tests simulate a specific biological receptor or target function and a qualitative decision ("hit" or "miss") is generated.

Once hits are identified, chemists perform an iterative process to synthesize and screen smaller, more focused libraries for lead optimization in an effort to improve compound activity toward a specific target. Using automated techniques, ultrahigh throughput can be obtained by the most advanced laboratories and tens of thousands of compounds can be screened in one day. In parallel studies, information is learned on a drug molecule's absorption, distribution (including an estimate of protein binding), metabolism, and elimination by sampling from dosed laboratory animals (called *in vivo* testing) and from working cells and/or tissues removed from a living organism (called *in vitro* testing since the cells are outside a living animal). These important tests are collectively referred to as ADME characteristics (Absorption, Distribution, Metabolism, and Elimination). A candidate compound that will potentially meet an important medical need receives an exhaustive review addressing all the key issues concerning its further development. Evaluation of the available data, competitive therapies, expected therapeutic benefit, market opportunities, and financial considerations all contribute to the final decision to grant development status to a particular compound. A multifunctional project team is assembled to guide the development efforts into the next phase—preclinical development. High-throughput bioanalysis (nonregulated) has been directly and indirectly involved in above activities and programs to facilitate and enhance accelerating discovery and early development phases.

5.1.1.2 Preclinical Screening and Development The preclinical development process largely consists of a safety evaluation (toxicity testing) and continued study into a drug candidate's metabolism and pharmacology. Both *in vitro* (mostly for screening purpose) and *in vivo* tests are conducted; many species of animals will be used because a drug may behave differently in one species than in another. An early assessment of dosing schedules in animal species can be determined, although human dosage regimens are not determined until the subsequent clinical trials in the next development phase.

Toxicology tests in preclinical development examine acute toxicity at escalating doses and short-term toxicity (defined as 2 weeks to 3 months), as well as the potential of the drug candidate to cause genetic toxicity. Today's research efforts attempt to utilize as few animals as possible and so more *in vitro* tests (screening) are conducted. The use of metabonomics for toxicity testing is making an impact on both drug discovery (to select a lead compound) and preclinical development (to examine safety biomarkers and mechanisms). Metabonomics is a technology that explores the potential of combining state of the art high resolution NMR (nuclear magnetic resonance) spectroscopy with multivariate statistical techniques. Specifically, this technique involves the elucidation of changes in metabolic patterns associated with drug toxicity based on the measurement of component profiles in biofluids (e.g., urine). NMR pattern recognition technology associates target organ toxicity with specific NMR spectral patterns and identifies novel surrogate markers of toxicity [6]. Also in preclinical development, the pharmacokinetic profile of a drug candidate is acquired. Pharmacokinetics is a specific, detailed analysis that refers to the kinetics (e. g., time course profile) of drug absorption, distribution, and elimination. The metabolites from the drug are identified in this stage. Definitive metabolism studies of drug absorption, tissue distribution, metabolism, and elimination are based on the administration of radiolabeled drug to animals.

It is important that the radionuclide is introduced at a position in the chemical structure that is stable to points of metabolism and conditions of acid and base hydrolysis. Pharmacology testing contains two major aspects—*in vivo* (animal models) and *in vitro* (receptor binding) explorations. Comparisons are made among other drugs in the particular collection under evaluation, as well as among established drugs and/or competitive drugs already on the market. More informative and/or predictive biomarkers are also identified and monitored from these studies. Detailed information about the drug candidate is developed at the proper time in preclinical development, such as the intended route of administration and the proposed method of manufacturing. In order to supply enough of the drug to meet the demands of toxicology, metabolism, and pharmacology, the medicinal chemistry and analytical groups work together to determine the source of raw materials, develop the necessary manufacturing process (scale up), and establish the purity of the drug substance and then product. The exact synthesis scheme and methodology needed to produce the drug are recorded in detailed reports. The pharmaceutics research group develops and evaluates formulations for the drug candidate. These formulations are assessed *in vivo* by the drug metabolism group. Quality and stability are the goals for this dosage form development effort. In the United States, after the active and inactive ingredients of a formulation containing the candidate compound have been identified and developed, a detailed summary called an Investigational New Drug (IND) Application is prepared. This document contains reports of all the data known to date on a drug candidate's toxicology, metabolism, pharmacology, synthesis, manufacturing, and formulation. It also contains the proposed clinical protocol for the first safety study in man. Up to date, part of early development programs may not be subjective to GLP regulations. However, more and more studies including animal toxicology had become regulated or GLP programs. It has been noted that a general trend is that data generation and data package for agencies' review should be in a reliable format good bioanalytical practice and/or GLP from regulated bioanalytical perspectives. USFDA, OECD, WHO, ICH, and other agencies and organizations have issued numerous guidance documents for industry implementations [6–12].

All of the information contained in an IND application is submitted to the United States Food and Drug Administration (FDA). Typically, thousands of pages of documents comprise this IND. The FDA reviews the information submitted and makes a decision whether or not the drug has efficacy and appears safe for study in human volunteers. The IND becomes effective if the FDA does not disapprove the application within 30 days. The drug sponsor is then approved and allowed to begin clinical studies in humans. When questions arise, the FDA responds to the IND application with a series of inquiries to be answered and a dialogue begins between the drug sponsor and the FDA.

5.1.1.3 Phase I–III Clinical Development Clinical trials are used to judge the safety and efficacy of new drug therapies in humans. Drug development is comprised of three clinical phases: Phase I–III prior to NDA and potentially with phase IV after conditional and full approval from regulatory agency. Each phase constitutes an important juncture, or decision point, in the drug's development cycle. A drug can be terminated at any phase for any valid reason. Should the drug continue its development, the return on investment (ROI) is expected to be high so that the company developing the drug can realize a substantial and often sustained profit for a period of time while the drug is still covered under patent.

Phase I safety studies constitute the "first time in man." The objective is to establish a safe dosage range that is tolerated by the human subject. These studies involve a small number of healthy male volunteers (usually 20–80) and may last a few months; females are not used at this stage because of the unknown effects of any new drug on a developing fetus. Biological samples are taken from these volunteers to assess the drug's pharmacokinetic characteristics. During a Phase I study, information about a drug's safety and pharmacokinetics is obtained so that well controlled studies in Phase II can be developed.

Phase II studies are designed to demonstrate efficacy, for example, evidence that the drug is effective in humans to treat the intended disease or condition. A Phase II controlled clinical study can take from several months to 2 years and uses from 200 to 800 volunteer patients. These studies are closely monitored for side effects as well as efficacy. Animal studies may continue in parallel to determine the drug's safety. A meeting is held between the drug sponsor and the FDA at the end of Phase II studies. Results to date are reviewed and discussion about the plan for Phase III studies is held. Additional data that may be needed to support the drug's development are outlined at this time and all information requirements are clarified. A month prior to this meeting, the drug sponsor submits the protocols for the Phase III studies to the FDA for its review. Additional information is provided on data supporting the claim of the new drug, its proposed labeling, its chemistry, and results from animal studies. Note that procedures exist that can expedite the development, evaluation, and marketing of new drug therapies intended to treat patients with life-threatening illnesses. Such procedures may be activated when no satisfactory alternative therapies exist. During Phase I or Phase II clinical studies, these procedures (also called "Subpart E" for Section 312 of the US Code of Federal Regulations) may be put into action.

Phase III studies are usually carefully designed and launched after evidence establishing the effectiveness of the drug candidate has been obtained in Phase II clinical studies, and the "End of Phase II" meeting with the FDA has shown a favorable outcome. These studies are large-scale controlled efficacy studies and the objective is to gain more data on the effectiveness and safety of the drug in a larger population of subjects. A special population may be used, for example, those having an additional disease or organic deficiency such as renal or liver failure. The drug is often compared with another drug used to treat the same condition. Drug interaction studies are conducted as well as bioavailability studies in the presence and absence of food. A Phase III study is a clinical trial in which the patients are assigned randomly to the experimental group or the control group. From 500 to 5000 volunteer patients are typically used in a Phase III study; this aspect of drug development can last from 2 to 3 years. Data obtained are needed to develop the detailed physician labeling that will be provided with the new drug. These data also extrapolate the results to the general population and identify the side effect profile and the frequency of each side effect. In parallel, various toxicology, carcinogenicity, and metabolic studies are conducted in animals. The cumulative results from all of these studies are used to establish statements of efficacy and safety of the new drug for NDA submission and approval.

Phase IV clinical trials are those studies conducted after a product launch to expand on approved claims, study the drug in a particular patient population, as well as extend the product line with modified and/or new formulations. A clinical study after the drug is sold may be conducted to evaluate a new dosage regimen for a drug, for example, fexofenadine (Allegra is an antihistamine sold by Aventis (Sanofi-Aventis in Bridgewater, NJ, USA)). The original clinical studies indicated that a dosage of 60 mg, given every 12 h, was adequate to control symptoms of allergies and rhinitis. Their product launch was made with this strength and dosage regimen. In response to competition from a once a day allergy drug, Aventis conducted Phase IV clinical vials (after the product was on the market) with different dosages and obtained the necessary data to show that a 180 mg version of Allegra could be taken once a day and relieve allergy symptoms with similar efficacy as 60 mg taken twice a day. Aventis then filed the clinical and regulatory documentation, and obtained approval to market a new dosage form of their drug. Another example of a postmarketing Phase IV clinical study is the investigation of whether or not sertraline (Zoloft an antidepressant drug) could be taken by patients with unstable ischemic heart disease. Results suggested that it is a safe and effective treatment for depression in patients with recent myocardial infarction or unstable angina.

There are hundreds and thousands of clinical samples from various sites to be generated for bioanalytical processing and analyses. A fully regulated bioanalytical process may be implemented throughout the processes of sample transport, storage, analysis for data package assembling, and submission. An overall bioanalytical function in support of drug discovery and development is given in Figure 5.2.

5.1.2 Basic Processes in Bioanalytical Method Development, Validation, and Sample Analysis

Bioanalytical methods involve a systemic evaluation of all the processes required to demonstrate that a particular technique is reliable for its intended purpose. Practically, this process consists of three to four phases each with its own predefined goals (Table 5.1). The first stage is method development where the aim is to perform feasibility studies, assess reagent availability, and so on, resulting in the construction of the validation plan. This phase is then followed by prestudy method validation, where the validation scope and plan is put into effect and is conducted normally

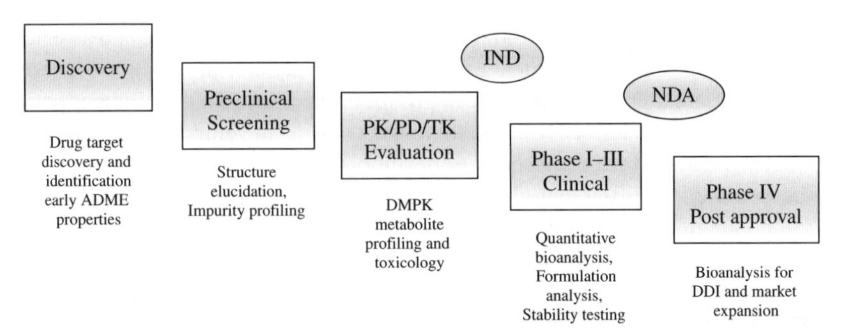

FIGURE 5.2 Bioanalytical functions in support of drug discovery and development.

utilizing validation samples (VS, but referred to as quality controls (QCs) during sample analysis). VS and QC are samples employing a test matrix, either matrix derived or a suitable surrogate, containing a known concentration (nominal) of analyte(s) that are treated as unknowns in an assay. The threefold goal of this phase is to (a) produce a body of data that proves that the method meets acceptable standards of performance; (b) formalize these data into an analytical report; and (c) draft a method (Lab Method or SOP) which is then taken forward sample analysis. Finally, during sample analysis QCs are incorporated to confirm that the method continues to perform consistently within specifications, thus allowing matrix derived data to be confidently accepted as valid. Performance parameters that are studied during the second stage of validation of bioanalytical methods would normally include: selectivity; sensitivity; calibration response; choice of quality control samples, analyte recovery, precision, accuracy; and reproducibility (incurred sample repeat analysis). For a fuller definition of these terms refer to Chapter 10. Bioanalytical method also requires a systematic study of analyte stability in calibration standards, quality

TABLE 5.1 Stages in Method Development, Validation, and Sample Analysis Processes

Stage	Main Purpose	End Results
Initial development	Assemble all of the components required; feasibility and validation	Go/no go decision
Development	Develop method and perform preliminary validation	Validation scope and plan
Prestudy validation	Run validation samples, derive acceptance criteria	A validated method with report and SOP
In-study validation	Quality Controls used in sample analysis, identify sampling issues, etc.	Valid bioanalytical data
Poststudy validation	Incurred sample repeat analysis, identify/confirm method robustness	Confirmation of reliable data

controls and study samples. Incurred sample repeat (ISR) analysis has become very critical to determining both method robustness and reliability of bioanalytical data in support of drug/product approval. This may be performed either in-study or poststudy validation (sample analysis). A bioanalytical method consists of two main components—sample preparation and detection of the compound(s).

5.1.2.1 ***Sample Preparation*** This is a technique/process used to clean up a sample before analysis and/or to concentrate a sample to improve its detection. When samples are biological fluids such as plasma, serum, or urine, this technique/process is described as bioanalytical sample preparation. The determination of drug concentrations in biological fluids yields the data used to understand the time course of drug action, or pharmacokinetics, in animals and man and is an essential component of the drug discovery and development process. Sample preparation prior to the chromatographic separation has three principal objectives:

- the dissolution of the analyte in a suitable solvent;
- removal of as many interfering compounds as possible; and
- preconcentration of the analyte.

Protein precipitation, liquid–liquid extraction, and solid phase extraction (SPE) are routinely used with and/or without automation.

5.1.2.2 ***Protein Precipitation*** In protein precipitation, acids or water-miscible organic solvents are used to remove the protein by denaturation and precipitation. Acids, such as trichloroacetic acid (TCA) and perchloric acid, are very efficient at precipitating proteins. The proteins, which are in their cationic form at low pH, form insoluble salts with the acids. Organic solvents, such as methanol, acetonitrile, acetone, and ethanol, although having a relatively low efficiency in removing plasma proteins, have been widely used in bioanalysis because of their compatibility with high-performance liquid chromatography (HPLC) mobile phases. These organic solvents that lower the solubility of proteins and precipitate them from solutions have an effectiveness, which is inversely related to their polarity.

A volume of sample matrix (one part) is diluted with a volume of organic solvent or other precipitating agent (three to four parts), followed by vortex mixing and then centrifugation or filtration to isolate or remove the precipitated protein mass. The supernatant or filtrate is then analyzed directly. Protein precipitation dilutes the sample. When a concentration step is required, the supernatant can be isolated, evaporated to dryness, and then reconstituted before analysis. This procedure is popular because it is simple, universal, inexpensive, and can be automated in microplates. However, matrix components are not efficiently removed and will be contained in the isolated supernatant or filtrate. In MS/MS detection systems, matrix contaminants have been shown to reduce the efficiency of the ionization process. The observation seen is a loss in response and this phenomenon is referred to as ionization suppression.

5.1.2.3 Liquid–Liquid Extraction Liquid–liquid extraction (LLE) is the direct extraction of the biological material with a water-immiscible solvent. The analyte is isolated by partitioning between the organic phase and the aqueous phase. The analyte should be preferentially distributed in the organic phase under the chosen conditions. For effective LLE some considerations should be taken such as

- The analyte must be soluble in the extracting solvent.
- A low viscosity of solvent to facilitate mixing with the sample matrix.
- The solvent should also have a low boiling point to facilitate removal at the end of the extraction.
- The use of pH control allows the fractionation of the sample into acid, neutral and basic components.
- A large surface area is important to ensure rapid equilibrium. This is achieved by thoroughly mixing using either mechanical or manual shaking or vortexing.

Generally, selectivity is improved by choosing the least polar solvent in which the analyte is soluble.

5.1.2.4 Solid Phase Extraction In SPE the analyte is retained on the solid phase while the sample passes through, followed by elution of the analyte with an appropriate solvent. SPE can be considered as a simple on/off type of chromatography. A typical SPE sorbent consists of a 40–60 μm silica particle to which has been bonded a hydrocarbon phase. The SPE is typically carried out using a five-step process—condition, equilibrate, load, wash, and elute. The solid phase sorbent is conditioned by passing a solvent, usually methanol, through the sorbent to wet the packing material and solvate the functional groups of the sorbent. The sorbent is then equilibrated with water or an aqueous buffer. Care must be taken to prevent the phase from drying out before the sample is loaded, otherwise variable recoveries can be obtained. Samples are diluted 1:1 with aqueous prior to loading to reduce viscosity and prevent the sorbent bed from becoming blocked. Aqueous and/or organic washes are used to remove interferences. Table 5.2 provides some basic information regarding the techniques used for sample preparation.

Detection of the compound—usually following chromatographic separation from other components present in the biological extract. The detector of choice is a mass spectrometer. Before a bioanalytical method can be implemented for routine use, it is widely recognized that it must first be validated to demonstrate that it is suitable for its intended purpose.

Bioanalytical method validation includes all of the procedures that demonstrate that a particular method used for quantitative measurement of analytes in a given biological matrix, such as blood, plasma, serum, or urine, is reliable and reproducible for the intended use. The fundamental parameters for this validation include accuracy, precision, selectivity, sensitivity, reproducibility, and stability. Validation involves documenting, through the use of specific laboratory investigations, that the performance characteristics of the method are suitable and reliable for the intended

TABLE 5.2 Typical Choices of Sample Preparation Techniques Useful in Bioanalysis

- Dilution followed by injection
- Protein precipitation
- Filtration
- Protein removal by equilibrium dialysis or ultrafiltration
- Liquid–liquid extraction
- Solid-supported liquid–liquid extraction
- Solid phase extraction (off-line)
- Solid phase extraction (on-line)
- Turbulent flow chromatography
- Restricted access media
- Monolithic columns
- Immuno-affinity extraction
- Combinations with some and/or all of the above

analytical applications. The acceptability of analytical data corresponds directly to the criteria used to validate the method.

In early stages of drug development, it is usually not necessary to perform all of the various validation studies. Many researchers focus on specificity, linearity, and precision studies for drugs in preclinical through Phase II (preliminary efficacy) stages. The remaining studies penetrating validation are performed when the drug reaches the Phase II (efficacy) stage of development and has a higher probability of becoming marketed product. But now, for pharmaceutical methods, guidelines from the United States pharmacopoeia (USP), International Conference on Harmonization (ICH), and Food and Drug Administration provide a framework for regulatory submission must include study on such fundamental parameters.

The following principles of bioanalytical method validation provide steps for the development of a new method.

- Parameters essential to ensure the acceptability of performance of a bioanalytical method validation are accuracy, precision, selectivity, sensitivity, reproducibility, and stability.
- A specific detailed description of the bioanalytical method should be written. This may be in the form of a protocol, study plan, report, and/or standard operating procedures (SOPs).
- Each step in the bioanalytical method should be investigated to determine the extent to which environmental, matrix material, or procedural variables from time of collection of materials up to the analysis and including the time of analysis may affect the estimation of the analyte in the matrix.
- Whenever possible, the same biological matrix as that intended in the samples should be used for validation purposes.
- The accuracy, precision, reproducibility, response function, and selectivity of the method with respect to endogenous substances should be established with

reference to the biological matrix. With regards to selectivity, there should be evidence that the substance being quantified is the intended analyte.

- The concentration range over which the analyte will be determined must be defined in the bioanalytical method, based on the evaluation of actual standard samples over the range, including their statistical variation. It is necessary to use sufficient number of standards to define adequately the relationship between concentration and response. The relationship between response and concentration must be demonstrated to be continues and reproducible. The number of standards to be used will be a function of the dynamic range and the nature of the concentration–response relationship. In many cases, five to eight concentrations (excluding blank) may define the standard curve.
- For bioanalytical methods, accuracy and precision must be accomplished by the analysis of replicate sets of the analyte samples of known concentration—quality control samples from an equivalent biological matrix.
- The stability of analyte in the biological matrix at intended storage temperature (s) should be estimated. In addition the influence of freeze–thaw cycles should be studied. The stability of the analyte at ambient temperature should be evaluated over a period of time that encompasses duration of typical sample preparation, sample handling, and analytical run time.
- The specificity of the assay methodology should be established using independent sources of the same matrix.
- Recovery should be reproducible at each concentration.
- Acceptance/rejection criteria for spiked, matrix-based calibration standards, and validation QC samples should be based on the nominal (theoretical) concentration of analyte(s).

In drug development, important decisions based on data obtained from bioanalytical methods are taken hence it is generally accepted that sample preparation and method validation are required to demonstrate the performance of the method and the reliability of the analytical results. The acceptance criteria should be clearly established in a validation plan, prior to the initiation of the validation study. Synopsis of validation parameter requirements is shown in Table 5.3.

5.2 QUALIFICATION OF PERSONNEL, INSTRUMENTATION, AND ANALYTICAL PROCEDURES

To be effective and functionally operational, bioanalytical laboratories should have three fundamental elements (personnel, facility/equipment, and analytical procedures) to meet technical and regulatory standards and excellence. These are clearly involved in the qualification/validation processes of personnel (analysts/scientists/laboratory supervisors), facility, analytical equipment and analytical/bioanalytical methodologies, and standard operating procedures.

TABLE 5.3 Synopsis of Validation Parameter Requirements

Parameter and/or Process	Requirements
Selectivity (matrix interference)	Review noninterference in at least six sources of matrix for non-LC–MS/MS assays For LC–MS/MS assays, determine matrix factor (MF) in six sources if the nonisotopically labeled IS is used. If isotopically labeled IS is used, demonstrate that IS-normalized MF is close to unity
Validation batches	Analyze at least three batches for accuracy and precision evaluation. At least one validation batch should be performed as large as the largest anticipated sample batch
QC samples	Concentration of QC samples should be Low QC—about three times of the LLOQ Mid QC—middle of the range (at about the geometric mean of low and high QC concentration) High QC—near the high end of the range ~75–85% of ULOQ Dilution QC—sufficient to cover highest anticipated dilution
QC acceptance criteria	Intra- and interbatch precision (%CV) and accuracy (%RE) should be QCs prepared at all concentration greater than LLOQ $\leq$15% QC prepared at LLOQ concentration $\leq$20%
Calibration standards	Include the following calibration standards: Minimum of six nonzero standards Matrix blank: matrix sample without internal standard Zero standard: matrix sample with internal standard
Standard acceptance criteria	Acceptance criteria for calibration standards are LLOQ standard $\leq$20% All other standards $\leq$15% At least 75% of standards should meet above criteria
Matrix blank	Interference in matrix blank should be $\leq$20% of LLOQ response
Recovery	Extent of recovery of analytes and IS should be consistent, precise, and reproducible Determine recovery at three concentration levels

(continued)

TABLE 5.3 (*Continued*)

Parameter and/or Process	Requirements
Stability	Perform the following stability experiments: Stock solution: minimum of 6 h at room temperature Post preparative (extracted samples/auto-sampler tray) Longer time from preparation through sample analysis (injection). Assess against fresh standards, except for auto-sampler reinjection reproducibility Benchtop: Stability at ambient temperature (or temperature used for processing of samples) to cover the duration of time taken to extract the samples (typically 4–24 h) Freeze–thaw: QC samples at minimum of two concentrations, three cycles, completed thawed, refrozen at least 12 h between cycles, at anticipated temperature of sample storage Long term: cover longest time from collection to final analysis for any sample in study Analyze three aliquots at low and high concentrations with fresh standard curves and compare against intended (nominal) concentrations. Long-term stability can be completed post validation

Quality Personnel (Scientists, Supervisors, and QA Auditors)

GLP work represents the public and repeatable foundation for a new drug entity's safety margin. The validated exposures of a drug in an animal model establish the threshold of highest tolerable dose. The difference in this exposure and the dose anticipated to demonstrate efficacy is the basis of the safety margin, the therapeutic index. The work of personnel who perform regulated/GLP bioanalysis represents a critical component of the entire organization's assessment of safety margin. That work must be unassailable. If necessary, it must be able to withstand scrutiny in a court of law. Training documentation for analysts must emphasize the importance of this work and specifically warn against any sort of fraud, backdating, or reporting false QC results. This might seem obvious, but you must anticipate and plan for preventing the weakest of defenses: "I wasn't trained," "I was rushed," or "I didn't have proper resources." On a more positive note, an analyst's ability to perform GLP-quality work is a mark of distinction. It provides breadth to a résumé and proves the analyst able to give a remarkable degree of attention to details and quality. It also offers him or her significant professional satisfaction: the ability to produce a highly refined bioanalytical method, as opposed to the fast, generic methods seen in the discovery phase of drug design. Bioanalytical laboratories should seek analysts who understand the accountability associated with GLP work, its intrinsic importance, and the personal opportunities it affords.

To be functionally and operationally effective, bioanalytical facilities should have a basic organizational structure which consists of management team, well-trained lead scientists, qualified laboratory analysts, quality control specialists, quality assurance unit (separate and independent from laboratory), and facility support staff. In the accountability structure, every individual is responsible for a process and entire process. Each process is 100% accountable for any errors or omissions determined by the subsequent step/process. However, each subsequent step/process is responsible for providing oversight to ensure that reasonable procedures are utilized by the proceeding step. A management team retains overreaching accountability for each step/process.

It must be noted that even for a relatively small facility, true 100% QC combined with metrics tracking, outsourcing management, SOP training, and SOP revisions justifies the services of a full-time employee. For instance, QC scientist, wherever possible, maintains a checklist system approved by a quality assurance unit (QAU) and other procedures to maintain documentation, chromatograms, electronic archival resources, and all other QC mechanisms for the work. Its is also highly recommended that bioanalytical laboratories do the same, creating a dedicated resource for this function and staffing it with an experienced GLP bio-analyst whose sense of the urgency and necessity of performing unassailable bioanalytical work is fully developed. Management is responsible for ensuring adequate staffing for proper surveillance of the work, as stated in the preamble to the 1987 GLP regulations. If the QAU is able to deliver 100% QC service by retaining an experienced analyst whose time is fully dedicated to QC activities (as perhaps originally envisioned in the preamble and code), then doing so would suffice. Nevertheless, to maintain a connection to the process we recommend the QC specialist perform some amount of bench work in addition to his or her QC responsibilities.

All GLP work requires a protocol, a study director, and a QAU. Without a QAU, a GLP group cannot exist. When drafting the master plan to build a regulated/GLP bioanalytical laboratory, the need to adequately staff the quality assurance unit is sometimes underestimated. The organization contemplating building or rebuilding the laboratory often tries to combine this new work function with its existing in-life GLP or good manufacturing practices (GMPs) regulatory groups, which is not a desirable approach. GLP bioanalysis requires complex and highly cross-referenced documentation, and its regulatory environment is fast evolving. Form 483s and guidances are highly technical and require a focused effort to capture in an applied audit. If resources are limited, alternatives to hiring a new auditor exist, but it is not recommended avoiding the hire by appointing a GMP or in-life GLP auditor in an "overtime" scenario.

In the 1987 GLP preamble, the FDA makes two important statements. The first is that the QAU auditor can be any scientist not involved in the study, and the second is that audits must be made in a timely fashion so that correction can be made. This would indicate that the agency recognizes that some firms lack sufficient resources to recruit new employees for the QAU position. Management might therefore staff QAU positions with existing employees, provided those employees are not directly involved in a current study. This is a far cry from the level of separation often called for by QAU groups. As applied today, the industry trend calls for the QAU to report directly to the senior council or separate upper management line away from R&D operation. This minimizes conflicts-of-interest, insulates the auditor from the work, and provides each auditor with an enhanced career opportunity to audit multiple areas of work: GMP, GLP in-life, GMP dose formulations, regulated bioanalysis, and so on. However, it limits the specialization of the auditor in highly complex and increasingly specialized technical applications. It has been strongly recommended that the QAU be adequately staffed to enable satisfactory specialization and turnaround time to support a new lab, requirements that might supersede a few additional layers of independence or auditor career track.

Audit turnaround time, a critical issue for all organizations, is particularly critical for a new laboratory. An audit of several months' delay could catastrophically affect a new group. If the QAU group takes several months to return its audit of a single study while a new group continues to produce several studies' worth of data, the QAU would, in so doing, place the entire organization at risk of regulatory noncompliance. For lack of feedback, the resultant safety margin data for the several studies produced would all potentially contain the same flaws as the first study. It is believed that speedy turnarounds are becoming standard for contract research organizations (CROs) and throughout the industry. Indeed, the 1987 preamble to the GLPs states that audits "must be conducted in a timely fashion so that corrections can be made." As the accepted time for the making of corrections becomes shorter, the need for fast audits becomes more pressing. To fulfill this need when internal resources are unavailable, you can outsource the work to one of the many organizations that provide auditing services. Alternatively, as allowed for in the 1987 preamble, organization can adopt a peer audit model whose participants are not involved in the study.

Before establishing a new group, it is highly recommended that organization engage an external auditor to perform a full gap analysis. For this, management should choose an individual consultant or firm specializing only in quality assurance. Many CROs see internal laboratories as competitors or customers and so, either way, have a conflict of interest in performing the gap analysis.

5.2.1 From Regulatory Perspectives: Personnel, Training, and Qualification

According to GLP regulation 21 CFR 58.29, each person engaged in the conduct of or responsible for the supervision of a nonclinical laboratory study should have the education, training, and experience or any combination of the above to allow that individual to perform the assigned duties. Additionally, the personnel who provide the training must be qualified to do so. Similarly in cGMP regulation 21 CFR 211.25 and 211.28, personnel engaged in the manufacture, processing, packing, or holding of a drug should have the education, training, and experience, or a combination of the three elements to enable that person to perform his or her assigned functions. In both GLP and cGMP regulations a qualified trainer on a continuing basis should provide the training. Personnel in a bioanalytical laboratory should have the required education, training, experience or a combination of the above to carry out their responsibilities. Job descriptions should be available and current for the personnel directly related to bioanalytical testing and a training plan for each of the job descriptions should be designed to allow for consistent and uniform training for the affected personnel. Qualified trainers should provide the training on a continuous basis and effectiveness of training should be monitored to ensure that the quality of personnel training is sustained. The bioanalytical laboratory management should ensure that the training programs are adequate, properly administered, and documented for all involved personnel (analysts and supervision) involved.

The company's training program should ensure that involved personnel are trained initially prior to performing a job function, are provided periodic training, and are verified to be qualified to conduct the tasks assigned to them. More importantly, training should be documented and adequate records should be maintained and available for regulatory inspections. It is customary for GLP/cGMP analytical and GLP bioanalytical laboratories to require the analyst's training curriculum to include SOPs, analytical instrumentation, and the required analytical techniques. However, due to the complex nature of the bioanalytical methods, most bioanalytical laboratories require study-specific training to be performed to further qualify the analysts used to perform the testing of specimens from human studies. These study-specific training/qualification records should be retained with the study data package to support future third party inspections. The study-specific training generally requires the analysts to prepare spiked calibration standards and QC samples and perform trial accuracy/precision test runs. If the results generated meet the preestablished acceptance criteria for the calibration curves and the QC samples, the analysts can then be considered as qualified to perform testing of the subject specimen samples to support the clinical study.

Quality Assurance Unit/Quality Control

The firm or institution providing bioanalytical services for clinical studies should implement a QA system to include all relevant aspects of bioanalytical testing as recommended in the FDA "Guidance for Industry: Bioanalytical Method Validation." The extensive role of the Quality Assurance Unit in this system and as delineated in both the GLP and cGMP regulations cannot be overemphasized. The GLP regulation for 21 CFR 58.35 assigns the QAU with the responsibility for monitoring each nonclinical laboratory study to assure that the facilities, equipment, personnel, methods, practices, records, and controls are in conformance with the regulations. The QAU has the authority to inspect each nonclinical laboratory study periodically to assure the integrity of the study and provide periodic notification to management noting any problems and the corrective actions taken. The QAU should ensure that deviations from approved protocols or procedures have been addressed with the appropriate authorization and documentation. The GLP regulations mandate that the QAU be a separate and independent entity from the entity engaged in the conduct and management of the study.

Some of the major roles for the Quality Control Unit (QCU) outlined in cGMP regulation 21 CFR 211.22 include the responsibility of the QCU in providing the assurance that the facilities, equipment, personnel, methods, records, controls, procedures, processes, and practices are in conformance with all applicable regulations, standards, and/or specifications. The QCU has the responsibility to review production records to assure that no deviations/errors have occurred or, if deviations/errors have occurred, that they have been fully investigated and documented. Additionally, the QCU has the authority to sample and release or reject raw materials, labeling, components and finished product upon review of all aspects relating to manufacturing, packaging, and control of the final product. The QCU has the oversight for the implementation and administration of a change control program including computer systems. On a timely basis and through self-audits, the QCU provides written reports to management to provide assurance that all operations are proceeding in a compliant manner. The QCU is an independent function and has separate reporting responsibilities from production.

The bioanalytical testing roles of the QAU in a bioanalytical laboratory should be akin to the roles played by QA (or QC) in a GLP or cGMP environment. The QAU will report to a position totally independent of the line management that is directly engaged in the clinical study. This unit will provide the line management the necessary oversight and assurance that all operations, procedures, and services are completed in conformance with appropriate regulations and standards. The roles and responsibilities of QA in a bioanalytical laboratory performing clinical studies should be delineated in a procedure and its responsibilities should include the monitoring of each clinical study to ensure that the systems (facilities, equipment, personnel, methods, records, controls, procedures, processes, and practices) supporting the conduct of the study are operating in a compliant manner. Written procedures (Standard Operating Procedures), bioanalytical sample analysis reports, and method validation reports should be reviewed and approved by QA. The independent authority of QA has become more relevant in light of FDA's and the public's concerns on drug safety and its role relative to the review/approval of documents, which were not previously under its scope is becoming a regulatory expectation (see examples in Table 5.4).

QA should have oversight in a formal system for handling of deviations, for example, planned and unplanned events, with the ultimate goal of ensuring that the deviation is appropriately investigated with a root cause assignment and a Corrective Action/Preventive Action (CAPA) taken to prevent its recurrence. The investigation should also determine whether the deviation has an adverse impact on the accuracy and validity of the study data individually and in its totality. Another quality system that should be implemented in a bioanalytical laboratory deals with a robust change control program to effectively manage changes to protocols, procedures, methods, and processes. Changes (including computerized systems) must be reviewed and approved by the QAU prior to implementation.

The QAU in a bioanalytical laboratory should undertake the responsibility of ensuring that all levels of bioanalytical operations are being conducted in a compliant manner and periodic reports should be prepared and issued. The QAU in a bioanalytical laboratory performing human studies should also perform internal selective audits of study data to ensure they are compliant with protocol and SOP requirements before the study report can be approved. Study reports should not be issued without this QA review and QA statement of compliance. As noted previously, QA review and

TABLE 5.4 Typical QAU Functions

Control of issuance and return of SOP	Auditing for compliance/ownership
Adequacy of the SOPs for their functions	Auditing/approval
Receipt/handling of specimens	Auditing
Freezer qualification/calibration	Approval and auditing and control
Documentation/record-keeping and record retention	Auditing/approval
Data and report	Audit and approval
Reprocessing of raw data	Audit for compliance
Preparation of standards and QCs	Audit for compliance
Adherence to SOPs/protocols	Audit for compliance
Investigations and event resolution, documentation of Deviations	Audit and approval
Change management/change control (CC)	Audit, approval, and ownership
GLP, FDA compliance training ownership or oversight reference standards certifications	Audit for compliance
Support regulatory inspections ownership instrument calibration/PM	Audit for compliance
Instrument qualification program protocol, report and CC approval	Audit for compliance
Computer system validation QA leadership	Audit for compliance
Analyst training program ownership or oversight	
BE report prior to submission	Audit and approval
BE final report (incl. bioanalytical)	Audit and approval
Vendor qualification	Audit and approval

verification/approval of the accuracy and reliability of data are becoming an industry practice. A summary of typical QA involvement in the day-to-day activities of a typical bioanalytical laboratory is summarized in Table 5.4. The management of a bioanalytical laboratory needs to review its operation and to optimize the roles and responsibilities of the QAU to ensure full compliance with current FDA expectations and to minimize risk. A Quality Manual, with senior management approval, may be developed to define the responsibilities, accountabilities, and authorities relative to the QAU and the implementation of quality systems [13].

5.2.2 Facility Design and Qualifications

The GLP regulations (21 CFR 58.31) require that the testing facility be of suitable size and construction to facilitate the conduct of the nonclinical laboratory study. This is also required in cGMP regulation 21 CFR 211.42 where any building and premises used in the manufacture, processing or holding of a drug product shall be of suitable size, construction, and location to facilitate cleaning, maintenance, and proper operations. Separate or defined areas should be available to prevent contamination or adverse effect on the study or drug product.

The bioanalytical laboratory area should be spacious and designed in a way that allows for efficient sample and personnel flow. For example, test specimens should efficiently flow from "storage freezers" to the "sample processing/extraction laboratory" to the "instrument laboratory." Storage space (freezers, refrigerators) for subject test samples, calibration standards, and QC samples should be adequate and properly secured to allow for efficient, storage, and retrieval. Separate areas for the receipt, verification, storage of incoming test specimens, and preparation of calibration standards and QC samples should be made available to avoid mix-ups and to ensure integrity of subject samples and standards. Utilities such as water, air, gas, and electricity should be adequate, stable, and uninterrupted. A brief summary of typical components for a Quality Manual is given in Table 5.5.

5.2.3 Equipment Design and Qualification

The equipment used in the generation, measurement, or assessment of data and the equipment used for facility environmental control in a nonclinical laboratory study should have the appropriate design and construction for its intended use and should be suitably located to facilitate operation, cleaning, inspection, and maintenance (GLP, 21 CFR 58.61). For the manufacture of drug products, the manufacturer is mandated by cGMP to ensure that the equipment used in the facility is constructed to be nonreactive, additive or absorptive, suitably validated or qualified, easy to clean and sanitize to prevent cross contamination (GMP, 21 CFR 211.63, 211.65). In the event of the GLP and cGMP regulations, the equipment (e.g., balances, LC, GC, GC/MS, LC–MS/MS, etc.) used to support the generation of laboratory results intended for BA/BE, PK/TK, and other regulated studies should likewise be of adequate design and construction for the conduct of the study. As is required in cGMP, the equipment (including storage chambers) in the bioanalytical laboratory should be qualified to demonstrate its suitability of use and proper performance. The same principles of

TABLE 5.5 Typical Content of a Quality Manual

1. Title page (including version number of Quality Manual, implementation date)
2. Approval pages
3. Quality policy and mission statement
4. Quality manual requirements/scope
5. Management responsibilities/review
6. Quality systems—QAU functions
7. Document control—record-keeping/retention
8. Contract review
9. Test specimen control, and tracking
10. Testing requirements
11. Equipment qualification, calibration and preventive maintenance
12. Data review and approval
13. Change control
14. Deviation and event management/resolution
15. Reporting requirements
16. Training and qualification of personnel
17. Internal audits
18. Attachments

equipment qualification relative to Installation, Operational, and Performance Qualifications (IQ, OQ, and PQ) could be adopted in the qualification of equipment in the bioanalytical laboratory. The laboratory equipment should be properly situated to allow ease of operation, cleaning, and maintenance. Where a software database is used to maintain and track calibration and preventive maintenance (PM) schedules and reports, the application needs to be validated to be in conformance with the regulations [14, 15]. Based on the review of current FDA 483 observations, most bioanalytical laboratories have not been cited for the lack of instrument/equipment/ freezer chambers qualification data; nevertheless, it is good business and good scientific practice to address this instrument qualification issue in a proactive manner to avoid any future potential compliance issues.

5.2.3.1 Maintenance and Calibration of Equipment The GLP regulations for conducting nonclinical laboratory studies in support of applications for research or marketing permits for products (21 CFR 58.63) and the cGMP regulations for the manufacture, processing, packing, or holding of drug products mandate that the respective laboratory, institution, or manufacturing facility implement a sound calibration and PM program (21 CFR 211.67, 21 CFR 211.68) to ensure that the validated equipment continue to operate properly, as intended. Similarly, the bioanalytical laboratory should adopt a calibration/PM program to ensure equipments (including storage chambers) are periodically calibrated and maintained to ensure continued compliant operation and performance of such equipments. A calibration/PM schedule should be used and followed and procedures must be in place to handle missed calibrations. Equipment with expired calibration dates or which has missed the calibration interval should be properly identified; for example, "Out of Calibration—

Do Not Use" and be taken out of service. Data from uncalibrated equipment are deemed to be unreliable and should not be reported for regulated studies. Of importance is the need for calibration procedures that clearly provide instructions to the laboratory personnel on how to calibrate and maintain the equipment. Where calibration is performed by outside service groups, the firm should have a copy of the contractor's calibration procedures, if applicable, and the calibration report should be reviewed and approved by laboratory management prior to placing the equipment back into service. Equipment calibration needs to be performed against traceable standards. Remedial actions need to be defined in the event of calibration and/or equipment failure.

5.2.3.2 ***Facility and Operation Qualification*** Sample receipt, storage conditions, laboratory layout, operational setting, laboratory information management system (LIMS), and instruments (hardware, firmware, and software) are essential elements for the data quality and integrity of bioanalytical laboratories. The entire process from initial sample generation/collection, shipment/transfer, storage conditions, sample preparation, sample analysis, data acquisition, data analysis/processing, data retention/security, and data archiving is underlined criticality to data delivery. It is inevitable that some of the steps may be beyond normal expectation. Laboratory information management system may assist laboratory operation to systematically establish the process workflow and monitor/capture every step within the processes.

LIMS provides systems for experiment design, sample specification, sample tracking and data recording, workflow management, process optimization and documentation, QA, and sharing of such data across facilities or projects. LIMS ensures the rigor of experimental data by linking it with associated laboratory analysts, QA/QC factors, characterizations, protocols, and related experiments and data. LIMS maintains a detailed pedigree for each sample by capturing processing parameters, protocols, stocks, tests, and analytical results for the sample's complete life cycle. Project and study data are also maintained to define each sample in the context of research tasks it supports. LIMS will be required for each analytical pipeline to track all aspects of sample handling. Scientists within bioanalytical laboratory facilities will conduct many thousands of experiments, each with hundreds to thousands of individual samples upon which several analytical measurements will be made. Although a number of LIMS are sold by commercial vendors, no single LIMS will be able to meet the large scale, varied needs of all bioanalytical facilities and projects. The broad range of experimental protocols used in the facilities and in the laboratories of principal investigators will require LIMS customizations flexible enough to meet constantly changing requirements (e.g., new experimentation, protocols, parameters, and data formats).

Throughput is vital to bioanalytical facilities, so care must be exercised in the design of systems critical to the facility's uptime. The core LIMS at each facility is just such a system. When LIMS is not operating, data cannot be processed and the facility cannot run. LIMS must be very robust, highly available, and secured in ways similar to an institution's critical information-technology systems. An external data query or database operation must not impact LIMS or operations. Databases assimilating a facility's data must be inaccessible to hackers/intruders, and the system and databases

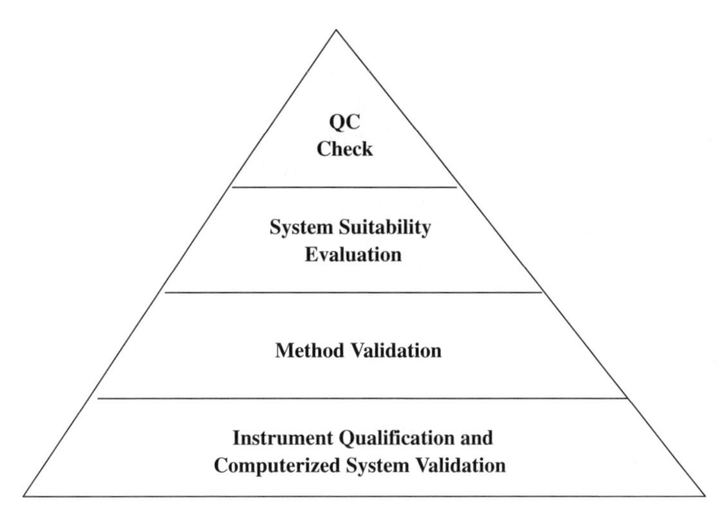

FIGURE 5.3 Basic components of data quality.

for recording data should be separate from those for sharing data. A working group should be established to examine existing and future needs of bioanalytical facility grantees and research centers. The group will assess and analyze existing LIMS as a prelude to adopting or creating a flexible and interoperable LIMS across a number of laboratory and center environments.

Analytical data generation is based upon a quality analytical procedure along with qualified equipment utilized during the analysis. Method validation process basically demonstrates the acceptability of data (e.g., quantitative aspect) from precision and accuracy perspectives. Analytical instrument qualification helps justify the continued/intended use of equipment, but it alone does not ensure the quality of data. Analytical instrument qualification is one of the four critical components of data quality. Figure 5.3 shows these components as layered activities within a Quality Triangle. Each layer adds to the overall quality. Instrument Qualification forms the base for generating quality data. The other essential components for generating quality data are the following: quality control check, system suitability evaluation, method validation and instrument qualification, and computerized system validation. These quality components are described below.

Analytical instrument qualification (AIQ) is documented evidence that an instrument performs suitably for its intended purpose and that it is properly maintained and calibrated. Use of a qualified instrument in analyses contributes to confidence in the veracity of generated data.

5.2.3.3 Analytical Instrument Qualification The following sections address in detail the analytical instrument qualification process. The other three components of building quality into analytical data—analytical methods validation, system suitability tests, and quality control checks—are not within the scope of this report [16]. Timing, applicability, and activities for each phase of analytical instrument qualification are clearly described within this report/article.

Qualification Phases Qualification of instruments is not a single, continuous process but instead results from many discrete activities. For convenience, these activities have been grouped into four phases of qualification. These phases are described as follows: (1) Design Qualification (DQ); (2) Installation Qualification (IQ); (3) Operational Qualification (OQ); and (4) Performance Qualification (PQ).

These qualification phases were used for AIQ because of their wide acceptance within the community of users, manufacturers, and quality assurance. Some of these qualification phases have their roots in manufacturing process validation. Note, however, that adoption of process validation terms does not imply that all process validation activities are necessary for AIQ. Some AIQ activities could arguably be performed within one or the other qualification phase. It is important that required AIQ activities be performed, but it should not be important under which qualification phase the individual activity is performed or reported. In any case, performing the activity is far more important than deciding where it belongs.

Design Qualification– The Design Qualification activity is most suitably performed by the instrument developer/manufacturer. Since the instrument design is already in place for the commercial off-the shelf (COTS) systems, the user does not need to repeat all aspects of DQ. However, users should ensure that COTS instruments are suitable for their intended applications and that the manufacturer has adopted a quality system for developing, manufacturing, and testing. Users should also establish that manufacturers and vendors adequately support installation, service, and training. Methods for ascertaining the manufacturer's design qualification and an instrument's suitability for its intended use depend on the nature of the instrument, the complexity of the proposed application, and the extent of users' previous interaction with the manufacturer. Vendor audits or required vendor-supplied documentation satisfy the DQ requirement. The required scope and comprehensiveness of the audits and documentation vary with users' familiarity with the instrument and their previous interactions with the vendor. Informal personal communications and networking with peers at technical or user group meetings significantly inform users about the suitability of instrument design for various applications and the quality of vendor support services. Informal site visits to other user and/or vendor facilities to obtain data on representative samples using the specified instruments also are a good source of information regarding the suitability of the instrument design for intended use. In many instances an assessment of the quality of vendor support, gleaned from informal discussions with peer users, significantly influences instrument selection.

Installation Qualification– Installation Qualification is a documented collection of activities needed to install an instrument in the user's environment. IQ applies to a new, preowned or an existing onsite—but not previously qualified—instrument. The activities and documentation associated with IQ are as follows:

- *System Description:* Provide a description of the instrument, including its manufacturer, model, serial number, and software version. Use drawings and flowcharts where appropriate.

- *Instrument Delivery:* Ensure that the instrument, software, manuals, supplies, and any other accessories arrive with the instrument as the purchase order specifies and that they are without damage. For a preowned or existing instrument, manuals, and documentation should be obtained.
- *Utilities/Facility/Environment:* Verify that the installation site satisfactorily meets vendor-specified environmental requirements. A commonsense judgment for the environment suffices; one needs not measure the exact voltage for a standard-voltage instrument or the exact humidity reading for an instrument that will operate at ambient conditions.
- *Network and Data Storage:* Some analytical systems require users to provide network connections and data storage capabilities at the installation site. If this is the case, connect the instrument to the network and check its functionality.
- *Assembly and Installation:* Assemble and install the instrument and perform any initial diagnostics and testing. Assembly and installation of a complex instrument are best done by the vendor or specialized engineers, whereas users can assemble and install simple ones. For complex instruments, vendor established installation tests and guides provide a valuable baseline reference for determining instrument acceptance. Any abnormal event observed during assembly and installation merits documenting. If the preowned or unqualified existing instrument requires assembly and installation, perform the tasks as specified here, and then perform the installation verification procedure described below.
- *Installation Verification:* Perform the initial diagnostics and testing of the instrument after installation. On obtaining acceptable results, the user and (when present) the installing engineer should confirm that the installation was successful before proceeding with the next qualification phase.

Operational Qualification– After a successful IQ, the instrument is ready for OQ testing. The OQ phase may consist of these test parameters.

- *Fixed Parameters:* These tests measure the instrument's nonchanging, fixed parameters such as length, height, weight, and so on. If the vendor-supplied specifications for these parameters satisfy the user, he or she may waive the test requirement. However, if the user wants to confirm the parameters, testing can be performed at the user's site. Fixed parameters do not change over the life of the instrument and therefore never need redetermining. *Note*: These tests could also be performed during the IQ phase and, if so, fixed parameters need not be redetermined as part of OQ testing.
- *Secure Data Storage, Backup, and Archive:* When required, secure data handling, such as storage, backup, and archiving should be tested at the user site according to written procedures.
- *Instrument Functions Tests:* Test important instrument functions to verify that the instrument operates as intended by the manufacturer and required by the user. The user should select important instrument parameters for testing according to

the instrument's intended use. Vendor-supplied information is useful in identifying specifications for these parameters. Tests should be designed to evaluate the identified parameters. Users, or their qualified designees, should perform these tests to verify that the instrument meets vendor and user specifications.

OQ tests can be modular or holistic. Modular testing of individual components of a system may facilitate interchange of such components without requalification and should be done whenever possible. Holistic tests, which involve the entire system, are acceptable in lieu of modular testing [11]. Having successfully completed OQ testing, the instrument is qualified for use in regulated samples testing/analysis. The extent of OQ testing that an instrument undergoes depends on its intended applications, therefore there is no specific offer regarding OQ tests for any instrument or application. Nevertheless, as a guide to the type of tests possible during OQ, consider these, which apply to a high-performance liquid chromatography unit: pump flow rate, gradient linearity, detector wavelength accuracy, detector linearity, column oven temperature, peak area precision, and peak retention time precision.

Routine analytical tests do not constitute OQ testing. OQ tests specifically designed to determine operation qualification should verify the instrument's operation according to specifications in the user's environment. OQ tests may not be required to be repeated at a regular interval. Rather, when the instrument undergoes major repairs or modifications, relevant OQ tests should be repeated to verify whether the instrument continues to operate satisfactorily.

Performance Qualification– After the IQ and OQ have been performed, the instrument's continued suitability for its intended use is proved through performance qualification. The PQ phase includes these following parameters:

(1) *Performance Checks:* Set up a test or series of tests to verify an acceptable performance of the instrument for its intended use. PQ tests are usually based on the instrument's typical on-site applications. Some tests may resemble those performed during OQ, but the specifications for their results can be set differently if required. PQ tests are performed routinely on a working instrument, not just on a new instrument at installation. Therefore, PQ specifications can be slightly less rigorous than OQ specifications. Nevertheless, user specifications for PQ tests should evince trouble-free instrument operation *vis-à-vis* the intended applications.

PQ tests should be performed independent of the routine analytical testing performed on the instrument. Like OQ testing, the tests can be modular or holistic. Since many modules within a system interact, holistic tests generally prove more effective by evaluating the entire system and not just the system's individual modules. Testing frequency depends on the ruggedness of the instrument and criticality of the tests performed. Testing may be unscheduled—for example, each time the instrument is used. Or it may be scheduled to occur at regular intervals; for example, weekly, monthly, and yearly. Experience with the instrument can influence this decision. Generally, the same PQ tests are repeated each time so that a history of the instrument's performance can be compiled and documented. Some system suitability tests or quality control checks that run concurrently with the test samples also imply that the instrument is performing suitably. However, though system suitability tests can supplement periodic PQ tests, they cannot replace PQ.

(2) *Preventive Maintenance and Repairs:* When PQ test(s) fail to meet specifications, the instrument requires maintenance or repair. For many instruments a periodic preventive maintenance may also be recommended. Relevant PQ test (s) should be repeated after the needed maintenance or repair to ensure that the instrument remains qualified.

(3) *Standard Operating Procedure for Operation, Calibration, and Maintenance:* Establish standard operating procedures to maintain and calibrate the instrument. Use a logbook, binder, or electronic record to document each maintenance and calibration activity.

A quick summary of instrument qualification is illustrated in Table 5.6.

5.2.3.4 Roles and Responsibilities

Users Users are ultimately responsible for the instrument operations and data quality. Users group includes analysts, their supervisors, and the organizational management. Users should be adequately trained in the instrument's use, and their training records should be maintained as required by the regulations.

Users should be responsible for qualifying their instruments. Their training and expertise in the use of instruments make them the best-qualified group to design the instrument test(s) and specification(s) necessary for successful AIQ. Consultants, validation specialists, and quality assurance personnel can advise and assist as needed, but the final responsibility for qualifying instruments lies with the users. The users must also maintain the instrument in a qualified state by routinely or regularly performing PQ.

Quality Assurance The quality assurance (QA) role in AIQ remains as it is in any other regulated study. QA personnel should understand the instrument qualification process, and they should learn the instrument's application by working with the users.

TABLE 5.6 Phase and Basic Requirements of Instrument Qualification

Design qualification	Defines the functional and operational *specifications*
Installation qualification	Establishes that the instrument is received as designed and specified, and that it is *properly installed*
Operational qualification	Demonstrates that the instrument *will function* according to the operational specifications
Performance qualification	Proves that an instrument will function in accordance with specifications appropriate to its *routine use*

Finally, they should review the AIQ process to determine whether it meets regulatory requirements and that the users attest to its scientific validity.

Manufacturer The manufacturer is responsible for DQ when designing the instrument. It is also responsible for validating relevant processes for manufacturing and assembly of the hardware and for validating software associated with the instrument as well as the stand-alone software used in analytical work. The manufacturer should test the assembled instrument prior to shipping to the user. The manufacturer should make available to the users a summary of its validation efforts and also the results of final instrument and software tests. It should provide the critical functional test scripts used to qualify the instrument and software at the user site. For instance, the manufacturer can provide a large database and scripts for functional testing of the network's bandwidth for laboratory information management system software. Finally, the manufacturer should notify all known users about hardware or software defects discovered after a product's release, offer user training and installation support, and invite user audits as necessary.

5.2.3.5 Software Validation Software used for analytical work can be classified into following categories: firmware, instrument control, data acquisition, and processing software, and stand-alone software.

Firmware The computerized analytical instruments contain integrated chips with low-level software (firmware). Such instruments will not function without properly operating firmware, and users usually cannot alter the firmware's design or function. Firmware is thus considered a component of the instrument itself. Indeed, qualification of the hardware is not possible without operating it via its firmware. So when the hardware; for example, analytical instrument, is qualified at the user's site, it essentially qualifies the integrated firmware. No separate on-site qualification of the firmware is needed. Any changes made to firmware versions should be tracked through change control of the instrument (see Section 5.2.3.7).

5.2.3.6 Instrument Control, Data Acquisition, and Processing Software

Software Software for instrument control, data acquisition, and processing for many of today's computerized instruments is loaded on a computer connected to the instrument. Operation of the instrument is then controlled via the software, leaving fewer operating controls on the instrument. Also, the software is needed for data–acquisition and postacquisition calculations. Thus, both hardware and software, their functions inextricably intertwined, are critical to providing analytical results.

The manufacturer should perform the DQ, validate this software, and provide users with a summary of validation. At the user site, holistic qualification, which involves the entire instrument and software system, is more efficient than modular validation of the software alone. Thus, the user qualifies the instrument control, data acquisition, and processing software by qualifying the instrument according to the AIQ process defined earlier.

Stand-Alone Software An authoritative guide for validating stand-alone software, such as LIMS, is available [12]. The validation process is administered by the software developer, who also specifies the development model appropriate for the software. It takes place in a series of activities planned and executed through various stages of the development cycle [12].

The software validation guidance document [15] indicates that user-site testing is an essential part of the software development cycle. Note, however, that user-site testing, though essential, is only part of the validation process for stand-alone software and does not constitute complete validation. Refer to the guide for activities needed to be performed at the user site for testing stand-alone software used in analytical work.

5.2.3.7 Change Control Changes to the instrument and software become inevitable as manufacturers or testing facility add new features and correct known defects. However, implementing all such changes may not always benefit users. Users should therefore adopt only the changes they deem useful or necessary. The Change Control process enables them to do this.

Change Control follows the DQ/IQ/OQ/PQ classification process. For DQ, evaluate the changed parameters, and determine whether the need for the change warrants implementing it. If implementation of the change is needed, first, one should install the changes to the system during IQ, and then evaluate which of the existing OQ and PQ tests need revision, deletion, or addition as a result of the installed change. Where the change calls for additions, deletions, or revisions to the OQ or PQ tests, one should follow the procedure outlined below.

- OQ: Revise OQ tests as necessitated by the change.
- Perform the revised OQ testing. If the OQ did not need revision, repeat only the relevant tests affected by the change. This procedure ensures the instrument's effective operation after the change is installed.
- PQ: Revise PQ tests as necessitated by the change.
- Perform the PQ testing after installation of the change if similar testing is not already performed during OQ. In the future, perform the revised PQ testing.

For changes to the firmware and the instrument control, data acquisition, and processing software, Change Control is performed through DQ/IQ/OQ/PQ of the affected instrument. Change Control for the stand-alone software requires user site testing of the changed functionality.

5.2.3.8 AIQ Documentation Two types of documents result from AIQ: Static and Dynamic.

Static Documents: Static documents are obtained during the DQ, IQ, and OQ phases and should be kept in a "Qualification" binder. Where multiple instruments of one kind exist, common documents should go into one binder or section, and documents specific to an instrument should go into that instrument's binder or section. During Change Control, additional documents can be placed with the static ones, but previous documents should not be removed. When necessary, such documents may be archived.

Dynamic Documents: Dynamic documents are generated during the OQ and PQ phase, when the instrument is maintained, or when it is tested for performance. Arranged in a binder or logbook, they provide a running record for the instruments and should be kept with them, available for review by any interested party. These documents may also be archived as necessary.

5.2.3.9 Instrument Categories Modern laboratories typically include a suite of tools. These vary from simple spatulas to complex automated instruments. Therefore, applying a single set of principles to qualify such dissimilar instruments would be scientifically inappropriate. The users are the most qualified to establish the level of qualification needed for an instrument [16]. Based on the level of qualification needed, it is convenient to categorize instruments into three groups: A, B, and C, as defined below. Each group is illustrated by some example instruments. The list of instruments provided below as illustration is neither meant to be exhaustive, nor can it provide the exact category for an instrument at a user site. The exact category of an instrument should be determined by the user for their specific instruments or applications.

Group A Instruments: Conformance of Group A instruments to user requirements is determined by visual observation. No independent qualification process is required. Example instruments in this group include light microscopes, magnetic stirrers, mortars and pestles, nitrogen evaporators, ovens, spatulas, and vortex mixers.

Group B Instruments: Conformance of Group B instruments to user requirements is performed according to the instruments' standard operating procedures. Their conformity assessments are generally unambiguous. Installation of Group B instruments is relatively simple and causes of their failure readily discernable by simple observations. Example instruments in this group include balances, incubators, infrared spectrometers, melting point apparatus, muffle furnaces, pH meters, pipettes, refractometers, refrigerator-freezers, thermocouples, thermometers, titrators, vacuum ovens, and viscometers.

Group C Instruments: Conformance of Group C instruments to user requirements is highly method specific, and the conformity bounds are determined by their application. Installing these instruments can be a complicated undertaking and may require the assistance of specialists. A full-qualification process, as outlined in this section, should apply to these instruments. Example instruments in this group might include the following: atomic absorption spectrometers, differential scanning calorimeters, densitometers, diode-array detectors, electron microscopes, elemental analyzers, flame absorption spectrometers, gas chromatographs, high-performance liquid chromatographs, inductively coupled argon plasma emission spectrometers, mass spectrometers, micro-plate readers, near infrared spectrometers, Raman spectrometers, thermal gravimetric analyzers, UV/Vis spectrometers, and X-ray fluorescence spectrometers.

5.2.3.10 ***Summary*** The purpose of the use of analytical instruments is to generate reliable data. Instrument qualification helps fulfill this purpose. No authoritative guide (direct or specific guidance) exists that considers the risk of instrument failure and combines that risk with users' scientific knowledge and ability to use the instrument to deliver reliable and consistent data. In the absence of such a guide, the qualification of analytical instruments has become a subjective and often fruitless document-generating exercise.

Taking its cue from the new FDA initiative, "Pharmaceutical GMPs for the 21st Century," an efficient, science- and risk-based process for AIQ was discussed at a workshop on analytical instrument qualification. Above description and discussion represent the distillate of deliberations on the complicated issues associated with the various stages of analytical instrument qualification. It emphasizes AIQ's place in the overall process of obtaining quality reliable data from analytical instruments and offers an efficient process for its performance, one that focuses on scientific value rather than on producing documents. Implementing such a process should remove ambiguous interpretations by various groups.

5.2.4 Analytical/Bioanalytical Method Qualification and Validation along with Related SOPs

Regulated bioanalysis, employed for the quantitative determination of drugs and their metabolites in biological matrices, plays significant roles in the evaluation and interpretation of data from bioequivalence, bioavailability, pharmacodynamic (PD), pharmacokinetic, and toxicokinetic studies. The quality of these studies, which are often used to support regulatory filings, is directly related to the quality of the underlying bioanalytical data. Reliability and quality of methods are critical to ensuring quality data delivery.

5.2.4.1 ***Analytical Method Validation*** Analytical methods validation is documented evidence that an analytical method does what it purports to do and delivers the required attributes. Use of a validated method should instill confidence that the method can generate test data with acceptable quality. Various user groups and regulatory agencies have defined procedures for method validation. Specific requirements regarding methods validations appear in many references on the subject [1–8]. Among some common parameters generally obtained during method validations are the following: accuracy, precision, sensitivity (LOD/LOQ), specificity, repeatability, linearity, recovery (extractability), analyte stability, and matrix effect.

System Suitability Tests: Typically conducted before the system performs samples analysis, system suitability tests verify that the system works according to the performance expectations and criteria set forth in the method, assuring that at the time of the test the system met an acceptable performance standard.

Quality Control Checks: Most analyses are performed using reference or calibration standards. Single- or multipoint calibration or standardization correlates instrument response with a known analyte quantity or quality. Calibrators/standards are generally prepared from certified materials suitable for the test. Besides calibration or standardization, some analyses also require the inclusion of quality control check samples, which provide an in-process assurance of the test's performance suitability.

The extent of system suitability tests or quality control checks varies for different types of analyses. For example, chemical analyses, which are largely subject to GMP regulations, may require more system suitability tests than bioanalytical work. The bioanalytical work, largely subject to GLP regulations, requires more quality control checks during sample analysis. In summary, AIQ and analytical method validation assure the quality of analysis before conducting the tests. System suitability tests and quality control checks assure the high quality of analytical results immediately before or during sample analysis.

5.2.4.2 Standard Operating Procedures Standard operating procedures are a must in both regulations (GLP 21 CFR 58.81, cGMP 21 CFR 211.100) and are the cornerstone of all required activities undertaken in the laboratory and manufacturing facility. SOPs are important for any organization to assure safety, efficiency, and consistency in the performance of a task.

As recommended in the FDA "Guidance for Industry Bioanalytical Method Validation" [3], there should be an approved procedure in place for every type of bioanalytical laboratory operation, to cover all aspects of analysis from the time the sample is collected and reaches the laboratory until the results of the analysis are reported. Examples of phases and requirements along with essential SOPs generally required in a bioanalytical laboratory are presented in 10Appendix C. SOPs should be written to be clear, easy to understand, and follow. All SOPs need to be reviewed annually to ensure the procedure does not become obsolete. As a regulatory requirement, training in new and revised procedures should be provided to affected personnel. Any revisions made to the current SOPs need to be managed by a proper change control procedure and system should be installed to ensure that only the most current versions of the SOPs are made available to all personnel. Personnel are not allowed to make any uncontrolled copies of the SOPs.

Readily available in every laboratory area should be copies of the SOPs relevant to the activities performed in that area. The following are examples of SOPs relating to laboratory activities:

(1) Test and reference items: Receipt and handling, labeling and traceability, identification, characterization, storage, measures to prevent cross contamination

(2) Equipment, materials, and reagents
(3) Laboratory operations
(4) Documentation: Control and handling of documentation, definition of raw data, data collection, preparation of the analytical report or final report, data storage, and retrieval

The conference report [16] makes specific reference to the requirement to have SOPs for run acceptance criteria, assay procedure, reintegration, and reassay. Of these, generation of procedures for objective and consistent reintegration of chromatographic peaks is, in particular, a contentious area. While automatic algorithms often work well, there are inevitable situations where baselines appear to have been set incorrectly by integration software and could readily be "corrected" by, for example, manual redrawing of a baseline. However, allowing operators to perform such actions may introduce unintentional or deliberate bias. An increasingly popular option is to never allow manual reintegration of spurious peaks; this can be approached by careful setting of appropriate integration parameters but may be problematic for methods with less than ideal chromatography. If manual integration is to be implemented, the process to be followed will need to be fully documented in the SOPs. It is essential that operators be fully trained and that any changes be peer reviewed in order to ensure consistency in an area with the potential for bias.

Laboratories should also anticipate the need to have SOPs covering the reanalysis of incurred samples to demonstrate assay reproducibility. An SOP for the investigation of anomalous results (sometimes also termed "out of specification" results) is also an emerging requirement. This SOP needs to address problems that are obvious according to predefined specifications (e.g., multiple batch failures) and to identify and investigate situations in which all specifications have been met but there are indications of problems that may affect data quality (e.g., contamination or strange pharmacokinetic profiles). Analysts need to be aware that such problems may exist even when defined specifications are met. The SOP should address the need to pinpoint the source of the problem if possible, assess the problem's impact on the study, and discuss procedures for eliminating or minimizing this impact. Typical examples of quality system procedures are presented in Table 5.7.

Reagents and Solutions The GLP regulations (21 CFR 58.83) clearly provide the requirements for the labeling of reagents and solutions with the appropriate identity, titer, concentration, storage requirement, and expiration date. While not implicit in cGMP regulations 21 CFR Part 211, it is a regulatory expectation that QC laboratory reagents and solutions are accurately labeled with the identity and concentration of reagent and solution and expiration date.

The bioanalytical laboratory should have and follow a written program for the preparation and use of reagents and solutions relative to standardization (if applicable), labeling, storage, documentation, and expiration. Controlled log books or worksheets can be used to document preparation of reagents and solutions and should be reviewed periodically during the conduct of the bioanalysis study by supervision

TABLE 5.7 Typical Examples of Quality System Procedures

Preparation, Approval, and Issuance of SOPs
1. Training, operation, calibration, and preventive maintenance of equipment
2. Bioanalytical methods development and validation preparation, acceptance criteria of calibration curves, QC samples, and batch runs system suitability requirements documentation/record-keeping of laboratory/study
3. Data reporting of study data review and approval of laboratory data change control, procedure monitoring of specimens storage chambers/freezers
4. Equipment validation/qualification issuance and control of logbooks handling of reagents and solutions receipt, verification and control of subject samples certification, handling, and storage of reference standards
5. Procedure to evaluate stability of drug in biological fluid and stock solutions
6. Determination/qualification of biological fluid for use as control blank
7. Handling of potential outliers, repeat testing events/deviation management and resolution computer and laboratory data system validation
8. Training of laboratory personnel

Note: Personnel are not allowed to make any uncontrolled copies of the SOPs.

for completeness and accuracy. Expiration dating of reagents and solutions should be adequately supported by stability studies.

Test and Control Article Characterization and Handling 21 CFR 58.105 and 21 CFR 58.107 of the GLP regulations delineate the requirements for the characterization and handling of the test and control articles used in a nonclinical laboratory study. The regulations state that the identity, purity, strength, and composition or other characteristics which will define the test and control articles should be determined for each batch and should be adequately documented. In the cGMP area, although not expressly stated, characterization and handling of reference standards is in its own way a quality system within the QC laboratory operations and a regulatory expectation. The QC laboratory responsible for the testing of drug products follows a written program for the qualification and use (including certification, storage, labeling, and expiration dating) of reference standards, which may include compendia standards, noncompendia standards, and in-house standards. It is good business and scientific practice for the bioanalytical laboratory engaged in the analysis of BA/BE study samples to implement a similar program for properly characterizing the standards for identity, strength, and purity. The characterization should be performed following approved procedures, documented, and reviewed by qualified personnel.

The control and tracking of the chain-of-custody records for subject test samples is one of the most important quality systems to be implemented in support of any FDA inspection. Detailed chain-of-custody records must be maintained, either via manual logbooks or a Laboratory Information Management System. These chain-of-custody records provide solid evidence to support the integrity of the subject and control samples and to support that they are handled in controlled environment/conditions defined in the prestudy validation report.

Records and Reports The reporting of nonclinical laboratory study results, storage and retrieval of records and data and records retention are well defined in GLP regulation 21 CFR 58.185 and 58.190 and cGMP regulation 21 CFR 21.195, respectively. At a minimum, the final report for a nonclinical study includes the following: (a) name and address of facility performing the study, study initiation date, and study completion date; (b) objectives and procedures as stated in protocol including changes to protocol; (c) statistical methods employed for analyzing the data; (d) test and control articles; (e) stability of test and control articles; (f) description of methods used; (g) description of test system used; (h) description of the dosage, dosage regimen, route of administration, and duration; (i) description of all circumstances that may have affected the quality or integrity of the data; (j) name of Study Director, names of other scientists and professionals and supervisory personnel involved in the study; (k) description of the transformations, calculations, or operations performed on the data, summary and analysis of the data, and conclusion from the analysis of data; (l) signed and dated reports of each individual involved in the study; (m) locations where all specimens, raw data, and final report are to be stored; QA statement and signed and dated final report by the Study Director. The Study Director makes corrections or additions to the final report in the form of an amendment, with the appropriate review, approval, and audit trail. All raw data, documentation, protocols, final reports, and specimens (except those obtained from mutagenicity tests and wet specimens of blood, urine, feces, and biological fluids) generated from the nonclinical laboratory study are to be retained and archived for expedient retrieval. The integrity of all raw data should be maintained in accordance with the requirements for the time period of their retention. Designated individuals should be responsible for the archives and access to the archives should be limited.

Since laboratory data from BA or BE studies play a major role in the approval of drug applications (NDA, ANDA) [17], it would be prudent for a firm or institution engaged in the conduct of bioanalytical testing to implement robust procedures for ensuring the quality, reliability, and integrity of all data it produces through adherence to good documentation practices. Some key elements of good documentation practices are summarized in Table 5.8. Similar to nonclinical laboratory studies generated under the GLP environment, all raw data, documentation, protocols, and final reports generated from the BA/BE laboratory study are to be retained and archived for expedient retrieval. The integrity of all raw data should be maintained in accordance with requirements for time period of the subject NDA or ANDA remains in effect at the minimum.

It should be noted that the FDA requires the retention of the lots of the generic drug product and the reference listed drug product to support preapproval inspection (PAI) with reconciliation of records for BE studies. Back-up subject samples and remaining test specimens are normally stored up to 6 months by contract bioanalytical laboratories after the completion of the study free of charge, and additional charges may be incurred for longer frozen storage at the request of the sponsor. It is recommended that the back-up samples be appropriately stored until NDA/ANDA approval at the minimum. It is further recommended that, for the extended storage of subject specimens, the corresponding long-term stability samples should be stored along with the

TABLE 5.8 Summary of Good Documentation Practices

1. Use controlled logbooks or prenumbered worksheets
2. Except for data generated by automated data collection systems, all other data should be directly documented in a logbook or worksheet at the time the activity is performed using indelible ink. Include units (grams, milligram, milliliter, liter, etc.) where applicable. The use of sticky notes, scrap paper, etc. does not constitute an official documentation record and therefore is not allowed
3. Supervisory review of logbooks or worksheets
4. Records should not be filled out in advance or predated, do not enter data at a later date based on memory, no backdating of entries. Enter "N/A" (not applicable) in defined spaces that are not used, including items, blocks or sections, except for space reserved for comments
5. No photocopies of original records or completed forms should be made, until after appropriate QA or supervisory review
6. Documentation using thermal paper should be photocopied after appropriate review and approval of data
7. Incorrect entries: (a) Cross-out entry with one line through the incorrect entry, enter the correct information above or adjacent to the incorrect entry; (b) sign and date the cross-out; and (c) do not discard, erase, delete, or render illegible an incorrect entry
8. Missing entries: (a) Reference the document from which the information is obtained, add information, initial and date the entry with the date on which the information was entered; (b) explain date discrepancy; and (c) obtain approval signature/date from the next level of management for missing entry
9. Missing explanation for correction to an entry—have the individual who originally corrected the incorrect entry provide the additional explanation. If the individual is not available, the supervisor who has approval authority for the document should make the explanation based on specific knowledge of the situation and then record the circumstances appropriately
10. Additional entries to completed documents—this should be done through the use of a memo or amendments to add information. This should be approved by QA
11. Rewriting documents—this practice should be restricted to exceptional situations (original document is illegible, or the incorrect form was used) and must be approved in advance in writing by QA
12. Voided documents—identify document as VOID. Provide the reason for invalidation of document. Sign/date voided document. Need QA written approval
13. Conflicting documents—when conflict occurs between two or more related documents, initiate a deviation investigation, and resolve the situation
14. Stamps used on documents—describe their usage for documents in the appropriate procedure. The user department should appropriately control stamps

subject specimens so that in the event repeat testing is required due to FDA questions or other concerns, long-term stability data can be collected to support the repeat testing. The typical contents of a bioanalytical report are summarized in Table 5.9.

There should be a designated group in the bioanalytical laboratory facility that is responsible for maintaining and archiving controlled documentation; for example, protocols, methods, BA/BE study reports, bioanalytical reports, etc. Typically, this responsibility is within the realm of the QAU. The controlled documents should be maintained in a safe, limited access area and controls must be implemented to protect

TABLE 5.9 Typical Contents of a Bioanalytical Report

1. Title of project and project number (if applicable)
2. Reference to BA/BE study title, protocol number, approval date of protocol (copy of protocol can be an attachment)
3. Reference to amendment(s) to protocol (copy of amendment can be an attachment)
4. Personnel (and titles) involved in the bioanalytical testing
5. Definitions and abbreviations
6. Method used with reference to method validation report (copy of method validation report can be an attachment)
7. Instrument used during the bioanalysis including freezers and refrigerators used for storage of calibration standards, QC standards and unknown samples including calibration dates of equipment
8. Documentation for preparation of solutions and standards (calibration standards and QC standards)
9. Certificate(s) of analysis for standards (reference standards and materials)
10. Documentation for receipt and verification of subject samples
11. Deviations (planned and unplanned) during the conduct of the testing
12. Raw data (chromatograms, spectra, printouts of gamma counter, etc.) for all runs (*Note*: All results should have an audit trail.)
13. Calibration curves and tabulated calibration curve data with acceptance criteria for all runs including rejected calibration standards
14. Tabulated QC data with acceptance criteria for all runs (including rejected QC standards)
15. Treatment of potential outliers
16. Handling of repeat testing
17. SOPs in effect at the time of the bioanalysis
18. Summary and conclusion
19. Signed QA statement
20. Signature page (including QA sign-off) indicating review and approval of bioanalytical report

the integrity of the documents. In addition to a robust procedure delineating the mandatory requirements for documentation, a procedure that describes the preparation and approval of the final bioanalytical report should be available and followed by the affected personnel.

5.2.4.3 Use of Consultants cGMP regulation 21 CFR 211.34 requires consultants to be qualified by the firm's QAU to ensure that they have the proper education, experiences, and qualifications to support the consulting assignments. An approved procedure that defines the selection, qualification, training and control of the use of outside consultants should be prepared to ensure that the work product from the consultants are acceptable to the FDA. A similar practice and policy should be established and implemented for GLP programs.

5.2.4.4 Study Director and Principal Investigator As discussed below, bioanalysis is typically conducted according to the principles of GLP but can only be claimed to be in full compliance with GLP regulations when it supports GLP toxicology studies. This has led to some confusion in the terminology used for the roles and responsibilities associated with bioanalytical projects. For (preclinical) GLP studies (e.g., bioanalytical support for toxicokinetic assessment), the role of the (bioanalytical) principal investigator is clearly defined in Organization for Economic Cooperation and Development (OECD) guidelines ("acts on behalf of the Study Director for the delegated phase and is responsible for ensuring compliance with the Principles of GLP for that phase"). Outside of this clear definition, the principles of GLP are applied and a senior bioanalytical scientist is appointed, for example, to be the equivalent of a "study director" for a validation study, or to be responsible for the bioanalytical component of a clinical study. However, most laboratories avoid using "principal investigator" in the context of clinical bioanalytical support, as this term is also used to describe the individual with overall responsibility for the clinical study. For bioanalytical validation studies, some labs use "study director" or "principal investigator" to denote the scientist with overall responsibility for the study, while others avoid these terms, as they could indicate that the study is a full GLP study (which it generally is not).

5.2.4.5 Protocols and Amendments For validation studies there is typically a separate plan or protocol issued prior to commencement of the validation experiments to describe in detail how the validation will be conducted. As for all such documents generated according to GLP principles, plans or protocols should be altered only by issuing an amendment. Validations of methods are typically regarded as separate studies, and consequently the validation plan would be expected to follow the principles of GLP in terms of not only describing experimental details but also addressing issues such as data analysis, reporting, and archiving. For bioanalytical support of preclinical and clinical studies, there is wide variability in whether a separate analysis plan to describe the analysis of samples from an individual study is generated. Anecdotal evidence suggests that this practice be more common for outsourced studies in Europe than in the United States. At this time, there does not appear to be any absolute regulatory requirement to generate such a plan, provided that key details about responsibility for any bioanalysis are provided in the main study plan or protocol.

5.3 REGULATORY COMPLIANCE WITH GLP WITHIN BIOANALYTICAL LABORATORIES

FDA bioanalytical guidelines [3] are applicable to bioanalytical method validation and sample analysis from bioequivalence, pharmacokinetic, and comparability studies in both human and nonhuman subjects, and they indicate that validation and analysis will be performed according to the principles of GLP.

GLP was developed as a consequence of inadequacies in preclinical studies, and as it is now defined, it applies to only nonclinical studies. As a consequence, it is self-evident that the application of a bioanalytical method to a toxicokinetic study should be performed in compliance with GLP. However, for validation of a bioanalytical method, the requirement to carry out these aspects in compliance with GLP is debatable. Indeed, in both UK and Japanese regulations, validation is considered a non-GLP activity, although there is a preference throughout the industry for validation to be performed in a GLP environment, following the principles of GLP.

The issue becomes more complex when one considers which studies must or can be performed in full compliance with GLP regulations. The FDA bioanalytical method validation guidance can be implemented in a GLP or non-GLP environment, but implementation of this guidance is not synonymous with GLP. Indeed, in the context of the FDA, bioanalysis is an integral part of non-GLP-based guidances, such as 21 CFR 320, Bioavailability and Bioequivalence Requirements (Drugs for Human Use) [17].

The FDA bioanalytical method validation guidance provides some further insight on this point: "The analytical laboratory conducting bioavailability and bioequivalence studies should 'closely adhere' to FDA's GLPs and to sound principles of quality assurance throughout the testing process." The legal claim "compliance with GLP" is replaced with "closely adhere to FDA's GLPs." Many laboratories have developed similar statements that do not claim compliance with GLP but state that the claim processes are closely related to it, for instance, "This study was carried out in laboratories that are GLP certified" or "This study was carried out in accordance with the principles of GLP." These terminologies are reflected in the Committee for Proprietary Medicinal Products (CPMPs) *Note for Guidance on Investigation of Bioavailability and Bioequivalence*, [18] which states that the bioanalytical part of bioequivalence trials should be conducted according to principles of GLP. While this is not the same as requiring such studies to be done in full compliance with GLP regulations, there is an expectation by the inspecting agencies that there will be close adherence to GLP, although a specific claim to compliance with GLP in the case of clinical studies cannot be made.

Regardless of guidances and GLP, the quality of any analytical data is a function of the need to ensure sample integrity and stability from the time it leaves the "subject" to the time it is analyzed using a validated and fully documented analytical procedure that is suited to the study. Documentation must be available to reconstruct, if necessary, all processes and procedures used to generate the final analytical result, from sample collection through laboratory analysis and generation of the final authorized study report. Although there are no specific regulations that address the conduct of bioanalytical testing for clinical studies, bioanalytical laboratories can draw from both the GLP and cGMP regulations in building their own quality systems, in ensuring the reliability, integrity, and accuracy of the data generated in support of the approval of the regulatory application. Compliance to applicable GLPs and/or cGMPs requirements in a bioanalytical laboratory is a wise business strategy and is a good scientific practice, as failure to do so would inevitably lead to additional costs, delays and, worst-case scenario, a nonapproval of a regulatory application.

5.4 JOINT-EFFORT FROM INDUSTRIES AND REGULATORY AGENCIES

Up to date, AAPS and FDA have sponsored and organized several workshops and seminars on specific topics to implement current guidelines and continue to improve/refine benchmark practices, and standards. The scope of these proceedings aim (but not limited to) at the following:

- Review the scope and applicability of bioanalytical principles and procedures for the quantitative analysis of samples from bioequivalence, pharmacokinetic, and comparability studies in both human and nonhuman subjects
- Review current practices for scientific excellence and regulatory compliance, suggesting clarifications, and improvements where needed
- Review and evaluate validation and implementation requirements for chromatographic and ligand-based quantitative bioanalytical assays, covering all types (sizes) of molecules
- Review recent advances in technology, automation, regulatory, and scientific requirements and data archiving on the performance and reporting of quantitative bioanalytical work and
- Discuss current best approaches for the conduct of quantitative bioanalytical work regardless of the size of the molecule analyzed

The third AAPS/FDA Bioanalytical Workshop, held May 1–3, 2006, in Arlington, VA, concluded with several recommendations to achieve the above goals and objectives. While the FDA guidance remains valid, the recommendations obtained during the workshop were aimed at providing clarification and some recommendations to enhance the quality of bioanalytical work. This publication (conference report) provides the clarification and recommendations obtained at the workshop with a view to achieve uniformity among the practitioners and users of quantitative bioanalysis for all types of molecules. Nonchromatographic Assay—Special Issues Differences Between Ligand-Binding Assays Supporting Macromolecule PK Analysis and Small Molecule Analysis by Chromatography-Based Assays. Table 5.10 proposes a benchmark process and acceptance criteria within bioanalytical laboratories.

The biopharmaceutical, pharmaceutical, biotechnology, and other life science industries have evolved rapidly in the last decade with remarkable developments in genomic, proteomics, and bioinformatics. This has created unique challenges and opportunities in the drug development arena. Biopharmaceuticals are predominantly potent protein and peptide-based therapeutic entities. Consequently, their blood concentration levels can be very low and difficult to quantify. Ligand-binding assays (LBAs) are highly sensitive assays with low lower limits of quantification and are typically the method of choice. These assays are increasingly used to provide PK–PD support, immunogenicity evaluation, vaccine potency determination, etc. The increased number of biological agents used as therapeutics (in the form of recombinant

TABLE 5.10 Benchmark Process and Acceptance Criteria within Bioanalytical Laboratories

Process or Criteria	Chromatographic Assays	Ligand-Binding Assays
Preparation of standards and QCs	Standards and QC samples may be prepared from the same spiking stock solution, provide the solution stability and accuracy have been verified. A single source of matrix may also be used, provided selectivity has been verified.	
Placement of samples	Standard curve samples, blanks, QCs, and study samples can be arranged as considered appropriate within the batch run, and support detection of assay drift over the run.	
Number of calibration standards in a batch run	Include with each analytical batch Blank matrix (sample without internal std.) Zero standard (matrix sample with internal std.) Nonzero calibration of six standards: a minimum of six standard points	Include with each analytical batch or plate Blank matrix Nonzero calibration standards: a minimum of six standard points. Can include anchor points (below LLOQ and above ULOQ)
Acceptance criteria for Calibration curves	Residual (absolute difference between the back calculated and nominal value) for each calibration standard meet the following limit: LLOQ standard <20% All other standard <15%	Residuals for each calibration standard should meet the following limits: LLOQ and ULOQ standards < 25% All other standards < 20% Any anchor points if used, are not to be included in the above acceptance criteria
	A minimum of 75% standards (at least six nonzero points) should be within the above limits for the analytical run to qualify. Values falling outside these limits can be discarded, provided for they do not change the established model.	

(continued)

TABLE 5.10 (*Continued*)

Process or Criteria	Chromatographic Assays	Ligand-Binding Assays
Number of QC samples in a batch	QC samples at the following three concentrations (within the calibration range) in duplicate with each analytical batch	
	Low: near the LLOQ (up to 3xLLOQ)	Low: above the second nonanchor standard, $\sim 3 \times$ LLOQ
	Medium: midrange of calibration curve	Medium: midrange of calibration curve
	High near the high end of range	High: below the second nonanchor standard at ~75% of ULOQ
	Each analytical batch should contain six or a minimum of 5% of total number of unknown samples. Add QCs in multiples of three concentrations (low, mid and high) when needed.	
Acceptance criteria for QCs	Allowed % deviation from nominal values	Allowed % deviation from nominal values
	QCs prepared at all concentrations greater than LLOQ <15%	QCs prepared at all concentrations greater than LLOQ and ULOQ <20%
	Low QC (if prepared at LLOQ) <20%	Low and high QC (if prepared at LLOQ or UOQ) > 25%
		In certain situations wider acceptance
		Criteria may be justified (e.g., when total error during assay validation approaches 30%
	At least 67% (four of six) of the QC samples should be within the above limits; 33% of the QC samples (not all replicates at the same level of concentration) can be outside the limits. If there are more than two QC samples at a concentration, then 50% of QC samples at each concentration should pass the above limits of deviation.	
Replicate analysis	In general, samples can be analyzed with a Single determination without replicate analysis if the assay method has acceptable variability as defined by the validation data.	Accuracy can generally be improved by replicate analysis. Therefore, duplicate analysis is recommended. If replicate analysis is performed, the same procedure should be used for samples and standards

	Duplicate or replicate analysis can be performed For a difficult procedure where high precision And accuracy may be difficult to achieve
Multiple analysis in a run	Samples involving multiple analytes in a run should not be rejected based on the data from one analyte failing the acceptance criteria.
Rejected runs	The data from rejected runs need not be documented, but the fact that a run was rejected and the reason for failure should be investigated and reported.

The issues of monitoring the effect of sample dilution are referred to in the FDA guidance. The guidance indicates that if the dilutions are conducted with like matrix (human plasma for human plasma), no within-study dilution matrix QC samples are necessary. However, the extent to which samples are allowed to be diluted should be tested during validation. If tested during validation, there is no need to run dilution QCs up to the tested dilution factor during sample analysis. If during sample analysis it is determined that the required dilution factor is greater than the extent tested during validation, an additional dilution factor should be evaluated during sample analysis. On the other hand, if the dilution is allowed and performed with an unlike matrix, QC samples should be diluted in the same manner as the study samples, and should be analyzed with the diluted samples. All diluted QCs should be created within the assay calibration range, and similar acceptance criteria as defined here should be applied, unless alternate specific criteria can be justified.

Multiple analyte assessment in a single analytical run was a topic of discussion. Although the FDA Guidance indicates that a run should not be rejected for the remaining analytes if one fails, it does not address how to assess and report all analyte concentration upon reanalysis of the failed analyte(s). In this regards, concentration from the first accepted run should be reported and if this analyte us repeated in a simultaneous assays when analyzing for different analytes, it is not necessary to quantify the already reported analytes. However, the source data from all acceptable runs, regardless of whether the concentrations from these runs were reported or not, should be retained.

proteins, monoclonal antibodies, vaccines, etc.) has prompted the pharmaceutical industry to review and refine aspects of the development and validation of bioanalytical methods for the quantification of these therapeutics in biological matrices in support of preclinical and clinical studies. Most of these methodologies are used in quantitative assays supporting pharmacokinetic and toxicokinetic parameters of the therapeutic agents. The methods that are primarily used in these evaluations are ligand-binding assays (LBAs or immunoassays as known as ELISA), where the specificity and selectivity of the assays depend on the interactions of other biological molecules, such as receptors, antibodies against the therapeutic candidates, and aptamers. The response observed in these methods is indirectly related to the concentration of the therapeutic, that is, the basis of the detection is an enzymatic or radiochemical response tied to a variety of binding interactions. There is no direct physicochemical property of a macromolecule that can be used in this determination (unlike for a small-molecule drug candidate). Because of the nature of these binding interactions, the dynamic range of the standard curves is narrow as well as nonlinear/sigmoidal in most cases.

Ligand-binding assays are used throughout many organizations attempting to discover or develop new chemical entities (NCEs). Besides the obvious size difference between small and macromolecule analytes, there are key structural differences. Small molecules typically are organic molecules whereas macromolecules are complex biopolymers. In addition, small molecules are prepared by organic synthesis while macromolecules are typically formed biologically. As a direct result of how macromolecules are produced, the reference standards tend to be heterogeneous, often because of posttranslational modification (e.g., glycosylation or phosphorylation). In contrast, small molecule reference standards are homogeneous with a high degree of purity. Generally, small molecules are often hydrophobic and macromolecules are often hydrophilic. While chemical stability is assessed for small molecules with relative ease, macromolecule stability assessment is generally more complex, requiring the evaluation of not only chemical and physical properties but also biological integrity (e.g., is receptor binding affinity maintained?). Macromolecules are endogenous and/or structurally similar to endogenous counterparts, while small molecules are generally xenobiotics, foreign, and not present in the sample matrix. The catabolism of small molecules is typically well defined, whereas for macromolecules few specifics are known and understood. Macromolecules typically have specific carrier proteins while small molecules can be generically bound to several endogenous proteins. Macromolecule therapeutic agents are produced in cell culture; hence, they are not characterized as rigorously as small-molecule drug candidates. There is a greater potential for lot-to-lot variability in purity and potency in these preparations. In many instances, a true "reference standard" may not be available; rather, a well-characterized material may be the only choice. It is critical to develop and validate the methods for macromolecules with the appropriate reference material used in the relevant study (e.g., the lot of material used in the validation may not be the same as the administered material in the clinical study). The reference material used in the clinical study may have different posttranslational modifications, which could result in the loss of binding activity/epitope for the capture or detection of molecule, making the method unsuitable for the intended purpose. At the least, an assessment of the appropriateness of the new lot should be conducted.

Because of these significant differences between small and macromolecule analytes, different technologies, such as LC–MS/MS for small molecules and LBAs for macromolecules, are often employed to determine drug levels for PK assessments. Method validations for these divergent methods should consider important differences including the basis of measurement, the detection modality, and whether a sample is measured directly in the matrix or extracted before analysis. The basis of measurement of LC–MS/MS is owed to the chemical properties of the analyte, while for LBAs, the measurement depends on a high-affinity biological binding interaction between the macromolecule analyte and another macromolecule(s) in the form of one or more capture/detection antibodies. Detection in LC–MS/MS methods is direct and typically results in a linear measured response, where higher concentrations of analyte have a proportional increase in response. In contrast, the measured response in LBAs is indirect and this results in a nonlinear, often sigmoidal, measured response. Owing to the characteristics of the assay system, the calibration standard curve range for an LC–MS/MS method is broad, often covering a few and even several orders of magnitude. In contrast, the calibration range for an LBA is typically limited to less than two orders of magnitude. These analyte differences, combined with the unique technologies used to measure analyte concentration, provide a strong rationale as to why consideration should be given to the need of employing some analyte-specific (small versus macromolecule) method validation guidelines. A benchmark process and acceptance criteria within bioanalytical laboratories are given in Table 5.10.

The considerations that pertain to matrix selection are one of the key differences between the assays developed for small-molecule analysis and the LBAs developed for the quantification of macromolecules. Small-molecule assays often include a preassay extraction, which is often helpful to alleviate problems from individual matrix variability. In addition, the use of either analog or stable isotope–labeled internal standards in liquid chromatography/mass spectrometry assays for small molecules normalizes the influence of matrix effects and system fluctuations. The inherent characteristics of macromolecular therapeutics make it difficult and often impossible to extract samples before analysis. LBAs used to quantify macromolecules, therefore, are often developed to measure analyte in complex matrices without extraction. A lot of macromolecular therapeutics are recombinant or modified variants of endogenous proteins. It is highly unlikely that most of the LBA reagents used will be able to distinguish between the therapeutic and the endogenous counterpart, which could affect the accuracy of measurement of the assay. In these cases, special considerations must be made for matrix selection and for analysis of data.

The matrices collected for bioanalysis include plasma, serum, urine, cerebrospinal fluid, synovial fluid, and homogenized tissue. The characteristics of the macromolecule can be affected by the methods used for sample preparation, the need for additives (anticoagulants, protease inhibitors, etc.) at the time of collection, the stability of the macromolecule during collection procedures (whole blood before separation of plasma or serum), and the postcollection processing and storage conditions (temperature, vial type, shipping, freeze–thaw cycles, etc.), so these characteristics must be evaluated during the method development phase. Assay format, sample collection conditions, and other factors may influence the choice of matrix in the assay (e.g., plasma is the preferred matrix for labile analytes because of the extended time needed for the preparation of serum and because of the presence of proteolytic enzymes). Spiked samples (ideally at the low and high concentrations) should be prepared in the same matrix as the anticipated matrix of the unknown study samples to evaluate the accuracy (recovery) of the method. In the absence of an endogenous component, simple spiked recovery studies using the nominal concentrations will be sufficient to qualify a matrix. The use of a stripped matrix (e.g., charcoal, immuno-affinity) or an alternative matrix (e.g., protein buffers, dialyzed serum) is not recommended but is necessary when no other strategy for quantification can be designed for measuring endogenous analytes. Regardless of the source of the matrix interference, validation samples (e.g., quality control QC samples used during the prestudy validation phase) must be prepared using the same type of neat, unaltered matrix as was used for the study samples for the determination of the assay's precision and accuracy. If there needs to be a matrix lot change during the course of study sample analysis, appropriate QC samples must be prepared to evaluate the comparability of the data obtained during the prestudy validation.

One major point of concern in discussing the method validations for these divergent technologies centers on standards and quality control acceptance criteria (e.g., the acceptable deviation from a nominal value expressed as a percentage). Current guidance recommends the 15/20 rule, where the first number, in this case 15%, is the acceptance criterion for all standards and QC samples with the exception of the lower limit of quantitation (LLOQ), where the acceptance criterion is increased to a 20% deviation. This rule was developed before routine use of LC–MS, where chromatographic methods were employed, but internal standards were analyte analogs and not stable isotopes. When the 15/20 rule was proposed, most PK assessments that used LBAs (e.g., radioimmunoassay) measured small molecules. The typical radioimmunoassay (RIA) used high-affinity polyclonal antibodies that were quite suitable to measure well-characterized homogeneous organic small molecules. In most of these small molecule RIAs, meeting the 15/20 challenge was achievable and it is recommended that the 15/20 rule be continued when LBAs are used for small molecule analysis. However, nearly all small molecule analysis performed today is by LC–MS or LC–MS/MS, often with the incorporation of a stable isotopic internal standard; as a result, assay precision has continued to be improved. In fact, the results of a method validation survey conducted for the 3rd AAPS/FDA Bioanalytical Workshop found that 89% of chromatography respondents used the 15/20 target. As a result of small molecule analysis moving to the LC–MS platform, LBAs are now almost exclusively used to measure macromolecules. While some LBAs continue to be developed and validated to meet the 15/20 rule, different criteria are sometimes necessary because of the heterogeneous nature of macromolecules, and the fact that other macromolecules (antibodies) are employed in the assay. In fact, the 3rd AAPS/FDA Bioanalytical Workshop survey found that only 23% of the LBA respondents follow the 15/20 rule. Instead, 53% of respondents used somewhere between 20/25 (42%) and 30/30 (2%) as their acceptance rule, while 23% used "other criteria." These "other criteria" could possibly include statistically based approaches that estimate in-study assay performance based on prestudy validation results.

During prestudy validation, method precision and accuracy are determined through the analysis of QCs (validation samples) prepared in a biological matrix equivalent to that anticipated for study samples. Because of the endogenous nature of some biopharmaceuticals, it may be necessary to deplete the matrix of the analyte or employ a "surrogate" matrix to evaluate method accuracy and precision. One proposed validation protocol four recommends that matrix be spiked at five or more validation sample concentrations that span the range of quantification (e.g., the anticipated LLOQ, ~three times LLOQ, mid geometric mean., high ~75% the upper limit of quantitation, or ULOQ, and finally the anticipated ULOQ).

As previously noted, the major sources of variability (imprecision and inaccuracy) differ based on technology. For LBAs, the interbatch variance component is usually a greater contributor to the overall variability than the intrabatch variance component. It is recommended that at least two independent determinations be made for each validation sample per assay run across a minimum of six independent assays runs (balance validation design). For example, 12 reportable values would result from 2 measurements across 6 independent assay runs. An appropriate statistical method should then be used to compute the summary statistics (e.g., each validation sample, the repeated measurements from all runs should be analyzed together). A detailed description of this approach has been described previously. For a method to be considered acceptable, it is recommended that both the interbatch imprecision (%CV) and the accuracy be expressed as absolute mean bias (%RE) being ±20% (25% at LLOQ and ULOQ). As an additional constraint to control method error, it is recommended that the target total error (sum of the absolute value of the %RE accuracy and precision %CV be less than ±30% and ±40% at the LLOQ and ULOQ, respectively). The additional constraint of total error allows for consistency between the criteria for prestudy method validation and in-study batch acceptance. In assessing the acceptability of a method, including total error, it is not appropriate to reject assay runs. All assay runs during the validation should be included in the computation of summary statistics. The only exception would be runs rejected for cause or in cases where errors are obvious and documented.

5.4.1 Ligand-Binding Assays In-Study Acceptance Criteria

The recommended standard curve acceptance criteria for macromolecule LBAs are that at least 75% of the standard points should be within 20% of the nominal concentration (%RE of the back-calculated values), except at the LLOQ and ULOQ where the value should be within 25%. This requirement does not apply to "anchor calibrators," which are typically outside the anticipated validation range of the assay and used to facilitate and improve "sigmoidal" curve fitting. LBAs measure the signal of a series of interactions that follow the law of mass action, resulting in a nonlinear and often sigmoidal standard curve. The response–error relationship is not constant (heteroscedastic); therefore, the highest precision does not necessarily coincide with the highest sensitivity. In general, it is highly recommended that results from multiple runs be used to estimate the response–error relationship. Because of the heteroscedastic nature of the response variance, a weighted, nonlinear, least-squares method is generally recommended for fitting concentration response data from LBAs. Four- and five-parameter logistic calibration models are often used to fit the LBA standard curves. Standard points outside of the range of quantification (anchor calibrators) are often used to assist in fitting these nonlinear regression models. In summary, it is recommended that at least three runs be used to establish the calibration model, with at least eight nonanchor standard points run in duplicate. The acceptance of the model must be verified by evaluating the relative bias between the back-calculated and nominal concentrations of the calibration standards. The use of the correlation coefficient is not recommended for confirmation of the regression model.

The recommended QC acceptance criteria for macromolecule LBAs includes the use of low, medium, and high (LQC, MQC, and HQC) QCs typically run in duplicate (e.g., 6 results = 3 concentrations × 2 reportable values per concentration), with assays being accepted based on a 4–6-20 rule. Exceptions to this criterion should be justified (e.g., prestudy total error data approaching 30%). At least four of the six QCs must be within 20% of the nominal value. In addition, at least one QC sample per concentration needs to meet this criterion. If additional sets of QCs are used in a run, then 50% of them need to be "in-range" at each concentration. The following are recommendations for the placement of the controls in relation to the standard curve range. The LQC should be placed above the second nonanchor standard, ~three times the LLOQ. The MQC is placed near the mid point (geometric not arithmetic mean) of the standard curve, while the HQC should be placed below the second nonanchor point high standard and/or ~75% of the ULOQ.

5.4.1.1 Calibration Curve and QC Ranges QC samples serve to monitor the performance of the methodology throughout the course of the analysis. They are the basis for demonstrating, as required in 21 CFR 320.29(a), that the analytical method is sufficiently accurate, precise, and sensitive to measure the actual concentrations achieved in the body. For studies involving pharmacokinetic profiles spanning all or most of the calibration curve, QC samples run in duplicate (or at least 5% of the unknown samples), spaced across the standard curve as per the FDA Guidance, are likely sufficient to adequately monitor method performance. For an analysis where the study data fall over a small percentage of the calibration curve range, it is possible that none of the QC concentrations is near the concentrations of the unknowns, thus limiting the monitoring power of the QC samples. If a narrow range of analysis values is known or anticipated before the start of sample analysis, it is recommended that either the standard curve be narrowed and new QC concentrations used as appropriate, or if the original curve is used, existing QC concentrations be revised or sufficient QC samples at additional concentration(s) added to adequately reflect the concentrations of the study samples. Narrowing of the standard curve and preparation of new QC samples requires only a partial validation to ensure adequate performance of the new curve and QCs. A full validation is not required. If a narrow range of analysis values is unanticipated, but observed after start of the sample analysis, it is recommended that the analysis be stopped and either the standard curve narrowed, existing QC concentrations revised, or QC samples at additional concentrations be added to the original curve before continuing with sample analysis. It is not necessary to reanalyze samples analyzed before optimizing the standard curve or QC concentrations.

The stability of the macromolecular therapeutic in the anticipated matrix and in conditions the sample will be subjected to should be demonstrated. In situations where an altered matrix is used for standard curve and QC preparation, stability samples must be prepared in the unaltered matrix. The experiments must mimic the conditions under which the study samples will be collected, stored, and processed. Stability types that need to be assessed include the stability of the analyte in blood when processed into plasma or serum; storage stability such as benchtop, short-term, and long-term storage at $-20°C$ and $-70°C$; and freeze–thaw stability. It is important to understand the physicochemical properties of the macromolecule during the stability evaluation. For instance, does a protease inhibitor cocktail need to be added during collection? Is the molecule more hydrophobic than others, which may warrant the use of highly proteinaceous buffers for storage? Formal stability experiments must be conducted with an established method during prestudy validation, but long-term stability experiments may extend into the in-study validation phase. It is important to note that a freshly prepared standard calibrator curve and QC samples or those that are within the acceptable expiration should be used as the reference for comparison of the stability samples. The acceptance criteria applicable to the QC samples may be used for stability evaluations, or other statistically appropriate methods may be used.

5.4.1.2 Carryover and Contamination Evaluation Contamination, carryover, or blank response from matrix or reagents can affect the accuracy and precision of quantitation at all concentrations. However, low concentration samples are mostly affected as a percentage of concentration. Care should be taken to minimize interference from all contamination factors and the interference should not significantly affect the accuracy and precision of the assay. Carryover does not necessarily involve only the next sample in the sequence. In fact, carryover from late-eluting residues on columns may affect chromatograms several samples later. Carryover from residues in rotary sampling/switching valves often appears later in the samples. Precautions should be taken to avoid contamination during sample collection and preparation. Carryover should be assessed during validation by injecting one or more blank samples after a high concentration sample or standard. The injector should be flushed with appropriate solvents to minimize carryover. If carryover is unavoidable for a highly retained compound, specific procedures should be provided in the method to handle known carryover. This could include injection of blanks after certain samples. Randomization of samples should be avoided, since it may interfere with the assessment of carryover problems. Contamination can be assessed by monitoring blank response in the presence of high concentration samples or standards. The assay platform (manual or automated), configuration of sampling and extraction method (e.g., manual, automated, on-line, or solid phase) in the assay should be taken into consideration when ascertaining contamination. There is no standard acceptable magnitude of carryover for a passing bioanalytical run. Carryover should be addressed in validation and minimized, and an objective determination should be made in the evaluation of analytical runs. During validation, the operator should assess the analyte response due to blank matrix while eliminating or minimizing other contaminations. The analyte response at the LLOQ should be at least five times the response due to blank matrix. For immunoassays, and if the analyte is present endogenously in the matrix, the blank response can exceed 20% of LLOQ, but the contribution should not interfere with the required accuracy in the measurement of the LLOQ. In such cases, specific procedures should be provided in the method to handle blank matrix response.

5.4.2 Determination of Metabolites during Drug Development

A draft FDA Guidance for Industry, entitled "Safety Testing of Drug Metabolites" was issued in June 2005 by the Center for Drug Evaluation and Research (CDER) [19]. There is general support from the pharmaceutical community for the idea that a more extensive characterization of the pharmacokinetics of unique and/or major human metabolites (UMMs) would provide greater insight into the connection between metabolites and toxicological observations. This information would be best generated by the use of rugged, bioanalytical methods applied at appropriate times in drug development. Characterization of UMMs should proceed using a flexible, "tiered" approach to bioanalytical methods validation. This tiered approach would allow metabolite screening studies to be performed in early drug development using bioanalytical methods with limited validation, with validation criteria increasing as a product moves into clinical trials. A tiered validation approach to metabolite determination would defer bioanalytical resource allocation to later in the drug-development timeline when there is a greater likelihood of drug success. As a minimum, the specifics of this tiered validation process should be driven by scientifically appropriate criteria, established a priority.

5.4.3 Incurred Sample Analysis

There are several situations where the performance of standards and QCs may not adequately mimic that of study samples from dosed subjects (incurred samples). Examples include metabolites converting to the parent species, protein-binding differences in patient samples, recovery issues, sample inhomogeneity, and mass spectrometric ionization matrix effects. These factors can affect both the reproducibility and accuracy of the concentration determined in incurred samples. While these effects are often characterized and minimized during method development using QC samples, it is important to ensure that they are under control when the method is applied to the analysis of incurred samples. A proper evaluation of incurred sample reproducibility and accuracy needs to be performed on each species used for Good Laboratory Practice (GLP) toxicology experiments. It is not necessary for additional incurred sample investigations to be performed in toxicology species once the initial assessment has been performed. Incurred sample evaluations performed using samples from one study would be sufficient for all other studies using that same species.

It is generally accepted that the chance of incurred sample variability is greater in humans than in animals, so the following discussion pertains primarily to clinical studies. The final decision as to the extent and nature of the incurred sample testing is left to the analytical investigator, and should be based on an in-depth understanding of the method, the behavior of the drug, metabolites, and any concomitant medications in the matrices of interest. There should be some assessment of both reproducibility and accuracy of the reported concentration. Sufficient data should be generated to demonstrate that the current matrix produces results similar to those previously validated. It is recognized that accuracy of the result generated from incurred samples can be more difficult to assess. It requires evaluation of any additional factors besides reproducibility upon storage, which could perturb the reported concentration. These could include metabolites converted to parent during sample preparation or LC–MS/MS analysis, matrix effects from high concentrations of metabolites, or variable recovery between analyte and internal standard. If a lack of accuracy is not a result of assay performance (e.g., analyte instability or interconversion) then the reason for the lack of accuracy should be investigated and its impact on the study assessed. The extent and nature of these experiments are dependent on the specific sample being addressed and should provide sufficient confidence that the concentration being reported is accurate. The results of incurred sample reanalysis studies may be documented in the final bioanalytical or clinical report for the study, and/or as an addendum to the method validation report.

In selecting samples to be reassayed, it is encouraged that issues such as concentration, patient population, and special populations (e.g., renally impaired) be considered, depending on what is known about the drug, its metabolism, and its clearance. First-in-human, proof-of-concept in patients, special population, and bioequivalence studies are examples of studies that should be considered for incurred-sample concentration verification. The study sample results obtained for establishing incurred sample reproducibility may be used for comparison purposes, and do not necessarily have to be used in calculating reported sample concentrations.

5.4.4 Documentation Issues

Although the current guidance for the documentation section remains valid, further issues are now addressed and details are provided by Viswanathan et al. [5] to facilitate effective and proper documentation. Records generated during the course of method validation and study sample analysis are source records and should be retained to demonstrate the validity of the method under the conditions of use, and to support the statements made in the report. This is necessary to enable the reconstruction of the laboratory events as they occurred, since the information generated by the individual laboratory might differ from what the sponsor/developer includes in the application.

5.4.4.1 Documentation at the Analytical Site

(1) Documentation of standard analyte can be done by Certificate of Analysis (CofA) or recertification of purity or stability data at the time of the use. In case of the internal standard, no specific CofA is necessary but the lack of interference between internal standard (IS) and analyte should be established.
(2) The source data (run preparation, extraction and run summary sheets, and chromatograms) of all analytical and validation runs, including failed runs, should be retained.
(3) Reintegrated chromatograms should be explicitly identified. The reason for reintegration and the mode of reintegration should be documented. The original and reintegrated chromatograms should be retained ideally as electronic records.
(4) Any problems during extraction and analysis (e.g., run interruption, clogging of columns) should be identified. The appropriate remedial action should be documented.
(5) In the case of multianalyte assays (simultaneous measurement of multiple analytes in each sample), when samples are reassayed only for one analyte (e.g., because the analyte failed to meet acceptance criteria in the original assay), the raw data collected for the other analytes should also be retained.

5.4.5 Analytical/Validation Reports

Analytical/validation reports should include the following:

(1) Summary table of all analytical runs analyzed. The tables should list the runs with run IDs, dates of analysis, whether runs passed or failed, reason for the failure, and any deviations from the validated method.
(2) Summary table of all validation runs analyzed. The tables should list the runs with run IDs, dates of analysis, whether runs passed or failed, and the reason for the failure. QC data from validation runs that only failed to meet QC acceptance criteria with no assignable cause for failure should be included in the precision and accuracy estimation.
(3) Deviations from SOPs and assay procedures, and significant unexpected events should be identified and their impact assessed.

5.4.6 Source Data Documentation

The actual conditions of use should be stated in documentation. For example, the source documentation for stability determinations during method validation should explicitly record experimental conditions such as storage temperature and duration, use of freshly prepared standard curves, and so forth. Such documentation is necessary to confirm that validation experiments support the storage conditions that existed during sample analysis. Modification of calibration response (deletion of individual standard points that exceed predefined acceptance limits or alteration of the standard curve range) and QC levels (adding QCs or shifting in the concentration range of the study samples) should be documented with sufficient detail to demonstrate that the changes were justified and/or followed established procedures. Regarding chromatographic methods, source documentation should include original and reintegrated chromatograms for accepted runs, along with the reason for changing integration parameters across a run or for individual samples within a run. Disabling electronic audit trails that record changes to integration parameters is not acceptable.

5.4.7 Final Report Documentation

A complete account of the performance of the bioanalytical method should be provided in the final report for both method validation experiments and study sample analysis. Although drug concentration data from the rejected runs need not be included in the final report, a brief description of the reasons and a tabular listing of rejected runs should be provided. The information provided would be helpful in the evaluation of the overall assay performance and acceptance of runs rejected and accepted. The final report should include a tabular listing of the actual QC results from all runs during method validation and accepted runs during study sample analysis. A table listing all reassayed samples; reason for reassay; and the values for original, reassay, and final should be included in the final report.

Currently, as described in the FDA Guidance, 5–20% of all chromatograms, including QCs samples and standards, must be submitted with an NDA or an ANDA filing. Because of the crucial nature of bioequivalence studies, the practice of submitting 20% of chromatograms from serially selected subjects should be continued for both NDAs and ANDAs. In general, representative chromatograms of typical analysis for other PK studies for NDAs should be sufficient for FDA submission. In circumstances where other PK studies are critical to the approval of the NDA, 20% of chromatograms may be requested for submission. However all original chromatograms and reintegrations should be retained at the site and available for audit if necessary.

Further reference is made to the FDA Guidance that describes the final report attributes in detail and remains generally applicable.

5.4.8 Stability Recommendation

Drug stability experiments should mimic conditions under which samples are collected, stored, and processed, as closely as possible. The experiments should be conducted in unaltered representative matrix, including the same type of anticoagulant. In cases where stripped or altered matrix is used for preparation of study calibration standards and/or QC samples, stability evaluation must be yet conducted in samples prepared in unaltered matrix. If a stabilizer is normally employed with incurred samples, it should be employed with the stability samples also.

Short-term stability experiments should be designed and conducted to cover the type of storage conditions that are to be expected for study samples. This generally includes an evaluation of minimum three freeze–thaw cycles, 4 h bench top, and refrigerated stability. During freeze–thaw stability evaluations, the freezing and thawing of stability samples must mimic the intended sample-handling conditions to be used during sample analysis. If study samples are to be stored on wet ice, for thawed periods greater than 4 h, then these conditions should be evaluated during validation as well. If during the sample analysis for a study, a sample was thawed through more than three cycles or if storage conditions changed and/or exceeded the sample storage conditions evaluated during method validation, stability must be established under these new conditions in order to demonstrate that the concentration values from these study samples are valid.

While short-term stability measurements are generated during method validation, long-term measurements are initiated during method validation, possibly evaluating analyte stability for a period of a few weeks, with the remaining long-term storage time points evaluated after method validation. This postvalidation data can then be added to the original validation data in the form of a validation report addendum or as a stand-alone stability report. Long-term stability should be evaluated at the expected storage conditions, including expected satellite storage temperature and duration (e.g., prior to shipment to the analytical laboratory). In consideration of this, there may be the need to include both −70°C and −20°C evaluations (e.g., when samples are stored under different conditions at the various study locations). Refer to section entitled "Separate Stability Experiments Required at −70°C if Stability Shown at −20°C" for additional discussion.

Stability evaluations should be performed against freshly prepared standard curves. When evaluating data generated from stability experiments, intended (nominal) concentrations should be used for comparison purposes. Additionally, to determine if the initial batch of stability samples are suitable for the subsequent stability experiments, a comparison with the initial day 0 or day 1 samples is recommended. If the measured concentration of the day 0 or 1 stability sample differs substantially from the intended concentration, this difference may be an indication that the bulk stability samples were not prepared correctly and preparing new bulk stability samples should be considered.

With respect to stock solutions prepared from certified reference standards, if the reference standard is within its expiration date when the stock solution is prepared, there is no need to prepare a new stock solution when the reference standard expires. When the stock solution exists in a different state (solutions versus solid) or in a different buffer composition (generally the case for macromolecules), the stability data on this stock solution should be generated to justify the duration of stock solution storage stability. In general, newer stock solutions within their established stability period (e.g., a solution with established 60-day stability used on day 55) should not be used to measure stability of an older solution (e.g., 120 days old). Although the newer stock may meet stability criteria for bioanalytical purposes, the chance of misinterpreting the stability of the older solution is high. The suggestion is to make a solution fresh from powder when determining the stability of any older stock.

An additional concern for LBAs is reagent stability. This includes, but is not limited to, antibodies and antibody conjugates (e.g., horseradish peroxidase, biotin, and avidin conjugates). Therefore, during method validation, documentation should be made of the conditions under which the principal reagents maintain sufficient stability to meet the basic requirements of assay performance. Some of these data will need to be generated by the sponsor in the case of proprietary reagents, whereas other stability data can be obtained from the manufacturer for commercially available reagents. When using the manufacturer's data, reagents must be stored as recommended by the manufacturer. If different conditions are used, the analytical investigator will need to generate the appropriate storage stability data of their own.

The evaluation of extract stability with a freshly prepared standard curve is not part of routine validation testing, but should be conducted as needed. In cases where extracted samples are stored before analysis (e.g., extracted samples are refrigerated for several hours or days before placement on the instrument), extract stability should be demonstrated for the storage temperature and duration. With regard to autosampler reinjection reproducibility, a freshly prepared standard curve is not necessary.

5.4.9 Matrix Effects for Mass Spectrometric-Based Assays

One phenomenon influencing mass spectrometry (MS)-based bioanalytical assays is matrix effect. Matrix effect is the suppression or enhancement of ionization of analytes by the presence of matrix components in the biological samples. Quantitative measurement of matrix effect provides useful information in validation of MS-based bioanalytical methods. The quantitative measure of matrix effect can be termed as Matrix Factor (MF) and defined as a ratio of the analyte peak response in the presence of matrix ions to the analyte peak response in the absence of matrix ions, for example,

$$\text{Matrix Factor} = \text{Peak response in presence of matrix ions}/\\ \text{absence of matrix ions}$$

An MF of 1 signifies no matrix effects. A value of less than 1 suggests ionization suppression. An MF of greater than 1 may be due to ionization enhancement and can also be caused by analyte loss in the absence of matrix during analysis. Internal standard-normalized MF is the MF of analyte divided by the MF for IS. The IS-normalized MF can also be obtained by substituting peak response with peak response ratio (analyte/IS) in the above equation for MF. Stable isotope–labeled IS minimizes the influence of matrix effects most effectively since the matrix effects observed for stable isotope–labeled IS are generally similar to those observed for the matching analyte. Analog IS may also compensate for matrix effects; however, the stable isotope–labeled internal standards are mostly effective and should be used whenever possible.

An absolute MF (or IS-normalized MF) of about 1 is not necessary for a reliable bioanalytical assay. However, highly variable MF in individual subjects would be a cause for the lack of reproducibility of analysis. To predict the variability of matrix effects in samples from individual subjects, determine the MF (or IS-normalized MF) for six individual lots of the matrix. The variability in matrix factors, as measured by the coefficient of variation, should be less than 15%. If the matrix is rare and hard to obtain, the requirement for assessing variability of matrix factors in six lots can be waived. Stable isotope–labeled internal standards help by normalizing MF to a theoretical value of 1, and thereby reduce the effective IS-normalized MF variability. When using stable isotope IS, it is not necessary to determine the IS-normalized MF in six different lots.

5.4.10 System Suitability

Scientifically qualified and properly maintained instruments should be used for implementation of bioanalytical methods in routine drug analysis. As part of qualifying instruments, performance of system suitability ensures that the system is operating properly at the time of analysis. System suitability checks are more appropriately used for chromatographic methods to ensure that the system is sufficiently sensitive, specific, and reproducible for the current analytical run. However, the system suitability tests do not replace the required run acceptance criteria with calibration standards and QC samples. System suitability tests, when appropriate, are recommended to ensure success, but are neither absolutely required nor do they replace the usual run acceptance criteria.

5.4.11 Reference Standards

Analytical reference standards are used for the preparation of calibration standards and QC samples. Reference standard lot numbers, purity, storage, stability, handling, and supporting documentation should be monitored and maintained. Reference standards should be used before their expiration or recertification dates. Some compounds used as internal standards or rare metabolites are available in very small amounts, and their certificates of analysis may not be available. If the full CofA is not available for rare metabolites, at a minimum, the documented purity information should be obtained. CofA or purity information of internal standards is not always necessary for the use of internal standards. When purity information is not available for the internal standard, it needs to be demonstrated that the internal standard does not interfere with the chromatography of the analyte(s) of interest. Macromolecular reference standards are often heterogeneous and may present unique comparability and stability considerations.

5.4.12 Validation Topics with No Consensus

This section describes the topics for which no consensus could be reached during the 3rd AAPS/FDA Bioanalytical Workshop. However, this certainly provides comments and emphasizes that further discussion and direction will be necessary.

Cross-Validation of Bioanalytical Methods When Using Different Anticoagulant Counter-Ions: There was recognition of a distinct difference between EDTA and heparin-containing plasma and that a bioanalytical method validated for one could not be used for the other without some revalidation of the method but no consensus was reached for the need for cross-validation when using the same anticoagulant with a different counter-ion. As an example, attendees could not agree on the degree of cross-validation necessary for a method validated using sodium-heparinized plasma when it was applied to a lithium-heparinized sample.

Cross-Validation Required When Using Different Strains or Sexes of a Species: No decisive arguments came forward in support of this activity although it was agreed that there could be some differences in biological matrix originating from the different strains or sexes. The general trend of the debate was that validation experiments to address such differences should not be considered the norm and should be performed when there are method-related concerns that can be attributed to a specific strain or sex-related difference.

Cross-Validation Required When Moving a Method between LC–MS/MS Instruments: Moving from different models of instrument (e.g., Sciex API4000 to a Sciex API-5000) would require cross-validation, but there was very little support for requiring cross-validation when switching between equivalent instruments. It was noted that a qualification experiment is usually performed by most analysts before samples are run on a given instrument and that these experiments were usually sufficient to allow the qualification of a new instrument for a specific assay.

5.4.13 Specific Criteria for Cross-Validation

The term cross-validation was used liberally throughout the 3rd AAPS/FDA Bioanalytical Workshop when considering all different types of changes in bioanalytical methods [5]. Cross-validation was discussed when using matrix from different species, when using matrix with different anticoagulants, when transferring methods to other laboratories, and when transferring methods to other analysts. However, no specific strategies, procedures, and acceptance criteria were discussed to adequately perform these cross-validation experiments. Cross-validation procedures and acceptance criteria need to remain flexible, considering the various bioanalytical situations where it would be required. Specific cross-validation criteria should be established *a priori* via a standard operating procedure.

5.4.14 Separate Stability Experiments Required at −70°C if Stability Shown at −20°C

A significant group of workshop attendees believed that biological sample stability shown at a given temperature (e.g., about −20°C), automatically defined sample stability at a lower temperature (e.g., about −70°C) based on Arrhenius principles of chemical reactivity. These principles dictate lower rates of chemical reactivity (e.g., analyte degradation) occurring at lower temperatures. The argument against assuming sample stability at lower temperatures was based on matrix degradation rather than chemical stability of the analyte. An effective (if anecdotal) argument was made during the meeting, which questioned the stability of biological matrix proteins at lower temperatures. It was argued that a lower temperature could cause denaturation or precipitation of matrix proteins and that this could affect protein binding or the ability to extract the drug from the matrix. There was general agreement that sample matrix will often have a different consistency depending on freezing conditions. Based on these arguments, there did seem to be some support for validating stability at lower storage temperatures even if stability has already been determined at a higher temperature. Additional stability at lower temperature should be required for macromolecules and may also be performed for small molecules as needed.

5.4.15 Stability Criteria for Stock Solution Stability

The need to characterize the stability of stock solutions was emphasized throughout the meeting and accepted as a core validation experiment. However, there was no agreement on the degree of degradation that defines acceptable stability. The consensus was that lower degradation in the standard acceptable ranges is desirable since these stock solutions are used for making other solutions and this error may be propagated in the concentrations reported for biological samples.

5.4.16 Acceptance Criteria for Internal Standards

The practice of placing precision criteria on internal standards as an additional run acceptance test was discussed. A highly variable internal standard can be an indication of an uncontrolled process during sample analysis, especially if the internal standard response is variable with incurred samples. It is recognized that the internal standard is present in a bioanalytical assay to compensate for variability of extraction in LC–MS/MS analysis. This is most likely to be the case when the internal standard is isotopically labeled. When assessing the impact of internal standard variability, it is important to determine, in cases of low internal standard response, that the assay continues to have the ability to accurately quantify at the LLOQ. No agreement was reached on the inclusion of internal standard criteria or on the magnitude of acceptable internal standard precision. However, if study samples or analytical runs are rejected or repeated based on internal standard response variability, objective criteria are necessary and need to be established a priority.

5.4.17 Summary

For quantitative bioanalytical method validation procedure and requirements, there was a relatively good agreement between chromatographic assays and ligand-binding assays. It was realized that the quantitative and qualitative aspects of bioanalytical method validation should be reviewed and applied appropriately.

(1) Some of the major concerns between the two methodologies related to the acceptable total error for precision and accuracy determination and acceptance

criteria for an analytical run. The acceptable total error for precision and accuracy for both the methodologies is less than 30. The 4–6-15 rule for accepting an analytical run by a chromatographic method remained acceptable while a 4–6-20 rule was recommended for ligand-binding methodology.

(2) The 3rd AAPS/FDA Bioanalytical Workshop clarified the issues related to placement of QC samples, determination of matrix effect, stability considerations, use of internal standards, and system suitability tests.

(3) There was a major concern and issues raised with respect to stability and reproducibility of incurred samples. This should be addressed for all analytical methods employed. It was left to the investigators to use their scientific judgment to address the issue.

In general, the 3rd AAPS/FDA Bioanalytical Workshop provided a forum to discuss and clarify regulatory concerns regarding bioanalytical method validation issues. The conference report revealed some of these concerns along with recommended implementations. In author's perspective, future Crystal City conference(s) will continue focusing on real world scenarios and further clarify issues and result in resolutions and consensus for execution in securing the integrity and reliability of data for regulatory submission and consideration.

In drug development, important decisions based on data obtained from bioanalytical methods are taken hence it is generally accepted that sample preparation and method validation are required to demonstrate the performance of the method and the reliability of the analytical results. The acceptance criteria should be clearly established in a validation plan, prior to the initiation of the validation study.

The following principles of bioanalytical method validation provide steps for the development of a new method:

- Parameters essential to ensure the acceptability of performance of a bioanalytical method validation are accuracy, precision, selectivity, sensitivity, reproducibility, and stability.
- A specific detailed description of the bioanalytical method should be written. This may be in the form of a protocol, study plan, report, and/or standard operating procedures.
- Each step in the bioanalytical method should be investigated to determine the extent to which environmental, matrix material, or procedural variables from time of collection of materials up to the analysis and including the time of analysis, may affect the estimation of the analyte in the matrix.
- Whenever possible, the same biological matrix as that intended in the samples should be used for validation purposes.
- The accuracy, precision, reproducibility, response function, and selectivity of the method with respect to endogenous substances should be established with reference to the biological matrix. With regards to selectivity, there should be evidence that the substance being quantified is the intended analyte.

- The concentration range over which the analyte will be determined must be defined in the bioanalytical method, based on the evaluation of actual standard samples over the range, including their statistical variation. It is necessary to use sufficient number of standards to define adequately the relationship between concentration and response. The relationship between response and concentration must be demonstrated to be continues and reproducible. The number of standards to be used will be a function of the dynamic range and the nature of the concentration–response relationship. In many cases, five to eight concentrations (excluding blank) may define the standard curve.
- For bioanalytical methods, accuracy and precision must be accomplished by the analysis of replicate sets of the analyte samples of known concentration—QC samples from an equivalent biological matrix.
- The stability of analyte in the biological matrix at intended storage temperature (s) should be estimated. In addition, the influence of freeze–thaw cycles should be studied. The stability of the analyte at ambient temperature should be evaluated over a period of time that encompasses duration of typical sample preparation, sample handling, and analytical run time.
- The specificity of the assay methodology should be established using independent sources of the same matrix.
- Recovery should be reproducible at each concentration.
- Acceptance/rejection criteria for spiked, matrix-based calibration standards, and validation QC samples should be based on the nominal (theoretical) concentration of analyte(s).

REFERENCES

1. Shah VP, Midha KK, Dighe S, et al. Analytical methods validation: bioavailability, bioequivalence and pharmacokinetic studies. Conference Report. *Eur J Drug Metab Pharmacokinet* 1991;16:249–255.
2. Shah VP, Midha KK, Findlay JW, et al. Bioanalytical method validation—a revisit with a decade of progress. *Pharm Res* 2000;17:1551–1557.
3. Food and Drug Administration. *Guidance for Industry: Bioanalytical Method Validation*. US Department of Health and Human Services, FDA, Center for Drug Evaluation and Research, Rockville, MD, 2001.
4. DeSilva B, Smith W, Weiner R, et al. Recommendations for the bioanalytical method validation of ligand-binding assays to support pharmacokinetic assessments of macromolecules. *Pharm Res* 2003;20:1885–1900.
5. Viswanathan CT, Bansal S, Booth B, et al. Quantitative bioanalytical methods validation and implementation: best practices for chromatographic and ligand binding assays. *J AAPS* 2007; *9* (Article 4): E30–E42.
6. Food and Drug Administration. *Draft Guidance for Industry: Safety Testing of Drug Metabolites*. US Department of Health and Human Services, FDA, Center for Drug Evaluation and Research, Rockville, MD, 2005.

7. International Conference on Harmonization. ICH Q2A: text on validation of analytical procedures. Fed Regist 1995; 60:FR 11260. Available at http://www.fda.gov/cder/guidance/ichq2a.pdf.
8. International Conference on, Harmonization. ICH Q2B: validation of analytical procedures: methodology. Fed Regist 1997; 62:FR 27463. Available at http://www.fda.gov/cder/guidance/1320fnl.pdf.
9. US, Department of Health and Human Services. *Draft Guidance for Industry: Analytical Procedures and Methods Validation, Chemistry, Manufacturing and Controls Documentation.* US Department of Health and Human Services, Food and Drug Administration, Rockville, MD, 2000.
10. United States Pharmacopoeial Convention. *United States Pharmacopoeia 26, National Formulary 21, <1225> Validation of Compendial Methods.* United States Pharmacopoeial Convention, Rockville, MD, 2003.
11. Rules and Guidance for Pharmaceutical Manufacturers and Distributors 2002. *Medicines Control Agency-MCA (Now Medicines and Healthcare Regulatory Agency-Medicines Section-MHRA)*, Her Majesty's Stationary Office, London, UK, 2002.
12. National Health and Welfare. *Acceptable Methods.* Drug Directorate Guidelines, National Health and Welfare, Health Protection Branch, Health and Welfare. Ottawa, Ontario, Canada, 1992.
13. Ocampo A, Lum S, Chow F. Current challenges for FDA-regulated bioanalytical laboratories for human (BA/BE) studies. Part I: Defining the appropriate compliance standards—application of the principles of FDA GLP and FDA GMP to bioanalytical laboratories. *Qual Assur J* 2007;11:3–15 (Wiley Inter-Science).
14. Furman WB, Layloff TP, Tetzlaff J. Validation of computerized liquid chromatographic systems. *J AOAC* 1994;77:1314–1318.
15. US Food and Drug Administration. *General Principles of Software Validation. Final Guidance for Industry and FDA Staff.* US Department of Health and Human Services, Food and Drug Administration, Rockville, MD, 2002.
16. Bansal SK, Layloff T, Bush ED, Hamilton M, Hankinson EA, Landy JS, Lowes S, Nasr MM, St. Jean PA, Shah VP. Qualification of analytical instruments for use in the pharmaceutical industry: a scientific approach. *AAPS J* 2004;5 (1): Article 22.
17. Food and Drug Administration. *21CFR320—Bioavailability and Bioequivalence Requirements (Drugs for Human Use).* US Department of Health and Human Services, FDA, CDER, Rockville, MD, 2002.
18. Committee for Proprietary Medicinal Products. *Note for Guidance on Investigation of Bioavailability and Bioequivalence.* CPMP, EWP, QWP-1401/98, Canary Wharf, London, 2001.
19. Food and Drug Administration. *Draft Guidance for Industry: Safety Testing of Drug Metabolites.* US Department of Health and Human Services, FDA, Center for Drug Evaluation and Research, Rockville, MD, 2005.

6

COMPETITIVENESS OF BIOANALYTICAL LABORATORIES—TECHNICAL AND REGULATORY PERSPECTIVES

6.1 TECHNICAL ASPECT OF COMPETITIVE BIOANALYTICAL LABORATORIES

The discovery and development of a new drug costs around $1 billion and it may take approximately 10 years for the drug to reach the marketplace. Drug discovery and development are the processes of generating compounds and evaluating all of their properties to determine the feasibility of selecting one new chemical entity (NCE) to become a safe and efficacious drug. Among many criteria, obtaining experimental pharmacokinetics (PK) data from laboratory animals in the nonclinical stage is critical to evaluating a drug candidate before it can be qualified to be tested in the clinical trials for safety and efficacy evaluation. A key parameter in pharmacokinetics is the plasma or tissue concentration of the new drug after its administration to laboratory animals. Therefore, developing an accurate and fast analytical method for measuring the concentrations of a compound in plasma or tissue is the first step in order to yield the PK of a compound. As the drug candidate moves down in the pipeline, the requirements for PK information differ at the different stages of drug discovery and development, leading to the introduction of "fit for purpose" analytical strategies that provide the appropriate level of bioanalytical support for the purpose at hand, while simultaneously minimizing resource expenditures. Due to its superior sensitivity and selectivity, LC–MS/MS has become the primary analytical technique used by most bioanalytical laboratories performing analyses for preclinical and clinical studies. The increased number of biological agents used as therapeutics

Regulated Bioanalytical Laboratories: Technical and Regulatory Aspects from Global Perspectives,
By Michael Zhou

(in the form of recombinant proteins, monoclonal antibodies, vaccines, etc.) has prompted the pharmaceutical industry to review and refine aspects of the development and validation of bioanalytical methods for the quantification of these therapeutics in biological matrices in support of preclinical and clinical studies. Most of these methodologies are used in quantitative assays supporting pharmacokinetic and toxicokinetic parameters of the therapeutic agents. Up to date, an alternative approach for dealing with macromolecules is to utilize ligand-binding assays (LBAs) to generate data for intended purposes.

Liquid chromatography coupled with tandem mass spectrometry (LC–MS/MS) has played an incredible role in drug metabolism and pharmacokinetics (DMPK) studies at various drug discovery and development stages since its inception to the pharmaceutical and biotechnology industries. This section will elaborate the most recent advances in sample preparation, separation, and the mass spectrometric aspects of high-throughput quantitative bioanalysis of drug and metabolites in biological matrices. Recently introduced techniques such as ultra-performance or ultra-speed liquid chromatography (UPLC or USLC) with small particles (sub-2 μm) and monolithic chromatography offer improvements in speed, resolution, and sensitivity compared to conventional chromatographic techniques. Hydrophilic interaction chromatography (HILIC) on silica columns with low aqueous/high organic mobile phase is emerging as a valuable supplement to the reversed-phase LC–MS/MS. Sample preparation formatted to 96-well and/or 384-well plates has allowed for semi-automation of off-line sample preparation techniques, significantly improving throughput. On-line solid phase extraction (SPE) utilizing column-switching techniques is rapidly gaining acceptance in bioanalytical applications to reduce both time and labor required for generating bioanalytical results. Extraction sorbents for on-line SPE extend to an array of media including large particles for turbulent-flow chromatography, restricted access materials (RAMs), monolithic materials, and disposable cartridges utilizing traditional packings such as those used in Spark Holland (Symbiosis) systems. In latter part of this section, recent studies of matrix effect in LC–MS/MS analysis and how to reduce/eliminate matrix effect in method development and validation will be also discussed.

Bioanalytical laboratories in the pharmaceutical and life science industries are constantly under pressure to reduce overall time for discovery and development. This is often accompanied with an increase in the number of biological samples requiring pharmacokinetic analysis and a decrease in the desired quantitation levels. Hyphenated techniques are examples of new tools adopted for developing fast and cost-effective analytical methods. Liquid chromatography tandem mass spectrometry (LC–MS/MS) has led to major breakthroughs in the field of quantitative bioanalysis since the 1990s due to its inherent specificity, sensitivity, and speed [1, 2]. It is now generally accepted as the preferred technique for quantitating small-molecule drugs, metabolites, and other xenobiotic biomolecules in biological matrices. The use of LC–MS/MS has grown exponentially in the last decade, due to its unmatched sensitivity, extraordinary selectivity, and rapid rate of analysis. Typical workflow of bioanalysis from sample generation to data delivery is summarized in Figure 6.1 (workflow).

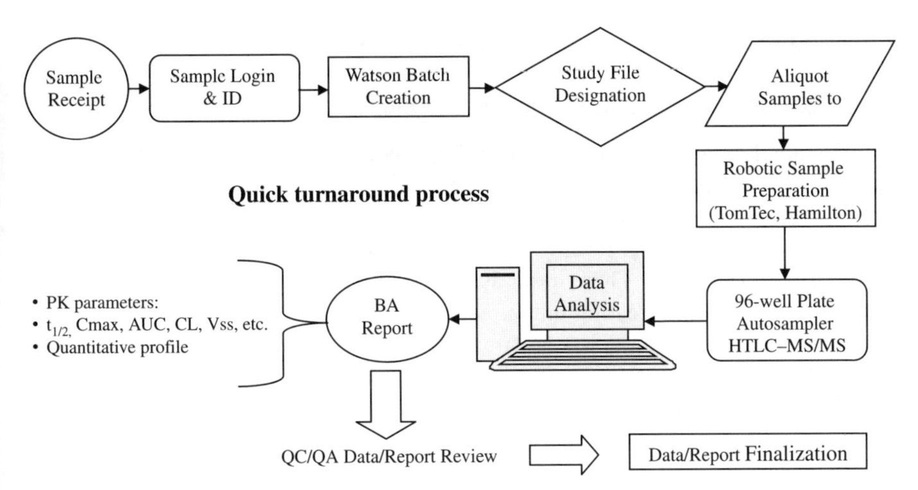

FIGURE 6.1 Typical workflow of bioanalysis from sample generation to data delivery.

Samples from biological matrices are usually not directly compatible with LC–MS/MS methodologies. Sample preparation has traditionally been done using such approaches as protein precipitation (PPT), liquid–liquid extraction (LLE), or SPE. Manual operations associated with these processes (in particular with LLE and SPE) are fairly labor intensive and time consuming. Parallel sample processing in 96-well format using robotic liquid handlers has significantly shortened the time that analysts have to spend in the laboratory for sample preparation. An alternative sample extraction method that has generated a lot of interest in recent years is the direct injection of plasma using an on-line extraction method. A major advantage of on-line SPE over off-line extraction techniques is that the sample preparation step is embedded into the chromatographic separation and thus eliminates most of the sample preparation time traditionally performed at the bench and potentially resulting in more variability from batch-to-batch and analyst-to-analyst performance.

Steep gradients and relatively short columns were first utilized in early applications of high-throughput LC–MS/MS assays to reduce run times. Better understanding of how matrix effects can compromise the integrity of bioanalytical methods has reemphasized the need for adequate chromatographic separation of analytes from endogenous biological components in quantitative bioanalysis using LC–MS/MS technique. New developments from chromatographic techniques such as UPLC and USLC with sub-2 μm particles and monolithic chromatography are showing promise in delivering higher speed, better resolution, and sensitivity for high-throughput analysis while minimizing matrix effects. LC–MS/MS applications for quantitative bioanalysis have been documented in hundreds of articles in just the past several years, and a number of reviews dealing with one or more aspects of quantitative LC–MS/MS bioanalysis have been published [3–6]. Within this section, publications related to technology development for throughput improvement, associated applications, and discussions of key developments in quantitative analysis from 2002 to 2009 will be reviewed and elaborated.

6.2 BIOANALYTICAL PROCESSES AND TECHNIQUES

6.2.1 Sample Generation, Shipment, and Storage

Prior to sample analysis from bioanalytical standpoint, it is very important to maintain the quality and integrity of samples where are to be generated from discovery, nonclinical and clinical programs regardless of nonregulated or regulated studies. Laboratory management regarding sample handling/custody is key to the success for bioanalytical operation. These processes generally include (but not limited to): sample receipt, login (creating unique identification numbers or ID), storage (adequate and acceptable conditions), and transfers/relocations from lab to lab and facility to facility, sample retention and/or disposal. Zhang et al. described an integrated sample collection and handling process for bioanalysis [7]. The sample handling process consumes a significant portion of the resources in a bioanalytical laboratory. One of the primary purposes for analytical automation is the possibility of analyzing a huge number of samples, either simultaneously or sequentially, and without any human intervention. When large amounts of samples are analyzed, mainly in routine analyses, the importance of the correct identification of each sample in order to obtain the appropriate correlation between the sample analyzed and the results obtained is obvious. At present, one of the efficient and reliable ways to identify a sample, among a number of them, from collection to result, is by using bar codes. A bar code can best be defined as an "optical Morse code." Series of black bars and white spaces of varying widths are printed on labels to uniquely identify items. The bar code labels are read with a scanner, which measures reflected light and interprets the code into number and letters that are passed on to a computer. All bar codes have start/stop characters that allow the bar code to be read from both left to right and right to left. Unique characters placed at both the beginning and end of each bar code, the stop/start characters provide timing references, symbol identification, and direction of reading information to the scanner. By convention, the unique character on the left of the bar code is considered the "start" and the character on the right of the bar code is considered the "stop." Besides the linear bar code, today's new bar codes are two dimensional, electronic or small computer chips, which sort data in inconspicuous places like a credit card. This technology has been included in clinical and bioanalytical analyzers where large amounts of samples are analyzed daily and when a correct assignation of the results obtained is mandatory.

Automation for sample handling meets the business needs of improving productivity and reduces the documentation required for compliance. Procedural elements involved in maintaining bioanalytical data integrity for good laboratory practices and regulated studies are discussed by James and Hill [8]. There elements can be divided into three areas. First, there are those ensuring the integrity of the analyte until analysis, through correct sample collection, handling, shipment, and storage procedures. Incorrect procedures can lead to loss of analyte via instability, addition of analyte through contamination or instability of related metabolites, or changes in the matrix composition that may adversely affect the performance of the analytical method. Second, the integrity of the sample identity needs to be maintained to ensure that the final result reported relates to the individual sample that was taken. Possible sources of error include sample mix-up or mislabeling, or errors in data handling. Finally, there is the overall integrity of the documentation that supports the analysis, and any prestudy validation of the method. This includes a wide range of information, from paper and electronic raw data, through standard operating procedures, analytical procedures, facility records, to study plans and final reports. These are critical to allowing an auditor or regulatory body to reconstruct the study. LIMS is a preferable database and handling system that maintains a detailed pedigree for each sample by capturing processing parameters, protocols, stocks, tests, and analytical results for the sample's complete life cycle. Project and study data also are maintained to define each sample in the context of research tasks it supports. LIMS will be required for each analytical pipeline to track all aspects of sample handling.

6.2.2 Sample Preparation

One strategy for high-throughput bioanalytical analysis is to use well-established instrumentation, rigorous, standardized techniques; and automation, wherever possible, to minimize or replace manual tasks. Automation results in greater performance consistency over time and in more reliable methods transfer from site to site. Automated 96-well plate technology is well established and accepted and has been shown to effectively replace manual operations. The 96-well instruments can execute automated off-line extraction and sample cleanups. Automated solid phase extraction, liquid–liquid extraction, and protein precipitation all can be performed in 96-well format and will be discussed in the following sections. Automation of the equipment used in daily laboratory activities has experienced massive development in recent years. Bioanalysis has required growing automation as a consequence of the exponential increment in the number of analyses that all areas of discovery and development phases—particularly nonclinical, clinical, environmental, industrial, and food areas—presently demand. Full automation of the analytical process is sometimes the best solution to laboratory overload as 24 h working days are possible without increasing personnel costs. Nevertheless, fully automated analyzers have—as a sometimes unsurpassable shortcoming—high acquisition costs with medium-to-high maintenance costs, mainly caused by the consumable material these analyzers use. For economic reasons, partial automation is, in most cases, the accepted solution. The simplest and least expensive partial automation affects the last steps of the analytical process, that is, measurement–transduction of the analytical signal and data collection and treatment. At present, commercial robotic apparatus and analytical instruments are automated, so these two aspects are addressed most times. Automated sample preparation (ASP) can be accomplished by using a batch or a serial approach. In the batch mode, multiple samples are prepared and then transferred to the analytical instrument for measurement. In the serial mode, samples are prepared one at a time and the ASP device is connected (integrated) with the analytical instrument only. From a temporal viewpoint, it is assumed that, when the ASP time is equivalent to or shorter than the analysis time, then a serial method is preferable. Serial methods are also more useful with samples where the assays are time dependent or when there is a stability question of the prepared analyte (e.g., with labile samples or matrices).

6.2.2.1 ***Off-Line Sample Preparation*** Adequate sample preparation is a key aspect of quantitative bioanalysis and can often be the cause of bottlenecks during high-throughput analysis. Sample preparation techniques in 96-well format have been well adopted in high-throughput quantitative bioanalysis with a number of applications [9–19]. The techniques that can use the format include LLE, SPE, and PPT. Typically, liquid transfer steps, including preparation of calibration standards (STD) and quality control (QC) samples as well as the addition of the internal standard (IS), were performed automatically using robotic liquid handling workstations for parallel sample processing (Hamilton STAR, etc. refer to robot liquid handler diagram in Figure 6.2). The increasing demand for high throughput causes a unique situation of balancing cost versus analysis speed as each sample preparation technique offers unique advantages. Dilute-and-shoot and protein precipitation are among one of the most popular and effective techniques due to their simplicity. Sample preparation with PPT is widely used in bioanalysis of plasma samples. The method has been extended to quantitation of drug and metabolites from whole blood. Koseki et al. developed a sensitive and specific LC–MS/MS method for the simultaneous determination of cyclosporin A (CsA) and its three main metabolites (AM1, AM4N, and AM9) in human blood [9]. Following protein precipitation, supernatant was directly injected onto the LC–MS system. Overall, PPT offers a generic and fast sample preparation technique that can be easily automated and reported by Ma et al. [10]. However, when analyzing supernatant from a plasma sample using PPT, salts and endogenous material are still present and can cause ion suppression or enhancement that will lead to higher or unexpected/undesirable variation from sample to sample.

Solid phase extraction has its commercial roots in the late 1970s. Since then, it has become a common and effective technique for extracting analytes from complex

FIGURE 6.2 Hamilton robot diagram for automated sample preparation.

samples. SPE prepares multiple samples in parallel (typically with 96-well format) and uses relatively low quantities of solvents, and the procedures can be readily automated with a broad range of sorbents (silica based or polymer based). Mixed-mode polymer-based sorbents (e.g., Waters Oasis MCX cartridge) were introduced in the late 1990s for the isolation of drugs with ionizable functional groups from biological fluids. The extraction procedures can be a generic protocol or can be optimized if better sample clean up is desired. The use of solid phase extraction often gives superior results to those with a PPT approach but may not be as cost effective as PPT due to the labor and material costs associated with the procedures. Mallet et al. [11] described a novel 96-well SPE plate that was designed to minimize the elution volume required for quantitative elution of analytes. The plate was packed with 2 mg of a high-capacity SPE sorbent that allows loading of up to 750 μL of plasma. The novel design permitted elution with as little as 25 μL solvent. Therefore, the plate can offer up to a 30-fold increase in sample concentration. The evaporation and reconstitution step that is typically required in SPE is unnecessary due to the concentrating ability of the sorbent. Yang et al. [12] developed a sensitive μElution SPE–LC–MS/MS method for the determination of $M + 4$ stable isotope labeled cortisone and cortisol in human plasma. In the method, analytes were extracted from 0.3 mL of human plasma samples using a Waters Oasis HLB 96-well μElution SPE plate with 70 μL methanol as the elution solvent. The lower limit of quantitation was 0.1 ng/mL and the linear calibration range was from 0.1 to 100 ng/mL for both analytes. Several related formats have recently appeared using miniaturized packed-bed formats as well as an adaptation of disk technology for the 96-well format. Ultimate goal for the 96-well approach is to add a higher level of parallel sample preparation using essentially a further miniaturized SPE format. By extracting multiple 96-well format SPE plates, up to four 96-well plates or 384 samples may be prepared and then analyzed within 24 h by one person with one LC–MS/MS system, which represents a significant improvement in sample throughput using commercially available equipment and supplies. Future developments should further increase throughput significantly.

Liquid–liquid extraction gives excellent sample clean up but poses engineering challenges for use in an automated high-throughput fashion. Several groups have developed different approaches to solve mixing and phase separation problems typically observed in a 96-well LLE method [13–18]. By using vigorous vortexing after well-controlled heat-sealing, or using repeated aspiration and dispensing by robotic liquid handler, common extraction solvents such as methyl *t*-butyl ether (MTBE) or ethyl acetate (EA) may be used in routine extraction of plasma, blood, or tissue samples. Wang et al. [13] developed and validated a 96-well LLE assay, using LC–MS/MS in the atmospheric pressure chemical ionization (APCI) mode for the simultaneous determination of two human immunodeficiency virus (HIV) protease inhibitors, lopinavir and ritonavir, in human plasma. The sample preparation consisted of liquid–liquid extraction with a mixture of hexane and ethyl acetate using 100 μL of plasma. The method was validated over the concentration ranges of 19–5300 ng/mL for lopinavir and 11–3100 ng/mL for ritonavir, respectively. Zhang et al. [14] presented a 96-well LLE method for measuring zotarolimus drug concentrations from drug-eluting stents in swine artery samples. The authors used 100% swine blood as the homogenization solution to improve the consistency of the extraction recovery and stability of the zotarolimus in tissue homogenates. Xu et al. [17] described a 96-well liquid–liquid back extraction LC–MS/MS method for the determination of a basic and polar drug candidate from human plasma samples. The analyte was extracted from plasma using MTBE first, followed by a back extraction from the organic phase into a small volume of acidified water. A linear range of 0.38–95.02 ng/mL was established for the method with good accuracy and precision. A similar approach was reported by Bolden et al. [18] in a liquid–liquid back-extraction (LLE) procedure for sample preparation of dextromethorphan (DEX), an active ingredient in many over-the-counter cough medicines, and dextrorphan (DOR), an active metabolite of DEX, in human plasma. After back extraction, the acidified water isolated from the back extraction was analyzed directly by LC–MS/MS, eliminating the need for a dry down step.

Combinations of sample preparation techniques have been developed to achieve desired sample extract purity with high throughput. Xue et al. [19] investigated a simplified protein precipitation/mixed-mode cation-exchange solid phase extraction (PPT/SPE) procedure. A mixture of acetonitrile and methanol along with formic acid was used to precipitate plasma proteins prior to selectively extracting the basic drug. After vortexing and centrifugation, the supernatants were directly loaded onto an unconditioned Oasis MCX µElution 96-well extraction plate, where the protonated drug was retained on the negatively charged sorbent while interfering neutral lipids, steroids or other endogenous materials were eliminated. Additional wash steps were deemed unnecessary and not performed prior to sample elution.

6.2.2.2 On-Line Sample Extraction Another approach is on-line SPE for the automated preparation of samples prior to LC–MS/MS analysis. This approach uses a commercial device that combines an auto-sampler and a solvent delivery unit to aliquot multiple liquid samples into a flowing stream of solvent. The solvent has preconditioned an in-line SPE cartridge. After conditioning, the SPE cartridge retains the targeted analytes while the relatively weak solvent elutes unretained salts and polar matrix components to waste. An empirically optimized sequence of increasingly stronger solvents is then used to further elute weakly retained unwanted sample components. A final elution with HPLC mobile phase elutes the targeted analytes off the SPE cartridge and onto an analytical HPLC column for LC–MS/MS analysis. The on-line SPE technique offers speed, high sensitivity by the preconcentration factor, and low extraction cost per sample, but typically requires the use of program controlled switch valves and column reconfigurations. However, the on-line technique can be fully automated offering real-time high-throughput sample analysis. Several generic approaches have recently been developed for on-line sample extraction coupled to LC–MS [7, 20–22]. Different extraction supports allowing direct injection of biological fluids or extracts in various applications were described in publications [23–39]. These extraction supports or sorbents include restricted-access media (RAM), large-size particle, monolithic material, and disposable cartridges. Most on-line SPE approaches use column switching to couple with the

FIGURE 6.3 Spark Holland Symbiosis system.

analytical columns. Various column dimensions can be configured for the fast analysis of drug and their metabolites in biological matrices at the nanogram per milliliter level or lower.

One commercial automated on-line SPE system is the Symbiosis system manufactured by Spark Holland (Figure 6.3). It includes an auto-sampler (Reliance), two binary HPLC pumps, an on-line SPE unit with two high-pressure solvent delivery pumps (HDPs) and a combined valve systems to direct fluid for different steps of SPE. At the beginning of each run, an on-line SPE cartridge is loaded into the unit. After a conditioning step with high organic solvent and an equilibrium step with low organic aqueous solution, a sample is injected onto the cartridge and washed with aqueous solution. Proteins and other matrix materials from the sample are removed during the washing step. Analyte of interest is then eluted onto the analytical column and detected by mass spectrometry or tandem mass spectrometries. During the sample elution step, a second sample is loaded to a new on-line SPE cartridge for the next analysis. In this parallel mode, the sample analysis cycle time approximates the LC run time without the time required for the SPE procedures. Since the on-line SPE cartridge is disposable and each sample uses a new cartridge, the carry over problem from the extraction cartridge is eliminated. A generic method for the fast determination of a wide range of drugs in serum or plasma has been presented for the Spark Holland system [40]. The method comprises generic solid phase extraction with HySphere particles, on-line coupled to gradient HPLC with tandem mass spectrometric detection. The optimized generic SPE–LC–MS/MS protocol was evaluated for 11 drugs with different physicochemical properties. Good quantification for 10 out of 11 of the pharmaceuticals in serum or plasma could be readily achieved.

The quantitative assays gave recoveries better than 95%, lower quantification limits of 0.2–2.0 ng/mL, acceptable precision and accuracy and good linearity over 2–4 orders of magnitude. Carry-over was determined to be in the range of 0.02–0.10%, without optimization. An approach for on-line introduction of internal standard for quantitative analysis was developed on the Spark Holland system [41]. In this approach, analyte and IS were introduced into the sample injection loop in different steps. Analyte was introduced into the injection loop using a conventional auto-sampler (injector) needle pickup from a sample vial. IS was introduced into the sample injection loop on-line from a microreservoir containing the IS solution. As a result, both analyte and IS were contained in the sample loop prior to the injection into the column. The authors demonstrated comparable accuracy and precision to those obtained using off-line IS introduction (e.g., IS and analyte were premixed before injection) while maintaining chromatographic parameters (e.g., analyte and IS elution time and peak width). This new technique was applied for direct analysis of model compounds in rat plasma using on-line SPE–LC–MS/MS quantification. On-line IS introduction allows for nonvolumetric sample (plasma) collection and direct analysis without the need of measuring and aliquoting a fixed sample volume prior to the on-line SPE–LC–MS/MS. The method enables direct sample (plasma) analysis without any sample manipulation and preparation. Alnouti et al. [33] reported another study with Symbiosis system connected to a Luna C_{18} analytical column or a Chromolith C_{18} monolithic column for analysis of two model compounds. Rat plasma spiked with the analytes was diluted with internal standard and injected directly onto the system. Method development including on-line SPE cartridge selection and extraction condition optimization was performed by the Symbiosis system automatically. The total cycle time of 4 min with the Luna C_{18} column was reported. The run time was reduced to 2 min per sample for the monolithic column without compromising the quality and validation criteria of the method. Koal et al. [34] developed a method for the quantitation of seven protease inhibitors and two nonnucleoside reverse transcriptase inhibitors in patient plasma samples. Only a sample dilution step was used to dilute samples and add internal standard before the analysis. Run time was 6.6 min for the first sample and only 3.3 min per sample thereafter.

On-line SPE with high flow rate has been achieved by using extraction columns packed with large diameter particles. The extraction flow rate is typically set to 4–6 mL/min. Sample extraction occurs with very high solvent linear velocity without significant backpressure. Turbulent flow chromatography (TFC) columns marketed by Cohesive Technologies (acquired by Thermo-Fisher Scientific) are widely used for this purpose. Turbulent flow in the extraction column results in rapid binding of small molecules to the absorbent while proteins being removed from the sample matrix. Minimum or no sample pretreatment is required and significant sample preparation time is saved. Chassaing et al. [28] demonstrated a parallel micro-TFC method to analyze pharmaceutical compounds in plasma. Plasma samples were mixed with an equal volume of internal standard solution and injected onto a parallel Aria TX-2 system equipped with microextraction columns. The narrow diameter of the TFC extraction column (0.5 mm i.d.) allowed the extraction flow rate to be reduced to only 1.25 mL/min. Special effort was made to lower the carry over from both auto-sampler and extraction column. The carry-over value was reduced to well below 0.2% for all six compounds used for method development. Smalley et al. [29] reported a method using turbulent flow chromatography to analyze Caco-2 cell-based permeability study samples. Ten compounds could be analyzed simultaneously in a cassette mode. The standard curve range for most compounds was 10–2500 nM with regression coefficients (r^2) greater than 0.99 for all compounds. The run time with individual sample was 6.5 min and was reduced to 3.5 min when Aria system equipped with a dual injection arm auto-sampler, dual injection ports, and multiplexed LCs were used.

Another commonly used on-line SPE sorbent material is restricted access material. With a small pore size, RAM works by eliminating the access of large molecules such as proteins to the inner surface of the particles. Small molecules can freely bind to the sorbent in the normal hydrophobic interaction mode. Proteins molecules quickly pass through the column and are washed out to waste. RAM columns have been used as the SPE and analytical column in the single column mode or coupled with another analytical column in column switching mode. Vintiloiu et al. [23] demonstrated the success of combining RAM with turbulent flow chromatography for on-line extraction of rofecoxib (Vioxx) in plasma samples. The on-line SPE was performed on a column packed by the researchers (Licrospher 60, RP-18 ADS, 40–63 μm diameter) at a loading flow rate of 5 mL/min. After on-line SPE, the analyte was eluted and separated on a monolithic column. The total run time for the analysis was 5 min per sample. The lower limit of quantitation was 40 ng/mL. The extraction method showed good recovery and robustness after more than 200 plasma sample injections. Kawano et al. [42] developed an on-line SPE method with methylcellulose-immobilized cation-exchange RAM to analyze basic drugs in plasma. Samples were injected onto the RAM exchange column at a flow rate of 3 mL/min with and 0.1% acetic acid and then eluted onto a C_{18} analytical column by fast gradient with acetonitrile and ammonium acetate buffer at 0.5 mL/min. The total cycle time was 7 min per sample run.

A polar functionalized polymer (Strata-X, Phenomenex) has been explored as the extraction support in an on-line SPE–LC–MS/MS assay [21]. This newly developed SPE column allows direct analysis of plasma samples containing multiple analytes. A gradient LC condition was applied to separate eight analytes that cannot be distinctly differentiated by MS/MS. With a run time of 2.8 min per injection using a Chromolith column as the analytical column, 300 direct plasma injections were made on one on-line SPE column without noticeable changes in system performance.

Beside fast chromatographic separation, monolithic phases have been investigated as extraction support for on-line SPE. Thanks to their high permeability, the extraction can be performed at a high flow rate without generating high backpressure. The flow remains laminar and is 5–10 times higher than the flow rates generally used with conventional supports. More details of using a monolith as the analytical column can be found in the separation section of this review article. Xu et al. [22] described an automated procedure using on-line extraction with monolithic sorbent for pharmaceutical component analysis in plasma by LC–MS/MS. A short monolithic C_{18} cartridge was used for high flow extraction at 4 mL/min. Plasma samples were subjected to protein precipitation first with acetonitrile, and the supernatant was diluted and loaded onto the monolithic cartridge. Sample elution was accomplished with narrow-bore LC–MS/MS system with a total analysis time of 4 min. A method for determination of Amprenavir (APV) and Atazanavir (AZV) in human plasma was developed with this approach. Very low carry-over on the order of 0.006% was demonstrated using the monolithic-phase-based method. The method has high recovery and good tolerance to matrix effect, which was demonstrated in 12 lots of plasma. The backpressure of the monolithic extraction cartridge remained unchanged after 450 sample injections. The performance of the monolithic-phase online extraction method was compared with that performed by an automated 96-well liquid–liquid extraction procedure, carried out using hexane and ethyl acetate as the extraction solvent. The results from both methods produced similar precision and accuracy.

Endogenous material from urine contains a great deal of amount of metabolic products that may present a significant challenge to assay developers and often require tedious sample preparation to remove the interfering small-molecules (coextractives). Method development for determining drug or metabolite concentrations from urine samples has been simplified with the implementation of on-line SPE. Because of its aqueous nature and lack of protein content, urine samples can be easily loaded onto and cleaned by on-line SPE cartridges. Barrett et al. [35] developed a sensitive method for quantitation of urinary 6β-hydroxycortisol (6β-HC) and cortisol using on-line SPE and LC–MS/MS. Human urine samples were injected directly onto an on-line solid phase extraction apparatus, Prospekt-2, followed by HPLC separation and LC–MS/MS detection. The lower limit of quantitation was 1 and 0.2 ng/mL for 6β-HC and cortisol, respectively.

The advantages of sequential automated system include analyte trace enrichment, unattended on-line sample preparation and analysis, and minimized adsorptive losses that often occur with off-line sample transfers and sample-handling procedures. Limitations include those aspects typical of any serial analysis, such as reduced sample throughput and sample stability problems caused by extended storage times in the auto-sampler. It should be noted that improved MS/MS sample throughput was demonstrated by excluding the analytical HPLC column, which effectively provided automated on-line flow-injection analysis of the SPE extracts. A summary table (Table 6.1) is shown for a comparison up to date within these particular applications. The total automation features of this approach and impressive detection limits that may be achieved make this sample-preparation strategy desirable in certain instances. Where automated sample preparation will go in the future can be anticipated by the foreseeable evolution of the necessities in analytical or bioanalytical laboratories, namely:

(1) Shortening of the time required for ASP, as required by the growing number of samples to be analyzed. This reduction in time can be achieved by the design and commercialization of both, general and dedicated devices assisted by different sources of energy.
(2) Integration of several steps involved in common ASP procedures in a single commercial device.

TABLE 6.1 A List of Commonly Used Extraction Supports for On-Line Extraction and Associated Applications

Compound	Extraction Support	Platform	Total Run Time	Low Limit of Quantitation
Rofecoxib (Vioxx) in rat plasma	Licrospher 60, RP-18 ADS, 0.76 × 50 mm, 40–63 µm	Home built	5 min	40 ng/mL
Compound A (proprietary) in rat plasma	C_{18} RAM-ADS (alkyl diol silica), 25 µm, 25 × 4 mm	Home built	8 min	1 ng/mL
Cyclosporin A, Tacrolimus, and Sirolimus in human blood	Cohesive Cyclone 50 × 1 mm polymeric column, 50 µm	Cohesive Technologies (Thermo-Fisher)	3 min	4.5 ng/mL for Cyclosporin A, 0.2 ng/mL for Tacrolimus, and 0.4 ng/mL for Sirolimus
Mycophenolic acid (MPA) and glucuronide metabolite (MPAG) in human plasma	Applied Biosystems Poros Perfusion column 30 × 1 mm	Cohesive Technologies (Thermo-Fisher)	5 min	50 ng/mL for MPA and 100 ng/mL for MPAG
MK-0767, a dual PPAR alpha/gamma agonist in human plasma	Cohesive Turboflow C_{18} column 50 × 1.0 mm, 50 µm	Cohesive 2300 HTLC Turboflow system	65 s	4 ng/mL
Dextrorphan and dextromethorphan in human plasma	Cohesive Cyclone C_{18} 50 × 1 mm, 50 µm	Cohesive 2300 HTLC Turboflow system	1.5 min	5 ng/mL
Multiple compounds in human plasma	Cohesive Cyclone 50 × 0.5 mm column, 50 µm	Cohesive Aria TX-2 system	3.7 min	1 ng/mL
Ten compounds in Caco-2 cell based permeability study samples	Cohesive Cyclone trap column 50 × 0.5 mm column, 50 µm	Cohesive Aria TX-2 system	3.5 min	10–2500 nM
Eight analytes (Indiplon, Verapamil et al.) in plasma	Phenomenex Strata-X SPE column 20 × 2.1 mm, 25 µm	Home built	2.8 min	1.95 ng/mL
Piritramide in human urine	Oasis HLB extraction column 25 µm, 20 × 2.1 mm	Home built	8.5 min	0.5 ng/mL

(*continued*)

TABLE 6.1 (*Continued*)

Compound	Extraction Support	Platform	Total Run Time	Low Limit of Quantitation
Terbutaline enantiomers in human plasma	Oasis HLB extraction columns 50 × 1.0 mm, 25 μm	Home built	5.5 min	1 ng/mL
(*R*)- and (*S*)-propranolol in rat plasma	Oasis HLB extraction columns 50 × 1.0 mm, 25 μm	Home built	10 min	2 ng/mL
Amprenavir and atazanavir in human plasma	Chromolith C_{18} 4.6 × 10 mm	Home built	4 min	2.77 ng/mL for atazanavir, and 4.50 ng/mL for amprenavir
Propranolol and diclofenac in rat plasma	C_{18} HD 10 × 2 mm	Symbiosis, Spark Holland	2 min for Chromolith column and 4 min for Luna column	1 ng/mL
Multiple anti-retroviral drugs in human plasma	HySphere C_{18} HD	Symbiosis, Spark Holland	3.3 min	6.5 ng/mL
Cortisol and 6β-hydroxycortisol in urine	HySphere C_{18} HD, 7 μm	Prospekt-2, Spark Holland	9 min	1 ng/mL for 6β-hydroxycortisol and 0.2 ng/mL cortisol
Talinolol in human plasma	C8 End Capped (10 × 2 mm) cartridge	Prospekt-2, Spark Holland	4.8 min	2.5 ng/mL
Brostallicin in human plasma	HySphere Resin SH cartridges (10 × 2 mm)	Prospekt-2, Spark Holland	8 min	0.124 ng/mL
Clozapine and metabolites in human serum	HySphere-C_{18}-HD, 7 μm, 10 × 2 mm and HySphere-CN, 7 μm, 10 × 2 mm	Prospekt-2, Spark Holland	2.2 min	0.15–0.3 ng/mL
8-Oxo-7,8-dihydro-2'-deoxyguanosine in human urine	Inertsil ODS-3 column, 5 μm, 50 × 4.6 mm	Home built	10 min	0.019 ng/mL

A "No" in this column indicates pretreatment was limited to sample transfer, internal standard addition, dilution, and centrifugation steps only.

(3) Broadening of robotics implementation—either workstations or robotic stations, as required—for unattended development of routine analyses.
(4) Miniaturization of ASP devices is a present trend that will grow in the future as a result of the continuously smaller volumes required in the subsequent steps of the analytical process.

6.3 ENHANCING THROUGHPUT AND EFFICIENCY IN BIOANALYSIS

There are a number of scientific techniques that provide significant advantages for bioanalysis in support of preclinical clinical programs, from sample preparation, through screening and throughput to the end of the process. Some of the typical techniques and commercially available platforms are listed below:

- Highly selective reaction monitoring (H-SRM), which eliminates chemical noise, lowers detection limits and reduces the likelihood of generating false positives.
- High field asymmetric waveform ion mobility spectrometry (FAIMS) for increased selectivity, which results in shorter run times because FAIMS separation occurs on a millisecond timescale instead of the chromatographic timescale.
- TurboFlow technology to minimize sample preparation by enabling direct injection into the LC–MS/MS system as well as to reduce matrix effects during bioanalysis.
- Aria TLX technology for multiplexing can quadruple MS throughput. Up to four parallel systems are synchronized to a single MS. Each system operates independently, permitting multiple methods to run simultaneously.
- LCQUAN quantitative software package for importing sequence information from external systems, method development, data review, data processing, data reporting, and exporting results to external systems within a 21 CFR Part 11 compliant environment.
- Watson LIMS for full bidirectional interface capability to analytical instruments, shipment and sample tracking through user-designed barcode labels, support for a wide range of PK/TK analyses, and the organization of study results in a unique document management system.
- QuickQuan software was originally developed to speed the pace of LC–MS/MS analysis for high-throughput screening (HTS) in drug discovery through intelligent automation. However, the same automated method optimization capabilities can readily be applied to method development for regulated bioanalysis.

Automated Optimization: One of the rate limiting steps in LC–MS/MS sample analysis is the time required to create accurate and sensitive instrument methods. To ensure the best possible performance, operators may spend a significant amount of time optimizing the mass spectrometer for compounds of interest. In laboratories where higher throughput is required, the time available for instrument optimization is minimal. Automated optimization packages are frequently utilized to increase throughput, but these solutions compromise assay sensitivity for the sake of speed.

Thermo Fisher Scientific developed QuickQuan HTS software to enable the optimization of multiple compounds, acquisition of chromatographic data, and generation of quantitative results in a single analytical run. Combined with the Thermo Scientific TSQ Quantum triple quadrupole mass spectrometers, QuickQuan allows automated LC–MS/MS analysis of chemically diverse compounds. Compound optimization parameters are stored to a database allowing for compounds to be analyzed in multiple assays. QuickQuan uses the database information to automatically generate acquisition and postacquisition methods needed for multiple analytical runs.

Faster Optimization Results: In the experiment outlined below, four compounds (Minoxidil, Imipramine, Paroxetine, and Nefazodone) were optimized using the automated infusion optimization procedure available in QuickQuan. Optimal gas flow rates and temperatures were determined though the TSQ Quantum EZ-Tune interfaces. A 5 ng/mL neat standard was prepared in 50/50 methanol to water and samples were analyzed through QuickQuan. After results were obtained from QuickQuan, the instrument parameters (gas flows, temperatures, and product ion energies) were optimized manually for these four compounds. The same sample was then injected using the manually optimized parameters under identical LC conditions and ion source positioning to compare intensity results between the two techniques. Manual optimization of multiple compounds required a longer time (approximately 30 min) when compared to the time required for QuickQuan to perform the procedure (approximately only 5 min). QuickQuan provided an equivalent optimization result to the manual procedure, in one-sixth the time. Manual optimization only showed a slight improvement in on-column response for one of the four compounds tested. In order to achieve this increase in signal, an extra 25 min of manual tuning was required. Also in comparison, the user did not need to tend to the TSQ Quantum instrument during the QuickQuan optimization procedure. In high-throughput environments where a large number of compounds are analyzed daily, the time required for manual optimization is extremely inefficient.

As these results indicate, QuickQuan enables automated compound optimization when sample throughput is a high priority. The bioequivalence process lends itself well to automation, and Thermo Scientific technologies such as QuickQuan have been designed specifically to foster automation in the laboratory. The QuickQuan application can perform compound optimization, acquisition, processing, and reporting of quantitative sequences using standard methods [43].

6.3.1 Chromatographic Separation

With the initial sample cleanup using SPE, LLE, or PPT, unwanted compounds can be present still in higher concentrations than the analytes of interest. A second stage of cleanup, typically involving LC separation, further separates analytes of interest from the unwanted compounds. Without this further separation, those unwanted and MS/MS unseen compounds present significant challenges. In the LC–MS interface, these compounds compete with analytes for ionization and cause inconsistent matrix effects that are detrimental to quantitative LC–MS/MS performance. Reversed-phase LC has been traditionally used for the quantitative LC–MS/MS. With increase of organic solvent (modifier) concentration in the mobile phase, the analyte retention decreases. However, one should be aware of the potential bimodal retention on the reversed-phase column due to the residual silanol groups. This bimodel retention may cause retention shift during the run or irreproducibility of the method. Among all the reversed-phase columns, of particular interest is the use of monolithic columns (available in C_{18} format) operated at a high flow rate. The chemistry and characteristics of the monolithic columns have been studied and reviewed [44–51]. Compared to a particulate column, the monolithic column has a reduced pressure drop but still maintains high separation efficiency. This is due to its unique bimodal pore structure, which consists of macropores (2 μm) and mesopores (13 nm) [52]. The mesopores provide the surface area for achieving adequate capacity while the macropores allow high flow-rates because of higher porosity, resulting in reduced flow resistance. Monolithic columns have become increasingly popular for use in ultrafast bioanalysis of drug candidates using tandem mass spectrometric detection [53–56]. Polar compounds often have poor retention on a reversed-phase column even with high aqueous mobile phase. Poor analyte on-column retention may result in detrimental matrix effects, which has been identified as one of the major reasons why bioanalytical LC–MS/MS methods fail. High aqueous content mobile phase is also not conducive to achieving the good spray conditions that are critical for sensitivity. Many drugs have basic functional groups, and acidic mobile phases are used and MS in the positive ion mode detects these compounds as the protonated ions. Ionization of polar compounds further decreases the analyte on-column retention on a reversed-phase column. To overcome this mismatch between reversed-phase LC and MS detection, other chromatographic materials were investigated to achieve better sensitivity and better on-column retention. Zirconia-based column may offer different retention mechanism from silica-based column and has been explored for quantitative bioanalytical LC–MS/MS application. More review and discussion are present in following sections.

6.3.1.1 Ultra-Performance Liquid Chromatography The use of smaller particles in packed-column LC is a well-known approach to shorten the diffusion path for a given analyte. As an approximation, the time required to achieve a given degree of resolution between two compounds decreases as the square of the particle diameter assuming everything else being constant [57]. Recent technology advances have made available reverse phase chromatography media with sub-2 μm particle size along with liquid handling systems that can operate such columns at much higher pressures. This technology termed ultra-performance liquid chromatography (UPLC) offers significant theoretical advantages in resolution, speed, and sensitivity for analytical applications, particularly when coupled with mass spectrometers capable

FIGURE 6.4 UPLC–MS/MS (TQ-S) ACQUITY system from Waters Corporation.

of high-speed acquisitions. The principles of and recent developments in UPLC were reviewed by Mazzeo et al. [43].

In 2004, Waters commercialized the ACQUITY UPLC system (Figure 6.4 Courtesy from Waters UPLC-MS/MS TQ-S ACQUITY as a current Model) that is able to work at pressures up to 1000 bar. Other manufacturers followed this approach, such as Jasco with the Xtrem LC capable of handling of pressures up to 1000 bar, and Agilent with 1200 Series Rapid Resolution LC system compatible with pressures up to 600 bar. Sub-2 μm particle columns have become available from almost all major column manufacturers. The strengths of UPLC technology promote the ability to separate and identify drug compounds with significant gains in resolution and sensitivity and marked reductions in the overall time of analysis. Since its introduction, ultra-performance liquid chromatography has served as a powerful analytical tool for high-throughput analysis. Al-Dirbashi et al. [58] reported a method for the determination of doxazosine in human plasma by UPLC–MS/MS. Plasma extracts after liquid–liquid extraction were separated on a C_{18} UPLC column packed with 1.7 μm particles. The total run time was 2 min. The calibration curve based on peak area ratio was linear up to at least 100 ng/mL, with a detection limit of 0.02 ng/mL. Wren et al. [59] investigated UPLC as an alternative to HPLC for the analysis of pharmaceutical development compounds. Data on three compounds were presented showing that significant reductions in separation time can be achieved without compromising the quality of separation. Apollonio et al. [60] assessed the separation of several commonly encountered amphetamine-type substances using the Acquity UPLC-Micromass Quattro Micro API MS system (Waters Corporation, USA). Using a polydrug reference standard and whole blood extracts, the authors successfully separated and identified amphetamine, methamphetamine, ephedrine, pseudoephedrine, phentermine, MDA, MDMA, MDEA, and ketamine in less than 3 min. In addition to the significant reduction in overall run time, all peaks exhibited acceptable resolution, indicating the capability to separate 5–11 peaks in 1.75 min. Shen et al. [61] conducted validation of an bioanalytical method for determination of desloratadine and 3-hydroxydesloratadine using UPLC in conjunction with mix-mode solid phase extraction. The dynamic range of the assay was from 0.025 to 10 ng/mL using 96-well solid phase extraction. The total run time was slightly over 2 min per sample. The approach of orthogonal extraction/chromatography and UPLC significantly improves assay performance while also increasing sample throughput for drug development studies.

Other direct comparison experiments using UPLC–MS/MS and HPLC–MS/MS have shown that the UPLC–MS/MS improved cycle time by 50–100% with increased sensitivity. Churchwell et al. [62] explored the differences in LC–MS performance by conducting a side-by-side comparison of UPLC for several methods previously optimized for HPLC-based separation and quantification of multiple analytes with maximum throughput. Sensitivity increases with UPLC, which were found to be analyte dependent, were as large as 10-fold and improvements in method speed were as large as fivefold under conditions of comparable peak separations. Improvements in chromatographic resolution with UPLC were apparent from generally narrower peak widths and from a separation of diastereomers not possible using HPLC. Yu et al. [63] developed a quantitative UPLC–MS/MS protocol for a five-compound mixture in rat plasma. A similar high-performance liquid chromatography/tandem mass spectrometry (LC–MS/MS) quantification protocol was developed for comparison purposes. Both UPLC–MS/MS and LC–MS/MS analyses were performed in both positive and negative ion modes during a single injection. Peak widths for most standards were 4.8 s for the HPLC analysis and 2.4 s for the UPLC analysis. There were 17–20 data points obtained for each of the LC peaks. Compared with the LC–MS/MS method, the UPLC–MS/MS method offered threefold decrease in retention time, up to 10-fold increase in detected peak height, with twofold decrease in peak width. Limits of quantification (LOQs) for both HPLC and UPLC methods were also evaluated. For UPLC–MS/MS analysis, a linear range up to four orders of magnitude was obtained with r^2 values ranging from 0.991 to 0.998. The LOQs for the five analytes ranged from 0.08 to 9.85 ng/mL. The carryover of the UPLC–MS/MS protocol was negligible and the robustness of the UPLC–MS/MS system was evaluated with up to 963 QC injections.

6.3.1.2 Monolithic Chromatography There is considerable interest to improve throughput by using monolithic columns because they exhibit higher separation efficiency at high flow velocities when compared to conventional LC columns [44–51, 64]. The structural characteristics of the monoliths used in chromatography and those of the conventional beds of particulate packing materials are very different. An important characteristic of monoliths is their high external porosity resulting from the structure of the network of through-macropores. Another is the structure of the stationary phase skeleton that consists in a network of small, thin threads of porous silica. The networks of the two phases twist around each other and provide the intricate structure of the monolithic medium. These two structural characteristics allow the combination of a low hydraulic resistance of the column to the stream of mobile phase and an enhancement of the rate of the mass transfers of the sample molecules through the beds of these continuous porous stationary phases. Consequently, the chromatographic behavior of monolithic columns differs markedly from that of the conventional columns packed with spherical particles. Two types of monolithic supports are currently available, namely organic polymers such as polymethacrylates, polystyrenes, or polyacrylamide and inorganic polymers based on silica and more recently on carbon and zirconia. In LC, monoliths made of silica produce better chromatographic performances than organic polymers. Monolithic silica columns developed from a sol–gel process [44, 64] have been commercialized by Merck (Darmstadt, Germany) and Phenomenex (Torance, CA, USA) under the brand names Chromolith™ and Onyx™, respectively. They possess macropores of 2 µm and mesopores of 13 nm. The main feature of silica rod columns is a higher total porosity, approximately 15% higher than those of conventional particulate HPLC columns. The resulting column pressure drop is therefore much lower, allowing operation at higher flow rates including flow gradients. Consequently, HPLC analysis can be performed much faster, as it is demonstrated by various applications. Because of the high permeability of monoliths, several columns can be coupled in series ($L > 1$ m) to generate high efficiency ($N > 100{,}000$ theoretical plates). However, the large internal column diameter (e.g., 4.6 mm in Chromolith and Onyx) of currently available monolithic silica columns is not fully compatible with MS and requires a high consumption of organic solvent. When coupled to an ESI interface, the use of splitter is required to keep the flow rate entering the source within optimum conditions.

A high-throughput LC–MS/MS method [47] using a Chromolith RP-18 monolithic column was developed for the determination of bupropion (BUP), an antidepressant drug, and its metabolites, hydroxybupropion and threo-hydrobupropion (TB), in human, mouse, and rat plasma. After semi-automated 96-well liquid–liquid extraction, analytes were separated with a mobile phase delivered isocratically at 5 mL/min, the eluate was split postcolumn to 2 mL/min and directed into a turbo-ionspray source. Chromatographic separation of bupropion and its metabolites was achieved within 23 s. The method was linear over a concentration range of 0.25–200 ng/mL for bupropion and threo-hydrobupropion, and 1.25–1000 ng/mL for hydroxybupropion. The monolithic column performance as a function of column backpressure, peak asymmetry, and retention time reproducibility was adequately maintained over 864 extracted plasma injections. Barbarin et al. [48] presented a high-throughput LC–MS/MS method for the determination of methylphenidate (MPH), a central nervous system stimulant, and its deesterified metabolite, ritalinic acid (RA) in rat plasma samples. A separation of these two compounds was achieved in 15 s by employing a 3.5 mL/min flow-rate, a porous monolithic column and a Turbo-Ionspray source compatible with relatively high flow-rates. Overall 768 protein-precipitated rat plasma samples (eight 96-well plates) containing both MPH and RA were analyzed within 3 h and 45 min. The calibration curves ranged from 0.1 to 50 ng/mL for MPH and from 0.5 to 50 ng/mL for RA. Baseline resolution of MPH and RA was consistent throughout analysis.

A monolithic column was directly compared with a conventional C_{18} column as the analytical column in method validation of a drug and its epimer metabolite [49]. Because the chosen drug and its epimer metabolite have the same selected reaction monitoring (SRM) transitions, chromatographic baseline separation of these two compounds was required or essential. Sample preparation, mobile phases, and MS conditions were kept the same in the column comparison experiment. The eluting flow rate for the monolithic column system was 3.2 mL/min (with 4:1 splitting) and for the conventional C_{18} column system was 1.2 mL/min (with 3:1 splitting). The monolithic column system had a run time of 5 min and the conventional C_{18} column system had a run time of 10 min. The methods on the two systems were found to be equivalent in terms of accuracy, precision, sensitivity, and chromatographic separation, but the monolithic column method increased the sample throughput by a factor of two.

The significantly improved separation speed by monolithic columns demanded higher throughput on sample extraction. An attractive approach using monolithic separation is to combine it with high-flow on-line extraction, which allowed for the fast extraction and separation of samples. Zeng et al. [50] used such an approach for multiple-component quantitative LC–MS/MS assays of drug candidates in biological fluids. An evaluation of the approach was performed using a mixture of fenfluramine, temazepam, oxazepam, and tamoxifen in plasma. A considerably reduced runtime was achieved while maintaining good chromatographic separations. A total cycle time of 1.2 min was achieved which included both sample extraction and subsequent monolithic column separation via column switching. A total of over 400 plasma samples were analyzed in <10 h in routine support of drug discovery programs. Zhou et al. [51] developed a high-throughput LC–MS/MS method that combined on-line sample extraction using turbulent flow chromatography with a monolithic column separation, for direct injection analysis of drugs and metabolites in human plasma samples. By coupling a monolithic column into the system as the analytical column, the method enables running "dual-column" extraction and chromatography at higher flow rates, thus significantly reducing the time required for the transfer and mixing of extracted fraction onto the separation column as well as the time for gradient separation. It was demonstrated that the total run time for this assay with a baseline separation of two analytes is less than 1.5 min.

6.3.1.3 Hydrophilic Interaction Chromatography Hydrophilic interaction chromatography (or hydrophilic interaction liquid chromatography, HILIC) is a version of normal phase liquid chromatography. Any polar chromatographic surface can be used for HILIC separations. Even nonpolar bonded silicas have been used with extremely high organic solvent composition, when the silica used for the chromatographic media was particularly polar for the application. Hydrophilic interaction liquid chromatography coupling with mass spectrometry has been gaining recognition as a valuable technique for analyzing polar molecules in biological matrix in recent years [52–55]. Polar compounds typically have very limited retention on reversed-phase (RP) columns. In order to separate the analyte from the matrix interference, reverse phase HPLC mobile phase with a very low organic content must be used. Sometimes trifluoroacetic acid or ion pair reagents have to be added into the mobile phase. When using ESI–MS, the very high aqueous mobile phase can cause low ionization efficiency. HILIC is a useful technique for the retention of polar analytes that offers a difference in selectivity compared to traditional reversed-phase chromatography. The highly volatile organic mobile phases used in HILIC provide increased electrospray ionization–mass spectrometry (ESI–MS) sensitivity. Although some column companies are marketing column specific for HILIC, most columns used with normal phase HPLC such as pure silica columns or cyano columns can be operated under HILIC conditions. The retention of an analyte on a HILIC column is determined by its polarity. Elution is driven by the water content in the mobile phase for better detection. HILIC often yields narrower peak, which further improves the signal noise ratio.

Eerkes et al. [52] developed a bioanalytical method using automated sample transferring, automated liquid–liquid extraction, and hydrophilic interaction liquid chromatography-tandem mass spectrometry for the determination of fluconazole in human plasma. After liquid–liquid extraction, the extract was evaporated to dryness, reconstituted, and injected onto a silica column using an aqueous-organic mobile phase. The chromatographic run time was 2.0 min per injection.

In a typical off-line sample preparation procedure using liquid–liquid extraction, solid phase extraction, or protein precipitation, the organic extracts need to be evaporated and reconstituted. The evaporation step could be very time consuming if the water content of the organic extracts is high. The use of HILIC could eliminate the evaporation and reconstitution steps that hamper improvement of throughput and automation. With the high organic mobile phase, samples can be dissolved in organic solvent and injected without the problem of mismatching with mobile phase and peak shape deterioration. Thus it is possible to have samples injected onto columns after protein precipitation, liquid–liquid extraction, or solid phase extraction without the steps of dry down and reconstitution. Xue et al. [53] developed and validated a single-pot LLE with HILIC–MS/MS method for the determination of Muraglitazar, a hydrophobic diabetes drug, in human plasma. After extraction with acetonitrile and toluene, the organic layer was then directly injected onto an LC–MS/MS system. Chromatographic separation was achieved isocratically on a Hypersil silica column with a mobile phase containing 85% of methyl *t*-butyl ether and 15% of 90/10 (v/v) acetonitrile/water and 0.3% trifluoroacetic acid. Postcolumn mobile phase of 50/50 (v/v) acetonitrile/water containing 0.1% formic acid was added. The standard curve, ranged from 1 to 1000 ng/mL, was fitted to a $1/x$ weighted quadratic regression model. The modified mobile phase was more compatible with the direct injection of the commonly used extraction solvents in LLE. Furthermore, the modified mobile phase improved the retention of Muraglitazar, a hydrophobic compound, on the normal phase silica column. In comparison with a reversed-phase LC–MS/MS method, this single-pot LLE, HILIC/MS/MS method improved the detection sensitivity by greater than fourfold based upon the LLOQ signal to noise ratio. Song et al. [54] demonstrated a similar approach of eliminating evaporation and reconstitution steps in 96-well LLE by using HILIC–MS/MS on silica column with high organic/low aqueous mobile phase. Omeprazole, its metabolite 5-OH omeprazole, and internal standard, desoxyomeprazole, were extracted from 0.05 mL of human plasma using 0.5 mL of ethyl acetate in 96-well plate format. A portion (0.1 mL) of the ethyl acetate extract was diluted with 0.4 mL of acetonitrile and 10 μL was injected onto a Betasil silica column. Mobile phase with linear gradient elution consists of acetonitrile, water, and formic acid. The flow rate was 1.5 mL/min with total run time of 2.75 min. The method was validated for a low limit of quantitation at 2.5 ng/mL for both analytes, respectively.

Deng et al. [55] coupled high-flow on-line reversed-phase extraction with normal phase on silica columns with aqueous-organic mobile phase LC–MS/MS to quantify drug candidates in biological fluids. The orthogonal separation effect obtained from this configuration considerably reduced matrix effects and increased sensitivity for highly polar compounds as detected by selected reaction monitoring. This approach also significantly improved the robustness and limit of detection of the assays. An evaluation of this system was performed using a mixture of albuterol and bamethan in rat plasma. The system has been used for the quantitation of polar ionic compounds in biological fluids in support of drug discovery programs.

6.3.2 Selective and Sensitive Detection

6.3.2.1 Mass Spectrometric Detection and System Multiplexing In addition to the high flow rate LC–MS/MS, another approach for achieving high throughput is to employ parallel analysis approach, either using multiple inlets in the mass spectrometer source and multiplexing LC unites into one MS. A four-channel multiplexed electrospray (MUX) ion source (Waters-Micromass, Manchester, UK) allows the nonstop introduction of the eluent from four LC columns and/or LC units into the mass spectrometer. Multiplexing, or parallel LC–MS/MS, is widely accepted as a way to increase bioanalysis throughput [56, 65, 66]. The concept of multiplexing originated from taking advantage of the time difference between the chromatography run time and the mass spectrometer data acquisition time. The mass spectrometer data acquisition time often occupies only a small portion of the total chromatographic run time. Most of time, the mass spectrometer is idle in waiting for the next sample to come. In multiplexing setups, multiple LC systems or columns are connected in parallel to a single mass spectrometer. Samples are introduced onto the LC systems in a staggered fashion so the analyte reaches the mass spectrometer from each LC system in a serial order without overlapping. Multiple samples can be analyzed within the same time period required for one sample to be analyzed with a single LC system. Hsieh et al. [56] reported their validation of an LC–MS/MS method with multiplexing HPLC. A Leap HTS Twin-PAL with two injection syringes was employed. With high speed on-line extraction using turbulent flow columns from Cohesive Technologies

(Thermo-Fisher Scientific), the method has a sample to sample cycle time as low as 0.4 min. Lindqvist et al. [65] made a system with three parallel HPLC columns and one mass spectrometer. Samples were injected by an auto-sampler and directed to three columns and the mass spectrometer by two six-port valves and one multiposition valve. The timing of the valve's action was controlled by the auto-sampler time program. More than 2.5 times throughput increase was achieved, as the "per sample" analysis time was decreased from 8 to 3 min (Figure 6.5).

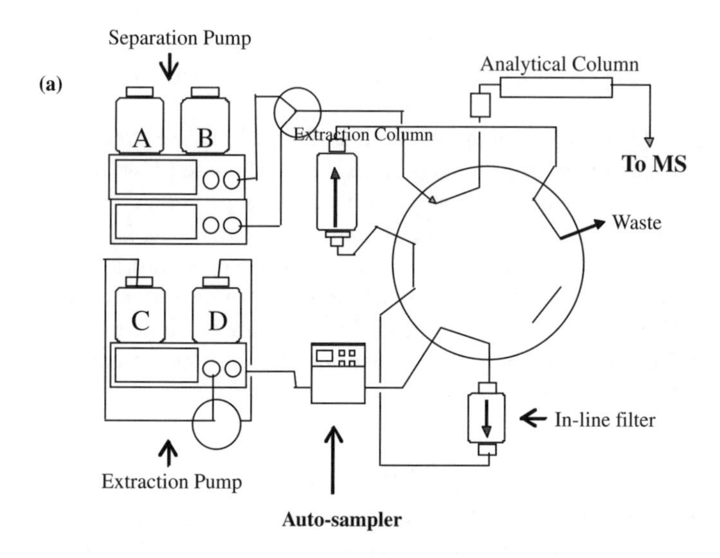

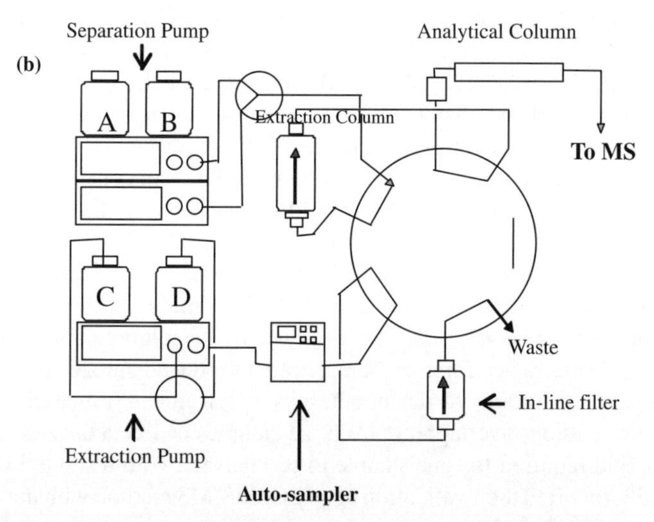

FIGURE 6.5 Schematic diagrams of the instrumental setup for the on-line extraction with a Chromolith cartridge (4.6 × 10 mm) as extraction column: (a) sample loading and extraction mode and (b) elution and separation mode.

Sample extraction, separation, and detection performed in a four-channel parallel format that resulted in an overall throughput of about 30 s/sample from plasma have been reported by Deng et al. [67]. After automated solid phase extraction, the extracted plasma samples were injected onto four parallel monolithic columns for separation via a four-injector auto-sampler. The use of monolithic columns allowed for fast and well-resolved separations at a considerably higher flow rate without generating significant column backpressure. This resulted in a total chromatographic run cycle time of 2 min on each 4.6 × 100 mm column using gradient elution. The effluent from the four columns was directed to a triple quadrupole mass spectrometer equipped with an indexed four-probe electrospray ionization source (Micromass MUX interface—Waters Corp.). The performance of this system was evaluated by extracting and by analyzing twelve 96-well plates (1152) of human plasma samples spiked with oxazepam at different concentrations. The good separation efficiency provided by this system allowed for rapid method development of an assay quantifying the drug candidate and its close structural analog metabolite. The method was cross-validated with a conventional LC–MS/MS assay.

High-field asymmetric waveform ion mobility spectrometry is another technology used by the bioanalytical practitioner to improve selectivity, sensitivity, and throughput. FAIMS separates ions at atmosphere pressure by transmitting a subset of ions and filtering out chemical background and isobaric endogenous interferences. It is installed between the atmospheric pressure ionization sprayer and mass spectrometer orifice. Ions formed from the ion source are carried by a stream of gas through a pair of closely spaced electrodes. A high frequency asymmetric waveform voltage (dispersion voltage, DV) and a DC voltage (compensation voltage, CV) are applied across FAIMS electrodes. The transmissions of ions are based on the mobility of the ions in the electric field, which can be adjusted by changing the compensation voltage. Ions from different compounds with same mass may have different mobility values. Thus FAIMS can separate background and interference that is not distinguishable by quadruple mass spectrometer. As an additional filter between HPLC and mass spectrometer, FAIMS technology can improve method selectivity, reduce noise, simplify HPLC condition, and shorten method development time. Kapron et al. [66] reported a study of analyzing an amine compound using FAIMS to eliminate its *N*-oxide metabolite interference. When analyzed using traditional LC–MS/MS method, this coeluting metabolite caused interference that made the analysis results unreliable. The interference was due to conversion of the metabolite to the drug molecule in the ion source. After applying FAIMS, this metabolite interference was successfully removed.

Due to their unpredictable character, matrix effects in quantitative analysis using LC–ESI–MS or LC–APCI–MS are a serious concern. It is clear that the use of real sample extracts is necessary already at an early stage of method development, as the matrix effect may have serious impact on the choice of the most appropriate sample pretreatment method, ionization method and mode, and even the most adequate mobile-phase composition. A study was performed where high-flow online reversed-phase extraction was coupled with normal phase chromatography on silica columns. Matrix effects were reduced considerably by this orthogonal separation configuration [67].

6.4 TECHNICAL CHALLENGES AND ISSUES ON REGULATED BIOANALYSIS

Trace level and quantitative determination of analytes in complex biological matrices has been always great challenge in data quality and integrity for regulatory review and approval processes. Some of the real world issues and situations are elaborated below:

6.4.1 Matrix Effect

Often described in the literature as other terms such as matrix ionization effect or ion suppression effect, matrix effect is a phenomenon observed when the signal of analyte can be either suppressed or enhanced due to the coeluting components that originated from the sample matrix. When a rather long isocratic or gradient chromatographic program is used in the quantitative assay, matrix effect may be not present at the retention time for an analyte. However, in the case of high-throughput LC–MS/MS analysis, matrix effect is one of the major concerns to be addressed in method development and validation, especially when analyte is not well separated from the LC-front.

Matrix effect could be introduced from a formulation agent used in toxicological studies. Larger et al. [68] observed strong ion-suppression in a preliminary pharmacokinetic study from a polysorbate cosolvent, which, if undetected, would have given highly erroneous pharmacokinetic results and possibly could have led to the inappropriate elimination of a promising drug candidate. Some excipients commonly used in formulations are polydispersed polymers, for which very limited pharmacokinetic information is available. Further investigation is needed to better understand the mechanisms of ion-suppression and the kinetics of the suppressing species to allow the development of new LC–MS/MS-based analytical strategies, which will not be subject to such ionization interferences.

One problem brought by matrix suppression effect is reduced sensitivity when analyte signal is suppressed [69, 70]. Detailed studies on matrix effects revealed that the ion suppression or enhancement is frequently accompanied by significant deterioration of the precision of the analytical method as demonstrated by Matuszewski et al. [71]. The authors studied the precision (%RSD) upon repetitive injection of postextraction spiked plasma samples as a function of the analyte concentration for a single lot and for five different lots of plasma. While for the single plasma lot the precision is acceptable, it may not be when different plasma lots are taken into account. Generally, matrix effect impacts more on the low end of calibration curve (e.g., LLOQ) than the mid range or high end of the curves.

When discussing matrix effects, it is useful to discriminate between ion suppression or enhancement by the matrix at one hand, and different matrix effects exerted by different sample lots at the other hand. A useful nomenclature was suggested by Matuszewski et al. [71] and is adopted here for discussion. The difference in response between a neat solution sample and the postextraction spiked sample is called the absolute matrix effect, while the difference in response between various lots of postextraction spiked samples is called the relative matrix effect. If no counteraction is taken, an absolute matrix effect will primarily affect the accuracy of the method, while a relative matrix effect will primarily affect the precision of the method.

The matrix effects are generally due to the influence of coeluting compounds on the actual analyte ionization process, that is, they happen well before the analyte ions enter the high vacuum of the mass analyzer. Matrix effects are known to be both compound and matrix dependent. It was demonstrated that matrix-induced ion suppression is especially important for early eluting compounds, while later eluting compounds are not affected as often. Suppression or enhancement effects may be exerted by any coeluting component entering the atmospheric-pressure ionization source via the liquid stream. Some mobile-phase additives are also known to suppress or enhance analyte response. Although such effects are sometimes called matrix effects, it appears useful to discriminate between effects due to the analytical system, for example, mobile-phase composition, source parameters, and effects due to the actual analyte matrix.

Absolute matrix effect can be easily detected by comparing the response obtained from a neat solution and that from a postextraction spiked sample. Difference in response may indicate either ion suppression or ion enhancement. To pinpoint the location of matrix peaks or affected region in the chromatogram by matrix effect, the analyte solution is usually postcolumn infused into the ion source while a blank matrix extract is injected through a column. For testing of relative matrix effect, samples from different sources or lots must be analyzed. Often plasma samples from different lots are spiked with analyte at the low end of calibration curve (e.g., samples at low quality control or lower level quantitation limit) and tested. Matuszewski et al. [72] described a simple experimental approach for studying and identifying the relative matrix effect in quantitative analyses by LC–MS/MS. It was shown that the variability of standard line slopes in different lots of a biofluid precision of standard line slopes expressed as coefficient of variation, CV (%), may serve as a good indicator of a relative matrix effect and, it is suggested, this precision value should not exceed 3–4% for the method to be considered reliable and free from the relative matrix effect liability. Endogenous phospholipids cause ion suppression in both positive and negative ESI modes, as observed, and must be removed or resolved chromatographically [70, 73]. It is suspected that one major contributor to matrix effects are Glycerophosphocholines (GPChos) because of their surfactant-like properties. A method was developed for detecting GPChos during LC–MS/MS method development [73]. The approach uses high energy in-source collisionally induced dissociation (CID) to yield trimethylammonium-ethyl phosphate ions (*m/z* 184), which are formed from mono- and disubstituted GPChos. The resulting ion is selected by the first quadrupole (Q1), and monitored without further fragmentation.

The best way to eliminate the influence of matrix effects on the accuracy and precision of a quantitative method is through the use of stable isotope labeled internal standards [72, 74]. It is important to add stable isotope labeled internal standards prior to sample pretreatment and/or extraction. In that way, they can correct or normalize for analyte losses during sample pretreatment as well as matrix-related suppression or enhancement during analyte ionization. Although it is generally believed that the use of an isotopically labeled internal standard corrects for almost all matrix effects, data reported for the bioanalysis of mevalonic acid indicate that this assumption needs to be demonstrated during method development and validation [75]. Wang et al. [76] reported that retention time difference between analyte and isotopically labeled internal standard could lead to the variability of a method's precision. In addition, mutual suppression or enhancement of responses of an analyte and its coeluting isotopically labeled internal standard has occasionally been reported [77]. However, calibration curves were linear if an appropriate IS concentration was selected for a desired calibration range to keep the response factors constant.

Obviously, limited availability and high costs have hampered the wide application of isotopically labeled internal standards. Advanced sample pretreatment methods can help in reducing or eliminating matrix effects [69, 78], and together with efficient chromatographic separation, one may eliminate the sample constituents responsible for the matrix effects. Alternatively, one may reduce or eliminate the influence that matrix effects have on the accuracy and/or precision of the method by one or a number of the following measures: change to a different MRM channel, change to another ionization methods, and/or change the mobile-phase composition [79, 80]. It has been shown that the precision of a method in which an analog internal standard is used, can be significantly improved by modifying the mobile-phase conditions in such a way that analyte and analog internal standard coelute [77, 81]. Regardless, stable isotopic internal standards have clearly become industry best practice for reliable quantitation.

6.4.2 Method Validation and Critical Issues during Sample Analysis

Once a bioanalytical method is developed, various tests (collectively called method validation) are conducted to prove that the method can be used for its intended application. The FDA guidance defines the relevant bioanalytical terms and acceptance criteria. In addition to the evaluation for precision, accuracy, and stability of the LC–MS/MS methods, an investigation of matrix effects should be performed. Analyte extraction recovery and potential carry-over of the extraction apparatus or autosampler should also be carefully examined. The pharmaceutical industry and regulatory agencies generally recognize and accept that the validation of a bioanalytical method is performed using standards and quality control samples in which the analytes have been fortified with a blank biological matrix. However, the incurred samples may or may not be the same as the fortified quality control samples. This difference may lead to a significant bias for the quantitative bioanalysis and must be carefully evaluated.

The drug and metabolite(s) are assumed to follow the same pharmacokinetic profile and therefore have roughly the same proportion for their concentration profiles. Quality control samples are therefore prepared in such a way that the concentration ratio for drug and metabolite is constant over the entire curve range. In many cases, however, this is not true and lopsided concentration profiles can be observed, especially for prodrugs that typically have much shorter half-life than the active drug. This phenomenon has been observed by Lee et al. [82]. Protein-binding of the drug candidate and protein could be significantly different for the quality control samples and incurred samples. This can also cause quantitation bias, depending upon the extraction method. Ke et al. used a chelating agent (disodium EDTA) to release protein-bound analytes during 96-well liquid–liquid extraction [83]. Yang et al. presented an interesting case study for diagnosis and trouble-shooting of this type of problem [84]. The method was validated using stable-isotope labeled internal standard with a LLE method using hexane as the extraction solvent. However, upon repeat analysis of the same samples, the concentration values increased to fivefolds of the original value. The concentration increased with each additional freeze/thaw cycle. It was found that this drug candidate has a strong protein-binding and hexane is not sufficient to release the drug candidate from the protein. Freeze/thaw cycles gradually denatured protein and weakened the binding, resulting in increase the free and extractable drug candidate concentration. PPT method (protein-binding was disturbed) was then used and consistent result was obtained. Two lessons learned here—A more selective extraction method (here LLE versus PPT) may not be the one mostly suitable for the study; stable-isotope labeled internal standard is not an automatic guarantee for a method suitable for routine sample analysis. The later point of view was also expressed by Jemal et al. [85]. Fura et al. also discussed the shift in pH of biological fluids during storage and processing as well as its impact on bioanalysis [86]. Compound adsorption to storage container often happens in sensitive LC–MS/MS assays for urine analysis, which causes nonlinear response and loss of sensitivity. The use of a zwitterionic detergent 3-3-cholamidopropyl-dimethylammonio-1-propanesulfonate (CHAPS) as an additive in urine can prevent analyte from adhering to surfaces during sample collection, storage, and preparation [87]. Due to the superior selectivity offered by the tandem mass spectrometric detection, fast LC–MS/MS has been frequently used to speed up the sample analysis. This can be achieved either by increasing the flow rate or by increasing the elution strength of the mobile phase. Increase of flow rate may slightly decrease the column efficiency but to a large extend the capacity factor is not sacrificed. Of course, one has to find suitable column amenable to the high flow rate, not always an easy task. Increasing elution strength will certainly decrease capacity factor and the analyte of interest may fall into the suppression or interference zones. How fast is too fast? Phase II metabolites (glucuronide, sulfate and glutathione-conjugation products) and pro-drugs are notoriously unstable in the LC–MS interface and are easily fragmented back to the drugs [88–90]. Without adequate chromatographic retention and separation between the drug and its Phase II metabolites or pro-drugs, the drug concentration can be easily over-estimated for the incurred samples. This bias will not be easily detectable since neither the calibration standards nor the quality control samples may contain these Phase II metabolites. A very fast LC using stronger elution strength would mask the interference peak with the analyte, causing an overestimation of the analyte's concentration. Generally in our organization, a fast and adequate quantitative LC–MS/MS analysis should have the retention time $\geq 2\ t_0$, at least 10 acquisition points per peak, and good separations from possible interferences or matrix effect. Also, the analysis cycle is limited by the auto-sampler speed as well as the adequate injector wash time in order to reduce carry-over.

6.4.3 Method Transfer

Development and establishment of bioanalytical methods—essential for any drug development program—is one area where it takes unpredictable time period and delays often occur, but with proper planning, could be alleviated. Timely method transfer plays an important role in expediting drug candidates through development stages. Method transfer is not a trivial task and requires careful planning and constant communication between the laboratory personnel involved in the transfer. It is often necessary to move bioanalytical methods from one laboratory to another over the course of a drug development program. It is therefore critical to understanding the challenges of method transfers and plan for this eventuality to ensure rapid transition between the different stages of drug development and to constrain one's predictable cost. Method transfers occur for a number of reasons. This may occur among crossfunctional groups, site to site within country or abroad. Quite often with typical situation—a sponsor has a method available in-house and requires a contract research organization (CRO) to increase throughput for the assay. In other instances, a sponsor must transfer assays to a different laboratory because of financial opportunity, logistical requirements, or dissatisfaction with services offered from internal or external suppliers or collaborators. Method transfers frequently occur between preclinical and Phase I clinical studies, or Phases I and II clinical studies. If in-house methods have been developed and analytical resources exist, a sponsor will often elect to analyze the preclinical and early clinical samples in order to maintain control over studies that support an IND submission or early clinical studies. Some of these studies involve smaller sample numbers and have fewer logistical issues than later-phase studies. If methods are transferred between Phases I and II, the majority of these will have already been validated—either by the sponsor or contracted laboratories—according to existing regulations and guidelines for validation of pharmacokinetic, immunogenicity, or cell-based neutralizing antibody assays, and in line with the needs of multiple programs. If assays are transferred between preclinical and clinical phases, it is often due to a sponsor's internal resources being limited or redirected to support other programs; consequently it may not be possible to manage the sample analysis. In some instances, transferred methods are in a partially evaluated or prevalidated state, as they have only been used in an unregulated capacity where such a method may be sufficient for providing exploratory data to support pharmacology, toxicology, drug metabolism, or formulation studies.

The challenges of method transfer apply to both small- and large-molecule programs. Large-molecule therapeutics tend to encounter these challenges more frequently, partly because compounds are often unique, having been modified in some manner to improve or extend their pharmacological activity or bioavailability. Companies with large-molecule candidates will have more developed analytical methods to detect their product utilizing specific reagents (e.g., antibody pairs), which will more frequently necessitate a method transfer. In addition, due to the immunogenic nature of biological therapeutics, bioanalytical programs for large molecules have additional complexity. Sponsors will require validated methods for both pharmacokinetic measurements (e.g., quantity of analyte in matrix) and immunogenicity measurements (e.g., the immune response to the drug product). Generally, issues that occur during bioanalytical method transfers can be grouped into the following potential categories: method development/establishment, performance expectations and information transfer/communication situations.

6.4.3.1 Method Development/Establishment The success of the method transfer will depend to a great extent on the method development scientists' experience and knowledge of applicable regulatory guidelines. Laboratories that have limited experience in developing assays for bioanalysis in preclinical or clinical studies may not appreciate or be completely aware of the requirements of a fully validated method, or even have relevant experience to resolve common issues that may arise. An example of this problem can be seen in the process by which analyte stability is established. A method might be transferred under the assumption that long-term stability of the molecule in relevant biological matrices has been established at either −20 or −70°C. When tested at the laboratory after being transferred it fails to meet stability acceptance criteria. Upon further investigation, it has often been discovered that the calibrator curve was not prepared immediately before analysis, or that stability had not been established for drug stock solutions used to prepare for the calibrators. Industry bioanalytical method validation consensus requirements are clear that stability must be established against calibrators that have been freshly prepared on the day of the evaluation. However, if the stability QCs were compared to freshly prepared calibrators, the stability evaluations would have failed. The degradation of the QCs below an acceptable limit would be masked by the use of frozen calibrators. This situation has been likened to "the shrinking yardstick" where your ruler is changing to the same extent as the item it is meant to measure. This results in a method with inadequate stability for the intended purpose and will require more method development before it can be used in a regulated environment.

For large molecule assays, selectivity testing is one aspect of method development that is often overlooked. Selectivity is defined as the ability of the method to detect analyte in the biological matrix of interest. It is negatively affected by matrix components that obstruct the assay's ability to detect the analyte of interest, either through binding directly with the analyte or through indirect effects on the assay system. Selectivity is evaluated by fortifying individual lots of matrix with the analyte at or near the lower limit of quantitation of the assay, and at a higher concentration (typically the high-quality control or HQC) and assessing the accuracy of recovery. Ideally, it should be possible to accurately measure analyte concentration in the individual lots. Conversely, a method for which many lots show poor recovery requires further development. For some methods, selectivity testing may have only been performed in a small number of lots or even a single pool of matrix before transfer. Testing in a pool of matrix obscures lot-to-lot differences and commonly results in further method development, sometimes after validation attempts fail.

CROs consider this a critical question to resolve before method transfer can be considered advisable.

CROs develop and validate numerous methods per year and possess standard operating procedures (SOPs) that govern how a method should be validated. A best practice before transferring a method from sponsor to CRO is to ensure that both parties' scientific teams communicate early and frequently about method characteristics. A sponsor should request a review of the CRO's method validation SOP or a summary of the evaluations performed. If possible, a face-to-face review should take place to discuss the way evaluations are performed with the CRO's team.

6.4.3.2 Performance Expectations Another aspect of a drug development program that can lead to an unsuccessful method transfer is failure to recognize the stresses to which methods are subjected in a high-throughput laboratory. A method that functions well when run infrequently by a single individual in a pharmaceutical or academic research laboratory may not demonstrate robust performance in a high-throughput environment where hundreds or thousands of samples are to be analyzed each week by multiple analysts. It is incumbent on the sponsor and CRO to outline their throughput expectations and explain how the method will be used. Thoroughly testing the robustness of the method prior to method transfer is required. Testing by multiple individuals over many days ascertains whether methods are able to withstand the rigors of the bioanalytical testing environment. CROs often use this approach to evaluate a method before formally validating the method.

6.4.3.3 Transfer of Information/Communications Less-than-optimal method performance will be compounded by a limited transfer of technical information between sponsor and the CRO performing the validation. Although it may seem obvious, thorough sharing of information during transfer is critical to success. Both individuals and organizations develop technical habits that differ slightly from other laboratories (e.g., methods of pipetting, storage of reagents, analyte stock solution formulation, use of specialized equipment, plate washers, etc.). Differences between the laboratories may exaggerate inherent method variability and cause a validation failure due to unacceptable accuracy and precision. In addition, laboratories source common reagents from several suppliers. Differences in the purity and preparation of chemicals among method developers, suppliers, and users are common. This may affect the performance of both large- and small-molecule methodologies, often in different and unpredictable ways. Furthermore, difficulties with reagent bridging and qualification are not restricted to the project validation phase and may be encountered throughout validation and sample analysis, in particular with ligand-binding assays. Securing the availability of antibodies, antibody conjugates, and other detection reagents for the foreseeable duration of the project is critical for immunochemical methods. For projects with timelines spanning several years, it is strongly recommended that a strategy for sourcing these reagents be adopted early to avoid potential problematic situations. Plans ensuring suitability of new reagent lots and/or manufacturers will be critical. However, it must be stressed that even slight changes in common buffers, kit solutions, columns and assay plates have been known to produce aberrant results. Even in evolving laboratory environments that adhere to significant current Good Manufacturing Practices (cGMPs)—like attention to event investigations, corrective action/preventative action and change control—a failure in a critical reagent can lead to delays both in validation and sample analysis. It is the responsibility of the method users to ensure reagents are received and stored appropriately.

Immunogenicity assays are equally influenced by the factors described above. In addition, project specific information for the validation is very important. Immunogenicity screening assays are statistically based. Samples are deemed to be positive or negative for antibodies to the drug product on the basis of their response in the assay system relative to a threshold set during validation. The threshold is calculated on the performance of matrices (typically around 50 lots) that have not been fortified with the surrogate antibody. The lots are chosen to mimic a study population unexposed to the drug and the threshold is usually set as the mean of the negative control run in the assays, plus a factor derived from the standard deviation of the lots tested during validation. Setting the threshold accurately for sample analysis relies upon selecting the correct lots of matrix for the validation. A best practice is to select lots of matrix that are from individuals as close to the study population as possible (e.g., age, weight, gender, diet, ethnicity, etc.). In some circumstances this is difficult, if not impossible, to achieve due to the availability of certain patient specific matrices. The clinical study protocol should be supplied to the end users at the time of method transfer, along with the biological matrix specifications.

Very importantly the sponsor (likely method establisher) and end user(s) collaborate to ensure that the pertinent details of the sponsor's method are clearly communicated at the time of transfer. As much as possible, detail should be provided in the method SOP or procedure document. Seemingly irrelevant information can be useful in clarifying procedures. Identifying key reagents for the method and testing different lots/suppliers ahead of time will ensure that the method will function with reagents from variable sources. When possible, assay performance should be tested with multiple individuals to ensure method transferability. Ideally, individuals from both the sponsor and end users should run the method side by side at the laboratory from which the method originated or, preferably, at the laboratory or CRO where the transfer is taking place. The geographical location of the two laboratories can in some instances make this an expensive proposition. However, this short-term investment is insignificant when compared to overall drug development costs. When the cost of travel is weighed against that of expending additional drug development time or resources, which typically run up to the hundred thousands per diem, it is a very worthwhile investment.

Incurred samples in method transfers given the significant financial pressures on drug development for meeting analytical milestones, a best practice for project managers is to ensure methods are validated at more than one site. This ensures an alternate or back-up laboratory is available should an issue arise with the capacity at a single bioanalytical site. Once a method is transferred into multiple sites the question of the comparability of results commonly arises. A great deal of attention has been focused recently on the use of incurred (in-study) samples for the assessment of bioanalytical method performance. The robustness and reliability of a method is assessed through examining the reproducibility of results for a portion of the study samples. While no official guidance has been delivered, the general consensus derived from meetings between industry representatives and the FDA held over the past several years is that small molecule/chromatographic methods are considered to have acceptable reproducibility if the repeat result of a minimum of two-thirds of tested samples is within ±20% of the original results. For macromolecule ligand-binding assays this criterion may be widened to ±30% due to the larger variability associated with these types of assays. While primarily aimed at determining the reproducibility of the method itself, this type of analysis can reveal nonmethodological issues such as technical errors, differences in matrices, and changes in reagents.

In some cases incurred sample analysis has been used as a means of evaluating the acceptability of a method transfer. At method validation completion, once the precision and accuracy has been established, incurred or spiked samples that have previously been analyzed at another location are analyzed at the validation site and evaluated for concordance. In a manner similar to incurred sample reanalysis, the results and the methods are deemed to be equivalent if they meet predetermined acceptance criteria (e.g., 2/3 of results must be within 20% of the original value). Delivering results from methods that meet acceptance criteria instills confidence that samples could be analyzed at either site with an equivalent outcome. However, in contrast to the situation described for in-study incurred sample analysis where the evaluation is for a method at a single site, an additional set of questions may arise from a method when data from two different sites fail acceptance criteria. Since both methods met acceptance for precision and accuracy and would be considered validated at each site, judging which of the two methods requires investigation is very difficult. Objective evaluation of cross-site data remains a topic of spirited discussion. The potential for divergent data may call into question the strategy of placing a single assay at multiple bioanalytical laboratories.

The comparability of the qualification samples at different sites may not be an issue, depending upon type of analyte under investigation. While for quantitative pharmacokinetic assays interlab accuracy is an absolute requirement, for some pharmacodynamic assays this stringency may be less critical and, in some cases, impossible to obtain. For example, with some biomarker assays sponsors may be less interested in knowing the absolute value of the measurements than the relative change in biomarker levels. The criteria for establishing concordance between laboratories under these circumstances would require a fit-for-purpose approach.

To have a successful transfer, the bioanalytical method itself must be robust and the equipment differences between the delivering and receiving parties should be carefully evaluated. Use of standardized automation equipment has shown to be advantageous during the method transfer. Unfortunately very limited information on method transfer can be found in the literatures [91, 92]. Typically in an organization, before the method transfer, scientists from both sites go through the method details very carefully. It has been found that even some apparently trivial differences such as the hood air flow speed, types of pipettors, brands of vortex, or centrifuge can sometime cause method performance difference. Every method generated in laboratories should contain a very detailed discussion about the method robustness and any potential factors that may impact the method performance.

Fast LC–MS/MS" for the quantitative bioanalysis can only be achieved after considering all aspects of the analysis cycle. LC–MS/MS practitioners should never take any doctrine for granted and should always ask the questions "How fast is too fast?" and "Can this validated method be used successfully for sample analysis." Only with these questions in mind, can we as bioanalytical practitioners always be aware of the potential pitfalls and culprits that could occur in any stage of the study. Otherwise, a seemingly perfect method can generate wasted data, and even worse, poor medicines that could impair public health. Hopefully, above critical review summarizes the advances in this field and points out the current critical view of quantitative bioanalytical LC–MS/MS. The average preapproval costs of drug development run at just above $800 million or roughly $183,000 per day over a 12 year timeframe. The loss of a day of patent life of a therapeutic has a considerable impact. It is therefore essential to adopt strategies for successful and efficient establishment and transfers of bioanalytical methods, which occur frequently in the drug development process. Good communication, close collaboration, and planning between developers and end users are the keys to their success. Likewise, the use of incurred samples offers a powerful tool for assessing the suitability of the transfer, but one whose outcomes must be carefully discussed and planned prior to use in order to avoid any delays resulting from unexpected or inconclusive results. Strategic relationships between developers and partners/collaborators facilitate these processes through the sharing of common processes and expectations.

From technical point of view, a review of up-to-date application is covered in several areas including sample preparation, separation, and detection. Although much of the emphasis is put on the first two areas, it should be noted that the progress in mass spectrometer designs over the years provided the basis for sensitive detection of ever more potent drug candidates from biological matrices. Without sensitivity gains, many of the commonly used approaches such as the "dilute and shoot" would not be practical. The results from many applications cited in this chapter have demonstrated that innovative chromatography technologies are reshaping the ways that separations are performed in high-throughput bioanalytical laboratories. Together with advancements made in laboratory automations like parallel sample processing, column-switching, and usage of more efficient extraction supports for SPE, they drive the trend toward less sample cleanup time in today's quantitative bioanalysis field. Importantly, the efficiencies are accomplished without compromising the quality of assay such as precision, accuracy, selectivity, and robustness. On the other hand, it is recognized that some of these techniques such as Spark Holland or UPLC systems need specialized equipments. Some of the materials like certain types of extract sorbents or specialized columns are not cost-effective yet to many users. Most of the techniques described in the article continue to be developing. For example, the achievement of small particle UPLC has not been fully extended from reversed phase to other types of stationary phase. Monoliths made of silica possess a limited pH range over which they are applicable (2–8). There is a need of more dimensions and different type of monolithic column, especially microbore monolithic columns so that highly efficient separation can be performed using less HPLC solvent. The separation efficiency of such columns can be optimized with improved fabrication. Nevertheless, further expansion or advancement of these techniques will be beneficial to bioanalytical scientists in either developing strategies for a new method or modernizing a high-throughput laboratory.

6.5 REGULATORY ASPECTS OF COMPETITIVE BIOANALYTICAL LABORATORIES

6.5.1 General Consideration

Within highly regulated industry, the-state-of-the-art equipment, modern technologies and competitive techniques may not be adequate to become competitive advantages. Strictly following regulation requirements and compliance have played increasingly roles in successful bioanalytical laboratories. Over the past decade, the validity and accuracy of data generated from bioanalytical laboratories in support of regulated nonclinical (animal) pharmacokinetics/toxicokinetics (PK/TK), bioavailability (BA)/bioequivalence (BE), pharmacokinetics/pharmacology (PK/PD) and other studies designed for new drug applications (NDAs)/abbreviated new drug applications (ANDAs) have been a focus of Food and Drug Administration (FDA) inspections as evident from published FDA inspectional findings. As a result, an important trend has been noted for the number of FDA 483 observations and untitled letters issued to several well-established bioanalytical laboratories. Although there are no definitive regulations that mandate how to conduct bioanalytical testing of specimens from animals to human subjects or PK/TK and BA/BE studies, from the regulatory perspective, it is the responsibility of the firm or institution performing the bioanalysis for PK/TK and BA/BE studies to implement sound quality systems to ensure the reliability, accuracy and integrity of the data it generates. This expectation is clearly defined within the FDA Guidance for Industry: Bioanalytical Method Validation [93]: "The analytical laboratory conducting pharmacology/toxicology and other nonclinical studies for regulatory submissions should adhere to FDA's Good Laboratory Practices (GLPs) (21 Code of Federal Regulations (CFRs) Part 58) and to sound principles of quality assurance throughout the testing process. The bioanalytical method for human BA, BE, PK, and drug interaction studies must meet the criteria in 21 CFR 320.29. The analytical laboratory should have a written set of standard operating procedures to ensure a complete system of quality control and assurance. The SOPs should cover all aspects of analysis from the time the sample is collected and reaches the laboratory until the results of the analysis are generated and reported. The SOPs also should include record keeping, security and chain of sample custody (accountability systems that ensure integrity of test articles), sample preparation, and analytical tools such as methods, reagents, equipment, instrumentation, and procedures for quality control and verification of results." The following sections illustrate how quality systems can be built and implemented in a bioanalytical laboratory by applying key elements of Good Laboratory Practices and current Good Manufacturing Practices, thus meeting current FDA/regulatory expectations.

6.5.2 Historical Perspective

The history of ensuring the quality and safety of drugs in the United States can be traced as early as 1848 when the Drug Importation Act [94] passed by Congress requiring the US Customs Service to inspect and stop, when appropriate, the entry of adulterated drugs from overseas. Since then, successive legislatures including the Pure Food and Drugs Act of 1906 [95], the Federal Food, Drug, and Cosmetic (FD&C) Act of 1938 [96] and Kefauver–Harris Amendments of 1962 [97] were enacted by Congress requiring new drugs, among other things, to demonstrate safety and efficacy prior to marketing. In 1977, FDA finalized the BE regulations [98]. The regulations were further amended in 1992, 1993, and 1999 and leading to the 2003 issuance of the FDA "Guidance for Industry: Bioavailability and Bioequivalence Studies for Orally Administered Drug Products" [99]. The regulations delineated the requirements for the submission of *in vivo* BA and BE data as a condition for the marketing of a new or generic drug product. BA/BE studies are conducted to support NDAs and ANDAs, respectively, and their supplements.

For NDAs, BA studies are generally conducted to support approval of a new chemical compound, new formulation, new dosage form, or a new route of administration. Bioavailability, as provided in FDA's 21 CFR 320(a) relative to BA and BE requirements, means the rate and extent to which the active ingredient or active moiety is absorbed from a drug product and becomes available at the site of action. For ANDAs and certain NDAs, BE studies are conducted to support approval of generic drugs or changes to existing drugs not eligible for ANDA submission. As defined in 21 CFR 320.1(e), BE means the absence of a significant difference in the rate and extent to which the active ingredient or active moiety in pharmaceutical equivalents or pharmaceutical alternatives becomes available at the site of drug action when administered at the same molar dose under similar conditions in an appropriately designed study. In other words, a typical BE study is conducted to show evidence that the drug product that is the subject of the application and the reference listed drug product exhibit a similar pharmacokinetic profile in a biological matrix, for example, plasma, serum, and/or urine.

The BE requirements as defined in 21 CFR Part 320 also provide guidelines relative to the design of BA/BE studies. There are two components in a BA/BE study, namely the clinical component and the analytical component. This part of chapter focuses on the analytical aspects of a BE study relative to the use of GLP and cGMP in bioanalytical laboratories. An important section of the BE requirements (21 CFR 320.29(a)) requires that the analytical method used in an *in vivo* BA or BE study to measure the concentration of the active drug ingredient or therapeutic moiety or its active metabolite(s) in the body fluids or excretory products or the method used to measure an acute pharmacological effect shall be demonstrated to be accurate and of sufficient sensitivity to measure, with appropriate precision, the actual concentration of the active drug ingredient or therapeutic moiety or its active metabolite(s) achieved in the body. The bioanalytical method should be validated to demonstrate its reliability for the intended use. The characteristics critical to the acceptability of the performance of the method and reliability of the results include: precision; accuracy; specificity; limit of quantitation; stability of stock solutions and of the analyte(s) in the biological matrix; and response function. Further details relative to the validation of a method are provided in the FDA Guidance for Industry entitled "Bioanalytical Method Validation" [93].

The regulations for conducting nonclinical laboratory studies in support of applications for research or marketing permits for products including food and color additives, animal food additives, human and animal drugs, medical devices for human use, biological products and electronic products are delineated in GLP regulation 21 CFR Part 58 [100]. The Federal Regulations for the manufacture, processing, packing or holding of drugs are embodied in cGMP regulation 21 CFR Part 210 [101], while 21 CFR Part 211 [102] delineates the regulations for the preparation of drug products for administration to humans and animals. There is, however, no single set of regulations directly applicable to the conduct of bioanalytical testing of samples obtained in human studies, such as BA, BE, PK, pharmacodynamic (PD) studies, and drug–drug interaction (DDI) studies. Applicable GLP, cGMP, sound Quality Assurance (QA) practices, and, to a certain extent, notable sections from Good Clinical Practices (GCPs) and Good Clinical Laboratory Practices (GCLP) are key practice ingredients to a bioanalytical laboratory in providing accurate and reliable data. These practices need to be adopted, enforced and audited in a bioanalytical testing environment to ensure that the analytical measurements are performed appropriately, reported accurately, and archived securely.

Due to the continued growth of drug development in the branded as well as generic sectors in the pharmaceutical industry, laboratory compliance in all GxP areas have become a major focus of attention by regulatory agencies including the Division of Scientific Investigation (DSI) and the other compliance branches of the FDA during the review of the applications as well as during site inspections of laboratories performing bioanalytical testing in support of ANDAs and NDAs. There are key elements present in both the regulations for GLP (21 CFR Part 58) and cGMP (21 CFR Parts 210 and 211) that a firm or institution providing bioanalytical laboratory services should employ to assure the quality and integrity of data generated in a BA or BE or PK study.

In 1967, the FDA's DSI was established with the following goals [103]:

(a) To audit and verify clinical trial data submitted to the FDA in support of applications to demonstrate the safety and efficacy, or BE, of drugs for human use
(b) To direct inspections of Institutional Review Boards (IRBs) for compliance with standards and regulations designed to protect the rights and welfare of human research subjects and

(c) To ensure that investigators, sponsors, and Contract Research Organizations that conduct nonclinical and clinical studies on investigational new drugs comply with United States laws and regulations covering GCP and GLP. The following chronologies for *in vivo* BE studies [104] provide historical events relative to FDA's inspection of bioanalytical laboratories performing human BA/BE studies:

Within 1979–1992: Inspection of BE studies of generic drugs on an ad hoc basis.

1993: Separate BE unit formed within DSI. The general approaches used by DSI in performing their inspection of human BA/BE studies, including bioanalytical laboratories can be found in the FDA "Compliance Program Guidance Manual (CPGM) 7348.001—Bioresearch Monitoring: Human Drugs—*In vivo* BE" [105]. Based on the review of FOI information and the authors' experiences, many FDA 483s and several untitled letters were issued to bioanalytical laboratories in the past few years as a result of DSI inspections. Ultimate objectives are to establish appropriate Quality Systems, Procedures, and Controls for Bioanalytical Laboratories.

6.5.3 Personnel—Training and Qualification

According to GLP regulation 21 CFR 58.29, each person engaged in the conduct of or responsible for the supervision of a nonclinical laboratory study should have the education, training, and experience or any combination of the above to allow that individual to perform the assigned duties. Additionally, the personnel who provide the training must be qualified to do so. Similarly in cGMP regulation 21 CFR 211.25 and 211.28, personnel engaged in the manufacture, processing, packing, or holding of a drug should have the education, training, and experience or a combination of the three elements to enable that person to perform his or her assigned functions. In both GLP and cGMP regulations, a qualified trainer on a continuing basis should provide the training. Personnel in a bioanalytical laboratory should have the required education, training, experience, or a combination of the above to carry out their responsibilities. Job descriptions should be available and current for the personnel directly related to BA/BE/PK testing and a training plan for each of the job descriptions should be designed to allow for consistent and uniform training for the affected personnel. Qualified trainers should provide the training on a continuous basis and effectiveness of training should be monitored to ensure that the quality of personnel training is sustained. The bioanalytical laboratory management should ensure that the training programs are adequate, properly administered, and documented for all involved personnel (analysts and supervisors) involved. The firm's training program should ensure that affected personnel are trained initially prior to performing a job function, are provided periodic training, and are verified to be qualified to conduct the tasks assigned to them. More importantly, training should be documented and adequate records should be maintained and available for regulatory inspections. It is customary for cGMP analytical laboratories to require the analyst's training curriculum to include SOPs, analytical instrumentation, and the required analytical techniques. However, due to the complex nature of the bioanalytical methods, most bioanalytical laboratories require study-specific training to be performed to further qualify the analysts used to perform the testing of specimens from human BA/BE or other studies. These study-specific training/qualification records should be retained with the study data package to support future third party inspections. The study specific training generally requires the analysts to prepare for spiked calibration standards and Quality Control samples and perform trial accuracy/precision test runs. If the results generated meet the preestablished acceptance criteria for the calibration curves and the QC samples, the analysts can then be considered as qualified to perform testing of the subject specimen samples to support the BA/BE and other studies.

6.5.3.1 Quality Assurance Unit The firm or institution providing bioanalytical supports for BA/BE and other studies should implement a QA system to include all relevant aspects of bioanalytical testing as recommended in the FDA "Guidance for Industry: Bioanalytical Method Validation" [93]. The extensive role of the Quality Assurance Unit (QAU) in this system and as delineated in both the GLP and cGMP regulations cannot be overemphasized. The GLP regulation for 21 CFR 58.35 assigns the QAU with the responsibility for monitoring each nonclinical laboratory study to assure that the facilities, equipment, personnel, methods, practices, records, and controls are in conformance with the regulations. The QAU has the authority to inspect each nonclinical laboratory study periodically to assure the integrity of the study and provide periodic notification to management noting any problems and the corrective actions taken. The QAU should ensure that deviations from approved protocols or procedures have been addressed with the appropriate authorization and documentation. The GLP regulations mandate that the QAU be a separate and independent entity from the entity engaged in the conduct and management of the study.

Some of the major roles for the Quality Control Unit (QCU) outlined in cGMP regulation 21 CFR 211.22 include the responsibility of the QCU in providing the assurance that the facilities, equipment, personnel, methods, records, controls, procedures, processes, and practices are in conformance with all applicable regulations, standards and/or specifications. The QCU has the responsibility for reviewing production records to assure that no deviations/errors have occurred or, if deviations/errors have occurred, that they have been fully investigated and documented. Additionally, the QCU has the authority to sample and release or reject raw materials, labeling, components, and finished product upon review of all aspects relating to manufacturing, packaging, and control of the final product. The QCU has the oversight for the implementation and administration of a change control program including computer systems. On a timely basis and through self-audits, the QCU provides written reports to management to provide assurance that all operations are proceeding in a compliant manner. The QCU is an independent function and has separate reporting responsibilities from management perspectives.

The bioanalytical testing roles of the QAU in a bioanalytical laboratory should be akin to the roles played by QA (or QC) in a GLP or cGMP environment. The QAU will report to a position totally independent of the line management that is directly engaged in the BA/BE and other studies. This unit will provide the line management the necessary oversight and assurance that all operations, procedures, and services are completed in conformance with appropriate regulations and standards. The roles and responsibilities of QA in a bioanalytical laboratory performing human BA/BE and other studies should be delineated in a procedure and its responsibilities should include the monitoring of each BA/BE study to ensure that the systems (facilities, equipment, personnel, methods, records, controls, procedures, processes, and practices) supporting the conduct of the study are operating in a compliant manner. BA/BE reports and method validation reports should be reviewed and approved by QA. The independent authority of QA has become more relevant in light of FDA's and the public's concerns on drug safety and its role relative to the review/approval of documents, which were not previously under its scope is becoming a regulatory expectation.

QA should have oversight in a formal system for handling of deviations, for example, planned and unplanned events, with the ultimate goal of ensuring that the deviation is appropriately investigated with a root cause assignment and a corrective action and preventive action (CAPA) taken to prevent its recurrence. The investigation should also determine whether the deviation has an adverse impact on the accuracy and validity of the study data individually and in its totality. Another quality system that should be implemented in a bioanalytical laboratory deals with a robust change control program to effectively manage changes to protocols, procedures, methods, and processes. Changes (including computerized systems) must be reviewed and approved by the QAU prior to implementation.

The QAU in a bioanalytical laboratory should undertake the responsibility of ensuring that all levels of bioanalytical operations are being conducted in a compliant manner and periodic reports should be prepared and issued. The QAU in a bioanalytical laboratory performing human BA/BE and other studies should also perform internal selective audits of study data to ensure they are compliant with protocol and SOP requirements before the study report can be approved. Study reports should not be issued without this QA review and QA statement of compliance. As noted previously, QA review and verification/approval of the accuracy and reliability of data is becoming an industry benchmark practice. The management of a bioanalytical laboratory needs to review its operation and to optimize the roles and responsibilities of the QAU to ensure full compliance with current FDA expectations and to minimize risk. A Quality Manual, with senior management approval, may be developed to define the responsibilities, accountabilities, and authorities relative to the QAU and the implementation of quality systems. An example of workflow between bioanalytical laboratory and quality assurance unit is given in Table 6.2.

6.5.4 Facility—Design and Qualifications

The GLP regulations (21 CFR 58.31) require that the testing facility be of suitable size and construction to facilitate the conduct of the nonclinical laboratory study. This is also required in cGMP regulation 21 CFR 211.42 where any building and premises

TABLE 6.2 Bioanalytical and Quality Assurance Workflow

GLP/Regulated Nonclinical and Clinical Studies:

- Study Plan or Protocol draft, reviewed and finalized
- Master Schedule Initiating Study completed and forwarded to QA
- Samples received (chain of custody) and stored securely
- BA memorandum initiated and completed
- LIMS set up, batch ID created and stored along with study files
- Sample Analysis starts
- In lab audit scheduled for all dose formulation and sample analysis (toxicology, clinical, etc.)
- Data compiled and reviewed daily (or on timely basis)
- Data internally QC'd within the BA Group
- Sample analysis completed; master schedule updated
- Report tables, binders, and draft report prepared for submission to QA
- Draft report reviewed and internally QC'd
- Master Schedule updated—to reflect data binders and report sent to QA
- Data and report audited by QA
- Review and audit observations prepared by QA
- Audit findings submitted to BA department for response/resolution of audit findings. If needed draft report will be updated
- Finalized audit and draft report sent to study director if necessary for comments
- Comments incorporated from/by study director if any
- Report issued; master schedule updated
- Report and all raw data organized, filed, and archived

used in the manufacture, processing, or holding of a drug product shall be of suitable size, construction and location to facilitate cleaning, maintenance, and proper operations. Separate or defined areas should be available to prevent contamination or adverse effect on the study or drug product. The bioanalytical laboratory area should be spacious and designed in a way that allows for efficient sample and personnel flow. For example, test specimens should efficiently flow from "storage freezers" to the "sample processing/extraction laboratory" to the "instrument laboratory." Storage space (freezers, refrigerators) for subject test samples, calibration standards, and QC samples should be adequate and properly secured to allow for efficient, storage, and retrieval. Separate areas for the receipt, verification, storage of incoming test specimens and preparation of calibration standards, and QC samples should be made available to avoid mix-ups and to ensure integrity of subject samples and standards. Utilities such as water, air, gas, and electricity should be adequate, stable, and uninterrupted.

6.5.5 Equipment Design and Qualification

The equipment used in the generation, measurement, or assessment of data and the equipment used for facility environmental control in a nonclinical laboratory study should have the appropriate design and construction for its intended use and should be suitably located to facilitate operation, cleaning, inspection, and maintenance (GLP, 21 CFR 58.61). For the manufacture of drug products, the manufacturer is mandated by cGMP to ensure that the equipment used in the facility is constructed to be nonreactive, additive or absorptive, suitably validated or qualified, easy to clean, and sanitize to prevent cross contamination (GMP, 21 CFR 211.63, 211.65). In the spirit of the GLP and cGMP regulations, the equipment (e.g., balances, LC, GC, GC/MS, LC–MS/MS, etc.) used to support the generation of laboratory results intended for BA/BE and other studies should likewise be of adequate design and construction for the conduct of the study. As is required in cGMP, the equipment (including storage chambers) in the bioanalytical laboratory should be qualified to demonstrate its suitability of use and proper performance. The same principles of equipment qualification relative to Installation, Operational, and Performance Qualifications (IQ, OQ, and PQ) could be adopted in the qualification of equipment in the bioanalytical laboratories. The laboratory equipment should be properly situated to allow ease of operation, cleaning and maintenance. Where a software database is used to maintain and track calibration and Preventive Maintenance (PM) schedules and reports, the application needs to be validated to be in conformance with the regulations [106, 107]. Based on the review of current FDA 483 observations, most bioanalytical laboratories have not been cited for the lack of instrument/equipment/freezer chambers qualification data; nevertheless, it is good business and good scientific practice to address this instrument qualification issue in a proactive manner to avoid any future potential compliance issues.

The GLP regulations for conducting nonclinical laboratory studies in support of applications for research or marketing permits for products (21 CFR 58.63) and the cGMP regulations for the manufacture, processing, packing, or holding of drug products mandate that the respective laboratory, institution, or manufacturing facility implement a sound calibration and PM program (21 CFR 211.67, 21 CFR 211.68) to ensure that the validated equipment continue to operate properly, as intended. Similarly, the bioanalytical laboratory should adopt a calibration/PM program to ensure equipments (including storage chambers) are periodically calibrated and maintained to ensure continued compliant operation and performance of such equipments. A calibration/PM schedule should be used and followed and procedures must be in place to handle missed calibrations. Equipment with expired calibration dates or which has missed the calibration interval should be properly identified; for example, "Out of Calibration—Do Not Use" and be taken out of service. Data from noncalibrated equipment are deemed to be unreliable and should not be reported. Of importance is the need for calibration procedures that clearly provide instructions to the laboratory personnel on how to calibrate and maintain the equipment. Where calibration is performed by outside service groups, the firm should have a copy of the contractor's calibration procedures, if applicable, and the calibration report should be reviewed and approved by laboratory management prior to placing the equipment back into service. Equipment calibration needs to be performed against traceable standards. Remedial actions need to be defined in the event of calibration and/or equipment failure.

6.5.6 Standard Operating Procedures

Standard Operating Procedures are a must in both regulations (GLP 21 CFR 58.81, cGMP 21 CFR 211.100) and are the cornerstone of all required activities undertaken in the laboratory and manufacturing facility. SOPs are important for any organization to assure safety, efficiency, and consistency in the performance of a task. As recommended in the FDA "Guidance for Industry Bioanalytical Method Validation," there should be an approved procedure in place for every type of bioanalytical laboratory operation, to cover all aspects of analysis from the time the sample is collected and reaches the laboratory until the results of the analysis are generated and reported. Examples of essential SOPs generally required in a bioanalytical laboratory are presented in Table 6.3. SOPs should be written to be clear, easy to understand and follow. All SOPs need to be reviewed annually to ensure the procedure does not become obsolete. As a regulatory requirement, training in new and revised procedures should be provided to affected personnel. Any revisions made to the current SOPs need to be managed by a proper change control procedure and system should be installed to ensure that only the most current versions of the SOPs are made available to users.

There should be a designated group in the bioanalytical laboratory facility that is responsible for maintaining and archiving controlled documentation; for example, protocols, methods, BA/BE and other study reports, and bioanalytical reports. Typically, this responsibility is within the realm of the QAU. The controlled documents should be maintained in a safe, limited access area and controls must be implemented to protect the integrity of the documents. In addition to a robust procedure delineating the mandatory requirements for documentation, a procedure that describes the preparation and approval of the final bioanalytical report should be available and followed by the affected personnel. A collection of good SOPs is a prerequisite for successful GLP compliance. Setting up an SOP system is often seen as the most important and time-consuming compliance task. Even without GxP regulations, classical quality assurance techniques, indeed good management, require standardized, approved, written working procedures. The successful implementation of SOPs requires

- sustained and enthusiastic support from all levels of management with commitment to establishing SOPs as an essential element in the organization and culture of the laboratory;
- SOP-based education and training of personnel so that the procedures are performed in the same way by all personnel;
- a sound SOP management system to ensure that current SOPs are available in the right place.

6.5.7 Laboratory/Facility Qualification Perspectives

Commercially available instruments and systems have already been tested by the vendors, which can be verified by audits. Therefore, why should a risk analysis methodology that is very effective for new designs and processes be dumped or foisted

TABLE 6.3 Essential SOPs Required for Bioanalytical Laboratories

1. Organization, roles, and responsibilities of GLP lab
2. Study Director role and responsibilities
3. Procedure for the control of incoming test specimen—receiving, storage, inventory, login/logout, and disposal—chain-of-custody check
4. Procedure for the control, qualification, calibration, and PM of laboratory freezers
5. Control, documentation, and control of control articles—qualification of reference standards
6. Control, documentation, and control of blank biological matrices
7. Control, storage, expiry dates, and destruction of chemicals, reagents
8. Procedure to Label test specimens, chemicals, and solutions
9. Preparation and control of Project Binder
10. Preparation of Standard and Quality Control samples
11. Preparation and qualification of stock standard solutions
12. Procedure for validation of bioanalytical methods—content, review, approval, and change control
13. Sample analysis plan—in study control
14. Chromatography Acceptance Criteria—procedure to review and accept HPLC. LC–MS/MS chromatographic data—peak integration, baseline review, and acceptance criteria
15. Calibration curves, quality control, and analytical runs acceptance criteria
16. Sample Reassay procedure, acceptance, and reporting
17. Significant figures round off procedure
18. System suitability tests requirements
19. Documentation and record-keeping and record retention practices—procedure to enter raw data
20. Laboratory verification, review, and approval of analytical raw data, calculation, derived data, and reports
21. Procedure to determine and Reassay of PK outliers
22. Procedure to investigate and resolve of significant recurring events/deviations—SOP for Event Investigation and Resolution
23. Backup and archiving of study data
24. Testing of incurred sample reanalysis (ISR)
25. Procedure to record deviations
26. Procedure to issue analytical report
27. Procedure to evaluate the stability of analyte in stock solution and biological samples
28. Procedure to evaluate recovery of analytes in biological fluid
29. Procedure for the cleaning of laboratory glassware
30. Instrument qualification, calibration, and PM program
31. Operation, maintenance, and calibration of various lab instruments:

For example, HPLC, LC–MS/MS, centrifuge, balances, autopipette, and robotic multiprobe dilutors

32. Computer system validation (CSV) and related CSV SOPs
33. Training and qualification requirements for bioanalytical personnel

on laboratories using mainly commercial systems? There are alternative and simpler risk analysis approaches that can be used for the commercial off-the-shelf and configurable COTS software applications used throughout laboratories. For example, there are also (1) hazard analysis and critical control points (HACCP) and (2) functional risk assessment (FRA) [108].

6.5.7.1 Testing Approach Versus Intended Purpose Throughout the Good Practice Guide (GPG), there appears to be an emphasis on managing regulatory risk. This is in contrast to the introductory statements in the GPG mentioned at the start of this column. From author's perspective, this is not desirable or recommended, and emphasis should be placed on defining the intended purpose of the system and hence functions of the instrument and software that are required first and foremost. Only then will you be able to assess the risk for the system based on the intended functions of the system.

The testing approach outlined for Qualification, Testing, and Release and verifying the operation of the PQ against user requirements, the following are usually performed:

- Verification of user SOPs
- Capacity testing (as required)
- Processes (between input and output)
- Testing of the system's backup and restore (as required)
- Security
- Actual application of the system in the production environment (e.g., sample analysis)

The risk classification is then plotted against the probability of detection to determine high, medium, and low priority of testing. A high-risk classification coupled with a low likelihood of detection determines the highest class of test priority. This probably encapsulates the overall approach of the guide in my view—regulatory rationale rather than business approach in contrast to the stated aims of the guide in the introduction. Using this approach, it is believed that laboratory personnel will be performing over complex and over detailed risk assessments forever for commercial systems that constitute the majority of laboratory systems. What the writers of the GPG have forgotten is that the FDA has gone back to basics with Part 11 interpretation [109]. Remember that the GMP predicate rules (21 CFR 211 and ICH Q7A for active pharmaceutical ingredients) for equipment/computerized systems state: *§211.63 Equipment Design, Size, and Location: Equipment used in the manufacture, processing, packing, or holding of a drug product shall be of appropriate design, adequate size, and suitably located to facilitate operations for its intended use and for its cleaning and maintenance* [110].

ICH Q7A *(GMP for active pharmaceutical ingredients), in §5.4 on Computerized Systems states in §5.42: Commercially available software that has been qualified does not require the same level of testing* [111].

As usual in the world, professional groups must have their own statement in how things should be done. The American Association of Pharmaceutical Scientists (AAPSs) is no exception and has produced a white paper titled "*Qualification of analytical instruments for use in the pharmaceutical industry; a scientific approach*." Of course, this is a different approach from GAMP perspectives [112].

In contrast to the GAMP GPG [113], which looks at laboratory equipment from the computer perspective, the AAPS document looks at the same issue from equipment qualification perspective. The AAPS white paper has devised three classes of instruments with a user requirements specification necessary to start the process as described in Chapter 5.

- *Group A Instruments*: Conformance to the specification is achieved visually with no further qualification required. Examples of this group are ovens, vortex mixers, magnetic stirrers, and nitrogen evaporators.
- *Group B Instruments*: Conformance to specification is achieved according to the individual instrument's SOP. Installation of the instrument is relatively simple and causes of failure can be easily observed. Examples of instruments in this group are balance, IR spectrometers, pipettes, vacuum ovens, and thermometers.
- *Group C Instruments*: Conformance to user requirements is highly method specific according to the guide. Installation can be complex and require specialist skills (e.g., the vendor). A full qualification is required for the following spectrometers: atomic absorption, flame absorption, ICP, MS and MS/MS, FT-IR, UV/VIS, and so on.

This approach is simpler but the only consideration of the computer aspects is limited to data storage, backup, and archive. Thus, this approach is rather simplistic from the computer validation perspective. Furthermore, the definition of IQ, OQ, and PQ is from the equipment qualification perspective (naturally) with operational release occurring after the OQ stage and PQ intended to ensure continued performance of the instrument. This is different from the GAMP GPG, which uses the computer validation definition of IQ, OQ, and PQ where PQ is end user testing and operational release occurs after the end of the PQ phase. It is a great challenge when two major publications may not agree on terminology for the same subject. However, the AAPS white paper is now the baseline document for the new proposed general chapter ⟨1058⟩ for the USP XXIX [114]. Consider the following issues that are not fully covered by the AAPS guide that will now be enshrined in a formal regulatory text:

- The scope of the guidance and proposed USP chapter [115] is limited only to commercial off-the-shelf analytical instrumentation and equipment.
- The three instrument groups are described along with suggested testing approaches to be conducted for each. However, in author's view, there is not sufficient definition of the criteria for placing instruments in particular groups.

- Group C instruments cover a wide spectrum of complexity and risk, and may have very diverse requirements. There is no specific allowance made within the approach for custom developed applications such as macros commonly found when operating spectrometers.
- The guide covers the initial qualification activities for analytical instruments but there is very little on the validation of the software that controls the instrument. There is little guidance on operational, maintenance, and control activities following implementation such as access control, change control, configuration management, and data back-up. How many spectrometers can laboratories name that do not have computer-controlled equipment and data acquisition?
- The proposed chapter uses the term "analytical instrument qualification" (AIQ) to describe the process of ensuring that an instrument is suitable for its intended use and/or application but the instrument is only a part of the whole computerized system. It is the computerized system that controls the whole—not the instrument.

6.5.7.2 Integrated Approach to Computer Validation and Instrument Qualification What it really needs for any regulated laboratory is an integrated approach to the twin problems of instrument qualification and computer validation. As noted by the GAMP GPG, the majority of laboratory and spectrometer systems come with some degree of computerization from firmware to configurable off-the-shelf software. The application software controls the instrument. If you qualify the instrument you will usually need the software to undertake many of the qualification tests with an option to validate the software at the time [116].

AAPS Analytical Instrument Qualification guide and GAMP laboratory GPG considered two examples as have been looked. They are looking at different parts of the same overall problem and coming up with two different approaches. No wonder if we do not take a considered and holistic view of the whole problem. For example, we use the same qualification terminology (IQ, OQ, and PQ) for both instrument qualification and computer system validation but they mean different things [117]. This fact is exemplified in the two guides. Therefore, the following guidance should be developed as a minimum:

Integrated terminologies usually cover both the qualification of the instrument and validation of the software. This must ensure that the laboratory is not separated from the organization or creates a profession of Lablish interpreters.

- Simple classification of laboratory equipment software based on the existing GAMP software categories to be consistent with the rest of the organization. The laboratory is not a unique part of a facility anymore than production is.
- Realistic life cycle(s) based on the further development of the simple system implementation life cycle (SILC) outlined in the GPG that reflect the different options that we face in the laboratory: from COTS to configurable COTS and where necessary customization of an application.

- Writing a specification or specifications to document both the instrument and the associated software functions. The equipment qualification requirements for traceable reference standards can also be devised for input into the user requirement specification (URS).
- Use of a simple to use but effective risk assessment methodology that reflects the majority of instrument and systems are commercial.
- Integrated and practical approaches to combined equipment qualification and computer validation to test and demonstrate that the system does what it is intended to do.

Qualification of laboratory equipment and validation of computerized laboratory systems are going into two different directions that lack an integrated approach. It is necessary to have an integrated approach that recognizes a combined approach to qualifying the instrument through the controlling software that also needs to be validated at the same time. This approach must harmonize the use of terminology and definitions while moving forward on this effort in this competitive marketplace.

6.6 ADVANCED/COMPETITIVE BIOANALYTICAL LABORATORIES

Thanks to significant advances in discovery science, there is a dramatic increase in the number of new compounds moving from discovery into life science product development. The goal of nonclinical testing is to eliminate poor candidates while moving quality candidate drugs into clinical development as quickly as possible. At any given time, there are 3000–4000 compounds in nonclinical testing that have made the cut from discovery into development. Ongoing incremental improvements in the discovery and selection process continue to occur, which should result in a higher rate of success for compounds entering the expensive clinical development phase.

Given today's time pressures and the high cost of discovery and development, it is more important than ever that supporting function/services, such as bioanalysis, are optimized for speed and success. Bioanalysis, simply stated, is the quantitative measurement of active drug(s) plus their metabolites in biological matrices; it is required for evaluation of a therapeutic entity throughout the entire drug development cycle. In the early stages of discovery, lead selection/optimization and nonclinical development, the focus is on science: creating and refining analysis methods that will accurately and efficiently assess the drug level in a variety of species and biological matrices. As development moves from nonclinical to clinical investigations, the bioanalytical support required also undergoes a transition. In the later stages of development (Phases II–IV), the challenges of sample transport and handling from complex, multisite, often multicountry trials, along with substantial data management needs, add a requirement for logistical excellence in addition to scientific capability.

Bioanalytical sciences such as feasibility and method development support critical pharmaceutical evaluations, including toxicokinetics and pharmacokinetics, bioavailability, drug–drug interaction studies, and other systematic assessments. Due to the sheer volume of work required in nonclinical and clinical product development, lead times must be carefully managed to keep projects on track. Today, many pharmaceutical and biotech companies utilize outsourcing to accomplish at least a portion of their bioanalytical needs. Outsourcing of bioanalytical projects may be done strategically or tactically and depends on business, financial, and technical factors or consideration.

Using outsourcing as a strategy helps manage the flow of projects entering or leaving product development due to toxicity, economics, or changing priorities. It may amortize the strengths of a particular outsourcing partner and provide additional scientific and operational excellence that can help provide information to make go/no-go decisions. Depending upon the specific type of project and compound, commercial bioanalytical laboratories can often quickly develop and validate scientifically robust methods, then smoothly integrate those methods into sample runs within weeks or sometimes even days of the initial project inquiry. Project initiation may begin immediately for established, validated assays, depending on the laboratory's capacity. Bioanalytical outsourcing often provides sponsors with more flexibility, global reach, and less investment in personnel and capital infrastructure, while expanding scope and capabilities that may not be available internally.

6.6.1 Strategy Versus Tactics

Strategic outsourcing generally involves a corporate commitment to a vendor partnership for value added benefits, with a focus on each partner's strengths as a long-term strategy. It entails a decision to leverage the sponsor's internal resources so that it adds maximum value for the discovery, development, and commercialization of new pharmaceutical or biological products. More commonly, capital- and/or labor-intensive development activities are outsourced, so the sponsor's focus is on the evaluation of incoming compounds and which entities are the most promising candidates for development. Benefits that may be gained from strategic partnerships include: effective leveraging of investment in methods and science across the life cycle of the product development program; consistent, customized services; enhanced communication; and specific scientific talent and expertise. Strategic outsourcing relationships generally yield financial benefits that range from maintaining tight control of fixed costs to leveraging fixed-cost investments toward activities that occur earlier in the discovery-development sequence—activities that are usually more of a unique "core" strength of the sponsor organization.

Companies that take a more tactical approach are often faced with short-term workloads that exceed their internal human resources or capacity. Tactical outsourcing is more "deliverable-focused" as opposed to value-added. Outsourcing the bioanalytical component helps them gain additional capacity when internal capacity is unavailable, streamline workflow and validate claims for additional internal resources. Often, these encounters are transactional one-time events to manage peak work volume and complete projects more quickly than the sponsor could accomplish internally. Neither the sponsor nor the contractor derive or amortize the longer-term benefits that may result from effectively working together over time. Of course, many companies find themselves somewhere along the continuum between strategic and tactical outsourcing and thereby derive some of the advantages and limitations of each approach. In either case, it may be important to evaluate which approach predominates in order to assess and choose a bioanalytical partner that will best meet a sponsor's unique needs.

6.6.2 Bioanalytical Laboratory Assessment

The process of contract laboratory selection starts with the sponsor's request for proposal (RFP). The RFP details the scope of work and includes all the pertinent information necessary for a contract bioanalytical laboratory to estimate the time, talent, and costs associated with the potential project. Internally, the sponsor should have an unambiguous understanding of the project scope (is it research-focused method development or production with thousands of samples?) and clearly communicate that to the laboratory under consideration. While the vocabulary may sound similar, meanings from organization to organization often differ, so expectations and terminology should be well defined in order to maximize efficiencies and cost effectiveness. A number of quantitative and qualitative elements should be examined when making a commercial bioanalytical laboratory assessment. The analysis should, of course, include assessments of capabilities, capacity, and scientific talent. A deeper investigation should reveal a clear understanding of the laboratory's quality processes, turn-around times, regulatory compliance, communication vehicles, and other underlying factors that ultimately contribute to a laboratory's "ease of use" factor and ability to consistently deliver over time.

6.6.3 Capacity

It is vital that method development, transfer, and validation be coupled with sufficient "production" capacity to keep a product on the critical development path. These activities are frequently conducted using different resources, so it is useful to evaluate the capacity of a laboratory in each area: (1) acquiring insight on whether any of the needed resources are assigned "double duty" (e.g., resource committed to work in both method development and production meaning sample analysis) and develop a clear understanding of the effectiveness of communication and interaction among resources and (2) discovering whether any bottlenecks exist in the process. There are always times when sponsor companies need more instruments or staff. Understanding what personnel, equipment, and facility resources are available and how they are managed, allocated and scheduled to projects is crucial. A commercial bioanalytical laboratory typically runs high sample volumes (particularly for later stage development) and is uniquely equipped to rapidly provide additional capacity. Typical questions that may help evaluate commercial laboratory capacity are listed below:

- How many samples does the analytical laboratory analyze per year?
- What is the laboratory's capacity to store samples or expand services on demand?

- How much of a laboratory's total capacity would the work being considered for placement occupy?
- What types of analytical equipment does the laboratory have? How often is the equipment calibrated and maintained? Are the technologies and systems compatible with the sponsor company's systems?
- How many type of test of interest? Does the contractor do in a week or a month or a year?
- How many specify type of equipment of interest? Are available and in what geographic locations?
- How quick are turnaround times for type of test or results?
- How flexible is the laboratory if a project is added or canceled?
- How are resources managed and scheduled? Describe the policies and processes in place for scheduling work. How does this differ for method development versus production work?
- Does the laboratory provide dedicated resource services to accommodate large capacity needs over longer periods of time?

6.6.4 Experience

No sponsor can afford to put a development program at risk. A commercial bioanalytical laboratory should have the talent and experience it takes to execute the project or it should decline the work: (1) examine the organization's peripheral experience in regulatory, quality systems, and information technology, (2) has the laboratory identified opportunities to improve quality and efficiency and installed methods to drive those changes?; and (3) look for an organization that is focused on amortizing the experience it has within and is willing and able to leverage that experience for a sponsor's program.

- Does the laboratory have experience with the class of compounds or technique of interest?
- How many methods are validated each year? How many are originated at the laboratory and how many involve technology transfer?
- How many scientists/technical experts will be devoted to the project in question? What are the policies and processes on how those experts are engaged and scheduled?
- What are the scientific credentials of the study director, study monitor or principal investigator assigned to the project?
- What is the training process for scientists or analysts?
- What is the staff turnover in each department?
- Do the study directors, study monitors, or principal investigator have the necessary experience relative to the project in question?
- Does the laboratory have experience in method transfer? Do they provide efficient systems for doing so?

6.6.5 Quality

Quality is paramount in pharmaceutical product development. How a bioanalytical laboratory approaches quality is a statement about its priorities. The FDA requires nonclinical bioanalytical work for regulatory submissions to be in compliance with GLP. While a recently communicated FDA guidance explicitly omits GLP compliance for bioanalytical work related to human clinical trials, the importance of these studies requires quality even beyond compliance. Quality systems should be "built in" to prevent errors or to catch and correct them in real time. If a laboratory's systems and procedures are well set up, quality control and quality assurance should not cause delays. It is important to look for process auditing and process controls that minimize time delays or escalation of costs.

- What quality systems, SOPs, GLP, or auditing procedures are in place?
- Is quality proactive or reactive and how is that demonstrated?
- What critical review procedures or peer-review steps are performed?
- Does the QA/QC process in the bioanalytical laboratory contribute or cause delays or bottlenecks to receipt of the deliverable?
- What regulatory expertise does the laboratory have and how does it keep abreast of the changing regulatory environment?
- How long has the laboratory operated in a regulatory environment?
- How many regulatory agency and/or client audits has the laboratory had and what were the results?
- What plans are in place or steps are being made to ensure electronic records and signatures comply with 21 CFR Part 11?
- What type of quality control talent is dedicated to each project?

6.6.6 Performance and Productivity Measures

Each study is ultimately unique, but the performance and track record of a commercial bioanalytical laboratory will provide insight and help set expectations for future work. Ensure that the laboratory has mechanisms in place to track performance.

- What are the on-time delivery performance metrics? How are they assessed against client expectations?
- How is customer satisfaction assessed?
- What is the laboratory's error rate (including error severity and corrective action plans)?
- How are errors tracked and followed up?
- What productivity measures does the laboratory track in development? Or in production? Are these shorter or longer than sponsor company internal metrics (if available)?
- How quickly does the laboratory develop and validate a method (on average)?

- How focused is the organization on maximizing efficiencies and minimizing costs?
- Does the promised turn-around time match the requirements of the sponsor?
- Where are the typical bottlenecks and slow points in any particular bioanalytical assessment? How likely are they to affect turn-around time?

6.6.7 Information Technology and Data Management

The information technology infrastructure and level of automation within a contract laboratory should be examined. Highly automated organizations may have replaced labor-intensive repetitive work, which may provide more consistency from run to run and/or result in more rapid delivery of cost-effective services. Contract commercial laboratories should also demonstrate efficient, robust data management, and data processing systems, particularly when data-intensive studies such as Phase II and III studies are under consideration. Strong science capabilities are moot if the contractor is unable to meet deadlines because reports or deliverables are stalled by data delays. It is likely to look for automated data processes from sample receipt to final deliverable. Highly IT-enabled bioanalytical units should be able to simultaneously maximize their ability to work efficiently, provide exceptional sample tracking and integrity, maintain superior regulatory compliance, minimize turnaround times (particularly from the collection of the last data point to the report), streamline client communication, and access to data and continuously address areas of inefficiency through analysis of business metrics.

- What functions have been automated and where has this resulted in efficiencies or reduction of development costs?
- What IT strategies are in place for study management (tracking, data transfer, processing, and accuracy)?
- Have IT systems been properly validated?
- Are data systems compatible with the sponsor systems?
- Are data systems capable of handling large quantities of data?
- How sophisticated are the data management processes?
- What sample management system is employed?
- What company-specific systems are there (e.g., LIMS-laboratory information management system) and what do they do?

6.6.8 Communication

Success in bioanalytical outsourcing is probably more dependent on good communication than any other possible single factor. Detailed communication plans are essential for meeting expectations and keeping projects on track. Some contract laboratories provide a single point of contact, while others allow the sponsor to contact anyone within the organization who is involved in the project. Most contractors provide a customized communication plan to meet sponsor requirements. The value of face-to-face interactions to establish a strong working relationship cannot be underestimated, particularly when deadlines are critically tight.

- What expectations exist for reports, formats, and/or timing?
- What elements comprise the communication plan and how is it executed?
- Who are the key contacts responsible for the project and how available are they?
- How flexible is the contract laboratory with regard to report formats, file types and style standards?
- With what frequency is communication between sponsor and contract laboratory maintained? Using what methods?
- How flexible is the laboratory with respect to communication protocols? How effective are their communication protocols when "customized" to fit a sponsor?
- How often are face-to-face client meetings scheduled or allowed?
- How is contingency plan handled?
- What (if any) direct communication links can or will be provided as part of the contract (e.g., secure email)?

6.6.9 Financial Stability

It is important to know that the contractor is financially viable and will be in business when the sponsor needs it, particularly when late-phase, multiyear studies are under consideration. In addition, it is also desirable to assess the laboratory's level of commitment to the field and to refining and extending its support services.

- What is the financial and business position of the bioanalytical laboratory?
- Does the laboratory have the financial means to invest in quality people, equipment, and facilities?

6.6.10 Ease of Use

The answers to these questions are extremely important as evaluation tools. However, a less readily defined element—"ease-of-use"—may be critical in the final decision. Services must be crisp and clean. Problems have to be quickly communicated and solutions to those problems should be recommended. Information must be free-flowing, but at a pace and frequency desired by the sponsor. Flexibility should be applied to accommodate the sponsor's shifting needs and priorities. Each sponsor will have a unique collection of requirements that add up to "ease-of-use" and the answer to that set of questions will factor prominently in the final selection of a contracting partner. In the end, a good outsourcing partner will provide a combination of capabilities, services, and processes that instill confidence and save the sponsor time and money.

6.6.11 Contracting Bioanalytical Services

Contracts are the tools used to establish a framework for the working relationship between the sponsor and the contract laboratory. A clear understanding of project scope and specifications, including details of the essential activities to be outsourced, is necessary. While a large number of pharmaceutical and biotechnology companies have accepted outsourcing as a strategy, a small but rapidly growing number of organizations have put processes in place to broadly and systematically identify, select and manage bioanalytical laboratory contractors. Both business and technical input are required for the contracting process to be fully successful.

In larger companies, the sponsor generally assembles a crossfunctional team to make the outsourcing decision that, at a minimum, includes the project manager, a bioanalytical scientist, a quality assurance representative and a contract (or purchasing/procurement) manager. In smaller organizations, a single individual and a quality assurance representative may be responsible for selecting a contract laboratory. Project managers responsible for overall clinical programs may be focused on big-picture logistical concerns (e.g., sample management or how soon data are available). Bioanalytical scientists, on the other hand, very often have worked on the proprietary methods that will be used. Science is paramount and they may be more focused on the experience of the scientists or technicians at the contract laboratory who will perform the work. The QA representative will focus attention on quality systems and processes and assess the laboratory's regulatory compliance program. The contract manager's focus may be on the protection of the sponsor's legal and financial interests (outlined in the contract), as well as the company's proprietary interests (delineated in a Confidential Disclosure Agreement). Balancing competing interests around the outsourcing decision can often be challenging.

The most successful outcomes, however, typically occur when the contracting group or the individual that contracts the service has worked in a bioanalytical laboratory and/or been very familiar with the processes and services in these organizations. This familiarity may help prevent "scope creep" and ensure a fair agreement on timelines and compensation. Sponsor companies are best served by a contracting individual or group that is able to make judgments on the bioanalytical component of drug development for the particular product(s), and understands the difficulty or simplicity of method development or transfer and the decisions that impact timelines.

6.6.12 The Contracting Process

Contracts may be simple or complex and are highly variable from one bioanalytical project to another. Contract pricing models vary according to the type of work under consideration and evaluation. Method development may involve hourly fees, based on an estimate of the time it will take to come up with a viable, scientifically robust assay. Production bioanalytical work may be priced at a fixed fee per sample, with a sliding or discounted fee schedule based on volume. Nearly any element of a project is fair game for negotiation, from method runs to equipment, or from the people assigned to the project to payment schedules.

Contracts may be negotiated in less than a day or taking much longer. One of the primary delays to quick study start-up is working through the legal verbiage that must necessarily go back and forth between the contract laboratory and sponsor. To a certain degree, master service agreements (MSAs) or preferred provider agreements (PPAs) address those bottlenecks, and facilitate the contracting process by covering common contractual elements including liability and confidentiality. MSAs and PPAs are particularly useful when strategic outsourcing is considered; the contract laboratory has a commitment for repeat business and can start prestudy work while project specific details are under negotiation. Fair prices and the costs of bioanalytical services require a certain amount of reflection and consideration. While it may be prudent to challenge high prices for the sake of saving money, for the sponsor using bioanalytical services strategically, it may be just as important to challenge very low prices. The acquisition and retention of good talent, the timely introduction of modern technology, and the implementation of adequate data systems are not sustained by low pricing.

Once a contract is signed, work is scheduled and initiated. Adequate supplies of test material and references should be provided to the contract laboratory with all the pertinent documentation well in advance of the study initiation date. Should the scope of the project change or should study delays or cancellations occur, the price quotation may change accordingly. Not all commercial contract bioanalytical laboratories are created equally. Each drug development program has its own unique requirements and those requirements change as the stages of development unfold. Identifying a laboratory with too narrow a match with one drug or stage of development will likely produce inefficiency later in the program as those needs to be changed. Conversely, identifying a laboratory with a broader range of capabilities may provide a less effective match with the needs of "today" or current task. Finding a laboratory with the breadth and depth to match the various talent, expertise, capacity, and speed requirements of a drug development program can be a daunting challenge. Method development may be fraught with technical difficulty in crafting a scientifically robust, consistent, transferable assay. Full-scale production bioanalysis may be more fast-paced, requiring quick turnaround, plenty of capacity, and logistical excellence. It is important to fully understand what present and future needs are or will be in order to find a complementary partner for outsourcing bioanalytical programs.

The right combination of capacity, experience, quality, performance, communication, and other attributes must be carefully assessed for "ease-of-use" in the evaluation of a contract laboratory. The contracting process is also an important consideration and must be factored into the overall planning process and timelines. A vigilantly nurtured bioanalytical outsourcing relationship has many benefits, including expanded resources, enhanced communication, global reach, and the flexibility to accommodate time-critical product development. Fully understanding the scope, depth, and breadth of what a commercial contract laboratory offers can provide the necessary information to make an informed decision and establish an outsourcing relationship that extends across time with a wide range of services for sponsor support.

6.7 APPLICATIONS AND ADVANCES IN BIOMARKER AND/OR LIGAND-BINDING ASSAYS WITHIN BIOANALYTICAL LABORATORIES

Protein biomarkers are used for various purposes in drug discovery, development and clinical diagnosis and prognosis. Method qualification/validation for protein biomarkers is typically applied to ligand-binding assays, although hyphenated mass spectrometric methods can be used as adjunct methodologies to confirm LBA specificity and provide valuable information during early discovery or demonstrative phases of a novel biomarker. Preanalytic variables of protein biomarkers, such as the purpose of the intended application, analyte(s), biological matrix, availability of reference standard, calibrator matrix, assay platform, and sample collection/handling, must be considered in any method development, qualification, and validation plan. Method qualification/validations for exploratory applications usually involve basic experiments for assay range finding, accuracy and precision, selectivity, specificity, and minimal stability. For advanced method validation, more rigorous tests with a wider scope are performed. These tests include additional patient population ranges, more runs on accuracy and precision from multiple analysts/reagent lots/instruments, selectivity and specificity tests using patient samples, and stability tests subjected to conceivable conditions over long-term use. Differences in biomarker method validation for drug development versus clinical diagnosis and issues of using developmental commercial kits are discussed. The codevelopment of biomarkers for drug development and diagnostics presents collaborative opportunities between the pharmaceutical and diagnostic sectors. Typically, panels of biomarkers (e.g., for cardiac, lipids, diabetes, and tumor markers) in serum or urine samples from patients are measured by auto-analyzers using FDA approved commercial kits. Recently, biomarkers have been recognized and accepted as useful tools for drug development [118, 119]. The appropriate application of biomarkers to preclinical and clinical drug development reduces time to market, hastens the attrition of undesirable candidate compounds prior to expensive phase III clinical trials, and guides dose selection with early indications of efficacy or toxicity [120–123]. There has been a paradigm shift from "trial and error" to mechanistic based, target-driven drug development using biomarkers to track the biological exposure–drug effect relationship. A long list of novel putative biomarkers, which are not included in routine clinical lab tests, has been generated from intense genomic and proteomic research. These biomarkers have been utilized in clinical trials of drug candidates for exploratory, demonstrative, or characterization applications as part of the process of developing mechanism-specific biomarkers toward the ultimate goal of surrogacy [124–127].

Development of novel biomarkers has notably increased in the areas of diabetes, cancer, rheumatoid arthritis, and cardiovascular disease. The use of novel biomarkers has become a prominent component of decision-making processes in drug development. They are used in *in vitro* and preclinical models and early clinical phase for quick hit and early attrition decisions [128–130]. In exploratory and demonstrative studies, pharmacodynamic correlations are typically unknown, data are used mainly for internal decision-making, and the output is generally not subject to regulatory review. The extent of method validation can thus be limited to a few basic components to expedite the process and preserve resources without unduly impacting commercialization, depicted as Exploratory Method Validation [131]. In contrast, the purposes of the characterization phase are to provide pivotal data to establish linkage to clinical outcome and to monitor patient progress upon treatment. Characterization of a novel biomarker in the translational phase requires data collection to show preclinical sensitivity and specificity and linkage to clinical outcomes in multiple clinical studies in humans. The purposes at this phase are different from those of the Exploratory or Demonstrative phase. The data are often used for critical decisions (such as supporting dose selection and patient stratification, demonstration of drug safety or efficacy, and differentiation of drug candidates), for submissions to be reviewed by regulatory agencies, or for postmarketing patient monitoring [132, 133]. In addition, the same method used for characterization would likely be used in the Qualification phase toward surrogacy confirmed over multiple drugs of similar mechanism and during surveillance studies. Therefore, biomarker characterization studies would require more intense rigor and cover a wider scope in Advanced Method Validation, with greater traceability and more detailed documentation than that of the previous phases to meet the study objectives in a defined context of its use. The knowledge of drug/protein target interactions in disease pathways also contributes to the concept of "personalized medicine" or "target therapy." Clinical biomarker assays are no longer simple results from an auto-analyzer with cutoff values that lead to the diagnosis of a certain disease. More significantly, gene-, protein-, or metabolite-based biomarker profiling of a patient can be used to identify and stage disease (diagnosis), decide treatment, and monitor progress and predict the outcome with certain confidence (prognosis). Mass spectrometric-based proteomic research has generated peptide maps of healthy versus patients with various disease types at different stages. Proposals have been made to use these patterns and the identified proteins from these peptides for early diagnosis and to monitor progress of treatment [134, 135]. Additionally, novel protein biomarkers linked to diseases or drug safety has been unveiled from this proteomic research [136–140]. With wider applications of biomarkers, especially novel biomarkers, in clinical diagnosis and prognosis and at various phases in drug development, questions arise about how the analytical laboratories (e.g., clinical or bioanalytical laboratories) should validate a method to be suitable for the intended applications. Methods for novel biomarkers are generally generated from the innovator laboratories in university or company settings, often using commercial kits developed "for research use only" (RUO) and therefore they may not be fully validated (Exploratory Method Validation). On the other hand, methods that are FDA approved or cleared would be subjected to rigorous validation tests similar to the Advance Method Validation. A fit-for-purpose method validation approach at various phases of biomarker application has been described by Lee et al. [131]. Here the discussion will focus on method validation of protein biomarker bioanalysis to meet the diverse purposes during drug development clinical trials and clinical diagnosis/prognosis. Advances have been made in the development of quantitative proteomics methodologies [141–146]. The performance of several of these methodologies has been recently assessed by a study organized by the

Association of Bimolecular Resource Facilities (ABRF) [147]. The analytical methods surveyed by this study fell into two broad categories: (1) gel-based methods and (2) MS-based methods (including both stable isotope-labeling and label-free techniques) [148]. The major conclusions from this study were that the methods are complex and require high levels of expertise for success. Hence, there was large laboratory-to-laboratory variability when the same technologies were used. As a result of this complexity and reliance on expertise, these methods are not commonly adopted for preclinical/clinical applications to study dose–response relationships in drug development. By virtue of the highly specific chemical information that is generated, MS-based techniques do play a major role in biomarker discovery, but validation and routine monitoring of protein biomarkers are most readily achieved by the use of ligand-binding assays [149, 150]. There may be instances, however, where LBA is not a viable option such as the case where no or inadequate affinity reagents are available. In these cases, MS-based approach may be a feasible alternative. An excellent review article was published by Lee and Hall [151], which describes the method validation of protein biomarkers in support of drug development or clinical diagnosis/prognosis. A depth of elaboration is shown regarding iterative processes of biomarkers method validation related to stages of discovery, characterization, and qualification. The flow diagrams are given depicting the process of biomarker development from discovery to surrogacy. Attrition of novel biomarkers is analogous to but not concurrent with drug candidate development, from a few hundred to a smaller potential panel and finally to a final few for a target mechanism. The rigor and scope of method validation are dependent on the intended applications. Lessons learned from exploratory applications contribute to the preanalytical planning, method refinement, and validation for advanced applications.

The application purpose of a biomarker and its method validation cannot be isolated without connection to various aspects in drug development. The decision process is interrelated with knowledge gained from disease pathways and patient data from exploratory and advanced studies. There is a need for the development of software to handle multiple biomarkers monitored during drug development. Optimal software would integrate multiple biomarkers' PD profiles with PK data to allow models to describe physiological compartments of exposure and effects on the disease and host biology. Preferably, knowledge of protein biomarkers is integrated with those of genomics, glycomics, and metabolomics. Biokinetics of the target, proximal, and distal biomarkers should be tracked in various patient populations. Thus, clinical validation of a biomarker goes through learning phases from continuously updated knowledge. The foundation of such knowledge rests upon reliable quantitative methods appropriately validated at each phase of application. LBA has been the major method in protein biomarker application. Other method types must be included for biomarker investigation to supply a complete picture of biology and chemistry for the biomarker's interaction with the drug candidate and proximal proteins. Western blot, 2D gel electrophoresis, MudPIT LC–MS/MS, and imaging have been used for biomarker research. Quantitative applications are being developed [148, 150, 152]. Technology integration of laser-microdissected cryostat sectioning, ProteinChip, gene microarray, immuno-histochemistry, multiplex binding assays, and hyphenated

MS methods (e.g., FACS–MS, MALDI–MS, and affinity–MS) will continue to impact biomarker discovery and the application to translational medicine. Since these technologies have evolved in a research environment, translation for the application to preclinical and clinical samples requires the cooperation of the scientists from both discovery and clinical realms. Bioassays coupled with LC–MS/MS to quantify reaction products can provide direct relevant chemical and biological information [153]. Affinity techniques can be coupled to MALDI and LC–MS/MS. Specific receptor proteins or antibodies can be covalently attached to a solid phase as a capture device for selective enrichment of low concentration biomarkers, and the eluent from the solid phase can subsequently be interrogated by MS for chemical information and quantification [154, 155]. Advances in molecular imaging will continue to contribute to diagnostic and drug development in tumor specific biomarkers [156, 157]. Flow cytometers can be used to select and enrich blood cell populations. The isolated cells can be manipulated and subsequently interrogated by LBA or MS methods [158, 159]. Multiple technological tools including cell-based, immuno-affinity, and biophysical methods will contribute to the integral knowledge of protein biomarker actions in target cells and biological fluids and the concomitant impact of drug intervention. Therefore, in addition to the need for integrating software, there is also a need for integrating methodologies. The development and validation of integrating software and technologies would enhance the overall knowledge of protein biomarkers and add greater support to understanding diseases and improving patients' lives.

As a whole, bioanalytical laboratories have been expanding to support drug discovery and development programs in terms of their capabilities (widening scope) and capacities (meeting high demands). The consideration of both technical and regulatory competitiveness is vital to successful implementation in delivering high quality and reliable data during the processes described above. In drug development, important decisions based on data obtained from bioanalytical methods are taken hence it is generally accepted that sample preparation and method validation is required to demonstrate the performance of the method and the reliability of the analytical results. The acceptance criteria should be clearly established in a validation plan, prior to the initiation of the validation study.

Bioanalysis and the production of pharmacokinetic, toxicokinetic, and metabolic data play a fundamental role in pharmaceutical research and development; therefore the data must be produced to acceptable scientific standards. For this reason and the need to satisfy regulatory authority requirements, all bioanalytical methods should be properly developed, validated, and documented. The lack of a clear experimental and statistical approach for the validation of bioanalytical methods has led scientists in charge of the development of these methods to propose a practical strategy to demonstrate and assess the reliability of chromatographic methods employed in bioanalysis. The aim of this chapter is to provide simple to use approaches with a correct scientific background to improve the quality of the bioanalytical method development and validation process. Despite the widespread availability of different bioanalytical procedures for low molecular weight drug candidates, ligand-binding assay remain of critical importance for certain bioanalytical applications in support of drug development such as for antibody and receptor. This chapter gives an idea about which criteria bioanalysis based on chromatographic-based assays and immunoassays should follow to reach for proper techniques and related acceptance criteria. Applications of bioanalytical methods in routine drug analysis also take into consideration for their simplicity (fit for purpose) and ease of operations (proper techniques and sound acceptance requirements). These various essential development and validation characteristics for bioanalytical methodology have been discussed with view to improving the standard and acceptance in this area of research.

REFERENCES

1. Bakhtiar R, Ramos L, Tse FLS. *Liq J Chrom Rel Technol* 2002;25:507–540.
2. Zhou S, Song Q, Tang Y, Naidong W. *Curr Pharm Anal* 2005;1:3–14.
3. Ackermann BL, Berna MJ, Murphy AT. *Curr Top Med Chem* 2002;2:53–66.
4. Jemal M, Xia Y-Q. *Curr Drug Metab* 2006;7:491–502.
5. Naidong W. *J Chromatogr B Analyt Technol Biomed Life Sci* 2003;796:209–224.
6. Hsieh Y, Korfmacher WA. *Curr Drug Metab* 2006;7:479–489.
7. Zhang NY, Rogers K, Gajda K, Kagel JR, Rossi D. *J Pharm Biomed Anal* 2000;23:551–560.
8. James CA, Hill HM. *AAPS J* 2007;9(2): E123–E127 (Article 14).
9. Koseki N, Nakashima A, Nagae Y, Masuda N. *Rapid Commun Mass Spectrom* 2006;20:733–740.
10. Ma J, Shi J, Le H, Cho R, Huang JC, Miao S, Wong BK. *J Chromatogr B Analyt Technol Biomed Life Sci* 2008;862(1-2): 219–226.
11. Mallet CR, Lu Z, Fisk R, Mazzeo JR, Neue UD. *Rapid Commun Mass Spectrom* 2003;17:163–170.
12. Yang AY, Sun L, Musson DG, Zhao JJ. *Rapid Commun Mass Spectrom* 2006;20:233–240.
13. Wang PG, Wei JS, Kim G, Chang M, El-Shourbagy TA. *J Chromatogr A* 2006;1130:302–307.
14. Zhang J, Reimer MT, Alexander NE, Ji QC, El-Shourbagy TA. *Rapid Commun Mass Spectrom* 2006;20:3427–3434.
15. Ji QC, Reimer MT, El-Shourbagy TA. *J Chromatogr B Analyt Technol Biomed Life Sci* 2004;805:67–75.
16. Zhang N, Yang A, Rogers JD, Zhao JJ. *J Pharm Biomed Anal* 2004;34:175–187.
17. Xu N, Kim GE, Gregg H, Wagdy A, Swaine BA, Chang MS, El-Shourbagy TA. *J Pharm Biomed Anal* 2004;36:189–195.
18. Bolden RD, Hoke SH 2nd, Eichhold TH, McCauley-Myers DL, Wehmeyer KR. *J Chromatogr B Analyt Technol Biomed Life Sci* 2002;772:1–10.
19. Xue YJ, Akinsanya JB, Liu J, Unger SE. *Rapid Commun Mass Spectrom* 2006;20:2660–2668.
20. Xu XS, Yan KX, Song H, Lo MW. *J Chromatogr B Analyt Technol Biomed Life Sci* 2005;814:29–36.
21. Zang X, Luo R, Song N, Chen TK, Bozigian H. *Rapid Commun Mass Spectrom* 2005;19:3259–3268.

22. Xu RN, Fan L, Kim GE, El-Shourbagy TA. *J Pharm Biomed Anal* 2006;40:728–736.
23. Vintiloiu A, Mullett WM, Papp R, Lubda D, Kwong E. *J Chromatogr A* 2005;1082:150–157.
24. Papp R, Mullett WM, Kwong E. *J Pharm Biomed Anal* 2004;36:457–464.
25. Ceglarek U, Lembcke J, Fiedler GM, Werner M, Witzigmann H, Hauss JP, Thiery J. *Clin Chim Acta* 2004;346:181–190.
26. Ceglarek U, Casetta B, Lembcke J, Baumann S, Fiedler GM, Thiery J. *Clin Chim Acta* 2006;373:168–171.
27. Zhou S, Zhou H, Larson M, Miller DL, Mao D, Jiang X, Naidong W. *Rapid Commun Mass Spectrom* 2005;19:2144–2150.
28. Chassaing C, Stafford H, Luckwell J, Wright A, Edgington A. *Chromatographia* 2005;62:17–24.
29. Smalley J, Kadiyala P, Xin B, Balimane P, Olah T. *J Chromatogr B Analyt Technol Biomed Life Sci* 2006;830:270–277.
30. Kahlich R, Gleiter CH, Laufer S, Kammerer B. *Rapid Commun Mass Spectrom* 2006;20:275–283.
31. Xia YQ, Liu DQ, Bakhtiar R. *Chirality* 2002;14:742–749.
32. Xia YQ, Bakhtiar R, Franklin RB. *J Chromatogr B Analyt Technol Biomed Life Sci* 2003;788:317–329.
33. Alnouti Y, Srinivasan K, Waddell D, Bi H, Kavetskaia O, Gusev AI. *J Chromatogr A* 2005;1080:99–106.
34. Koal T, Sibum M, Koster E, Resch K, Kaever V. *Clin Chem Lab Med* 2006;44:299–305.
35. Barrett YC, Akinsanya B, Chang SY, Vesterqvist O. *J Chromatogr B Analyt Technol Biomed Life Sci* 2005;821:159–165.
36. Bourgogne E, Grivet C, Hopfgartner G. *J Chromatogr B Analyt Technol Biomed Life Sci* 2005;820:103–110.
37. Calderoli S, Colombo E, Frigerio E, James CA, Sibum M. *J Pharm Biomed Anal* 2003;32:601–607.
38. Niederlander HA, Koster EH, Hilhorst MJ, Metting HJ, Eilders M, Ooms B, de Jong GJ. *J Chromatogr B Analyt Technol Biomed Life Sci* 2006;834:98–107.
39. Hu CW, Wang CJ, Chang LW, Chao MR. *Clin Chem* 2006;52:1381–1388.
40. Schellen A, Ooms B, van de Lagemaat, D, Vreeken R, van Dongen WD. *J Chromatogr B Analyt Technol Biomed Life Sci* 2003;788:251–259.
41. Alnouti Y, Li M, Kavetskaia O, Bi H, Hop CE, Gusev AI. *Anal Chem* 2006;78:1331–1336.
42. Kawano S, Takahashi M, Hine T, Yamamoto E, Asakawa N. *Rapid Commun Mass Spectrom* 2005;19:2827–2832.
43. Smalley J, Xin B, Olah TV. *Rapid Commun Mass Spectrom* 2009;23(21): 3457–3464.
44. Cabrera K. *J Sep Sci* 2004;27:843–852.
45. Li R, Dong L, Huang J. *Anal Chim Acta* 2005;546:167–173.
46. Wang G, Hsieh Y, Cui X, Cheng KC, Korfmacher WA. *Rapid Commun Mass Spectrom* 2006;20:2215–2221.
47. Borges V, Yang E, Dunn J, Henion J. *J Chromatogr B Analyt Technol Biomed Life Sci* 2004;804:277–287.
48. Barbarin N, Mawhinney DB, Black R, Henion J. *J Chromatogr B Analyt Technol Biomed Life Sci* 2003;783:73–83.

49. Huang MQ, Mao Y, Jemal M, Arnold M. *Rapid Commun Mass Spectrom* 2006;20:1709–1714.
50. Zeng H, Deng Y, Wu JT. *J Chromatogr B Analyt Technol Biomed Life Sci* 2003;788:331–337.
51. Zhou S, Zhou H, Larson M, Miller DL, Mao D, Jiang X, Naidong W. *Rapid Commun Mass Spectrom* 2005;19:2144–2150.
52. Eerkes A, Shou WZ, Naidong W. *J Pharm Biomed Anal* 2003;31:917–928.
53. Xue YJ, Liu J, Unger S. *J Pharm Biomed Anal* 2006;41:979–988.
54. Song Q, Naidong W. *J Chromatogr B Analyt Technol Biomed Life Sci* 2006;830: 135–142.
55. Deng Y, Zhang H, Wu JT, Olah TV. *Rapid Commun Mass Spectrom* 2005;19: 2929–2934.
56. Hsieh S, Tobien T, Koch K, Dunn J., *Rapid Commun Mass Spectrom* 2004;18:285–292.
57. Giddings JC. *Dynamics of Chromatography, Part I, Principles and Theory*. Marcel Dekker, New York, 1965.
58. Al-Dirbashi OY, Aboul-Enein HY, Jacob M, Al-Qahtani K, Rashed MS. *Anal Bioanal Chem* 2006;385:1439–1443.
59. Wren SAC, Tchelitcheff P. *J Chromatogr A* 2006;1119:140–146.
60. Apollonio LG, Pianca DJ, Whittall IR, Maher WA, Kyd JM. *J Chromatogr B Analyt Technol Biomed Life Sci* 2006;836:111–115.
61. Shen JX, Wang H, Tadros S, Hayes RN. *J Pharm Biomed Anal* 2006;40:689–706.
62. Churchwell MI, Twaddle NC, Meeker LR, Doerge DR. *J Chromatogr B Analyt Technol Biomed Life Sci* 2005;825:134–143.
63. Yu K, Little D, Plumb R, Smith B. *Rapid Commun Mass Spectrom* 2006;20:544–552.
64. Ikegami T, Tanaka N. *Curr Opin Chem Biol* 2004;8:527–533.
65. Lindqvist A, Hilke S, Skoglund E. *J Chromatogr A* 2004;1058:121–126.
66. Kapron JT, Jemal M, Duncan G, Kolakowski B, Purves R. *Rapid Commun Mass Spectrom* 2005;19:1979–1983.
67. Deng Y, Wu JT, Lloyd TL, Chi CL, Olah TV, Unger SE. *Rapid Commun Mass Spectrom* 2002;16:1116–1123.
68. Larger PJ, Breda M, Fraier D, Hughes H, James CA. *J Pharm Biomed Anal* 2005;39:206–216.
69. Muller C, Schafer P, Stortzel M, Vogt S, Weinmann W. *J Chromatogr B Analyt Technol Biomed Life Sci* 2002;773:47–52.
70. Shen JX, Motyka RJ, Roach JP, Hayes RN. *J Pharm Biomed Anal* 2005;37:359–367.
71. Matuszewski BK, Constanzer ML, Chavez-Eng CM. *Anal Chem* 2003;75:3019–3030.
72. Matuszewski BK. *J Chromatogr B Analyt Technol Biomed Life Sci* 2006;830: 293–300.
73. Little JL, Wempe MF, Buchanan CM. *J Chromatogr B Analyt Technol Biomed Life Sci* 2006;833:219–230.
74. Stokvis E, Rosing H, Lopez-Lazaro L, Schellens JH, Beijnen JH. *Biomed Chromatogr* 2004;18:400–402.
75. Jemal M, Schuster A, Whigan DB. *Rapid Commun Mass Spectrom* 2003;17:1723–1734.
76. Wang S, Cyronak M, Yang E. *J Pharm Biomed Anal* 2007;43:701–707.

77. Liang HR, Foltz RL, Meng M, Bennett P. *Rapid Commun Mass Spectrom* 2003;17:2815–2821.
78. Souverain S, Rudaz S, Veuthey JL. *J Chromatogr A* 2004;1058:61–66.
79. Constanzer ML, Chavez-Eng CM, Fu I, Woolf EJ, Matuszewski BK. *J Chromatogr B Analyt Technol Biomed Life Sci* 2005;816:297–308.
80. Mallet CR, Lu Z, MazzeoJr., *Rapid Commun Mass Spectrom* 2004;18:49–58.
81. Kitamura R, Matsuoka K, Matsushima E, Kawaguchi Y. *J Chromatogr B Biomed Sci Appl* 2001;754:113–119.
82. Lee J, Son J, Lee M, Lee KT, Kim D-H. *Rapid Commun Mass Spectrom* 2003;17:1157–1162.
83. Ke J, Yancey M, Zhang S, Lowes S, Henion J. *J Chromatogr B* 2000;742:369–380.
84. Yang L, Wu N, Clement RP, Rudewicz PJ. *J Chromatogr B* 2004;799:271–280.
85. Jemal M, Schuster A, Whigan DB. *Rapid Commun Mass Spectrom* 2003;17:1723–1734.
86. Fura A, Harper TW, Zhang H, Fung L, Shyu WC. *J Pharm Biomed Anal* 2003;32:513–522.
87. Tang YQ, Tollefson JA, Beato BD, Weng N. *AAPS Annual Meeting,* Salt Lake City, UT, October 26-30, 2003.
88. Jemal M, Xia YQ. *Rapid Commun Mass Spectrom* 1999;14:422–429.
89. Naidong W, Lee JW, Jiang X, Wehling M, Hulse JD, Lin P.P. *J Chromatogr B* 1999;735:255–269.
90. Naidong W, Jiang X, Newland K, Coe R, Lin PP, Lee JW. *J Pharm Biomed Anal* 2000;23:697–704.
91. Pan J, Song Q, Shi H, King M, Junga H, Zhou S, Naidong W. *Rapid Commun Mass Spectrom* 2004;18:2549–2557.
92. Leahy M, Felder L, Goytowski K, Shou W, Weng N. *52nd ASMS Conference on Mass Spectrometry and Allied Topics,* Nashville, TN, May 23–27, 2004.
93. US Department of Health and Human Services, Food and Drug Administration, Center for Drug Evaluation and Research, Center for Veterinary Medicine. Guidance for Industry: Bioanalytical Method Validation. May 2001.
94. Drug Importation, Act. 9 Stat 237, June 26, 1848.
95. Pure Food and Drugs, Act. 34 Stat 768, June 30, 1906.
96. Federal Food, Drug and Cosmetic (FDC), Act. United States Code (U.S.C.) Title 21, Chapter 9, June 25, 1938.
97. Kefauver-Harris Amendments, Public Law 87-781. 76 Stat 788-789, October 10, 1962.
98. Bioavailability and Bioequivalence, Requirements. 21 CFR Part 320, 2006.
99. US Department of Health and Human Services, Food and Drug Administration, Center for Drug Evaluation and Research, Center for, Veterinary. *Qual Assur J* 2007;11:3–15. DOI: 10.1002/qaj. Medicine. Guidance for Industry: Bioavailability and Bioequivalence Studies for Orally Administered Drug Products—General considerations. March 2003.
100. Good Laboratory Practice for Nonclinical Laboratory, Studies. 21 CFR Part 58, 2005.
101. Current Good Manufacturing Practice in Manufacturing, Processing, Packing or Holding of Drugs. 21 CFR Part 210, 2006.
102. Current Good Manufacturing Practice for Finished, Pharmaceuticals. 21 CFR Part 211, 2006.

103. Food and Drug Administration, Center for Drug Evaluation and Research, Division of Scientific Investigations homepage on the Internet. Rockville, MD, updated September 21, 2006, cited September 22, 2006. About DSI, about 4 screens. Available at http://www.fda.gov/cder/Offices/DSI/aboutUs.htm.
104. LePay D.GLP, bioequivalence, and human subject protection programs. Drug Information Association. Office of Compliance, 1998. Available at http://www.fda.gov/cder/present/dia698/diafda4/diafda4 (PPT#291,6, InVivo Bioequivalence Program).
105. Food and Drug Administration. *Compliance Program Guidance Manual. Bioresearch Monitoring: Human Drugs: In Vivo Bioequivalence,* Chapter 48, October 1999 (CPGM 7348.001).
106. Electronic Records, Electronic, Signatures. 21 CFR Part 11, 2006.
107. US Department of Health and Human Services, Food and Drug Administration, Center for Drug Evaluation and Research, Center for Veterinary, Medicine. Guidance for Industry: General Principles of Software Validation. January 2002. *Qual Assur J* 2007;11:3–15. DOI: (Copyright John Wiley & Sons, Ltd).
108. McDowall RD. *LCGC Eur* 2006;19(5): 274–282.
109. FDA Guidance for Industry on Part 11 Scope and Application, 2003.
110. FDA Current Good Manufacturing Practice for Finished Pharmaceutical Products, 21 CFR 211.
111. ICH Q7A Good Manufacturing Practice for Active Pharmaceutical Ingredients, 2000.
112. Good Automated Manufacturing Practice (GAMP) guidelines version 4, International Society of Pharmaceutical Engineering, Tampa, FL, USA, 2001.
113. GAMP Forum Good Practice Guide—Laboratory Systems; International Society for Pharmaceutical Engineering, Tampa, FL, USA, 2005.
114. United States Pharmacopeia XXIX, 2006.
115. Pharmacopoeal Forum.<1058> Analytical Equipment Qualification, January 2005.
116. McDowall D. *Qual Assur J* 2005;9:196–227.
117. Bansal SK, et al. Qualification of analytical instruments for use in the pharmaceutical industry: a scientific approach, American association of pharmaceutical scientists. *AAPS PharmSciTech* 2004;5(1): 1–8 (Article 22).
118. FDA, Innovation or Stagnation? Challenge and opportunity on the critical path to new medical products, 2004.
119. Zerhouni E. *Science* 2003;302:63–72.
120. DiMasi JA, Hansen RW, Grabowski HG. *J Health Econ* 2003;22:151–185.
121. Colburn WA. *J Clin Pharmacol* 1997;37:355–362.
122. Levy G. *Clin Pharmacol Ther* 1994;56:356–358.
123. Jadhav PR, Mehta MU, Gobburu JVS. *Am Pharm Rev* 2004;7:62–64.
124. Wagner JA, Williams SA, Webster CJ. *Clin Pharm Ther* 2007;81:104–107.
125. Pepe MS, Etzioni R, Feng Z. *J Natl Cancer Inst* 2001;93:1054–1061.
126. Bjornsson TD. *Eur Pharm Rev* 2005;1:17–21.
127. Wagner JA. *Dis Markers* 2002;18:41–46.
128. Lee JW, Weiner RS, Sailstad JM, Bowsher RR, Knuth DW, O'Brien PJ. *Pharm Res* 2005;22:499–511.
129. FDA Using Disease, Placebo, and Drug Prior Knowledge to Improve Decisions in Drug Development and at FDA, 2006.

130. Lee JW, Smith WC, Nordblom GD, Bowsher RR. Validation of assays for the bioanalysis of novel biomarkers. In: Bloom JC, Dean RA,editors. *Biomarkers in Clinical Drug Development*, Marcel Dekker, New York, 2003, pp. 119–149.
131. Lee JW, Devanarayan V, Barrett YC, Allinson J, Fountain S, Keller S, Weinryb I, Green M, Duan L, Rogers JA, Millham R, O'Brien PJ, Sailstad J, Khan M, Weiner RS, Ray C, Wagner JA. *Pharm Res* 2006;23:312–328.
132. Kummar S, Kinders R, Rubinstein L, Parchment RE, Murgo AJ, Collins J, Pickeral O, Low J, Steinberg SM, Gutierrez M. *Nat Rev Cancer* 2007;7:131–139.
133. Stoch SA, Wagner JA. *Int J Pharm Med* 2007;21:271–277.
134. Roessler M, Rollinger W, Mantovani-Endl L, Hagmann ML, Palme S, Berndt P, Engel AM, Pfeffer M, Karl J, Bodenmuller H, Ruschoff J, Henkel T, Rohr G, Rossol S, Rosch W, Langen H, Zolg W, Tacke M. *Mol Cell Proteomics* 2006;5: 2092–2101.
135. Manne U, Srivastava RG, Srivastava S. *Drug Discov Today* 2005;10:965–980.
136. Pan S, Shi M, Jin J, Albin RL, Lieberman A, Gearing M, Lin B, Pan C, Yan X, Kashima DT, Zhang J. *Mol Cell Proteomics* 2007;6(10): 1818–1823.
137. Rikova K, Guo A, Zeng Q, Possemato A, Yu J, Haack H, Nardone J, Lee K, Reeves C, Li Y, Hu Y, Tan Z, Stokes M, Sullivan L, Mitchell J, Wetzel R, Macneill J, Ren JM, Yuan J, Bakalarski CE, Villen J, Kornhauser JM, Smith B, Li D, Zhou X, Gygi SP, Gu TL, Polakiewicz RD, Rush J, Comb MJ. *Cell* 2007;131:1190–1203.
138. Han WK, Bonventre JV. *Curr Opin Crit Care* 2004;10:476–482.
139. McIntosh MW, Liu Y, Drescher C, Urban N, Diamandis EP. *Clin Cancer Res* 2007;13:4422–4428.
140. Guo A, Villen J, Kornhauser J, Lee KA, Stokes MP, Rikova K, Possemato A, Nardone J, Innocenti G, Wetzel R, Wang Y, MacNeill J, Mitchell J, Gygi SP, Rush J, Polakiewicz RD, Comb MJ. *Proc Natl Acad Sci USA* 2008;105:692–697.
141. Lambert JP, Ethier M, Smith JC, Figeys D. *Anal Chem* 2005;15:3771–3788.
142. Washburn MP, Wolters D, YatesJr., *Nat Biotechnol* 2001;19:242–247.
143. Kirkpatrick DS, Gerber SA, Gygi SP. *Methods* 2005;35:265–273.
144. Everley PA, Bakalarski CE, Elias JE, Waghorne CG, Beausoleil SA, Gerber SA, Faherty BK, Zetter BR, Gygi SP. *J Proteome Res* 2006;5:1224–1231.
145. Everley PA, Krijgsveld J, Zetter BR, Gygi SP. *Mol Cell Proteomics* 2004;3:729–735.
146. Bakalarski CE, Haas W, Dephoure NE, Gygi SP. *Anal Bioanal Chem* 2007;389: 1409–1419.
147. Turck CW, Falick AM, Kowalak JA, Lane WS, Lilley KS, Phinney BS, Weintraub ST, Witkowska HE, Yates NA. *Mol Cell Proteomics* 2007;6:1291–1298.
148. Wehr T. *LC–GC*, 2007;25:1.
149. Veenstra TD. *J Chromatogr B* 2007;847:3–11.
150. Sahab ZJ, Semaan SM, Sang QA. *Biomark Insights* 2007;2:21–43.
151. Lee JW, Hall M. *J Chromatogr B* 2009;877:1259–1271.
152. Anderson L, Hunter CL. *Mol Cell Proteomics* 2006;5:573–588.
153. Everley PA, Gartner CA, Haas W, Saghatelian A, Elias JE, Cravatt BF, Zetter BR, Gygi SP. *Mol Cell Proteomics* 2007;6:1771–1777.
154. Niederkofler EE, Tubbs KA, Gruber K, Nedelkov D, Kiernan UA, Williams P, Nelson RW. *Anal Chem* 2001;73:3294–3299.

155. Lee JW,and Hall M. *J Chromatogr B* 2009;877:1259–1271.
156. Stephen RM, Gillies RJ. *Pharm Res* 2007;24:1172–1185.
157. Leong FJ, Leong AS. *J Postgrad Med* 2004;50:62–69.
158. Mitsui K, Doi H, Nukina N. *Methods Enzymol* 2006;412:63–76.
159. Ghattas H, Darboe BM, Wallace DL, Griffin GE, Prentice AM, Macallan DC. *Trans R Soc Trop Med Hyg* 2005;99:675–685.

7

SPONSOR AND FDA/REGULATORY AGENCY GLP INSPECTIONS AND STUDY AUDITS

The most important challenges facing regulated bioanalytical laboratories are strict regulatory requirements and pharmaceutical and life science consolidation. Worldwide regulatory bodies increasingly require more detailed safety control of drug candidates during development stages and ultimately to ensure safety of final products and avoid potential product recalls. Over the past decade, the validity and accuracy of data generated from bioanalytical laboratories in support of bioavailability (BA)/bioequivalence (BE), pharmacokinetics/pharmacology (PK/PD), and other studies designed for new drug applications (NDAs)/abbreviated new drug applications (ANDAs) have been a focus of Food and Drug Administration (FDA) inspections as evident from published FDA inspectional findings. As a result, an important trend has been noted for the number of FDA 483 observations and untitled letters issued to several well-established bioanalytical laboratories and analytical facilities.

While most pharmaceutical and life science companies are subjected to routine inspection by the FDA district offices, inspection of bioanalytical laboratories performing analyses for regulated animal toxicology and human BA/BE studies is generally conducted and directed by the Division of Scientific Investigations (DSIs), which is within the Office of Compliance in Center of Drug Evaluation and Research of the FDA. DSI is an FDA division with the special mandate to inspect preclinical, clinical, and nonclinical laboratories, study sites, and contract research organizations to ensure the validity and accuracy of bioanalytical data.

The regulations for conducting nonclinical laboratory studies in support of applications for research or marketing permits for products including food and color additives, animal food additives, human and animal drugs, medical devices for human use, biological products and electronic products are delineated in GLP regulation

Regulated Bioanalytical Laboratories: Technical and Regulatory Aspects from Global Perspectives,
By Michael Zhou

21 CFR Part 58. The Federal Regulations for the manufacture, processing, packing, or holding of drugs are embodied in cGMP regulation 21 CFR Part 210, while 21 CFR Part 211 delineates the regulations for the preparation of drug products for administration to humans and animals. There is, however, no single set of regulations directly applicable to the conduct of bioanalytical testing of samples obtained in human studies, such as BA, BE, PK, pharmacodynamic (PD) studies, and drug–drug interaction studies. Applicable GLP, cGMP, sound Quality Assurance (QA) practices and, to a certain extent, notable sections from Good Clinical Practices (GCPs) and Good Clinical Laboratory Practices (GCLPs) are key practice ingredients to a bioanalytical laboratory in providing accurate and reliable data. These practices need to be adopted, enforced, and audited in a bioanalytical-testing environment to ensure that the analytical measurements are performed appropriately and reported accurately.

Due to the continued growth of drug development in the branded as well as generic sectors in the pharmaceutical industry, laboratory compliance in all GxP areas have become a major focus of attention by regulatory agencies including the Division of Scientific Investigation and the other compliance branches of the FDA during the review of the applications as well as during site inspections of laboratories performing bioanalytical testing in support of ANDAs and NDAs. There are key elements present in both the regulations for GLP (21 CFR Part 58) and cGMP (21 CFR Parts 210 and 211) that a firm or institution providing bioanalytical laboratory services should employ to assure the quality and integrity of data generated in nonclinical and clinical studies.

As the FDA compliance expectations have become clearer in the past few years, many US and international bioanalytical laboratories are in the process of reassessing their systems, SOPs, and controls in an effort to comply and thereby minimize future FDA inspectional issues.

7.1 GLP VERSUS BIOMEDICAL RESEARCH MONITORING AND MUTUAL ACCEPTANCE OF DATA FOR GLOBAL REGULATIONS AND INSPECTIONS

FDA uses Compliance Program Guidance Manuals (CPGMs) to direct its field personnel on the conduct of inspectional and investigational activities. The CPGM's describes in details of different forms based upon FDA's Bioresearch Monitoring Program. The purpose of each program is to ensure the protection of research subjects and the integrity of data submitted to the agency in support of a marketing application and approval. FDA requires that sponsors of FDA-regulated products submit evidence of their product's safety in research and/or marketing applications. These products include food and color additives, animal drugs, human drugs and biological products, human medical devices, diagnostic products, and electronic products. FDA uses these data to answer questions regarding:

- The toxicity profile of the article
- The observed no adverse effect dose level in the test system
- The risks associated with clinical studies involving humans or animals

- The potential teratogenic, carcinogenic, or other adverse effects of the article
- The level of use that can be approved

The importance of proper safety testing to the product approval process prompted the agency to begin its Toxicology Laboratory Monitoring Program (TLMP). To ensure the quality and integrity of the safety data submitted to it in support of the approval of any application for a research or marketing permit, FDA issued regulations specifying standards for adequate safety testing, prepared an inventory of domestic and foreign toxicology laboratories engaged in safety testing, conducted training sessions for agency investigators to develop proficiency in evaluating testing facilities, and instituted a compliance program that provided for periodic inspections of the testing facilities. These steps have been completed. In the Federal Register of December 22, 1978 (43 FR 60013), FDA issued final GLP regulations. The regulations, which were codified as 21 CFR Part 58, became effective on June 20, 1979. In the Federal Register of April 11, 1980 (45 FR 24865), FDA amended §58.113(b) to delete the requirement that reserve samples of test and control article-carrier mixtures be retained.

FDA maintains an inventory of testing facilities that have been inspected for compliance with the GLP regulations. The inventory listing is updated quarterly to accommodate laboratories that have either ceased or begun work on products regulated by the agency. FDA has conducted 17 training courses at its National Center for Toxicological Research in Jefferson, AR, to provide training in Good Laboratory Practice and the associated laboratory inspection techniques as well as "hands-on" exercises in toxicology experimentation. Finally, in 1976 the agency began an inspection program to assess laboratory compliance with the GLP regulations. Program features include biennial surveillance inspections to assess compliance with the procedural provisions of the GLP regulations and data audit inspections to assess the accuracy of the information contained in final study reports as required by §58.185. The importance of nonclinical laboratory studies to FDA's public health decisions demands that they be conducted according to scientifically sound protocols and with meticulous attention to quality.

GLP inspections 58 CFR 11(e.g., bioanalytical/analytical laboratories) usually includes (but not limited to)

(1) Organization and personnel
(2) Appropriate facility
(3) Qualified equipment
(4) Written procedures (SOPs)
(5) Reagents and solutions
(6) Animal care
(7) Test and control article
(8) Study protocol or plan
(9) Conduct of study (documentation)

(10) Master schedule
(11) Records and reports
(12) Data management and archiving

Types of GLP inspections may be categorized as follows:

Facility Based: Procedures (SOPs), Personnel Management, QA, Staff (including support), Training, Equipment (Procedures, Qualification/maintenance computer systems, etc.), Reagents/solutions, Facility design, Master schedule, and so on.

Study Based: Protocol/plan, Sponsor, Personnel roles (Study director, Principal investigator, Study personnel, QA, etc.), Test system, Test and control articles, QA inspections/audits, Study records/data, Study report, and so forth. Frequently asked questions for the inspections are: SOP table of contents, Company FDA inspection history, Copies of FDA483s and Responses (FDA does not issue GLP certificates), Equipment list, Floor plan, Org charts, and possibly other pieces of information.

Bioanalytical laboratory audits associated with nonclinical (animals) and clinical (human) studies for data submission usually covered by GLP regulations, FDA Guidance for Industry: BMA May 2001, GCP, cGMP, GCLP, and other related regulations, principles (OECD), and guidelines (ICH, WHO, etc.) are given in Table 7.1.

The FDA compliance program is one of four agency-wide Bioresearch Monitoring Compliance Programs. Regulations that govern the proper conduct of clinical studies establish specific responsibilities of sponsors for ensuring: (1) the proper conduct of clinical studies for submission to the Food and Drug Administration and (2) the protection of the rights and welfare of subjects of clinical studies. The specific regulations are found in 21 CFR 312 (CBER and CDER), 21 CFR 812 (CDRH), and 21 CFR 511.1(b) (CVM). The specific responsibilities of sponsors of clinical studies include obligations to

(1) obtain agency approval, where necessary, before studies begin;
(2) manufacture and label investigational products appropriately;
(3) initiate, withhold, or discontinue clinical trials as required;
(4) refrain from commercialization of investigational products;
(5) control the distribution and return of investigational products;
(6) select qualified investigators to conduct studies;
(7) disseminate appropriate information to investigators;
(8) select qualified persons to monitor the conduct of studies;
(9) adequately monitor clinical investigations;
(10) evaluate and report adverse experiences;
(11) maintain adequate records of studies;
(12) submit progress reports and the final results of studies.

TABLE 7.1 Typical Areas of Audits for Regulatory Compliance

- Audit of investigative site operations to determine adequacy of documentation and procedures, and to ensure your trial will be carried out in a controlled manner, per protocol, and in compliance with regulations
- Audit of toxicology laboratories (in-life) and studies to assess GLP compliance and accuracy of reports
- Audit of ongoing and completed studies (Phases I–IV) at investigative sites to determine site and monitor compliance with applicable regulations (21 CFR Parts 50, 54, 56, and 312/812), FDA Compliance Programs, the study protocol, and Standard Operating Procedures and guidelines, as well as to ensure documentation is appropriate for a regulatory inspection
- Audit of Phase I units and PK/PD studies
- Audit of Institutional Review Board operations and documentation for compliance with Protection of Human Subjects and Institutional Review Board regulations (21 CFR 50 and 56) and for responsiveness to the sponsor's needs
- Audit of sponsor central file record archives
- Audit of clinical study data listings
- Audit of clinical and nonclinical study reports
- Audit of sponsor or contract bioanalytical laboratory systems to assess sample assay performance and adherence to established procedures, industry standards, and FDA guidance for method validations, sample assays, and data reporting
- Audit of bioanalytical method validations, data, and final reports

Sponsors may transfer responsibility for any or all of these obligations to Contract Research Organizations (CROs). *Note*: The medical device regulations (21 CFR 812) do not define or delineate responsibilities for CROs. Under the regulations such transfers of responsibility are permitted by written agreement. Responsibilities that are not specified in a written agreement are not transferred. When operating under such agreements, the CROs are subject to the same regulatory actions as sponsors for any failure to perform any of the obligations assumed. Monitors are employed by sponsors or CROs to review the conduct of clinical studies to assure that clinical investigators abide by their obligations for the proper conduct of clinical trials. A Sponsor–Investigator is an individual who both initiates and conducts an investigation, and under whose immediate direction the investigational article is administered, dispensed, or implanted. The requirements applicable to a sponsor–investigator include both those applicable to an investigator and a sponsor.

Bioequivalence studies are generally performed to support abbreviated new drug applications and new drug applications. In ANDAs, approvals of generic versions of innovator drug products are normally based on results of bioequivalence studies. In NDAs, bioequivalence studies are frequently conducted to link to-be-marketed formulation(s) with the clinical trial formulation(s), and to support approval of a new formulation or a new route of administration of a marketed drug. The Bioequivalence Regulations (21 CFR 320) of January 7, 1977 and the amendments of April 28, 1992, April 28, 1993, and January 5, 1999 stated the requirements for submission of *in vivo* bioavailability and bioequivalence data as a condition of marketing a new (e.g., new chemical compound; new formulation, new dosage form, or new route of administration of a marketed drug) or generic drug. 21 CFR 320 also provided general guidance concerning the design and conduct of bioavailability/bioequivalence studies. However, it should be noted that bioequivalence studies conducted to support ANDAs involve testing of already approved drug entities and therefore, generally do not require an investigational new drug (IND) application. The Food and Drug Administration does not require bioequivalence studies on pre-1938 drug products.

Bioequivalence studies involve both a clinical component and an analytical component. The objective of a typical bioequivalence study is to demonstrate that the test and reference products achieve a similar pharmacokinetic profile in plasma, serum, and/or urine. Bioequivalence studies usually involve administration of test and reference drug formulations to 18–36 normal healthy subjects, but patients with a target disease may also be used. Formulations to be tested are administered either as a single dose or as multiple doses. Sometimes formulations can be labeled with a radioactive component to facilitate subsequent analysis. In a bioequivalence study, serial samples of biologic fluid (plasma, serum, or urine) are collected just before and at various times after dose administration. These samples are later analyzed for drug and/or metabolite concentrations. The study data are used in subsequent pharmacokinetic analyses to establish bioequivalence. This program will cover both the clinical and analytical components of bioequivalence studies. In some situations the clinical and analytical facilities for a study may be part of the same organization and therefore may be covered by one District. In other situations, the two facilities may be located in different Districts. For the purpose of this program, the District where the clinical facility is located will be referred to as the Clinical Component District, and the District where the analytical facility is located will be referred to as the Analytical Component District.

From a global perspective, Mutual Acceptance of Data (MAD) was established in 1981 under the Organization for Economic Cooperation and Development (OECD) Council Act. It states that data generated in an OECD member state in accordance with OECD Test Guidelines and Principles of Good Laboratory Practice (GLP) will be accepted in other member countries for assessment purposes and other uses relating to the protection of health and the environment. Nonmember states can gain acceptance into the system if they demonstrate that their test facilities produce safety data of comparable rigor and quality. This removes the need for costly and labor-intensive retesting. It also makes it possible for producers in OECD countries to have safety tests for their chemicals done in MAD-compliant non-OECD states. With OECD–MAD membership, biomedical companies can enjoy better market access and cost savings when exporting to the other member countries. Preclinical research data from GLP-compliant companies will be accepted and recognized in these countries. This removes the need for costly and labor-intensive duplicative testing overseas. Among the OECD–MAD members (30 nations) are many developed countries such as Australia, Japan, the United Kingdom, and United States, which are key biomedical export markets for other nonmember countries. Participation in the MAD system begins with provisional adherence by the nonmember who must accept data generated under MAD conditions from member countries.

The OECD is not a supranational organization, but rather a forum for discussion where governments express their points of view, share their experiences, and search for common ground. If Member countries consider it appropriate, an accord can be embodied in a formal OECD Council Act, which is agreed at the highest level of OECD, the Council. In general, there are two types of Council Act. A Council Decision, which is legally binding on OECD Member countries, and a Council Recommendation, which is a strong expression of political will. In the area of chemicals, for example, there is a Council Act relating to the Mutual Acceptance of Data. The testing of chemicals is labor intensive and expensive. Often the same chemical is being tested and assessed in several countries. Because of the need to relieve some of this burden, the OECD Council adopted a Decision in 1981 stating that data generated in a Member country in accordance with OECD Test Guidelines and Principles of Good Laboratory Practice shall be accepted in other Member countries for assessment purposes and other uses relating to the protection of human health and the environment. The 1981 Council Decision sets the policy context agreed by all OECD Member countries which established that safety data developed in one Member country will be accepted for use by the relevant registration authorities in assessing the chemical or product in another OECD country, that is, the data does not have to be generated a second time for the purposes of safety assessment. A further Council Act was adopted in 1989 to provide safeguards for assurance that the data are indeed developed in compliance with the Principles of GLP. This Council Decision-Recommendation on Compliance with GLP establishes procedures for monitoring GLP compliance through government inspections and study audits as well as a framework for international liaison among monitoring and data-receiving authorities. A 1997 Council Decision on the Adherence of nonmember countries to the Council Acts related to the Mutual Acceptance of Data in the Assessment of Chemicals sets out a stepwise procedure for non-OECD countries with a significant chemical industry input to take part as full members in this system.

From worldwide regulatory compliance and enforcements, there are several recognized and noticeable agencies and organizations such as US FDA, US EPA, EMEA (European Medicines Evaluation Agency), MHLW (Ministry of Health, Labor, and Welfare in Japan), MHRA (Medicines and Healthcare Products Regulatory Agency in UK), OECD, and ISO. Needless to say, it is great regulatory strategies to minimize the burden on the regulated industry whilst maintaining the required levels of public safety. Businesses within the FDA regulated industries, such as pharmaceutical, biotechnology, and medical devices, must think of more than just making a profit. Today, companies within the natural sciences industry must meet best practices (*GxP*) regulations to ensure that the entire development and manufacturing process behind a product meet requirements of compliance and quality.

7.2 PURPOSES AND BENEFITS OF REGULATORY INSPECTIONS/AUDITS

The investigator will determine the current state of GLP compliance by evaluating the laboratory facilities, operations, and study performance as outlined in Parts III, C, and D of the program (Compliance Program Manual). Some of the typical items and areas are usually considered to be audited for quality system and compliance implementation as follows:

(1) *Organization Chart:* If the facility maintains an organization chart, obtain a current version of the chart for use during the inspection and submit it in the EIR.

(2) *Facility Floor-Plan Diagram:* Obtain a diagram of the facility. The diagram may identify areas that are not used for GLP activities. If it does not, request that appropriate facility personnel identify any areas that are not used for GLP activities. Use during the inspection and submit it in the EIR (Facility-GLP: Label areas for GLP or non-GLP or sharing).

(3) *Master Schedule Sheet:* Obtain a copy of the firm's master schedule sheet for all studies listed since the last GLP inspection or past 2 years and select studies as defined in 21 CFR 58.3(d). If the inspection is the first inspection of the facility, review the entire master schedule. If studies are identified as non-GLP, determine the nature of several studies to verify the accuracy of this designation. See 21 CFR 58.1 and 58.3(d). In contract laboratories determine who decides if a study is a GLP study. (QA—need all studies in the current 2 years).

(4) *Identification of Studies*:

   (a) *Directed Inspections*: Inspection assignments will identify studies to be audited.

   (b) *Surveillance Inspections*: Inspection assignments may identify one or more studies to be audited. If the assignment does not identify a study for coverage, or if the referenced study is not suitable to assess all portions of current GLP compliance, the investigator will select studies as necessary to evaluate all areas of laboratory operations. When additional studies are selected, first priority should be given to FDA studies for submission to the assigning Center. *Note*: Studies performed for submission to other government agencies, for example, Environmental Protection Agency, National Toxicology Program, and National Cancer Institute, will not be audited without authorization from the Bioresearch Monitoring Program Coordinator (HFC-230). However, this authorization is not necessary to briefly look at one of these studies to assess the ongoing operations of a portion of the facility.

7.2.1 Criteria for Selecting Ongoing and Completed Studies

(a) Safety studies conducted on FDA-regulated products that have been initiated or completed since the last GLP inspection.

(b) Safety studies that encompass the full scope of laboratory operations.

(c) Studies are significant to safety assessment, for example, carcinogenicity, reproductive toxicity, and chronic toxicity studies.

(d) Studies in several species of animals.

The investigator is encouraged to contact the Center for guidance on study selection.

Ongoing Studies: Obtain a copy of the study protocol and determine the schedule of activities that will be underway during the inspection. This information should be used to schedule for inspections of ongoing laboratory operations, as well as equipment and facilities associated with the study. If there are no activities underway in a given area for the study selected, evaluate the area based on ongoing activities. GLP—make sure always protocol is in study binder before starting it.

Completed Studies: The data audit should be carried out as outlined in Part III, D. If possible, accompany laboratory personnel when they retrieve the study data to assess the adequacy of data retention, storage, and retrieval as described in Part III, C 10. QA—make sure a good retrieve procedure.

7.2.2 Areas of Expertise of the Facility

Testing facilities may conduct one or more types of studies within the scope of GLP regulations. The field investigator should identify in the EIR the types of studies conducted at the facility using the following broad categories:

(1) Physical–chemical testing
(2) Toxicity studies
(3) Mutagenicity studies
(4) Tissue residue depletion studies
(5) Analytical and clinical chemistry testing
(6) Other studies (specify—clinical department operation may be also inspected)

7.2.3 Establishment Inspections

The facility inspection should be guided by the GLP regulations. The following areas should be evaluated and described as appropriate.

7.2.4 Organization and Personnel (21 CFR 58.29, 58.31, 58.33)

Purpose: To determine whether the organizational structure is appropriate to ensure that studies are conducted in compliance with GLP regulations, and to determine whether management, study directors, and laboratory personnel are fulfilling their responsibilities under the GLPs regulations.

Management Responsibilities (21 CFR 58.31): Identify the various organizational units, their roles in carrying out GLP study activities, and the management responsible for these organizational units. This includes identifying personnel who are performing duties at locations other than the test facility and identifying their lines of authority. If the facility has an organizational chart, much of this information can be determined from the chart.

(a) Determine if management has procedures for assuring that the responsibilities in 21 CFR 58.31 can be carried out and

(b) Look for evidence of management involvement, or lack thereof, in the following areas:

Assigning and Replacing Study Directors:

- Control of study director workload (use the Master Schedule to assess workload).
- Establishment and support of the Quality Assurance Unit (QAU), including assuring that deficiencies reported by the QAU are communicated to the study directors and acted upon.
- Assuring that test and control articles or mixtures are appropriately tested for identity, strength, purity, stability, and uniformity.
- Assuring that all study personnel are informed of and follow any special test and control article handling and storage procedures.
- Providing required study personnel, resources, facilities, equipment, and materials.
- Reviewing and approving protocols and standard operating procedures (SOPs).
- Providing GLP or appropriate technical training.

Personnel (21 CFR 58.29): Identify key laboratory and management personnel, including any consultants or contractors used, and review personnel records, policies, and operations to determine if:

- Summaries of training and position descriptions are maintained and are current for selected employees.
- Personnel have been adequately trained to carry out the study functions that they perform.
- Personnel have been trained in GLPs.
- Practices are in place to ensure that employees take necessary health precautions, wear appropriate clothing, and report illnesses to avoid contamination of the test and control articles and test systems.
- If the firm has computerized operations, determine the following:
  - Who was involved in the design, development, and validation of the computer system?
  - Who is responsible for the operation of the computer system, including inputs, processing, and output of data?
  - Whether computer system personnel have training commensurate with their responsibilities, including professional training and training in GLPs?
  - Whether some computer system personnel are contractors who are present on-site full time, or nearly full time. The investigation should include these contractors as though they were employees of the firm. Specific inquiry may be needed to identify these contractors, as they may not appear on organization charts.

Interview and observe personnel using the computerized systems to assess their training and performance of assigned duties.

Study Director (21 CFR 58.33): Assess the extent of the study director's actual involvement and participation in the study. In those instances when the study director is located off-site, review any correspondence/records between the testing facility management and quality assurance unit and the off-site study director. Determine that the study director is being kept immediately apprised of any problems that may affect the quality and integrity of the study.

Assess the Procedures by Which the Study Director:

- Assures the protocol and any amendments have been properly approved and are followed.
- Assures that all data are accurately recorded and verified.
- Assures that data are collected according to the protocol and SOPs.
- Documents unforeseen circumstances that may affect the quality and integrity of the study and implements corrective action.
- Assures that study personnel are familiar with and adhere to the study protocol and SOPs.
- Assures that study data are transferred to the archives at the close of the study.

EIR Documentation and Reporting: Collect exhibits to document deficiencies. This may include SOPs, organizational charts, position/job descriptions, and CVs, as well as study-related memos, records, and reports for the studies selected for review. The use of outside or contract facilities must be noted in the EIR. The assigning Center should be contacted for guidance on inspection of these facilities. (GLP-QA: Updated organization chart, job description, CV, and training record (established), master schedule to design who is responsible for the study (established).

7.2.5 Quality Assurance Unit (QAU; 21 CFR 58.35)

Purpose: To determine if the test facility has an effective, independent QAU that monitors significant study events and facility operations, reviews records and reports, and assures management of GLP compliance.

QAU Operations (21 CFR 58.35(b–d)): Review QAU SOPs to assure that they cover all methods and procedures for carrying out the required QAU functions, and confirm that they are being followed. Verify that SOPs exist and are being followed for QAU activities including, but not limited to, the following:

- Maintenance of a master schedule sheet.
- Maintenance of copies of all protocols and amendments.
- Scheduling of its in-process inspections and audits.
- Inspection of each nonclinical laboratory study at intervals adequate to assure the integrity of the study and maintenance of records of each inspection.

- Immediately notify the study director and management of any problems that are likely to affect the integrity of the study.
- Submission of periodic status reports on each study to the study director and management.
- Review of the final study report.
- Preparation of a statement to be included in the final report that specifies the dates inspections were made and findings reported to management and to the study director.
- Inspection of computer operations.

In addition, it is also important to (1) verify that, for any given study, the QAU is entirely separate from and independent of the personnel engaged in the conduct and direction of that study; (2) evaluate the time QAU personnel spend in performing in-process inspection and final report audits; and (3) determine if the time spent is sufficient to detect or identify problems in critical study phases and if there are adequate personnel to perform the required functions.

Note: The investigator may request the firm's management to certify in writing that inspections are being implemented, performed, documented, and followed up in accordance with this section (refer to 58.35(d)).

EIR Documentation and Reporting: Obtain a copy of the master schedule sheet dating from the last routine GLP inspection or covering the past 2 years. If the master schedule is too voluminous, obtain representative pages to permit headquarters review. When master schedule entries are coded, obtain the code key. Deficiencies should be fully reported and documented in the EIR. Documentation to support deviations may include copies of QAU SOPs, list of QAU personnel, their curriculum vitae (CVs) or position descriptions, study-related records, protocols, and final reports. (QA—needs to make sure all those exist).

7.2.6 Facilities (21 CFR 58.41–58.51)

Purpose: Assess whether the facilities are of adequate size and design. The following items need to be checked:

- Facility Inspection.
- Review environmental controls and monitoring procedures for critical areas (e.g., animal rooms, test article storage areas, laboratory areas, handling of bio-hazardous material, etc.) and determine if they appear adequate and are being followed.
- Review the SOPs that identify materials used for cleaning critical areas and equipment, and assess the facility's current cleanliness.
- Determine whether there are appropriate areas for the receipt, storage, mixing, and handling of the test and control articles.
- Determine whether separation is maintained in rooms where two or more functions requiring separation are performed.

- Determine that computerized operations and archived computer data are housed under appropriate environmental conditions (e.g., protected from heat, water, and electromagnetic forces). Facility-IT: needs to check this part.

EIR Documentation and Reporting: Identify which facilities, operations, and SOPs were inspected. Only significant changes in the facility from previous inspections need be described. Facility floor plans may be collected to illustrate problems or changes. Document any conditions that would lead to contamination of test articles or to unusual stress of test systems.

7.2.7 Equipment (21 CFR 58.61–58.63)

Purpose: To assess whether equipment is appropriately designed and of adequate capacity and is maintained and operated in a manner that ensures valid results.

Equipment Inspection: Assess the following:

- The general condition, cleanliness, and ease of maintenance of equipment in various parts of the facility.
- The heating, ventilation, and air conditioning system design and maintenance, including documentation of filter changes and temperature/humidity monitoring in critical areas. (Facility: do we have temperature chart for all GLP area).
- Whether equipment is located where it is used and that it is located in a controlled environment, when required.
- Nondedicated equipment for preparation of test and control article carrier mixtures is cleaned and decontaminated to prevent cross contamination.
- For representative pieces of equipment check the availability of the following.
- SOPs and/or operating manuals.
- Maintenance schedule and log.
- Standardization/calibration procedure, schedule, and log.
- Standards used for calibration and standardization. (GLP: SOP for instruments established (ChemStation is in approval process).) Some maintenance are scheduled by facility or outsourced vendor as whole (such as balance, etc.), make sure that good log system is in place; logbooks are there, more information or title page need to be updated as given in Table 7.2.
- For computer systems, assess that the following procedures exist and are documented.
- Validation study, including validation plan and documentation of the plan's completion.
- Maintenance of equipment, including storage capacity and backup procedures.
- Control measures over changes made to the computer system, which include the evaluation of the change, necessary test design, test data, and final acceptance of the change.

TABLE 7.2 Typical Logbook Information with a Logbook Record

- Name of the equipment
- Name of the manufacturer
- Model or type for identifications
- Serial number
- Date equipment was received in the laboratory
- Condition when received (new, used)
- Details of checks made for compliance with relevant calibration or test standard specification
- Date equipment was placed in service by the laboratory
- Current location in the laboratory, if appropriate
- Copy of manufacturer's operating instruction(s)
- Details of maintenance carried out
- History of any damage, malfunction, modification, or repair
- Person responsible for the equipment

- Evaluation of test data to assure that data are accurately transmitted and handled properly when analytical equipment is directly interfaced to the computer.
- Procedures for emergency backup of the computer system (e.g., backup battery system and data forms for recording data in the event of a computer failure or power outage). (IT—needs to check this part; QA—should have kept all change control documents, also make sure that they are completed and indexed/archived for easy retrieval.)

EIR Documentation and Reporting: The EIR should list which equipment, records, and procedures were inspected and the studies to which they are related. Detail any deficiencies that might result in contamination of test articles, uncontrolled stress to test systems, and/or erroneous test results.

7.2.8 Testing Facility Operations (21 CFR 58.81)

Purpose: To determine if the facility has established and follows written SOPs necessary to carry out study operations in a manner designed to ensure the quality and integrity of the data.

- Review the SOP index and representative samples of SOPs to ensure that written procedures exist to cover at least all of the areas identified in 58.81(b).
- Verify that only current SOPs are available at the personnel workstations.
- Review key SOPs in detail and check for proper authorization signatures and dates, and general adequacy with respect to the content (e.g., SOPs are clear, complete, and can be followed by a trained individual).
- Verify that changes to SOPs are properly authorized and dated and that a historical file of SOPs is maintained.

- Ensure that there are procedures for familiarizing employees with SOPs.
- Determine that there are SOPs to ensure the quality and integrity of data, including input (data checking and verification), output (data control), and an audit trail covering all data changes.
- Verify that a historical file of outdated or modified computer programs is maintained. If the firm does not maintain old programs in digital form, ensure that a hard copy of all programs has been made and stored.
- Verify that SOPs are periodically reviewed for current applicability and that they are representative of the actual procedures in use.
- Review selected SOPs and observe employees performing the operation to evaluate SOP adherence and familiarity.

EIR Documentation and Reporting: Submit SOPs, data collection forms, and raw data records as exhibits that are necessary to support and illustrate deficiencies (GLP-QA: some SOPs need to be reviewed/approved, for example, Incurred sample reanalysis (high), ChemStation user (high), reagent (high), data QC (moderate), significant number (moderate), spreadsheet validation (low).)

7.2.9 Reagents and Solutions (21 CFR 58.83)

Purpose: To determine that the facility ensures the quality of reagents at the time of receipt and subsequent use.

- Review the procedures used to purchase, receive, label, and determine the acceptability of reagents and solutions for use in the studies.
- Verify that reagents and solutions are labeled to indicate identity, titer or concentration, storage requirements, and expiration date.
- Verify that for automated analytical equipment, the profile data accompanying each batch of control reagents are used.
- Check that storage requirements are being followed.

7.2.10 Animal Care (21 CFR 58.90)

Purpose: To assess whether animal care and housing is adequate to minimize stress and uncontrolled influences that could alter the response of test system to the test article.

- Inspect the animal room(s) housing the study to observe operations, protocol and SOP adherence, and study records. Refer to IOM 145.2 prior to inspecting subhuman primate facilities.
- Determine that there are adequate SOPs covering environment, housing, feeding, handling, and care of laboratory animals, and that the SOPs and the protocol instructions are being followed.
- Review pest-control procedures and documentation of the chemicals used.

- Identify individuals responsible for the program. When a contractor provides the pest control, determine if someone from the laboratory accompanies the exterminator at all times.
- Determine whether the facility has an Institutional Animal Care and Use Committee (IACUC). Obtain and submit a copy of the Committee's Standard Operating Procedures and the most recent committee minutes to verify committee operation.
- Determine that all newly received animals are appropriately isolated, identified, and their health status is evaluated.
- Verify that treatment given to animals that become diseased is authorized by the study director and documented.
- For a representative sample of animals, compare individual animal identification against corresponding housing unit identification and dose group designations to assure that animals are appropriately identified.
- For a representative sample of animals, review daily observation logs and verify their accuracy for animals reported as dead or having external gross lesions or masses.
- Ensure that animals of different species, or animals of the same species on different projects, are separated as necessary.
- Verify that cages, racks, and accessory equipment are cleaned and sanitized, and that appropriate bedding is used.
- Determine that feed and water samples are collected at appropriate sources, analyzed periodically, and that analytical documentation is maintained.

EIR Documentation and Reporting: The EIR should identify which areas, operations, SOPs, and studies were inspected and document any deficiencies.

7.2.11 Test and Control Articles (21 CFR 58.105–58.113)

Purpose: To determine that procedures exist to assure that test and control articles and mixtures of articles with carriers meet protocol specifications throughout the course of the study, and that accountability is maintained.

- Characterization and Stability of Test Articles (21 CFR 58.105)—The responsibility for carrying out appropriate characterization and stability testing may be assumed by the facility performing the study or by the study sponsor. When test article characterization and stability testing is performed by the sponsor, verify that the test facility has received documentation that this testing has been conducted.
- Verify that procedures are in place to ensure that
    - The acquisition, receipt and storage of test articles, and means used to prevent deterioration and contamination are as specified.

  - The identity, strength, purity, and composition, (e.g., characterization) to define the test and control articles are determined for each batch and are documented.
- The stability of test and control articles is documented.
- The transfer of samples from the point of collection to the analytical laboratory is documented.
- Storage containers are appropriately labeled and assigned for the duration of the study.
- Reserve samples of test and control articles for each batch are retained for studies lasting more than 4 weeks.

7.2.12 Test and Control Article Handling (21 CFR 58.107)

Purpose: Determine that there are adequate procedures for

(1) documentation for receipt and distribution;
(2) proper identification and storage;
(3) precluding contamination, deterioration, or damage during distribution.

- Inspect test and control article storage areas to verify that environmental controls, container labeling, and storage are adequate.
- Observe test and control article handling and identification during the distribution and administration to the test system.
- Review a representative sample of accountability records and, if possible, verify their accuracy by comparing actual amounts in the inventory. For completed studies, verify documentation of final test and control article reconciliation.
- Mixtures of Articles with Carriers (58.113)—If possible, observe the preparation, sampling, testing, storage, and administration of mixtures. Verify that analytical tests are conducted, as appropriate, to:
  - Determine uniformity of mixtures and to determine periodically the concentration of the test or control article in the mixture, and
  - Determine the stability as required under study conditions.
  - Verify that the results are reported to the study director. When the sponsor performs analytical testing, concentration data may be reported to the study director as absolute or relative (percent of theoretical) values.

EIR Documentation and Reporting: Identify the test articles, SOPs, facilities, equipment, and operations inspected. Review the analytical raw data versus the reported results for accuracy of reporting and overall integrity of the data that are being collected. Where deficiencies or other information suggest a problem with the purity, identity, and strength of the test article or the concentration of test article mixtures, document and obtain copies of the analytical raw data and any related reports. Consideration should be given to collecting a sample as described in Compliance Program Manual Part III, G.

7.2.13 Protocol and Conduct of Nonclinical Laboratory Study (21 CFR 58.120–58.130)

Purpose: To determine if study protocols are properly written and authorized, and that studies are conducted in accordance with the protocol and SOPs.

7.2.14 Study Protocol (21 CFR 58.120)

Purpose: To review SOPs for protocol preparation and approval and verify they are followed.

- Review the protocol to determine if it contains required elements.
- Review all changes, revisions, or amendments to the protocol to ensure that they are authorized, signed, and dated by the study director.
- Verify that all copies of the approved protocol contain all changes, revisions, or amendments. (GLP-QA: For validation, please have protocol and PI assigned, need to keep signature and initial document for all personnel involved in GLP studies).
- Conduct of the Nonclinical Laboratory Study (21 CFR 58.130)—Evaluate the following laboratory operations, facilities, and equipment to verify conformity with protocol and SOP requirements for the following.

7.2.15 Test System Monitoring

Purpose: To review data generation, analytical process and study documentation as follows.

- Recording of raw data (manual and automated).
- Corrections to raw data (corrections must not obscure the original entry and must be dated, initialed, and explained).
- Randomization of test systems.
- Collection and identification of specimens.
- Authorized access to data and computerized systems. (GLP-IT: security reason, for desk computer, need screen saver; for LC–MS/MS, log off when being away; for HPLC–UV, set up ChemStation lock out time; for vendor/engineer, set up and use an account only access local computer.)

EIR Reporting and Documentation: Identify the studies inspected and, if available, the associated FDA research or marketing permit numbers. Report and document any deficiencies observed. Submit, as exhibits, a copy of all protocols and amendments that were reviewed.

7.2.16 Records and Reports (21 CFR 58.185–58.195)

Purpose: To assess how the test facility stores and retrieves raw data, documentation, protocols, final reports, and specimens. Reporting of Study Results

(21 CFR 58.185)—Determine if the facility prepares a final report for each study conducted. For selected studies, obtain the final report, and verify that it contains the following:

- The required elements in 21 CFR 58.185(a)(1–14), including the identity (name and address) of any subcontractor facilities and portion of the study contracted, and a description of any computer program changes.
- Dated signature of the study director (21 CFR 58.185(b)).
- Corrections or additions to the final report are made in compliance with 21 CFR 58.185(c).
- Storage and Retrieval of Records and Data (21 CFR 58.190).
- Verify that raw data, documentation, protocols, final reports, and specimens have been retained.
- Identify the individual responsible for the archives and determine if delegation of duties to other individuals in maintaining the archives has occurred.
- Verify that archived material retained or referred to in the archives is indexed to permit expedient retrieval. It is not necessary that all data and specimens be in the same archive location. For raw data and specimens retained elsewhere, the archives index must make specific reference to those other locations.
- Verify that access to the archives is controlled and determine that environmental controls minimize deterioration.
- Ensure that there are controlled procedures for adding or removing materials. Review archive records for the removal and return of data and specimens. Check for unexplained or prolonged removals.
- Determine how and where computer data and backup copies are stored, that records are indexed in a way to allow access to data stored on electronic media, and that environmental conditions minimize potential deterioration.
- Determine to what electronic media such as tape cassettes or ultrahigh capacity portable discs the test facility has the capacity of copying records in electronic form. Report names and identifying numbers of both copying equipment type and electronic medium type to enable agency personnel to bring electronic media to future inspections for collecting exhibits.

EIR Documentation and Reporting: Provide a brief summary of the facility's report preparation procedures and their retention and retrieval of records, reports, and specimens. If records are archived *off-site*, obtain a copy of documentation of the records that were transferred and where they are located. Describe and document deficiencies. (QA–GLP laboratory: ensure master schedule is completed, and then check out all the items listed above for completed studies; QA—makes sure archiving system and data retrieval procedure are ready; GLP laboratory—needs logically store and archive all e-data, and delete some data from operation system according SOPs; IT—checks data backup and server facility is safe.)

7.2.17 Data Audit

In addition to the procedures outlined above for evaluating the overall GLP compliance of a firm, the inspection should include the audit of at least one completed study. Studies for audit may be assigned by the Center or selected by the investigator as described in Compliance Program Manual Part III, A. The audit will include a comparison of the protocol (including amendments to the protocol), raw data, records, and specimens against the final report to substantiate that protocol requirements were met and that findings were fully and accurately reported.

For each study audited, the study records should be reviewed for quality to ensure that data are

- *Attributable:* The raw data can be traced, by signature or initials and date to the individual observing and recording the data. Should more than one individual observe or record the data, that fact should be reflected in the data.
- *Legible:* The raw data are readable and recorded in a permanent medium. If changes are made to original entries, the changes:
  - must not obscure the original entry;
  - indicate the reason for change;
  - must be signed or initialed and dated by the person making the change.
- *Contemporaneous:* The raw data are recorded at the time of the observation.
- *Original:* The first recording of the data.
- *Accurate:* The raw data are true and complete observations. For data entry forms that require the same data to be entered repeatedly, all fields should be completed or a written explanation for any empty fields should be retained with the study records. (GLP laboratory—seems to be mostly established, but needs pay more attention.)

7.2.18 General

Determine if there were any significant changes in the facilities, operations, and QAU functions other than those previously reported.

Determine whether the equipment used was inspected, standardized, and calibrated prior to, during, and after use in the study. If equipment malfunctioned, review the remedial action, and ensure that the final report addresses whether the malfunction affected the study.

- Determine if approved SOPs existed during the conduct of the study.
- Compare the content of the protocol with the requirements in 21 CFR 58.120.
- Review the final report for the study director's dated signature and the QAU statement as required in 21 CFR 58.35(b)(7).

Protocol Versus Final Report: Study methods described in the final report should be compared against the protocol and the SOPs to confirm those requirements were met. Examples include, but are not limited to, the following:

- Selection and acquisition of the test system, for example, species, source, weight range, number, age, date ordered, and date received.
- Procedure for receipt, examination, and isolation of newly received animals.
- Methods of test system identification, housing, and assignment to study.
- Types and occurrences of diseases and clinical observations prior to and during the study, as well as any treatments administered.
- Frequency and methods of sampling and analysis for contaminants in feed and water.
- Feed and water contaminant levels did not exceed limits specified in the protocol.
- Types of bedding and pest control materials do not interfere with the study.
- Preparation and administration of test and/or control articles.
- Analysis of test article/test article carrier mixtures.
- Observation of the test system response to test article.
- Handling of dead or moribund animals.
- Collection and analysis of specimens and raw data.
- Necropsy, gross pathology, and histopathology.

7.2.19 Final Report Versus Raw Data

The audit should include a detailed review of records, memorandum, and other raw data to confirm that the findings in the final report completely and accurately reflect the raw data. Representative samples of raw data should be audited against the final report. Examples of types of data to be checked include, but are not limited to, the following:

- Animal body weight records.
- Food and water consumption records.
- Test system observation and dosing records.
- Records for the analysis of uniformity, concentration, and stability of test and control article mixtures.
- Protocol-required analyses (e.g., urinalysis, hematology, blood chemistry, ophthalmologic exams).
- Necropsy and gross pathology records.
- Histopathology, including records for tissue processing and slide preparation.
- Conformity between "interim" reports (e.g., a report on the first year of a multi-year study) and the final report.
- A representative number of animals from selected dose groups (initially the control and high) should be traced from receipt through final histopathologic examination.

Evaluate for the following:

- The accuracy of individual test system identification.
- Review trends in each parameter measured or observed and note inconsistencies.
- Explanations for reported data outliers.
- Conformity between *in vivo* test system observations and gross pathology observations.
- Conformity between gross pathology observations and histopathologic examinations.

Documentation of unforeseen circumstances may have affected the quality and integrity of the study.

7.2.20 Specimens Versus Final Report

The audit should include examination of a representative sample of specimens in the archives for confirmation of the number and identity of specimens in the final report.

- EIR Documentation and Reporting.
- All studies audited should be identified by
    (1) Sponsor's name.
    (2) Study title (and any study numbers).
    (3) Study director's name.
    (4) Test article name and/or code.
    (5) Test system.
    (6) The date the protocol was signed by the study director.
    (7) In-life starting and ending dates.
    (8) The date the study director signed the final report and if available.
    (9) The FDA research or marketing application number.
    (10) Describe and document any significant deficiencies in the conduct or reporting of the study. Obtain a copy of the narrative study report and protocol.

7.2.21 Refusal to Permit Inspection

Field investigators should refer to IOM Section 514 for general guidance on dealing with refusal to permit inspection. Test facility management should be advised that the agency will not consider a nonclinical laboratory study in support of a research or marketing permit if the testing facility refuses to permit inspection [58.15(b)]. Under this program, partial refusals, including refusal to permit access to, and copying of the master schedule (including any code sheets), SOPs, and other documents pertaining to the inspection will be treated as a "total" refusal to permit inspection.

7.2.22 Sealing of Research Records

Whenever a field investigator encounters questionable or suspicious records under any Bioresearch Monitoring Program, and is unable to copy or review them immediately, and has reason for preserving the integrity of those records, the investigator is to immediately contact by telephone HFC-230 for instructions. Refer to IOM 453.7 for more information.

7.2.23 Samples

Collection of samples should be considered when the situation under audit or surveillance suggests that the facility had, or is having, problems in the area of characterization, stability, storage, contamination, or dosage preparation. The investigator should contact the assigning Center and the Division of Field Science (HFC-140) before samples are collected. If the field investigator collects a sample, a copy of the methodology and reference samples from the sponsor or the testing facility must also be obtained. The copy will be sent to HFC-140 for designation of a laboratory to perform the sample analysis according to instrument capabilities and the availability of the selected district laboratory. The investigator should contact HFC-140 at (301) 827-7605 for specific disposition/handling instructions. The investigator must use the appropriate 7-digit code for product collected and provide complete documentation with each sample for possible legal/administrative actions.

Samples may include physical samples of (1) carrier, (2) test article, (3) control article, (4) mixtures of test or control articles with carriers, and (5) and specimens including wet tissues, tissue blocks, and slides (see above regarding Center and HFC-140 authorization and instructions).

Following are some of the typical observations cited for corrections and improvements on a continuing basis:

(1) *Laboratory-GLP:* Sample management system need to be established to quick retrieve a sample, and quickly provide a list of samples in a storage area. Software or databases are being explored with IT.

(2) *Facility-GLP:* Sample emergency transfer plan or procedures need to be established. Clearly define who will be called, where to be transferred, and what documents to be filled.

(3) *Laboratory-GLP:* Raw data need to be stored in a safe place after daily work (file cabinets or locked place), not just put it on the desk.

(4) *IT-GLP:* Security role setting for analyst and administrator need to be checked against SOPs.

(5) *QA-GLP:* Study audit for long-term study may need to be audited more than once. (Right now, at least one in-life audit is performed for each sample analysis study.)

(6) *QA-GLP:* No annual training for GLP compliance (firm should contact outside training service, and obtain training record/certificate).

(7) *Facility:* Security for cleaning service people, nobody escort them and they can reach most of the areas.

(8) *Facility-GLP:* Freezer alarm systems need to be routinely tested and documented.

(9) *Facility:* No business continuity plan. What happen if disaster happens to the testing facility? Where the data (paper and e-data) will be transferred? Where the samples will be transferred (to Fisher, contacted)?

(10) *QA-GLP:* Instrument validation documents, QA need to check if they have all documents, check their completeness, and indexed for easy retrieval.

(11) *QA-Management:* Some reference chemical testing data were obtained using non-GLP facility/method such as CMC.

(12) *QA:* No SOP for GxP training.

(13) *HR:* Job description is not version controlled.

(14) *GLP-IT:* Screen saver for security reason.

Please note that there are numerous cases of observations associated with different levels of understanding, and corrective action and preventative action (CAPA) plans.

7.3 TYPICAL INSPECTIONS/AUDITS AND THEIR OBSERVATIONS

Regulatory inspections and audits may include (but not limited to): study/project based, facility-based voluntary and nonvoluntary processes. Main objectives are as follows:

- To verify the quality and integrity of scientific data from bioequivalence studies submitted to the Center for Drug Evaluation and Research (CDER).
- To ensure that the rights and welfare of human subjects participating in drug testing are protected.
- To ensure compliance with the regulations (21 CFR 312, 320, 50, and 56) and promptly follow-up on significant problems, such as research misconduct or fraud.

Generally, unannounced visits to nonclinical, clinical, and/or analytical investigator will be impractical. Appointments to inspect may be made by telephone, but the time interval until the actual inspection should be kept as short as possible. Any undue delay of the inspection on the part of the clinical or analytical investigator should be reported immediately to GLP and Bioequivalence Investigations Branch (GBIB, HFD-48). In the event that an unannounced visit becomes necessary, this will be addressed and clearly stated in the inspection assignment memo sent to the appropriate District office(s).

GLP inspections are to verify the quality and integrity of nonclinical studies used to protect human subject safety toxicity studies and other nonclinical studies critical to the drug development process general toxicology, carcinogenicity, reprotoxicity, safety pharmacology, toxicokinetics studies pivotal to review decisions IND, NDA, BLA. Regarding clinical studies, these typically involve BE study types—*in vivo* studies clinical trial versus to-be-marketed, new salt/ester, new formulation, new combination (NDAs), generic versus innovator (ANDAs); *in vitro* studies—biowaivers based on the Biopharmaceutics Classification System (e.g., nasal aerosols and sprays for local action). The nature of these inspections is clinical, analytical, statistical, and clinical study portions. Facility types include CROs, sponsors, academic institutions, individual labs, and hospitals (domestic and foreign sites). Inspection type consists of either Routine studies pivotal to approval decisions or "For Cause" problems with study conduct or reported results or complaint follow-up.

Inspection of clinical and analytical facilities should include review of the practices, qualifications of personnel, and procedures utilized during conduct of the clinical study and analytical testing. The clinical and analytical data provided by the GBIB should be compared with the records at the clinical and analytical study sites. Analytical testing procedures, including controls, should be reviewed by the GBIB scientist or FDA field chemist for scientific soundness with relation to the data submitted in the NDA/ANDA. Here are some of typical concerns observed from general audits and inspections (Table 7.3).

7.4 REGULATORY CHALLENGES FOR BIOANALYTICAL LABORATORIES

7.4.1 Introduction

Due to its superior sensitivity and selectivity, LC–MS/MS has become the primary analytical technique used by most bioanalytical laboratories performing analyses for human bioavailability/bioequivalence studies [1, 2]. The compliance requirements in developing, validating, and using LC–MS/MS have been published in several Food and Drug Administration guidance documents, the Compliance Program Guidance Manual, Good Laboratory Practice regulations, and various publications [3–9].

In the FDA Guidance for Industry: Bioanalytical Method Validation, FDA clearly states that "... The bioanalytical method for human BA, BE, PK, and drug interaction studies must meet the criteria in 21 CFR 320.29. The analytical laboratory should have a written set of standard operating procedures to ensure a complete system of quality control and assurance. The SOPs should cover all aspects of analysis from the time the sample is collected and reaches the laboratory until the results of the analysis are generated and reported. The SOPs also should include record keeping, security and chain of sample custody (accountability systems that ensure integrity of test articles), sample preparation, and analytical tools such as methods, reagents, equipment, instrumentation, and SOPs for quality control and verification of results" [5].

Therefore, it is evident from this Guidance that it is an FDA expectation that all bioanalytical laboratories are required to implement proper systems, SOPs, and controls to ensure the accuracy and validity of the data reported. Absent specific requirements articulated by FDA, it is reasonable to extrapolate general principles from all applicable GLP, current Good Manufacturing Practice (cGMP), Good Clinical Practices regulations, and guidances.

TABLE 7.3 Typical Issues Found during Audits/Inspections

Procedures

- Written standard operating procedures were inaccurate, incomplete, ambiguous, and contained no documentation of origin, review, and approval
- Failure to follow written SOPs as required by 21 CFR 58.XXX

Internal quality audits

- Quality audits are inadequate to assure that the quality system is in compliance with the established quality system requirements and to determine the effectiveness of the quality system
- Failure to document the dates and observations of quality audits

Personnel

Laboratory controls

- Failure to demonstrate a consistent way in handling the receipt and storage of standards, samples, and other chemicals/reagents (did not follow appropriate SOPs)
- Expired standards were used in the calibration of equipment and/or in quantification of analyte(s). Working solutions were not properly labeled or documented in laboratory notebooks/forms or other records in that the data did not bear complete information, including the analyst or preparer's identity, solution designation, concentrations, and expiry dates

Laboratory equipment qualification

- Qualification and validation arrangements for system were poor with a lack of formal process, acceptance criteria, testing procedures, records, reviews, error handling, formal reporting, and signing off
- Failure to ensure that all inspection, measuring, and test equipment is suitable for its intended purpose and its capable of generating valid results

Computer/software validation

- Failure to provide complete records of computerized system installation and validation
- Lack of evidence in support of validating computer data integrity

Methods performance

- Failure to provide records and justification of batch run(s) failing acceptance criteria (high rate of failure than normal, e.g., 20%)
- Failure to follow established laboratory method procedures for routine analysis

Corrective and preventive actions

- Failure to verify or validate CAPA to ensure that such action is effective and does not adversely affect accuracy and reliability of data
- Failure to maintain adequate procedures for implementing CAPA such as analyzing data to identify existing and potential rootcause
- Failure to investigate other laboratory procedures/process and deficiencies that might have been associated with rootcause

Raw data

- Failure to maintain laboratory records for all the laboratory testing performed
- Failure to provide relevant SOPs handling laboratory raw data (generating, processing, storing, and archiving)

TABLE 7.3 (*Continued*)

Records

- Failure to retain all control and laboratory records to assure the integrity and completeness of laboratory data
- Failure to maintain records of changes to documents

Electronic records

- Failure to verify audit trail whiling revising the data file
- Failure to demonstrate internal polices in e-data handling process

Documentation

- Failure to provide all necessary supporting documentation pertaining to study
- Failure to implement SOPs relating to documentation

Examples of noncompliance from inspections/audits

Clinical studies

Examples of noncompliance:

(1) Subjects not receiving the test or reference drug formulation according to the study randomization codes
(2) Biological samples compromised by improper identification, handling, or storage
(3) Failure to report adverse experiences, such as vomiting, and diarrhea, which may affect absorption and elimination of drugs
(4) Inadequate drug accountability records
(5) Inadequate medical supervision and coverage
(6) Significant problems/protocol deviations/adverse events not reported to the sponsor
(7) Failure to adhere to the inclusion/exclusion criteria of the approved protocol
(8) Inadequate or missing informed consent for participating subjects
(9) Any other situation in which the health and welfare of the subjects are compromised

Analytical testing

Examples of noncompliance:

(1) Inconsistencies between data reported to FDA and at the site
(2) Inadequate or missing validation of assay methodology with respect to specificity (related chemicals, degradation products, metabolites), linearity, sensitivity, precision, and reproducibility)
(3) Failure to employ standard, scientifically sound quality control techniques, such as use of appropriate standard curves and/or analyte controls that span the range of subjects' analyte levels
(4) Failure to include all data points, not otherwise documented as rejected for a scientifically sound reason, in determination of assay method precision, sensitivity, and accuracy
(5) Samples are allowed to remain for prolonged periods of time without proper storage
(6) Failure to maintain source data, for example, source data written on scrap paper and/or discarded in trash after transferring to analytical documents
(7) Lack of objective standard for data acceptance of calibration standards and quality controls
(8) Unskilled personnel performing analytical procedures
(9) No traceable documentation of analytical findings
(10) Inadequate or no written procedures for drug sample receipt and handling
(11) Inadequate or missing standard operating procedures

While most pharmaceutical companies are subjected to routine inspection by the FDA district offices, inspection of bioanalytical laboratories performing analyses for human BA/BE studies is generally conducted and directed by the Division of Scientific Investigations, which is within the Office of Compliance in Center of Drug Evaluation and Research of the FDA. DSI is an FDA division with the special mandate to inspect preclinical, clinical, and nonclinical laboratories, study sites, and contract research organizations [10, 11].

As the FDA compliance expectations have become clearer in the past few years, many US and international bioanalytical laboratories are in the process of reassessing their systems, SOPs, and controls in an effort to comply and thereby minimize future FDA inspectional issues. These proactive efforts could result in redefining and further elevating the standards of "industry practices" in many critical quality systems and controls. Examples of the quality systems impacted by this enhancement process are listed below: (1) events/deviations investigation, resolution, and management; (2) document change control; (3) data review and approval process; (4) record-keeping and documentation practices; (5) instrument, equipment, and computer validation and qualification; (6) independent quality assurance oversight and control; and (7) method validation practices.

In this section, detailed analyses are made for some of the examples of laboratory practices routinely used in LC–MS/MS that could become FDA 483 observations today. In order to increase the potential for future successful FDA inspections, bioanalytical laboratories should review the current enforcement trends and revise or enhance their systems, SOPs, and controls accordingly.

7.4.2 Analysis of Current FDA Inspection Trends

Based on overall review of recent FDA 483 observations along with authors' expertise, several general themes can be found, which are summarized below. These points should be carefully evaluated and considered by the management of bioanalytical laboratories to ensure their systems, SOPs, and controls meet FDA expectations for a compliant bioanalytical laboratory.

(1) Adverse trends observed during method validation and/or production batch runs (and their impact on data accuracy and validity) need to be properly evaluated and documented before accepting and reporting any BA/BE data.

(2) A documented and scientifically sound investigation for significant recurring events or deviations needs to be performed to address unused, unreported, or rejected data.

(3) Investigation of significant events or deviations and adverse trends needs to focus on the identification of assignable cause/root cause and the implementation of appropriate CAPA plan to prevent recurrence.

(4) The validity or acceptability of the production batch runs needs to be assessed against the performance characteristics of the bioanalytical method generated during method validation. If substantial performance gaps are observed, the

validity of the BA/BE data could be questioned if the current investigation does not further support the results.

(5) If a certain population of the data set generated in support of a BA/BE study are considered to be inaccurate or invalid, the BA/BE conclusion could become questionable if there is not a proper investigation to support this data set as an isolated population, and if the data set has potential material impact on other passing data.

(6) All validated methods, after they have not been used for an extended length of time (e.g., over 1 year), need to be evaluated and requalified or revalidated as required before they can be used to support new BA/BE studies. This evaluation will ensure that the validated method will continue to meet current FDA expectations and industry standards.

7.4.3 Discussion and Analysis of Specific Potential FDA 483 Observation Issues

Based on collective experience from bioanalytical laboratory operational perspectives, as well as review of publicly available FDA 483s and Inspection Reports obtained under the Freedom of Information Act, the most common recent FDA 483 observations are in the following areas: (1) method validation; (2) acceptability of batch runs; (3) events/deviations investigations/resolution; (4) test specimen accountability; and (5) document change control. A list of potential FDA objectionable observations and the scientific and compliance basis of each of these potential observations is presented below. This information should allow the bioanalytical laboratory management to perform appropriate risk assessment to determine the proper level of enhancement to be made to their current systems, SOPs, and quality controls.

7.4.4 Method Validation Issues

The requirements for method validation for bioanalytical methods used in support of human studies have been an ongoing topic of discussion between bioanalytical laboratories and the FDA. This dialogue has led to the issuance of the May 2001 FDA Guidance for Industry—Bioanalytical Method Validation [5], which was developed based on two workshops in 1992 and 2000 jointly sponsored by the industry and the FDA [3, 4, 12]. A high percentage of the recent FDA 483 observations issued to bioanalytical laboratories are related to method validation issues, as highlighted in the examples (observations) provided below.

7.4.4.1 Observation 1 The accuracy and precision data in the method validation report is incorrectly stated; data from many runs performed during the prestudy validation were omitted, which fails to demonstrate accuracy and precision of the method.

7.4.4.2 Compliance/Scientific Basis The requirement to include all pertinent accuracy and precision data in the determination of accuracy and precision of the method is clearly stated in the May 2001 FDA Guidance for Industry—Bioanalytical Method Validation:"... Reported method validation data and the determination of accuracy and precision should include all outliers ..." [5]. This requirement is also stated in FDA's Compliance Program Guidance Manual 7348.001 (October 1, 1999) entitled "Bioresearch Monitoring—*In vivo* Bioequivalence" as,"... failure to include all data points, not otherwise documented as rejected for a scientifically sound reason, in determination of assay method precision, sensitivity and accuracy ..." [6].

However, excluding failed runs from the determination of accuracy and precision of the method during validation has been a common finding at bioanalytical laboratories and has resulted in many recent FDA 483 citations and observations, particularly when there is no rationale documented justifying the exclusions. Many bioanalytical laboratories routinely follow their SOP on chromatographic run acceptance to exclude failed method validation runs when the calibration standards and quality control (QC) samples do not meet the SOP acceptance criteria. However, if these method validation runs were rejected due to the variability inherent in the method, then, according to the above-referenced FDA documents, these data should be included in the determination of the "actual" accuracy and precision of the method. Therefore, bioanalytical laboratories need to review all method validation runs, passed or failed, to ensure that there is a valid reason to support any exclusion. Unless the failed runs are due to isolated and nonmethod (inherent variability) related issues, the data should be included. The rationale for exclusion should be documented, and appropriately evaluated and approved to support the firm's decision and for consideration by customers and regulatory authorities when required.

7.4.4.3 Observation 2 The bioanalytical laboratory fails to demonstrate analyte stability—recovery after freezing and thawing, postpreparative, and stock solution stability did not employ an unfrozen (freshly prepared) reference (calibration standards).

7.4.4.4 Compliance/Scientific Basis The May 2001 FDA Guidance for Industry—Bioanalytical Method Validation states"... All stability determinations should use a set of samples prepared from a freshly made stock solution of the analyte in the appropriate analyte-free, interference-free biological matrix ..." [5]. While the FDA Guidance was published several years ago, nevertheless, many bioanalytical laboratories are still collecting stability data using stored frozen standards. This practice is based on the rationale that the firm only needs to demonstrate "relative stability", as their practice is to prepare calibration standards and QC samples in bulk and store them with the test specimen samples under the same conditions. This is based on the view that, even if the analyte in the frozen standards degrades during storage, it would not affect the accuracy and precision of the results generated, since the analyte in the test specimen samples would degrade at the same rate. However, the FDA guidance requires the firm to demonstrate "absolute stability" so that the capability and suitability of the method can be fully understood. As a result, many FDA 483 observations issued to bioanalytical laboratories have been due to the use of frozen standards and samples in the determination of stability.

While the FDA guidance recommends the traditional approach of comparing analytical results of stored standards to that of the fresh standards, FDA also allows other valid statistical approaches using confidence limits for the evaluation of analyte stability to be used [13, 14].

7.4.4.5 ***Observation 3*** The use of stability data generated from other methods or studies (e.g., HPLC–UV) to support the stability of a revised or new method (e.g., LC–MS/MS) is without proper justification.

7.4.4.6 ***Compliance/Scientific Basis*** Most bioanalytical methods developed and validated by commercial contract bioanalytical laboratories can remain in active use for many years in order to support generic BA/BE and other studies. After the initial validation, methods could continue to be optimized and/or modified. As recommended by the FDA guidance, partial validation is required to support these changes. However, the FDA guidance does not further detail the partial validation requirements under various change conditions as presented. The FDA guidance only states that"... partial validation can range from as little as one intra-assay accuracy and precision determination to a nearly full validation" [5]. As a result, it has been observed that the extent of partial validation performed varies significantly among laboratories for similar changes. Many bioanalytical laboratories have considered that analyte stability (mainly for long-term storage stability) would not be affected by the test method used, and therefore, stability studies would not need to be repeated. However, due to the complexity and inherent variability of the biological matrix, one could argue that the stability duration established by a less specific method, such as HPLC–UV could be longer than that established by a much more specific method, such as LC–MS/MS. Therefore, before relying on old stability data to support a new method, proper scientific justification with supporting data needs to be generated and documented.

7.4.4.7 ***Observation 4*** Failure to use the same method as the validated method—for example, use of manual aliquots instead of automated aliquots, changes made in the method without supportive validation data demonstrating that these changes do not compromise the accuracy and precision of the data generated.

7.4.4.8 ***Compliance/Scientific Basis*** As discussed under Issue no. 3, from the time of initial implementation of a method, it is possible that the method could undergo optimization and/or modification. Many bioanalytical laboratories cited by the FDA attempted to justify that partial validation was not required since, in their view, the methodological changes were minor. Although the justification for the immediate change may be adequate, it is important to consider the cumulative effect of all prior changes when assessing the need for revalidation. The FDA could challenge the accuracy and validity of the data generated by the modified method unless proper supporting validation data are generated. In order to avoid this potential compliance issue, a formal document change control SOP should be instituted and controlled by the Quality Assurance Unit, to ensure that the decision for revalidation will be made based on the evaluation of all the changes made since the last complete validation. Based on previous experience in auditing bioanalytical laboratories, oftentimes change control and change management are the weak links in the entire quality system program. Change control is a critical control element in a cGMP environment and it is evident from the current trend that this requirement is gaining more and more attention from the FDA in the bioanalytical laboratory environment.

7.4.4.9 Observation 5 The bioanalytical laboratory fails to demonstrate the absence of matrix effect when using LC–MS/MS as the bioanalytical method in the human BA/BE and other studies.

7.4.4.10 Compliance/Scientific Basis Due to the complexity and inherent variability of the biological matrix, materials from the extracted matrix could affect the extent of ionization of the analyte and consequently may suppress and/or enhance the ionization process, potentially compromising the accuracy and precision of the test results generated using LC–MS/MS [2, 15–17]. It should be noted that matrix effect is unique to LC–MS/MS and is one of the major unknown parameters that needs to be evaluated during method development and validation, before the method can be considered as "suitable for its intended use". This issue has been discussed in the FDA guidance and many other publications and will be further discussed. Bioanalytical laboratories often evaluate the matrix effect during method development, and many have not further demonstrated the absence of matrix effect during method validation. However, the development records did not adequately document efforts and data to support the absence of matrix effect. As a result, without this documentation, any event or adverse trend of highly variable recovery and/or loss of sensitivity observed could be challenged during FDA review as potentially due to a matrix effect.

Bioanalytical laboratories often evaluate the matrix effect during method development, and many have not further demonstrated the absence of matrix effect during method validation. However, the development records did not adequately document efforts and data to support the absence of matrix effect. As a result, without this documentation, any event or adverse trend of highly variable recovery and/or loss of sensitivity observed could be challenged during FDA review as potentially due to a matrix effect.

7.4.4.11 Observation 6 The composition of the spiked QC and calibration standards are substantially different from that of the test specimen (e.g., the presence of excessive aqueous or organic solvent in the final spiked matrix samples).

7.4.4.12 Compliance/Scientific Basis While it is not explicitly stated in the FDA guidance, it is good science to perform method validation using analyte spiked into the same matrix as the test specimen in order to ensure that the method validation data are truly representative of the actual test article conditions. It will be ideal if the spiked matrix does not contain any solvent or water. However, due to inherent solubility limitation of many analytes in biological fluid, direct spiking is impossible, in many cases. Therefore, stock and diluted spiking solutions in organic or aqueous mixtures are generally used. Bioanalytical laboratories should have an SOP to define how the spiking validation samples are to be prepared, and an absolute limit of organic solvent and the aqueous content needs to be established. Generally, the presence of organic solvent in the spiked samples should be about 2% and in any case, should not be more than 5%.

7.4.4.13 Observation 7 The size of the batch run during method validation is much less than the actual production or sample analysis run, or runs were accepted exceeding the validated run time.

7.4.4.14 Compliance/Scientific Basis It is considered a current industry standard (as well as an FDA expectation) that the method validation accuracy and precision runs should contain enough test samples to mimic the actual run time of a production batch run, as defined in the test protocol for the designated BA/BE study. The ruggedness of the method can therefore be further demonstrated. However, this has not been the practice in the past, and the FDA could challenge this deficiency during study-specific inspection, particularly when the ruggedness issue was observed for production runs. In order to maintain the same run time as the production runs, "dummy" extracted matrix blank samples are usually added to the method validation run sequences in order to extend the run time.

7.4.5 Batch Runs Acceptance Criteria Issues

7.4.5.1 Issue 1 Batch runs were accepted when more than 50% of QC samples at each concentration level did not meet the acceptance criteria.

7.4.5.2 Compliance/Scientific Basis This has been a common recent FDA 483 observation for several bioanalytical laboratories, even though the requirement of having 50% of QC samples at each concentration level meeting the acceptance criteria is not explicitly defined in the May 2001 FDA guidance. Nevertheless, this observation makes good scientific sense, as adopting this additional requirement would further ensure the accuracy and precision of the data generated for the entire concentration range of the method. Many bioanalytical laboratories are revising their SOPs for the acceptance of calibration curves, QC samples, and batch runs to include this requirement in order to avoid future FDA challenge.

7.4.5.3 Issue 2 Analytical runs that originally failed when results of QC samples were outside the acceptance limits were reprocessed by excluding selected calibration standards until the QC samples passed. The order of rejection of calibration standards was not consistent; the calibration curve would not pass if proper rejection order were followed.

7.4.5.4 Compliance/Scientific Basis As described in the FDA guidance, it is an acceptable practice to drop some calibration standards in the construction of the regression line based on not meeting the preestablished acceptance criteria [5].

However, the FDA guidance does not provide the procedural details regarding the order of rejection of calibration standards. As a result, the practices of some bioanalytical laboratories became a subject of recent FDA 483 observations. Consequently, many bioanalytical laboratories have updated their SOPs to provide a consistent and detailed guidance regarding the order of rejection of calibration standards.

7.4.5.5 Issue 3 Based on a review of recent FDA 483 observations issued to bioanalytical laboratories, below is a list of additional batch runs acceptance issues that could be found objectionable to the FDA during data review:

- Acceptance of batch runs data with recurring highly variable internal standard response without proper investigation and justification to support the accuracy and validity of the data reported.
- Acceptance of batch runs data when excessive poor chromatography, such as tailing, spikes, or split peaks, was observed.
- Acceptance of batch runs data with inconsistent manual integration without proper justification to support the accuracy and validity of data generated.
- Acceptance of batch runs data with traveling peaks and/or variable retention times without proper justification.
- Acceptance of batch runs data that were interrupted for a considerable period of time, with no data to support the accuracy and validity of the data generated after the runs were resumed.
- Acceptance of study data when there is an adverse trend observed; for example, the percentage of rejected runs, rejected QC samples and calibration standards are excessive and not consistent with the method validation data. As a result, there is no justification to support the accuracy and validity of the other reported data.
- Acceptance of batch runs when anomalous results were rejected as PK outliers and replaced by repeated assay results without proper justification to support that the outlier results were, indeed, PK outliers.
- Acceptance of batch runs with significant interference observed for blanks, zero standards, and predose samples for a large population of the batch runs.
- Acceptance of batch runs with numerous repeated results based on the sponsor's requests without proper justification. Analytical runs were accepted when regression type was changed during the analysis of study samples without proper cross-validation.
- Acceptance of batch runs data when most repeated results were significantly different from initial results; for example, the entire run was repeated due to a sample being out of regression range, or a repeated diluted sample is significantly different from the initial result. As a result, validity of the reported data is questionable.
- Acceptance of batch runs data with many results generated using inconsistent manual integrations.

Inconsistencies in the manual integration of chromatograms, in those identical peaks of some standards and QC samples were reintegrated using different integration methods. If re-integration was performed consistently across all samples within a run, some runs would not meet the run acceptance criteria.

7.4.5.6 Compliance/Scientific Basis It is evident from the list of objectionable issues presented above that bioanalytical laboratories are not only expected to produce the raw data to support the reported data but also that the laboratories must also be able to produce the proper documentation to justify the exclusion of the unused data with the documented scientifically sound rationale. The laboratories should also be able to demonstrate that the exclusion of these unused data would not compromise the accuracy and validity of the reported data. The bioanalytical laboratories should have proper documentation to avoid any perception of "selective reporting."

Due to the inherent variability of the LC–MS/MS method and the complexity of the biological matrix, the observations described above could occur during any routine analysis even if the method had been properly validated. However, if the magnitude and frequency of these events become significant (e.g., becoming an adverse trend), then it would become a potential FDA inspectional issue. Therefore, in order to minimize this possibility, it is highly recommended for bioanalytical laboratories to perform a thorough internal review of BA/BE data generated using the "quality indicators" approach to identify and resolve potential questions regarding the accuracy and reliability of reported data. The quality indicators can be developed using the objectionable conditions presented above to ensure the study data are free of these objectionable conditions. A list of example quality indicators that can be used to facilitate an effective review of the LC–MS/MS method performance is provided in Table 7.4 and is further discussed in the section "FDA Inspection Readiness Preparation."

7.4.6 Events/Deviations Investigation/Resolution Issues

7.4.6.1 Issue 1 Acceptance of batch runs data when there is an adverse trend or recurring significant deviation observed and no investigation performed to identify its root cause, which would demonstrate that the passing reported data were, in fact, accurate and valid. For example,

Ten consecutive batch runs (except one) were rejected due to massive failures of in-study QC samples; one batch run was accepted with marginally passed QC samples. Selective acceptance of passing data generated from batch runs when subsequent investigation concluded that the robotic sample preparation instrument used was malfunctioning at the time of use.

Acceptance of batch runs data when there is a significant adverse trend observed. Failure to investigate when reassays of incurred samples showed significant variability, in that the original and repeated results differ significantly without proper justification.

7.4.6.2 Compliance/Scientific Basis It is evident from the examples of objectionable laboratory practices provided above that the bioanalytical laboratories are

TABLE 7.4 Summary of Quality Indicators to be Used in Evaluating LC–MS/MS Chromatographic Data

Run Acceptability: Determine whether the run was rejected or accepted with a supported, valid reason; determine if the rejection rate of runs, quality controls, and/or calibration standards is excessive; and identify any adverse trends

QC and STD Acceptability: Determine if the calibration standard was rejected in the correct order as described in the SOP; observe to see if there is an overall trend of the percentage deviated from nominal when quality controls and calibration standards marginally meet the acceptance criteria, and so on.

Internal Standard Variability: Determine if the change in internal standard responses is indicative of potential loss of system sensitivity, instrument malfunction, sample extraction problem, recovery problem, matrix effect, and so on.

System Suitability: Determine if the system suitability meets the acceptance criteria; determine if there are documented stability data to support the suitability of the system in completing the batch runs, and so on.

Interrupted Run: Evaluate the data to ensure that there are no excessive time gaps within a batch run that exceeded the allowable limit required by the SOP, and so on. If excessive time gaps are observed, determine if the test system is requalified prior to restarting the run.

Potential Interference: Evaluate the magnitude and frequency of potential interference observed in blanks, zero-standards, and predose samples.

Poor Chromatography: Determine the extent and impact of merging peaks, split peaks, spikes, traveling peaks, shift in retention time, and shift in baseline on the accuracy and validity of the data generated.

Missing Peak: Determine the frequency of missing peak events as it may be indicative of potential systemic sample extraction problem, and so on that would discredit the entire batch run.

Consistency in Data Reporting: Detect any systemic practices, which may be construed as selective reporting of data, inconsistent manual integration, and rejection of calibration standards.

Reported Below Limit of Quantitation (BLQ) Results: Determine the frequency and magnitude of samples with BLQ results, as this could be an indication of a potential issue with the analytical method and/or sample preparation error. Performing manual integration in an inconsistent manner is one of the most common FDA inspectional issues. Due to the need to assay an extended concentration range, the noise level at lower calibration ranges would be much higher than that at the higher concentration range. As a result, manual integration needs to be performed in order to improve the accuracy. However, if these manual integrations were frequently performed in an inconsistent manner, it would become a potential compliance issue.

expected to perform a thorough and well-documented investigation for recurring significant events or deviations. This is a current "hot" compliance topic. The expectation to perform formal laboratory investigation is parallel to the cGMP requirement for pharmaceutical analytical laboratories when an out-of-specification or aberrant test result is generated. It cannot just be replaced with repeated testing data without appropriate justification based on a formal investigation driven by an SOP. However, based on author's experience, performing formal or rigorous laboratory

TABLE 7.5 Summary of Critical Control Elements in an Event/Deviation Investigation SOP

Implement an SOP system, driven by QA, to document and track event investigations and to evaluate all events prior to acceptance of the BA/BE studies to include the following elements:

(1) Define the scope/nature of the events
(2) Conduct proper documented scientifically sound investigation
(3) Identify root cause or most probable cause
(4) Perform impact assessment for other reported data potentially implicated by these events
(5) Establish appropriate corrective actions and preventive actions, properly implemented and tracked
(6) Record final disposition of the study data affected and determine its impact on the accuracy and validity of the clinical study
(7) Follow SOP to close out investigation for single event and all events at the end of the study

investigations for these types of issues has not been the past practice at many bioanalytical laboratories. Due to the complexity and inherent variability of the biological matrix, many bioanalytical laboratories had allowed the replacement of "outlier" results with repeated assays of the test specimen, usually following SOPs that did not require a formal documented investigation. As a result of the FDA's concerns, many bioanalytical laboratories have started to review and revise their SOPs related to event investigation, resolution, and management. An effective investigation SOP should cover the control elements as summarized in Table 7.5.

Since the conclusion of the BA/BE study is not based on a single data point, but rather on a statistical evaluation of all the data points generated, the final closure of the study should not occur until all individual events can be evaluated in their totality, ensuring that all reported data are supportable and would not be compromised by these recurring events.

7.4.7 Test Specimen Accountability Issue

7.4.7.1 Issue 1 The bioanalytical laboratory fails to maintain adequate and accurate sample records. For example, a record system fails to document the removal or return of test samples to and from the frozen storage areas, fails to document the identity of the individual(s) who removed or returned the samples from the storage area, and/or fails to note the date and time of this removal or return.

It is a GLP requirement to document the chain of custody records for all test articles used. Similar control is expected for the specimen samples within bioanalytical laboratories to ensure that the test specimen and control samples are stored and handled within the validation conditions, as defined in the stability studies during prestudy validation. Therefore, the laboratory documentation should be designed in such a manner to provide a traceable audit trail.

7.4.8 Recommendations to Support an Effective FDA Inspection Readiness Preparation

In light of the current trends seen in recent FDA 483 observations issued to bioanalytical laboratories, it is recommended that the companies/institutions should thoroughly evaluate the state of existing controls and practices to assure the accuracy and reliability or test data and conclusions. As part of inspection readiness, the companies/institutions should perform a study-specific "mock" inspection and 100% data review using the "quality indicators" approach to ensure that all systems, practices, data, and documentation are ready to support a successful FDA study-specific inspection. As discussed in Issue no. 3 under "Batch Run Acceptance," these quality indicators were developed based on recent FDA 483 observations to ensure that all common potentially objectionable conditions can be detected and corrected, prior to the FDA inspection. Examples of FDA pock audits and quality indicators are summarized in Table 7.6.

To support an effective mock FDA study-specific inspection, a well-planned audit strategy should be developed to ensure that the scope of the audit coverage will address both the study data issues and the related quality system issues. The critical elements of a sample audit plan are mentioned earlier and given in Table 7.5. In addition to reviewing of hard-copy raw data, review of electronic data should be performed so that the potential chromatographic problems (e.g., peak tailing, peak splitting, baseline drifting, reintegration, etc.) and unreported data not reflected in printouts can be detected and properly evaluated.

7.5 HANDLING AND FACILITATING GLP OR GxP AUDITS/ INSPECTIONS

The Food and Drug Administration Bioresearch Monitoring Program (BIMO) oversees FDA-regulated research by performing site visits to clinical investigators, sponsors, Institutional Review Boards (IRBs), nonclinical animal laboratories, and analytical/bioanalytical laboratories. Site visits help to assure that human subjects and animals are protected from undue hazards and to verify that research data supporting new human and animal product approvals are reliable.

FDA conducts inspections to determine if investigators are in compliance with FDA regulations. Inspections can be announced or unannounced. Most inspections are routinely performed to verify data submitted to FDA (e.g., at sites enrolling the largest number of subjects), but can also occur as a result of a complaint made to FDA, due to sponsor concerns, as a result of a review division request within FDA, or based upon current and ongoing public health issues. The inspections must ensure that product development is carried out in a way that produces the reliability and quality of data and processes. They must also make sure that all key GLP, GCP, and cGMP systems at the development and commercial sites are up to current standards and regulations. The primary reasons for a preapproval inspection may be related to one of the following aspects:

TABLE 7.6 Typical Study-Specific FDA Mock Inspection Plan

(A) Review of study-specific records

(1) Copy of the final analytical method validation protocols and analytical reports used in support of the BA/BE studies
(2) Copy of the final bioanalytical reports submitted to the FDA as reference material
(3) Analytical raw data to support the above studies, including reassay and rejected data (notebooks, worksheets, freezer logs, chromatograms, computer printouts, etc.)
(4) All pertinent specimen receiving records in the laboratory
(5) All pertinent freezers/refrigerators in/out logs to support the chain-of-custody of specimens
(6) All pertinent monitoring, calibration, and qualification records for freezers
(7) All pertinent records for the preparation and qualification of quality control, calibration standards, and stability samples
(8) All pertinent records to support the suitability of the reference standards used
(9) All pertinent instrument calibration/qualification records
(10) All pertinent computer validation records
(11) All pertinent training and qualification records for analysts review of quality systems

(B) Review of organization/facility-specific information/records

(1) Organization, responsibility, and staffing of Bioanalytical Laboratory—Quality Assurance Unit
(2) Receipt, handling, and control of specimens
(3) Specimen freezer control—inventory, calibration, qualification, and monitoring
(4) Record-keeping/documentation/record retention practices
(5) SOPs for the review and approval of analytical raw data—laboratory check and QA check
(6) SOP to handle, document, and report results of repeat testing
(7) SOP to control the reprocessing/reintegration of sample specimens
(8) SOP to prepare calibration, quality control, and stability samples
(9) SOPs and requirements for prestudy and in-study validations
(10) SOPs and acceptance criteria to accept calibration curves, QC samples, and batch runs
(11) Control and qualification of reference standards
(13) SOP to evaluate the stability of analyte in biological fluids and stock solutions
(14) Quality audit SOP for bioanalytical data and reports
(15) Instrument calibration/qualification program
(16) Computer/laboratory data system validation program
(17) Analyst training and qualification program

New Drug Application: The New Drug Application is the registration file sent to the FDA for the marketing of a new drug product in the United States.

Biologics License Application: The Biologics License Application (BLA) is the equivalent of an NDA for biopharmaceutical and biotechnology products.

Marketing Authorization Application: The Marketing Authorization Application (MAA) is the registration file submitted to the relevant national authorities of EU member states or the EMEA as part of an application to market a new product in the European Union.

It is highly recommended that mock audit be conducted for awareness and general understanding of how an audit/inspection is carried out along with scope and purpose of the audit/inspection. Some of the examples for the preparation of regulatory inspections are as follows:

- Audit of investigative site operations to determine adequacy of documentation and procedures, and to ensure your trial will be carried out in a controlled manner, per protocol, and in compliance with regulations
- Audit of ongoing and completed studies (Phases I–IV) at investigative sites to determine site and monitor compliance with applicable regulations (21 CFR Parts 50, 54, 56, and 312/812), FDA Compliance Programs, the study protocol, and Standard Operating Procedures and guidelines, as well as to ensure documentation is appropriate for a regulatory inspection
- Audit of Phase I units and PK/PD studies
- Audit of Institutional Review Board operations and documentation for compliance with Protection of Human Subjects and Institutional Review Board regulations (21 CFR 50 and 56) and for responsiveness to the sponsor's needs
- Audit of sponsor central file record archives
- Audit of clinical study data listings
- Audit of clinical and nonclinical study reports
- Audit of sponsor or contract bioanalytical laboratory systems to assess sample assay performance and adherence to established procedures, industry standards, and FDA guidance for method validations, sample assays, and data reporting
- Audit of toxicology laboratories and studies to assess GLP compliance and accuracy of reports

7.5.1 General Preparation for an Inspection

Proper preparation work is vital for ensuring that your company successfully passes an inspection. As far as for a preapproval inspection (PAI), it is highly recommended that preparation work start at least 12 months before the date when you plan to contact the relevant authority. An initial meeting should be held to define the responsibilities of individual PAI team members and milestones must be set in order to monitor the progress of preparation work. PAI protocols for drug products and active pharmaceutical ingredients have been created to systematically identify the strengths, opportunities, weaknesses, and threats associated with organization, personnel, quality systems, facilities, utilities, water systems, clean steam, HVAC, gases, equipment, key raw material suppliers, manufacturing process synthesis, sieving, milling, micronization, analytical testing, test method validation, method transfer, stability, laboratory equipment, microbiological testing, contractors/suppliers, and inspection issues. These protocols enable you to identify any potential shortcomings during the early stages of preparation work, usually 6–12 months before the submission of a new drug, and give you the opportunity to resolve the issues or finalize pending work and reports in good time. The FDA is also increasingly focusing on the validation of computer and/or computerized systems and compliance with 21 CFR Part 11 on a risk-based evaluation. It is vital that bioanalytical laboratories are well prepared in these areas.

Companies should be aware that preparing for a PAI is not a last-minute activity. It is clear from current warning letters that SOPs are still extremely important. One of the most likely reasons for this is the FDA's new systematic approach. Common observations for SOPs relating to key issues such as deviation, OOS, change control and training are that they are often too convoluted or general and sometimes workers are not given sufficient training in how to implement them. Many warning letters also cite incomplete document history files or outdated SOPs. Generic SOPs are available to help you avoid such pitfalls and only need to be adapted slightly to suit your company. It is also advisable to organize comprehensive training for the personnel who are to participate in a regulatory authority inspection. Indeed, it is widely recognized that personnel who behave in an inappropriate manner or provide misleading information can cause inspectors to look into areas where they previously had no concerns. The same applies to the misinterpretation of questions and the use of phrases that should be avoided such as "I think," "I suppose," "I guess," "normally," "usually," "occasionally," or "mostly." It is vital that all personnel who could potentially come into contact with an FDA investigator undergo proper training on how they should conduct themselves during an inspection. It is important to remember that there is a whole range of do's and don'ts that must be followed in all circumstances. Troubleshooting and Gap analysis conducted by the PAI team provides a further useful tool. This "organized brainstorming" meeting is designed to list all potential strengths, opportunities, weaknesses, and threats. Everyone taking part in the meeting must be able to speak openly and honestly when discussing where they personally see potential risk areas or weaknesses. Senior management are often surprised by how much is already known within an organization, but is not communicated to the people who are able to change and improve issues that may have caused problems in the past but were never resolved. It is crucial that you closely examine the findings of any previous inspections and the commitments that were given to the agency as inspectors often use these as a starting point for their work. Shortly before the inspection, a key contacts list should be compiled detailing the names, positions, phone numbers, mobile phone numbers, and email addresses of the most experienced personnel. This will ensure that information flows rapidly and smoothly during the hectic inspection period.

Audits by third parties do not have to be stressful events. The level of preparation within your organization will make a significant difference in your level of success. Below are some helpful hints and techniques that will enable any organization to perform smoothly during an FDA inspection. Is being audited an opportunity to tell your story? Is it thought within your organization that "all of this quality stuff is in the way of doing business?" It must be understood that the single most important element for a successful audit/inspection outcome is the existing quality culture within a company. Substantial compliance to quality systems can only be derived from the diligent execution of all requirements defined within the Quality Management System as a matter of doing business, each and every day. What you are doing in an audit is communicating and demonstrating how you run your business within the framework of your Quality Management System. How well you do this makes a difference. In order to execute a third party audit, company needs to first prepare by defining how an audit wants to be executed within a Third Party Audit Procedure. The procedure should, at a minimum, define:

(1) How to Receive Inspectors
- Verifying Credentials
- Communications to Executive Staff & Employees
- Opening Presentation

(2) Duties of key individuals
- Receptionist
- Audit Support Team Members

(3) Audit Room and Control Room Set-up
(4) Conducting Tours
(5) Daily Audit Summaries

The Information Book: Prepare an Information Book for the Auditor containing standard information that will be needed during the audit. The book should contain

(1) Company Overview
(2) Organizational Charts
(3) Facility Demographics
  (a) Headcount
  (b) Facility diagram
(4) Documentation
  (a) Site/Corporate SOP Matrix
(5) Recent Field Actions (make sure you have information that spans the scope of their inspection)
(6) CAPA and Complaints information for past 6 months to 1 year

Employee Preparation: Next, identify "who" within your organization will participate in the audit and in what way will they participate. It is important to understand that you only get one chance to "tell the story." At a minimum you should identify the following:

(1) Facilitators—The primary host(s) for the auditor and are responsible for the oversight of all facets of audit execution
(2) Scribes and Runners
(3) Subject Matter Experts—Individuals who are focused on specific processes each and every day. They are specially trained and prepared for audit situations

(4) Control Room Staff
(5) Employees
(6) Management Team

Employee preparation is essential. Each and every employee does not need to know the details of the execution and goings on within the audit, but they do have to be prepared and know they play a role within the audit. Employees should be prepared to answer key questions such as:

(1) What is the Quality Policy? Ensure each employee understands and can speak to what the policy means to them. Use the Quality Policy Training.
(2) How do you know what's required of you in your job? As defined within my job description. As I have been trained to do. By following approved procedures and work instructions. Know how to access procedures and work instructions. Bring up errors in procedures and instructions now.
(3) Is your training current? Ensure your training is current. If there are gaps, complete the training now and maintain it!
(4) What if I don't know the answer? "I am not sure but I will get the information for you."

Inspection Procedure: Prior to inspection day, it is important to understand what the third party can and will look at while on site. The FDA, in particular, is somewhat prescriptive in what information they can and cannot access. The FDA has authority to inspect all equipment, materials, products, labeling, and surrounding documentation including but not limited to SOPs, clinical study, and quality systems documentation. They do not have the authority to inspect certain documentation such as personnel data, financial statements, R&D records, audits, and management reviews. Understanding how the auditor will determine the compliance and effectiveness of your Quality System can also help you prepare your staff in advance. The Auditor will plan to examine the system from either the "bottom up" or from the "top-down" as depicted below: A general inspection approach is a "top-down" approach where the auditor will be looking at the firm's "systems" for addressing quality before they actually look at specific quality problems. In the "top-down" approach, they will "touch bottom" in each of the subsystems by sampling records, rather than working their way from records review backwards towards procedures. What will they be looking for?

- To ensure that your CAPA system procedures have been defined and documented.
- To determine if appropriate nonconformance sources of product and quality problems have been identified.
- To determine if these sources of product and quality information are trended and if any unfavorable trends have been identified.
- To challenge the quality data information system and verify that the data received by the CAPA system are complete, accurate, and timely.
- To verify that appropriate statistical methods are employed to detect recurring quality problems.

- To make sure of timelines for approval and execution follow a logical sequence.
- To determine if failure investigation procedures are followed.
- To determine if appropriate actions have been taken for significant product and quality problems identified from nonconformance data sources.
- To determine if corrective and preventive actions were effective and verified or validated prior to implementation.
- To ensure that corrective and preventive actions for product and quality problems were implemented and documented.
- To determine if information regarding nonconforming product and quality problems and corrective and preventive actions has been properly disseminated, including to Management Review.
- To establish objective evidence that Executive Management is engaged.

Investigators should take the following actions when notified that an on-site inspection by FDA is going to occur:

- Determine the nature (e.g., "for cause," routine, etc.) and the scope of the audit. What protocols will be reviewed? How long will the FDA inspector(s) be on-site?
- Routine inspections are generally scheduled within ten (10) working days of the initial contact and cannot be postponed without sufficient justification.
- Inform the following groups (as applicable) when initially contacted by FDA so that each party can prepare for the visit, as necessary:
  - Research team and ancillary support services (e.g., pharmacy, nursing, Medical Information Management, etc.)
  - Department Chair and College/Center officials
  - Office of Responsible Research Practices (ORRPs) staff supporting the IRB or Institutional Laboratory Animal Care and Use Committee
  - University Laboratory Animal Resources (ULAR) staff
  - OSU Sponsored Program Officer, Sponsor, and/or Contract Research Organization
  - Office of Research Compliance (ORC) if the inspection is "for cause."
- Schedule an appropriate room for the auditors. Provide the FDA inspector(s) with office supplies, access to a copier and fax machine, and a list of staff contact names and telephone/pager numbers.
- Prereview all research records (medical and regulatory), and make them available to the inspector(s) at the time of the site visit. Investigators must allow the FDA to access, copy, and verify any case history and/or drug or device administration records made by the research team or others.
- At the conclusion of the site visit, the FDA inspector conducts an exit interview with the principal investigator and associated research team members. A written "Notice of Inspection" (Form FDA 483) is typically generated if deficiencies are found.

- Forward a copy of postaudit communications and/or FDA Form 483 to the research team, Department Chair, and IRB or IACUC, as applicable. Consult with ORRP staff as needed for assistance in responding to audit findings.

For the FDA or regulatory inspector, far from being an intruder, has a legal right to be in a facility or laboratory, and the results of his or her inspection will depend largely on three steps that the firm should already have taken. In short, it is assumed that the facility has (1) designed and installed effective quality assurance systems and procedures; (2) implemented an internal audit program to ensure that the desired results are being obtained on an ongoing basis; and (3) made arrangements to guarantee that regulatory inspectors will be courteously received and adequately looked after during their visits, and that appropriate follow-up action will be taken at the conclusion of each visit. Many companies manage steps one and two very nicely but then stumble at step three. This is unfortunate, as there is unquestionably a need for policies and procedures that will bring a high level of professionalism to the task of hosting visitors.

Assuming that a firm's GLP or GxP/quality assurance program is adequate, there are several steps that management must take to supplement the three steps that have already been delineated. Specifically, the firm must (1) accept the inevitability of inspections and assume that they will occur at the most inopportune time; (2) learn as much as possible about the why, how, who, when, and what of inspections and about the legal rights of the firm; (3) develop a corporate policy and a set of procedures governing inspections, and include positions on potentially controversial issues; (4) designate and train inspection coordinators to serve as hosts at each facility; and (5) analyze the results of each inspection to determine how the program can be strengthened for the future.

To a large extent, success or failure is dependent on the skills and capabilities of the inspection coordinator/moderator and the standard of a company's backroom organization. It is important to have at least two separate rooms set aside for the inspection. The front room is used by the inspectors and the second room, often referred to as the "war room" or backroom, is used to ensure that documents are quickly and efficiently retrieved and guarantee the smooth flow of both personnel and documents to the inspection room. The war room/backroom coordinator and his or her assistant must have clearly defined roles. The same applies for the inspection front room, where the coordinator moderates the inspection and the inspection observer (scriber) takes notes. Each individual request from the inspector should be written down on an inspection request form that notes the date, time, inspector's name, subject, and the requested information. This form should then be taken to the backroom for preparation or investigation by a designated person (runner). Documents such as validation master plans, protocols and reports, batch records, SOPs, and other key quality documents should be collected prior to the inspection and made available in the backroom. It is extremely important that documents are brought to the attention of the inspectors in their original form only (with original signatures) or alternatively as copies that are officially sanctioned in line with your document control procedure.

7.5.2 Why Are Audits/Inspections Needed and Conducted?

Preparations for hosting regulatory inspectors must begin with an understanding of the purpose of inspections. The objective in this case is expressed quite clearly in FDA's Inspection Operations Manual. The emphasis here is on ensuring the fitness for use of the product and securing the correction of deficiencies found in regulatory audits. The *Investigations Operations Manual* (IOM) is the primary source regarding Agency policy and procedures for field investigators and inspectors. This extends to all individuals who perform field investigational activities in support of the Agency's public mission. Accordingly, it directs the conduct of all fundamental field investigational activities. Adherence to this manual is paramount to assure quality, consistency, and efficiency in field operations. The specific information in this manual is supplemented, not superseded, by other manuals and field guidance documents. Recognizing that this manual may not cover all situations or variables arising from field operations, any significant departures from IOM established procedures should have the concurrence of district management. For 2009, the IOM contains some important changes that clarify or present new procedures. For instance, additional FDA 483 guidance was added; OEI maintenance instructions reference FMD-130; and additional information related to food, drug, and veterinary products was incorporated in Chapter 5; and current import terminology was updated in Chapter 6. Appendix C—Blood values was completely revised. As with each new edition of the IOM, please take time to review sections of the IOM for changes, which may apply to your work. Since December 1996, the IOM has been posted on ORA's Internet Website. The entire IOM is available there, with all graphics included. Future updates to the IOM will be performed periodically during the year to this on-line version. Remember, whether reviewing the "hard copy" or the "on-line" version of the IOM, the most recent version is the document of record.

7.5.3 Written Policy in Place

Given the aforementioned objective, together with the obvious requirement that a company be prepared to explain and defend its internal procedures, it becomes evident that a firm's inspection-response program must encompass certain key elements—the most important of which are written policies and procedures on issues pertaining to inspections. Translation of such matters into writing ensures that all relevant policies will be well thought out and that senior members of company management will approve both the policies and their implementation. In addition, standard operating procedures in written form can be readily communicated to all who are involved in hosting regulatory inspectors, thus ensuring uniform application of such policies throughout the firm. Finally, written policies and procedures provide a basis for periodic review as regulations and other factors change through the years.

In formulating its corporate policy toward inspections, the firm should make it clear that there will be full cooperation within the law with all properly constituted inspections. Full cooperation means just that—it includes complete responses to valid questions, a lack of defensiveness, an openness to the investigator's point of view, and, in general terms, projection of a professional and businesslike attitude. The written policy must, in addition, clearly state who is responsible for coordinating inspections in each facility; this can be defined in terms of a specific position in the organization or by the name of an individual. The procedures must then delineate the method that is used to select and train such personnel.

It should be noted, however, that no matter how well trained inspection coordinators may be, there will come a time when some help will be needed from others within the organization. Thus a firm should establish procedures to ensure that representatives of the corporate quality, regulatory affairs, and legal departments are immediately informed that an inspection is taking place and are told what the purpose of that inspection is. These individuals and other corporate staff and management personnel may have to be available for consultation; indeed, in some companies, quality assurance and regulatory affairs management personnel actually come to their plants to take an active part in the inspection process. Obviously, the more thorough a firm has been in preparing for inspections, the less necessary it will be for such individuals to be come actively involved in this way.

7.5.4 Positions on Controversial Issues

A good approach toward reducing dependence on staff groups is to arrive at corporate positions on potentially controversial inspection related issues. Failure to adequately consider such issues in advance can result in the firm's relinquishment of information to which the regulatory agency is not entitled. On the other hand, an unnecessarily restrictive position on an issue can cause a firm to appear uncooperative or result in a charge or refusal in part to permit an inspection. Although a firm should consult with its legal consul on the logistic or legal issues involved in such matters, some simple rules can be suggested.

Lists of Records: Above all, lists of records should be prepared, and these lists should be divided into three categories: (1) those to which the agency is entitled by law; (2) those that need not be shown from a legal standpoint but could be shared if they are clearly in the best interests of the firm; and (3) those that would not be shown under any circumstances. The location and form of the records in the first two categories should be indicated in these lists so that the coordinator can readily retrieve any of the records. At times, however, an effort should be made to limit the quantity of records delivered to the inspector. For example, if the goal is to demonstrate that the firm is involved in interstate commerce, FDA needs not review all of last year's distribution records. If however, one is concerned about the adequacy of the data-retrieval record system, an in-depth review may be appropriate. In either case, a note should be made as to which records were seen by the investigator.

Copying Records: Most firms allow FDA access to company copying machines if the agency wishes to make a reasonable number of copies of appropriate records. Inspectors may offer to pay for copies, but this is seldom required unless the number of copies to be made is quite large. It may be preferable to have someone make copies for the inspectors, so that a second copy can be made and kept for attachment to the inspection coordinator's report. Some companies request that investigators initial and date the back of each original that is copied.

Proprietary Information: All sensitive information should be marked confidential, and this fact should be called to the attention of the inspector so that there is a reasonable chance that the information will not be disclosed under the Freedom of Information Act. Such claims can be made in advance or at the time of the inspection, but too many claims of confidentiality tend to invalidate the whole process.

Samples of Items: Most companies are willing to provide FDA with a reasonable quantity of samples of small items at no charge. A common limit beyond which payment may be requested is $50.00. Most companies also make it a policy to retain duplicate samples for their own use. The inspector should be asked what testing will be done on the samples so that this testing can be duplicated in the firm's laboratory. A firm might also ask for the results to FDA testing so that a comparison can be made with company data.

Photographs on Inspections: The question of a policy on photographs is complex and widely debated. A firm should nonetheless formulate some policy on this issue, because sooner or later an investigator will wish to take photographs during an inspection. More and more companies are clearly stating, as part of their initial interview, that a "no-photographs" policy exists.

Questioning Nonmanagement Personnel: Regulatory inspectors may wish to question nonmanagement employees, and this can become a concern. A good policy is to state, "No questions on the job if there is any risk of interference with the task being performed." A firm can offer to have an individual temporarily relieved so that a discussion can take place away from the job site. However, it should be made clear that questions outside the individual's sphere of responsibility will be intercepted by the inspection coordinator, who should be present at all such interviews.

7.5.5 The Inspection Coordinator

The role of the inspection coordinator is a highly critical one. For this reason, each large facility should have a primary inspection coordinator and at least two backup personnel. In a smaller firm, the manager usually carries the responsibility of inspection coordination, but a backup person should be designated in the event that the manager cannot be present at an inspection. In larger facilities, the best candidate for the primary role is the plant's quality assurance manager/director. Many firms insist that the plant manager coordinate the inspection, but this individual may not be sufficiently familiar with the details of day-to-day plant operation. The responsibilities of the inspection coordinator seem quite simple but are in fact highly complex. The coordinator's first duty is to stay with the inspector at all times. One exception to this rule might occur when the inspector is involved in a lengthy review of records, for which purpose he or she must remain in a conference room. Even in this situation, however, the coordinator should be available to explain or interpret items whenever necessary.

The coordinator must also provide the firm with a consistent approach toward problems and procedures; in this way, he or she can minimize the chance that contradictions will arise. It is also quite useful to have a second person accompany the inspector at all times; a good choice for this duty is one of the backup coordinators. This individual is valuable in that he or she can leave the inspection group for various purposes—for example, to obtain information that is not at hand or to inform others of the progress of the inspection. Key managers of various departments should also be involved; however, if too many people are present, the inspector may feel that management is unduly concerned with the outcome of the inspection or that the firm lacks confidence in its coordinator. The coordinator must always remember that he or she is the company, as far as the investigator is concerned; indeed, a firm's attitude toward regulatory compliance will be judged largely on the basis of the coordinator's attitude. The coordinator must therefore be able to speak with authority and confidence, and yet, when necessary, be willing to say, "I don't know; I'll arrange to get the information for you," or simply, "That is beyond my authority."

The coordinator must also interact with other company employees—those working in the plant as well as those in staff groups. Questions posed by the inspector must be guided to people with both the knowledge and the verbal skills to answer them, and inappropriate questions must be challenged immediately. The coordinator must communicate inspection results on a continuing basis; then, once the inspection has been completed, he or she should summarize the results more formally.

Any individual who hopes to assume all of these responsibilities must possess knowledge and skills in several areas. Above all, he or she must thoroughly comprehend the Federal Food, Drug, and Cosmetic Act and its relevance to the products being inspected. An understanding of the good laboratory practices and other regulations is equally important. Finally, the coordinator must understand both FDA's inspection authority and the approach that the agency takes during inspections; to do so, he or she can attend internal or outside seminars and peruse the FDA Inspection Operations Manual, which is available under the Freedom of Information Act. Other important areas of knowledge include (1) company policies and procedures, particularly those relating to potentially controversial issues; (2) all aspects of laboratory operation and of the quality assurance system (particularly important here is the rationale for the choice of a certain approach when the regulations permit options by using words such as adequate or sufficient); and (3) inspection do's and don'ts.

Many coordinators have found that a good way to prepare for an inspection is to put together a booklet or a reference folder—one that contains "refresher" type information and other material that may be required during the inspection. Such information can also be invaluable if it becomes necessary for one of the firm's backup personnel to host the inspector. It is recommended that the following items be included:

- A card with the company's name, address, and phone number.
- A copy of the current annual report.
- The names and phone numbers of company regulatory and legal personnel.

- A simple line drawing of the manufacturing process used for each product. A simple facility-layout drawing may also be useful.
- Samples of current labels, wrappers, quarantine, and release stickers.
- Sample finished product release form.
- A list of components for each product. A list of the locations of (a) all appropriate records (consider indexing by GMP section); (b) reserve sample storage; (c) any appropriate operating permits licenses and so on; (d) device master files and the quality control manual; and (e) other important operational reference data.
- Brief biographical data on the plant manager, the quality assurance manager, and other key management personnel.
- A copy of sampling instructions.
- The name of the pest-control contractor, and a copy of the contract, the licensing status, and the chemical used.
- Notes on any unusual activity that is going on in the plant—for example, construction work.
- A copy of the GLP regulations.
- Contracts with custom sterilizers or with other custom operators.
- Data retrieval operating procedures.
- Records of past inspections (e.g., FD 483 forms, or 483s notices of observations, establishment inspection reports, and company responses).
- Data that may be required to show that an effective internal-audit program has been instituted.

Undoubtedly, each firm can add to this list on the basis of its own experience. It should be noted, however, that if a firm's inspection coordinators prepare for reference folders, a decision must be made as to which items will be given to inspectors, which might be shown to inspector, and which will be for the exclusive use of the coordinator. Another key area of preparation should involve everyone in the firm who might reasonably to contact the inspector. In short, all such individuals must be acquainted with the procedures to be followed during the inspection, and all must understand the role of the inspection coordinator. It is also useful to familiarize these individuals with a list of inspection do's and don'ts. Items such as the following are often included in such lists:

DO:

- Tell the receptionist in advance whom to call (both principal and alternate personnel) when a regulatory inspector appears.
- Call, or have an assistant call the plant/facility manager as well as other key managers at this time.
- Call the second individual involved with inspections—for example, the person who will be with the inspector at all times.
- Examine the credentials of the inspector, and establish the purpose of the inspection (e.g., routine, for cause, or survey or PAI).

- Immediately inform corporate management and staff that an inspection is under way, and tell these individuals what the purpose of the inspection is.
- Work out a rough schedule for the visit so that key personnel can be made available.
- Assume a friendly, cooperative attitude—but do not overdo it. Avoid creating an adversary relationship.
- Project an attitude of confidence and professionalism.
- Be professional and confident.
- Balance cooperation with wariness.
- Review pertinent plant/facility policies, for example, policies pertaining to cameras, recorders, wearing apparel, and safety equipment during the opening interview.
- Make sure you understand all questions and their contexts before you answer the inspector.
- Respond to requests appropriately.
- Wait for the investigator(s) to make specific requests before providing records, samples, labels, and the like.
- Focus on the positive aspects of a situation rather than getting defensive about negative aspects.
- Take immediate corrective action when appropriate. Ask to have such action included in the establishment inspection report.
- Take complete notes as you go along, and polish these up at each opportunity.
- Obtain duplicate copies of any documents taken.
- Obtain duplicates of any sample that is taken and get a receipt for all such samples. In addition, try to ascertain the purpose of each sample, and ask to receive the test results (e.g., analytical or microbial data).
- Avoid uncontrolled interchanges between personnel and the inspector.
- Take every issue seriously.
- Make a complete write-up of the inspection.
- Follow up to see that all written comments (e.g., 483s) and verbal comments have been addressed and resolved.

DON'T:

- Guess, lie, deny the obvious, or make misleading statements.
- Become argumentative or at worst hostile.
- Engage in unconstructive argument. Avoid win/lose or legalistic confrontations.
- Get uptight. Fear is inappropriate, although genuine concern may be expressed.
- Tell the investigator(s) that an inspection is not possible that day because the owner or person who is in charge is on vacation, and suggest she/he or they return next week.

- Say that something is impossible or could not happen. This is a red flag that challenges the inspector to find a way it could happen.
- Offer other materials and information that might relate to another matter pending with FDA but are unrelated to the request.
- Volunteer information unless that information is clearly in the best interest of the company.
- Respond to questions that are improper or outside your area of expertise or authority.
- Attempt to answer "what if" questions and other hypothetical questions.
- Threaten to contact the investigator's boss if conflict develops.
- Comment on the quality of the inspection.
- Offer to buy lunch!

Although the reasons for most of these recommendations are self-evident, the last item warrants some additional comment in as much as it has some interesting implications. If a facility is located at some distance from commercial eating establishments, it is perfectly appropriate for the firm to offer an inspector the paid use of the company cafeteria. The concern here is more with uncontrolled exposure of the inspector to situations in which chance remarks might be overheard and misinterpreted. To avoid such situations, one can take the inspector to a separate table or eating area. Normally, however, the inspector will want to use this time to update notes, call the office, review the file, or perform other activities, in which case he or she will leave the facility during lunch hour.

In the same vein, it is highly recommended that clip-on badges be given to all visitors—not only to inspectors. Firms should also review all appropriate safety and sanitation requirements as part of the opening interview. Inspectors are usually very sensitive to internal rules in these areas.

7.5.6 Follow-Up Procedures

One final task of the inspection coordinator is to provide appropriate follow-up and response. This should begin during the exit interview, which involves the plant manager, the inspector the coordinator, and other key management personnel. This meeting is highly critical to the whole process; thus, attentiveness, tact, responsiveness, and good note taking are musts.

At the meeting, the investigator reviews the results of the inspection(s); in addition, he or she may leave a 483 if potentially violative conditions have been observed. If corrections have already been made, these should be called to the attention of the inspector. A request to note the corrective action on the 483 can be made, but the inspector is under no obligation to comply with such a request. Finally, any misunderstandings about the facts of an observation must be cleared up at this point. Alternately, a firm may "agree to disagree" and may want to communicate the disagreement to the FDA district office. A good coordinator will resolve issues of this type during the inspection, thus avoiding confrontations in the exit interview.

After the inspector leaves, the coordinator should prepare a detailed report, including:

- The date, times, and purpose of the inspection. (Attach copies of FDA 482 for "Notice of Inspection"—the 483 if any, all sample receipts, and the like).
- The areas toured and the individuals contacted.
- The questions asked by the inspector and the responses given.
- The documents viewed. Attach duplicates of items copies.
- The inspector's spoken comments.
- All actions taken as a result of the inspection.

The inspection should be discussed with appropriate corporate staff groups; topics stressed should include the inspection's emphasis and any comments that were made. Calling together the appropriate line and staff managers to critique the inspection will also be of value. Such a meeting gives the firm an opportunity to further improve its compliance status and its handling of inspections in future.

A formal response should be made to the FDA district office on all 483s. Corrective actions taken should be described, and if there is disagreement with either the facts of an observation or the significance of these facts, such disagreement should be noted. It might be beneficial for a firm to request that its response be filed with the 483 and that this response be released if a request for the 483 is made under Freedom of Information regulations. It should be remembered that one of the stated purposes of an inspection is to secure correction of deficiencies. Prompt response on a 483, while not legally required, can help convince the agency of the firm's good intentions and can thus reduce the chances that further regulatory action to be taken.

FDA, too, does some follow-up. For example, a detailed establishment-inspection report (EIR) is prepared by the agency and placed in the company file; this report is also available for release under FOI regulations. Some companies routinely request copies of this report—often through a third party, so that the company can determine how thoroughly proprietary material has been purged. Immediate requests for establishment-inspection reports are not recommended, however, since such requests alert other companies and organizations to the fact that an inspection has taken place. The question is one of balancing the risk of alerting competitors or the media to the inspection against the benefit of having the material. However, the firm should definitely obtain a copy of the establishment inspection report if it is discovered that someone else has already requested one.

7.5.7 Summary

As one (firm) develops its own plans for handling and facilitating regulatory audits, one can obtain further assistance from a variety of sources. Quality assurance and regulatory compliance personnel from other firms, for example, are almost always willing to discuss their policies and procedures. A number of GLP or GxP quality assurance consultants are also available, and these individuals can be very helpful. A manual covering this important area can be purchased through the organization that publishes the Pink Sheet and related newsletters, and the FDA Inspection Operations Manual can be obtained as well. In summary, it could be stated that through preparation, development of written policies and procedures governing the handling of inspections, and care in the selection and training of coordinators are the keys to a successful inspection.

There are various kinds of good laboratory practice monitoring authorities (MAs) in the world. Some countries have only one MA, while others, including Japan, have more than one MA. In addition, each MA has its own relationship with regulatory authorities (RAs), receiving authorities (RcAs) and industry based on the internal regulatory systems. There are eight GLP MAs in Japan. This number is probably the largest in the world. Efforts have been made to establish a close link among MAs and to apply and implement GLP programs in an efficient, effective, and consistent way, namely: (i) interministerial meeting on GLP. It is essential to establish a system for information exchange and decision making when there are a number of MAs such as in Japan. To this end, the interministerial meeting on GLP has been set up as a means for MAs, RAs, and RcAs to share information on Organization for Economic Cooperation and Development and foreign countries and make national decisions as a whole country; and (ii) joint training program. With the goal of training inspectors and minimizing differences in inspections among MAs, a joint training program has been started including joint visits to test facilities (TFs) and participation in evaluation committees at other MAs. Currently, pharmaceuticals, medical devices, and other life science related products are internationally developed and marketed. In this context, the OECD MAD and the activities of the OECD working group on GLP have gained paramount importance for both industry and governments.

In the United States, laboratories testing products regulated by the Food and Drug Administration or the Environmental Protection Agency (EPA) must operate within the guidelines published by these agencies and codified as law in the *Code of Federal Regulations* (CFRs) and *Federal Register* (FR). These guidelines are known as current Good Manufacturing Practice, Good Laboratory Practice, and Good Clinical Practice. cGMP guidelines are codified under 21 CFR Parts 210 and 211 for finished pharmaceuticals, and 21 CFR Parts 225 and 226 for Type A medicated articles and finished feeds regulated by FDA. GLP guidelines are codified under 21 CFR Part 58 for nonclinical laboratory studies under FDA. The EPA regulates GLP through 40 CFR Part 160 for products/laboratory analysis regulated by the *Federal Insecticide, Fungicide, and Rodenticide Act of 1996* (*FIFRA*) and 40 CFR Part 792 160 for products/laboratory analysis regulated by the *Toxic Substances Control Act of 1976* (*TSCA*). Global acceptance for GCPs written by the International Conference on Harmonization of Technical Requirements for Registration of Pharmaceuticals for Human Use (ICH) have been published, and thus are codified under 62 FR 25692 for products/laboratory analysis regulated by FDA. For clinical laboratories, GCPs are codified under 21 CFR Part 493, the *Clinical Laboratories Improvement Act of 2003* (*CLIA*). GCPs can often be interpreted for the needs of each business concern. If work is being done in these areas, it is recommended that the quality assurance department be consulted to determine which GxP regulations apply.

GLP, cGMP, and GCP interpretations for analytical laboratories contain many common elements and have now become collectively known in the pharmaceutical industry as "GxP." Grouping these guidelines as GxP compliance has created a harmonized norm and helped many companies address the varied interpretations that exist within the industry. In terms of instrumentation, it is arguable whether cGMP or GLP is stricter in their requirements, simply because their interpretation is left up to the business concern and each company interprets the guidelines to suit their own business practices. Setting up policies to comply with both cGMP and GLP as GxP guidelines eliminates this issue within many companies.

Whether automated or not, GxP guidelines require analytical methods to be scientifically valid prior to implementation. The fact that automation can run more precisely or more accurately than manual methods is irrelevant. Documentation must exist for regulatory inspection that clearly demonstrates that the automated or manual method is appropriate for the phase of development and is sufficiently accurate, precise, and robust. Even though the compendial methods published by United States Pharmacopoeia (USP), American Society for Testing and Materials (ASTMs), and Association of Analytical Communities (AOAC) have been validated and fully tested prior to publication, it is still up to the user to demonstrate and document that the method is performing properly on their equipment.

The system is definitely efficient because thirty member countries (OECD) and three nonmember countries can attain a higher level of compliance through international GLP-related agreements. US FDA has effectively worked with global regulatory agencies/organizations, and industry focus/working groups to shape and refine the regulations for twenty-first century and beyond.

REFERENCES

1. Lee MS, Kerns E.H. *Mass Spectrom Rev* 1999;183–4:187–279.
2. James CA, Breda M, Frigerio E. *J Pharm Biomed Anal* 2004;35(4):887–893.
3. Shah VP, Midha KK, Dighe S, McGilveray IJ, Skelly JP, Yacobi A. et al. Analytical methods validation: bioavailability, bioequivalence, and pharmacokinetics studies *J Pharm Sci* 1992;81:309–312.
4. Shah VP, Midha KK, Findlay JW, Hill HM, Hulse JD, McGilveray IJ. et al. Bioanalytical method validation: a revisit with a decade of progress. *Pharm Res* 2000;17:1551–1557.
5. US Department of Health and Human Services, Food and Drug Administration, Center for Drug Evaluation and Research, Center for Veterinary Medicine. Guidance for Industry: Bioanalytical Method Validation, May 2001.
6. Food and Drug Administration Bioresearch monitoring: human drugs: in vivo bioequivalence. *Compliance Program Guidance Manual* (CPGM 7348.001), Chapter 48, October 1999. Available at http://www.fda.gov/ora/compliance_ref/bimo/7348_001/foi48001.pdf.
7. Food and Drug Administration. Bioavailability and Bioequivalence Requirements, 21 CFR Part 820, 2005.

8. US Department of Health and Human Services, Food and Drug Administration, Center for Drug Evaluation and Research, Center for Veterinary Medicine, (2001). Guidance for Industry: Bioavailability and Bioequivalence Studies for Orally Administered Drug Products—General Considerations, March 2003.
9. Food and Drug Administration. Good Laboratory Practice for Nonclinical Laboratory Studies, 21 CFR Part 58, 2005.
10. Food and Drug Administration, Center for Drug Evaluation and Research, Division of Scientific Investigations homepage on the Internet. Rockville, MD updated 21 September 21, 2006; cited September 22, 2006. About DSI; about 4 screens. Available at http://www.fda.gov/cder/Offices/DSI/aboutUs.htm.
11. LePay D.GLP, bioequivalence, and human subject protection programs. In Drug Information Association, Office of Compliance, 1998. Available at http://www.fda.gov/cder/present/dia698/diafda4/dia fda4.PPT.
12. Miller KJ, Bowsher RR, Celniker A, Gibbons J, Gupta S, Lee JW. et al. Workshop on bioanalytical methods validation for macromolecules: summary report. *Pharm Res* 2001;189:1373–1383.
13. Timm U, Wall M, Dell D. *J Pharm Sci* 1985;749:972–977.
14. Kringle R, Hoffman D, Newton J, Burton R. *Drug Info J* 2001;35:1261–1270.
15. Weng N, Halls TDJ. Systematic troubleshooting for LC–MS/MS. *Pharm Technol* 2002;151:22–26, 49.
16. King R, Bonfiglio R, Fernandez-Metler C, Miller-Stein C, Olah T. *J Am Soc Mass Spectrom* 2000;11:942–950.
17. Karns H.T. *Am Pharm Rev* 2003;61:106–107.

8

CURRENT STRATEGIES AND FUTURE TRENDS

Advances in many different disciplines have occurred to change the way drug discovery and development is performed today as compared to even 5 years ago. These advances include sequencing of the human genome; identifying of more drug target through proteomics; advances in the fields of combinatorial chemistry, high-throughput screening, and mass spectrometry; and improvements in laboratory automation and overall throughput in bioanalysis. The end result of these process improvements is that compounds can now be synthesized faster than ever before. These greater numbers of compounds are being quickly evaluated for pharmacological and metabolic activities utilizing high-throughput automated systems, with the ultimate objectives of bringing a drug product to market in a shorter timeframe to suffice unmet medical needs.

Regulatory bodies have drafted numerous guidance documents to ensure that proper and sufficient research and development need to be considered and performed as part of data generation and regulatory requirements for submission and approval. The concept and processes of regulated environment have greatly enhanced the quality and integrity of analytical and bioanalytical laboratories in support of research and development programs.

Regulated Bioanalytical Laboratories: Technical and Regulatory Aspects from Global Perspectives,
By Michael Zhou

8.1 STRATEGIES FROM GENERAL LABORATORY AND REGULATORY PERSPECTIVES

New drug discovery and development becomes more difficult and more expensive every year. The larger pharmaceutical companies continue to seek new leads outside of their own research laboratories in order to control R&D costs, or to extend their portfolio. This is particularly true in the biopharmaceutical area, since most traditional pharmaceutical companies have had limited direct experience in this field and prefer to license-in such technology. Alternatively, the smaller biotechnology company may be acquired by the larger partner. In the same week in May 2007, two major US companies both published their intent to invest further in biotechnology incubator companies (Pfizer, Inc.) or the purchase of biotechnology companies with late-stage products, especially vaccines (Merck & Co.).

In all of these cases, the new product leads will be of increased value to the acquiring partner if the principles of Good Laboratory Practice (GLP) have been followed by the originating scientists and research laboratories. In any drug development program, but especially with the acquisition and further development of any new technology by a major industrial partner, the term "time to market" becomes all important, since each month lost in the laboratory is a month when the product is not earning revenue to pay back the ever-increasing costs of development and testing. The drug development and testing timeline are being extended by more stringent regulatory requirements, especially in the preclinical testing stages. Time to market will be extended, with negative commercial impact, if critical early research work and drug testing have to be repeated under more strictly controlled conditions in order to comply with the GLP and GxP regulations.

These regulations, in particular with GLPs were introduced first in the United States in 1979 and revised in 1987. They were intended to assure the quality and integrity of any preclinical safety or efficacy data, especially animal testing data, which would be submitted in support of an application (IND) to start testing a drug in human subjects, or for approval to market the new drug on completion of the clinical trials. The regulations cover all key areas of nonclinical laboratory work, including facilities and operations, personnel qualification and training, study design and execution, data recording and archiving, and quality assurance procedures. The same requirements may apply in those laboratories that are testing samples derived from clinical studies in some countries; the concept of Good Clinical Laboratory Practice has been introduced, as a supplement to GLP and to Good Clinical Practice regulations.

University laboratories, spin-off companies, and smaller drug development companies, which therefore undertake to apply GLP to their work as soon as it is required, are those which can attract greater attention from potential industrial partners. Alternatively, if the smaller new companies prefer to proceed to further independent development before approaching "Big Pharma," they can gain a more rapid acceptance of their experimental results by the regulatory bodies, leading to earlier qualification of the drug for clinical studies. Contract laboratories with Good Laboratory Practice offering preclinical testing services will have to be able to demonstrate to their clients and the regulatory bodies that their operations are fully in GLP compliance. Companies planning to extend their work beyond the bounds of North America will benefit from a better understanding of the OECD principles of GLP and their practical applications.

Code of Federal Regulations, Title 21, Part 58, 2006 Edition (21 CFR 58)—The regulations laid down by other countries, especially the members of the European Community, are based upon guidelines developed by the Organization for Economic Cooperation and Development (OECD) in 1980. These are very similar to 21 CFR 58 and have been updated regularly. Guidelines on safety testing and analytical methods issued by the International Conference on Harmonization (ICH) are also reviewed on a continued basis. Those rules which are most commonly cited by FDA inspectors for noncompliance are highlighted. Since compliance is finally determined by official inspections of the laboratory organization and operations, and by audits of data and reports, special attention is paid to documentation and the preparations for internal and external inspections and audits. The recommendations presented in this Guide have been distilled from a large body of information and opinion. The most recent guidelines issued by the FDA and other regulatory authorities to industry and to their inspection staff have been accessed. Opinion and advice collected from other regulatory consultants, including ex-FDA personnel, has been reviewed. With this Guide, the task of GLP compliance is made easier to understand and to achieve successful GLP/regulated bioanalytical laboratories.

Industry was challenged when the first Good Laboratory Practice regulations were introduced in 1978 and industries are still faced with a number of challenges. The challenges of today are again the consequences of change (e.g., technologies, globalization, etc.); this time not in terms of implementing new regulations, but in terms of applying the regulations in the modern day environment, an environment that is constantly changing and evolving. The GLP principles, despite being a little dated remain fundamentally sound and so can still be applied in the modern day environment. On the one hand, industry must consider the principles before applying new science, technology and strategy, while on the other hands, the GLP monitoring authorities (MAs) should recognize the impact of change on the quality of data that should allow for a pragmatic approach without compromising compliance. Advances in sciences have led to new study types, new study designs, increased interest in biological entities, increased scientific methodology, and the ability to measure more parameters.

The Good Clinical Laboratory Practices (GCLPs) concept possesses a unique quality, as it embraces both the research and the clinical aspects of GLP. The development of GCLP standards encompasses applicable portions of 21 CFR Part 58 (GLP) and 42 CFR Part 493 (Clinical Laboratory Improvement Amendments, CLIAs). Due to the ambiguity of some parts of the CFR, the GCLP standards are described by merging guidance from regulatory authorities as well as other organizations and accrediting bodies, such as the College of American Pathologists (CAP), and the International Organization for Standardization 15189 (ISO). The British Association of Research Quality Assurance (BARQA) took a similar approach by combining Good Clinical Practice (GCP) and GLP in 2003. The GCLP standards were developed with the objective of providing a single, unified document that encompasses IND sponsor requirements to guide the conduct of laboratory testing for human clinical trials. The intent of GCLP guidance is that when laboratories adhere to this process, it ensures the quality and integrity of data, allows accurate reconstruction of experiments, monitors data quality, and allows comparison of test results regardless of performance location. By recognizing these standards as the minimum requirements for optimal laboratory operations, the expectation is that GCLP compliance will ensure that consistent, reproducible, auditable, and reliable laboratory results from clinical trials can be generated for clinical trials implemented at multiple sites. A corollary of this infrastructure is that the data will be produced in an environment conducive to study reconstruction, enable prioritization between candidate product regimens, and guide rational decision making for moving products forward into advanced clinical trials.

The GCLP standards were developed to bring together multiple guidance and regulatory information, as they apply to clinical research and to fill a void of a single GCLP reference for global clinical research laboratories with regard to laboratories that support clinical trials such as those that perform protocol-mandated safety assays, process blood, and perform immune monitoring assays for candidates on a product licensure pathway. To maintain a GCLP environment for a clinical trial, it is critical that all of the key GCLP elements are in place and operational. These elements include organizations and personnel, testing facilities, appropriately validated assays, relevant positive and negative controls for the assays, a system for recording, reporting and archiving data, a safety program tailored to personnel working in the laboratory, an information management system that encompasses specimen receipt/acceptance, storage, retrieval and shipping, and an overall quality management plan. The most appropriate way to ensure compliance with GCLP guidance is to audit laboratories. Because key decisions regarding the advancement of products are based on laboratory-generated data obtained from specimens collected during the trials, GCLP compliance is critical. Such compliance will assist laboratories in ensuring, accurate, precise, reproducible data are produced that guarantee sponsor confidence, and stand under regulatory agency review. Here are examples of recent guidance documents for industry (Table 8.1).

8.2 STRATEGIES FROM TECHNICAL AND OPERATIONAL PERSPECTIVES

Over the course of the past several years, the need for increased throughput analytical methods in support of drug discovery and development operations has dramatically increased. New nicety methods, which can provide reliable quantitation of target analytes in biological fluids, are now necessary to keep pace with biological screening. To date, the only successful approaches to direct high-speed quantitation of large numbers of diverse compounds are based on high-performance liquid chromatography/tandem mass spectrometric (LC–MS/MS) techniques. Key challenges to this task include sample preparation, the automated development of

TABLE 8.1 Examples of Recent Guidance Documents for Industry

Title and Format	Subject	Type	Issue Date
Pharmacokinetics in Patients with Impaired Renal Function—Study Design, Data Analysis, and Impact on Dosing and Labeling (PDF—151 KB)	Clinical Pharmacology	Draft	3/17/2010
S9 Nonclinical Evaluation for Anticancer Pharmaceuticals (PDF—170 KB)	International Conference on Harmonization—Quality	Final	3/5/2010
Non-Inferiority Clinical Trials (PDF—565 KB)	Clinical/Medical	Draft	2/26/2010
Adaptive Design Clinical Trials for Drugs and Biologics (PDF—424 KB)	Clinical/Medical	Draft	2/25/2010
Submission of Documentation in Applications for Parametric Release of Human and Veterinary Drug Products Terminally Sterilized by Moist Heat Processes (PDF—69 KB)	CMC—Microbiology (Chemistry, Manufacturing, and Controls)	Final	2/25/2010
Contents of a Complete Submission for the Evaluation of Proprietary Names (PDF—306 KB)	Labeling	Final	2/5/2010
Assessment of Abuse Potential of Drugs (PDF—138 KB)	Clinical/Medical	Draft	1/26/2010
The Use of Mechanical Calibration of Dissolution Apparatus 1 and 2—Current Good Manufacturing Practice (cGMP) (PDF—38 KB)	cGMPs/Compliance	Final	1/26/2010
M3(R2) Nonclinical Safety Studies for the Conduct of Human Clinical Trials and Marketing Authorization for Pharmaceuticals (PDF—295 KB)	International Conference on Harmonization—Joint Safety/Efficacy (Multidisciplinary)	Final	1/20/2010
Guidance to Pharmacies: Compounding Tamiflu Oral Suspension in Advance to Provide for Multiple Prescriptions (PDF—114 KB)	Procedural	Draft	1/11/2010
Planning for the Effects of High Absenteeism to Ensure Availability of Medically Necessary Drug Products (PDF—47 KB)	Procedural	Draft	1/7/2010
Q4B Evaluation and Recommendation of Pharmacopoeial Texts for Use in the ICH Regions • Annex 5: Disintegration Test General Chapter (PDF—32 KB)	International Conference on Harmonization—Quality	Final	12/22/2009

(*continued*)

TABLE 8.1 (*Continued*)

Title and Format	Subject	Type	Issue Date
• Annex 8: Sterility Test General Chapter (PDF—32 KB)[24]			
Q4B Evaluation and Recommendation of Pharmacopoeial Texts for Use in the ICH Regions	International Conference on Harmonization—Quality	Draft	12/16/2009
• Annex 11: Capillary Electrophoresis (PDF—90 KB) General Chapter			
• Annex 12: Analytical Sieving General Chapter (PDF—313 KB)			
Addendum to ICH S6:[29]			
Preclinical Safety Evaluation of Biotechnology—Derived Pharmaceuticals S6(R1) (PDF—160 KB)[30]	International Conference on Harmonization—Safety	Draft	12/16/2009
PET Drug Products—Current Good Manufacturing Practice (cGMP) (PDF—399 KB)	cGMPs/Compliance	Final	12/9/2009
Patient-Reported Outcome Measures: Use in Medical Product Development to Support Labeling Claims (PDF—295 KB)	Clinical/Medical	Final	12/8/2009
Assay Development for Immunogenicity Testing of Therapeutic Proteins (PDF—161 KB)	Chemistry, Manufacturing, and Controls (CMC)	Draft	12/3/2009

methods, and the rapid analysis of many samples. Accomplishing these tasks require the use of generic methods, which can be applied across many structurally diverse compounds. The state-of-the-art in high-speed LC–MS/MS should be taken into consideration with the aim of describing the most practical and effective approaches currently in use.

Combinatorial synthesis and high-speed analoging methodologies are now generating thousands of biologically active compounds per year. While extremely efficient high-throughput primary screening methodologies exist which can support this level of compound generation, the need to assess secondary parameters, such as physiochemical and ADME (absorption, distribution, metabolism, and excretion) properties has become critical. In fact, the need to determine whether a given chemical series possesses the appropriate "drug-like" qualities has become a key factor in library generation, particularly when these data are used to make synthesis decisions [1]. These secondary parameters have several characteristics that define challenges particular to them. First, endpoint determination is problematic. Many ADME-related assays involve direct transformation of the target compound rather than some perturbation in a biological process requiring direct measurement of the target compound. Thus, nonanalyte-specific detection methods such as those employed in traditional high-throughput screening (HTS) are ineffective. Second, many of these *in vitro* assays involve complex biological reagents requiring the target compound level to be measured against a background of other small molecules. One factor in common with its HTS progenitors is the reliance on consistent formats such as the 96-well microtiter plates and high capacity liquid handling. Using the methodologies developed for traditional HTS, the generation of samples is no longer an issue with screens developed for the determination of ADME parameters. These factors combine to make the rapid analysis of biological fluids for target compounds a rate-limiting step in this process. The present most effective and widely used method to address this disparity between sample generation and endpoint determination is by the development and implementation of high-throughput LC–MS/MS (HT LC–MS/MS). The state-of-the-art in HT LC–MS/MS for discovery support constitutes a completely different technique from either of the ancestors referred to in its abbreviated name. At its inception, LC–MS/MS became the tool of choice for the good laboratory practice/regulated analysis of biological fluids due to its superior selectivity and sensitivity for small molecule pharmaceutical compounds. Each of these highly compound-specific custom methods were developed and validated in various biological matrices to support toxicology and clinical studies. An LC–MS/MS method must possesses a high level of specificity, accuracy, and precision at a low enough lower limit of quantitation (LLOQ) to satisfy public regulatory authorities to be of use for compounds under full development. In a discovery setting, however, analysis speed becomes an equally important figure of merit. Recent advances in high-speed HPLC and MS technologies have allowed LC–MS/MS to become a critical tool for early discovery. HPLC techniques used for LC–MS/MS have evolved from a means to separate various components in a mixture to a rapid means to introduce clean material into the ionization source of the mass spectrometer. This change transformed the HPLC process away from chromatography in the separation science sense of the word, to a means of rapid on-line sample clean up prior to its introduction into the MS source. In a similar vein, the mass spectrometric portion of the technique deviates from the more classical form of the art in that the spectra produced are not of the nature, complexity, or quality of traditional mass spectrometry. In fact, HT LC–MS/MS has continuously evolved away from long, high theoretical plate separations, optimized lens settings, and unit (or better) resolution. This evolution has been driven solely by the need of discovery scientists to provide large numbers of quick and reliable measurements. Simply, a "discovery" measurement must be able to produce data of useable quality at the same velocity as the cutting edge synthetic and screening methodologies used to produce new compounds. The current state-of-the-art in HT LC–MS/MS for early discovery support of high-throughput *in vitro* screens as well as development applications is reviewed and elaborated below. Discussions will center on applications of linear quadrupole instruments in quantitative applications. As many of the approaches are very novel, relatively little has been published in the literature. This chapter/section attempts to offer a sense of the general trends in the field of HT LC–MS/MS. The more important elements of the technique are compared and contrasted, along with some analysis of future developments.

Bioanalytical laboratories typically involve in the processes of sample receiving (shipment, receipt, chain of custody, storage, etc.), sample preparation (extraction, cleanup, etc.), sample analysis (qualitative and quantitative determination of analytes of interest), data processing, analysis, and reporting. It is imperative that integrity and quality of the data be delivered, documented, and archived for decision-making processes in discovery and development programs.

8.3 BIOLOGICAL SAMPLE COLLECTION, STORAGE, AND PREPARATION

8.3.1 Sample Collection and Storage

The complex nature of biological matrices requires that sample preparation be an integral part of bioanalytical methods. One of the challenges faced in plasma sample collection, storage, and extraction is the instability of drugs, metabolites, and prodrugs in biological samples. Compound stability in plasma may be affected by enzymes and/or pH of the biological samples, anticoagulants, storage temperature, and freeze–thaw cycles [2–18]. While the degradation of a drug during sample collection and storage causes underestimation of the drug concentration, the degradation of the metabolite or prodrug may cause overestimation of the drug concentration. Acylglucuronides are probably the most commonly encountered problematic metabolites in bioanalysis. Acylglucuronides tend to be unstable and hydrolyze to release the original aglycone (the drug) under neutral and alkaline conditions and elevated temperatures [2, 3], although different acylglucuronides have been shown to have different rates of hydrolysis [3]. The mildly acidic pH of 3–5 tends to be the most desirable pH region for minimizing the hydrolysis of acylglucuronides in biological samples. The common recommendation is that blood samples should be immediately cooled on ice after collection, and then plasma should be separated using a cooled centrifuge within 10 min after blood collection. The plasma samples are then stored frozen at $-70°C$. During sample processing for analysis, the plasma samples should be kept cooled on ice and the aliquotted portions immediately buffered to lower the pH to 3–5. The lactone is another commonly encountered metabolite functional group which could be bioanalytically challenging since the lactone may be converted to its open-ring hydroxy acid drug. Systematic studies of the effects of pH and temperature on the stability of the lactone metabolites of hydroxy acid drugs show that the sample pH should be adjusted to 3–5 in order to minimize the hydrolysis of the lactone metabolite back to the drug or vice versa [4, 5].

It should be noted that if a drug, metabolite, or prodrug is very unstable, it may be necessary to add a stabilizing reagent directly to the blood before obtaining plasma samples. It has thus been reported that citric acid added directly to the blood, in the amount of 5 mg/mL of blood, provides the lower pH required for stabilization of pH-sensitive labile compounds without causing the undesirable gelling of the blood [6]. Another important finding is that the pH of the biological sample may change during storage or processing and thus cause unexpected degradation of drugs or metabolites [7, 8]. Fura et al. found that pH of plasma samples may be significantly higher than the normally expected value of approximately 7.4 depending on the conditions of sample storage and processing [7]. Initial addition of appropriate amounts of citrate or phosphate buffers to the plasma sample can be used to maintain the pH of plasma during storage or processing [8].

Metabolites with *O*-methyloximes [9] and carbon–carbon double bonds [10] can undergo *E* to *Z* isomerization due to nonoptimal pH or exposure to light. Epimerization is another potential cause for sample instability [11, 12]. A sample that contains a thiol drug and its disulfide metabolite may also cause an analytical challenge due to the potential for the conversion of the thiol to the disulfide or vice versa [13]. The type of anticoagulant used during blood sample collection may affect the stability of drugs or their metabolites. Evans et al. compared stability of hormones in human blood samples collected with EDTA, lithium heparin, sodium fluoride, and potassium oxalate [14]. They showed that most of the hormones were stable in blood collected with EDTA or fluoride at 4°C. A variety of chemical agents, such as methyl acrylate, sulfate, thiosulfate, fluoride, borate, phosphate, paraoxon, eserine, and organophosphate reagents, have been used to stabilize analytes in biological matrices [3, 8, 15–18].

Another important aspect of plasma sample storage is the minimization of clot formation that could impede the facile aliquotting of the plasma sample for analysis. The clogging of pipetting tips is mainly caused by the formation of fibrinogen clots when the plasma samples are stored at − 20°C and undergo freeze–thaw cycles [19]. The problem of fibrinogen clots can be overcome by storing plasma samples at − 80°C [19]. Berna et al. employed an interesting approach to solving the clot formation problem by the use of 96-well polypropylene filter plates to collect, store frozen, and then filter plasma samples prior to bioanalysis [20]. Sadagopan et al. reported that the failure rate for transfer of EDTA plasma, by automated workstation or manually, is less than that of heparinized plasma [21]. During analysis of a large number of samples, the task of manually uncapping and recapping the sample tubes could take a significant amount of time and cause physical stress. As a solution to this problem, Teitz et al. reported the direct transfer of plasma samples from pierceable-capped tubes into a 96-well plate using a robotic system [22].

8.3.2 Sample Preparation Techniques

As stated above, the final form of the sample prior to introduction into the analysis system is critical and impacts the system throughput. In this section, primary focus is based on the analysis of samples generated from bioanalytical methods relating to sample preparation techniques commonly used in the industry. Clearly, the sample preparation procedure used is dependent on the complexity of the sample matrix. Several approaches are currently in routine use across the industry.

8.3.2.1 Protein Precipitation While hardly a new development, protein precipitation is by far the simplest, most generic method in use. The resulting samples are somewhat crude, but the method is rapid and broadly applicable. The most common procedure involves the addition of an organic solvent (acetonitrile, methanol, or alike) to the sample in volumes adequate to remove a majority of protein with a minimal sample dilution, typically 3:1 (*v/v*). Higher volume ratios (5:1) of organic solvent will further precipitate most inorganic salts. While yielding cleaner samples, the high solvent strength of the sample solution can result in poor chromatographic performance [23] and in further sample dilution. Lower precipitation agent volumes using a variety of inorganic modifiers are also effective at protein removal [24], but liquid-handling operations can be more cumbersome with the reduced sample volumes. An inherent disadvantage of protein precipitation as a sample preparation method is that it generally requires a manual centrifugation step, making automation of this process difficult. Filtration may be used as an alternative to centrifugation, but nonspecific binding to the filtration apparatus may add to the assay imprecision. In any case, the crude nature of the samples produced often requires the selectivity of tandem MS as the detection step. Protein precipitation is currently in wide use for discovery and development purposes due to its generic nature, high recovery, and simplicity.

8.3.2.2 Solid Phase Extraction Solid phase extraction (SPE) has been used for many years in applications where removal of endogenous materials from the sample matrix is critical. In addition to removing proteins from the sample matrix, other endogenous materials can be removed using the selectivity of the SPE cartridge. Many phases and cartridge formats have been developed to optimize recovery and sample clean up. In terms of high-throughput applications, the most significant development in recent years was the introduction of the 96-well SPE plate format [25]. This technique has been successful in the assay of toxicology and clinical samples, which focus on a target compound. SPE plates have been effectively applied in discovery settings, in situations where programs are working with compound series with minimal structural diversity. In applications where compound physiochemical properties are similar, this technique can be employed to rapidly give clean samples with high recovery. This generally requires the use of large volume (1–5 mL) elution coupled with evaporation and low volume reconstitution. SPE has proven to be amenable to automation, despite the need for multiple steps and evaporation. Improvements in the SPE technique have been gained through the use of membrane materials. These materials are considerably more uniform and require significantly lower volumes for washing and elution. As a result, membrane-based 96-well plates are much more amenable to automation [26]. In situations where compounds of diverse structure and physical properties are to be analyzed, SPE is not the technique of choice. In this situation, optimization of the washing and elution solvents is often necessary for adequate recovery and desirable reproducibility. While automation solutions exist, SPE is inherently slow due to the need for multiple washes, elution, evaporation, and reconstitution steps. In addition, SPE is the most expensive sample preparation technique with the cost of 96-well extraction plates of the order of $100/plate. When thousands of samples are to be assayed, this has a substantial impact on cost/sample. In general, this technique is best employed in the late discovery/early development stages where sample throughput requirements are lower and more thorough sample clean up is required. Regardless, SPE has great potential in future development and improvement thanks to its unique selectivity.

8.3.2.3 Liquid–Liquid Extraction Liquid–liquid extraction using 96-well platform has been developed for high-throughput bioanalysis [27, 28]. While effective in these applications, it remains to be seen how broadly applicable this approach will be. Like SPE, optimization of the technique for maximum recovery and sample clean up is a compound-specific phenomenon. Issues surrounding phase separation, solvent selection and sample pH need to be optimized for each individual compound prior to sample preparation, making the technique more appropriate for late discovery/early development applications. Liquid–liquid extraction (LLE) has been widely considered as a simple and robust sample preparation technique in bioanalytical sample preparation. When extracting ionizable compounds, most analysts/bioanalysts adjust the pH of the sample to achieve fully unionized compounds. Usually, a generally accepted rule is applied to adjust the pH of the aqueous phase, known as the $pK_a \pm 2$ rule, depending on the acid/base characteristics of the analyte(s). By taking a closer look at the general equations that describe the extraction behavior of ionizable compounds, it is desirable to extend this pH adjustment rule by taking the distribution ratio and the volume of both liquid phases into account. By choosing an extraction pH based on this extended rule, the selectivity of the extraction can be influenced without loss of recovery. As a measure of this selectivity, two equations were proposed to indicate the ability of the extraction system to discriminate between two compounds. Also, milder extraction pH can be used for pH labile analytes. To use this new rule quantitatively, a new calculation method for the determination of the distribution ratio may be derived. These calculations are based on normalized recoveries making this method less susceptible to errors in absolute recovery determination. The proposed equations were supported by demonstrating that careful pH adjustment can lead to higher selectivity. The main conclusion was that a closer look at the extraction pH in bioanalytical methods extends the possibilities of obtaining a higher selectivity or the possibilities of extracting pH labile analytes at milder pH conditions without loss of recovery [29].

8.3.2.4 Direct Injection/Column Switching Techniques The concept of direct plasma/biofluid injection has sparked interest in recent years with obvious reasons and advantages. The idea that samples can be collected and analyzed directly is an extremely attractive alternative to complex (or any) sample preparation steps. Two approaches have been used, both to the same degree of success. The first approach involves the use of chemically modified stationary phases that allow plasma proteins to pass through the column largely unretained. Termed restricted access media (RAM), this material/approach has only recently been used as a sample clean up step prior to MS analysis [30]. In this application, analysis times were less than 5 min for a single target analyte. Another approach to direct injection involves the use of large particle size media in 1 mm internal diameter (ID) columns run at extremely high (3–4 mL/min) flow rates to create turbulent or semiturbulent flow. Under these conditions, proteinaceous materials from the sample solution pass rapidly through the column and are diverted to waste, while the analyte of interest is subsequently retained and eluted using gradient chromatography and MS detection [30–32]. Although advantageous from a sample preparation standpoint, both of these approaches allow a maximum of a few hundred injections before column performance and system pressure become problematic. Thus, direct injection technologies are best applied in the later stages of discovery, particularly where only a few hundred samples are to be assayed with less variation and better reproducibility.

8.3.3 Off-Line Sample Extraction

Off-line sample extraction methods, namely solid phase extraction, liquid–liquid extraction, and protein precipitation, are widely used in quantitative LC–MS/MS bioanalysis [33–47], as briefly discussed above. The multiple steps of the three modes of off-line extraction have been automated in 96-well format to different degrees using robotic liquid handlers, as described in details in an excellent book solely dedicated to bioanalytical sample preparation [34]. Using modular 96-well SPE plates populated with sorbent cartridges of different chemistries, Jemal et al. reported a strategy for the fast development of SPE procedures in a standardized, comprehensive manner [35]. Classical LLE methods, while considered to be less automation-friendly than SPE, may yield cleaner extracts as evidenced by less matrix effect and less tendency for clogging LC columns. A variant of LLE, solid-supported LLE, which is more automation-friendly, has also been used [34]. Protein precipitation, considered as a "dilute-and-shoot" technique, is one of the simplest sample preparation methods in bioanalysis. The other attraction to the protein precipitation technique is its universality since the same basic procedure can be applied to extract almost any analyte. However, the disadvantage of this technique is that the extract obtained is not as clean as that obtained with SPE and LLE as may be evidenced by the matrix effect and soiling of the LC system and mass spectrometer. Therefore, automated SPE, LLE, and on-line extraction (described below) are considered to be the preferred techniques, especially at a stage when a drug candidate enters the development phase, where sensitive, accurate, precise, and rugged bioanalytical methods are required for routine analyses of large batches of clinical samples.

8.3.4 On-Line Sample Extraction

On-line sample preparation, such as protein precipitation, is considered as another "dilute-and-shoot" method since there are no sample preparations except for aliquotting the samples, adding the internal standard, and centrifugation. However, the main difference between these two methodologies is that on-line sample extraction provides a cleaner extract with reduced chance for matrix effect and less soiling of the LC–MS system. The unique feature of probably the most commonly used on-line technique is the use of an on-line extraction column packed with large particles of a stationary phase material (typically $>20\,\mu m$), which is used in conjunction with a very high flow rate of the mobile phase (typically 3–5 mL/min). This translates to a very fast linear speed of 10–17 cm/s when a narrow bore column (typically 1 mm i.d.) is used. The combination of the fast flow and large particle sizes allows for the rapid removal of proteins with simultaneous retention of the small-molecule analyte of interest. The principles and numerous applications of this fast-flow on-line extraction technique have been presented in a number of publications [48–58], including an excellent review article [51]. Both homemade and commercially available on-line extraction systems have been used. Early on, a simple, homemade on-line system, based on a single extraction column for both extraction and analysis, was used [48, 51]. However, this approach gives little or no chromatographic separation. Then a system that incorporated an analytical column in-line with the extraction column was used to achieve adequate chromatographic separation [51, 52]. Further improvement to this type of homemade configuration was made by using two parallel extraction columns as well as two parallel analytical columns [53]. Large particle-size extraction columns commonly used include polymeric and silica-based sorbents, which work based on reversed phase, ion exchange, or mixed (combination of ion exchange and reversed phase) mode of separation. Fully integrated commercial systems are available which provide very desirable features such as the ability to achieve isocratic focusing while eluting from the extraction column in order to retain the eluted analyte as a sharp and well-defined plug at the head of the analytical column [54, 55]. Such a system can be used not only for low-level quantification of drugs [54] but also for identification of trace levels of metabolites in biological fluids via focusing multiple injections onto the analytical column prior to elution into the mass spectrometer [55].

A second category of extraction columns consists of columns that are packed with sorbent materials known as restricted access media [56–58]. RAM designates a family of sorbents that allows direct injection of biological fluids by limiting the accessibility of interaction sites within the pores to small molecules only. Macromolecules are excluded and interact only with the outer surface of the particle support coated with hydrophilic groups, which minimizes the adsorption of matrix proteins. RAM can be classified by the way of protein exclusion mechanism. Macromolecules can be excluded by a physical barrier due to the pore diameter or by a diffusion barrier created by a protein (or polymer) network at the outer surface of the particle. RAM can further be classified by subdividing RAM sorbents with respect to their surface chemistry. RAM-based on-line extractions have been used either in single-column mode, with the RAM column serving both as the extraction and analytical column, or in conjunction with a second analytical column. The main advantage of the one-column system is simplicity. Hsieh et al. reported the application of such a system using a single RAM column at a regular flow rate (1.0 mL/min for a column of 4.6×50 mm, 5 μm) for drug analysis in plasma [58]. A disadvantage of a one-column system is that only limited chromatographic separation of analytes can be achieved. On the other hand, a dual-column system consisting of one extraction column and one analytical column allows the achievement of superior chromatographic resolution by the use of analytical columns of various modes of chromatographic separation. For example, separation and quantification of enantiomers can be achieved by simply coupling a chiral column to the extraction column [59–61]. A recent review paper was published by Cassiano et al., who elaborated the RAM application for direct injection of biological samples [62]. The fundamental improvements in the preparation of tailored RAMs and diversity of applications with various phases were presented. Insights into diminishing the matrix effect by the use of RAM supports in methods by LC–MS/MS and into the low number of methods for enantiometric separations by direct injections of biological matrix samples are addressed.

Recently, on-line extraction LC–MS systems using monolithic alkyl-bonded silica rod columns for quantitative analysis have been reported [63–65]. Successful quantitative bioanalyses have been accomplished by using the monolithic column in a single-column mode, where the column serves as both the extraction and analytical column, or by using it as an analytical column in conjunction with a different extraction column. While monolithic column based on-line extraction looks promising, there is a need for more practical experience with this approach, and for the availability of more variety of monolithic columns, both in terms of column dimensions and stationary phase types.

8.4 STRATEGIES FOR ENHANCING MASS SPECTROMETRIC DETECTION

MS Instrumentation: The key to high-speed bioanalysis is the tandem mass spectrometers at the end of the instrumentation train. The high degree of selectivity afforded by the use of even a single mass analyzer allows compromises to be made in both the sample preparation and separation stages, which translate directly into time savings. In the past, these were the primary bottlenecks in the process. Modern instrumentation has been developed which is rugged, simple to use, and increasingly affordable. This section reviews the key elements of linear quadrupole-based instrumentation used in high-throughput bioanalysis.

Tandem Versus Single MS: When selecting MS instrumentation suitable for any given application, a subject worthy of serious thought concerns the use of tandem (MS/MS) or single (MS) quadrupole instruments for high-throughput sample analysis. Other instrumentation options exist (time-of-flight and/or ion traps instruments), but they are relatively new and untried for high-throughput applications and are beyond the scope of this review. While the single versus tandem question is sometimes debated, the appropriate choice is dependent on the complexity of the sample matrix, the throughput and signal-to-noise ratio required in the samples of interest, and the budget available. Arguments for the use of single MS center on cost and ease of use. Some *in vitro* experiments such as the Caco-2 model generate relatively clean buffer samples with little endogenous interference and a single MS instrument is more than adequate for this type of application. However, for more complex sample matrices, the selectivity of MS/MS is often required to support high-throughput bioanalysis. The techniques necessary for the removal or separation of the interferences observed when using MS are usually more complex than the operation of more sophisticated MS/MS instrumentation. When supporting many different biological screens, the up-front cost of purchasing MS/MS instrumentation is minimal compared to the time spent developing strategies to overcome MS limitations. Thus, given adequate financial resources, the clear choice points to

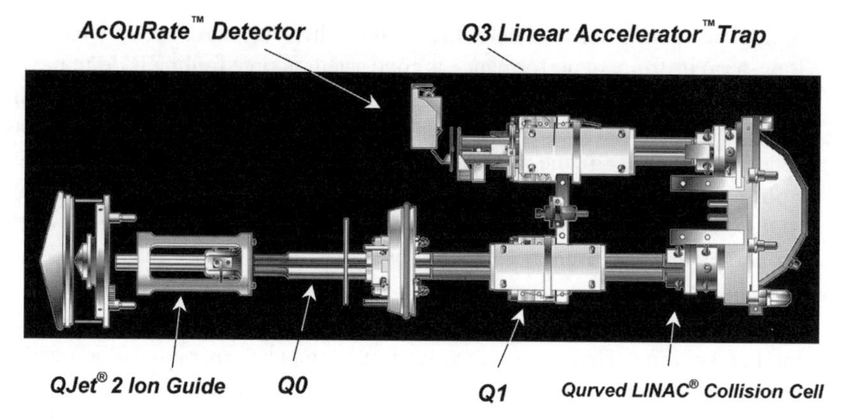

FIGURE 8.1 QTrap5500 ion rail with a curved Q2 collision cell (Courtesy Applied Biosystems).

the use of tandem MS instrumentation. This issue has been discussed in more detail elsewhere [66]. New advances in tandem MS/MS design provide users with more rugged and sensitive system and enable scientists to expedite bioanalysis with possible "direct and shoot" approach (Figure 8.1—API 5500 System).

Ionization Method: The two ionization methods most commonly used for LC–MS/MS are atmospheric pressure chemical ionization (APCI) and heated, pneumatically assisted ESI. These two atmospheric pressure ionization sources are both ideal for the introduction of chromatographic effluents and have been extensively reviewed [67, 68]. Of the two, ESI is the most widely used for high-throughput applications. ESI has several key characteristics that make it the technique of choice for applications involving the analysis of thousands of samples. First, a success rate (minimum height response for a neat standard injection) for structurally diverse compounds is generally somewhat higher than with APCI for the compounds assayed in bioanalytical laboratories ($n > 1000$). This fact is attributed to the softer ionization conditions in ESI, as more labile analytes often fragment and do not give a molecular ion in APCI. While only slightly more successful than APCI, discovery and development programs focused on a particular series may not be detected with adequate sensitivity using APCI, sharply limiting the utility of the method. It is important to note that the converse of this situation may also be true and that these effects may be compound specific. Second, ESI is a concentration-dependent detection method. While this means that small volume samples can be assayed with the same sensitivity as larger ones, the primary advantage in high-throughput analysis is that the sample stream can be split pre-source with no detection limit penalty. Thus, HPLC separations can be carried out at high (1 mL/min) flow rates and split down to readily volatilized flows before entering the MS source. In addition to enhanced detection success, the bulk of endogenous material from the biological sample does not enter the source. This has huge implications where long-term instrument performance is concerned. Source fouling is dramatically reduced, which enables detection sensitivity to remain far more stable over the course of a thousand sample analytical run. By contrast, the APCI source is mass sensitive, requiring all the material to enter the source for maximum sensitivity. Splitting an APCI flow stream will result in reduced sensitivity by a factor equal to the split, so source fouling becomes an issue. An excellent treatment of the properties of these sources is offered by Covey [69].

MS Method Optimization: In the past, MS/MS optimization required time-consuming sample infusion, parameter identification, and manual MS/MS method creation. This process involved many manual operations that resulted in sharply limited throughput. Typically, manual instrument tuning for a single compound requires approximately 10 min/compound. To address the challenge offered by the analysis of many individual compounds, approaches based on automated samplers and MS instrument advances have emerged which greatly streamline this process. Modern instruments allow for fast scan times and are able to cycle through different lens settings over the course of a flow injection peak, allowing for multiparameter data to be collected "on the fly." An example where this capability is used [70]. In this particular application, collision energy was cycled through increasing values, generating multiple analyte fragments. The optimum product/fragment ion pair and collision energy can then be selected and associated with a "generic" set of ion lens settings. Generic instrumentation settings differ only in collision energy and are chosen to give adequate sensitivity for diverse structures. Significant improvements in throughput are gained by avoiding extensive tuning procedures in a first pass. In most cases, this approach reduces instrument optimization time to approximately 60 s/compound. The cycle time for data collection promises to be further reduced with the advent of the multiple probe auto-sampler, which allows this task to be performed with extremely high-throughput [71].

8.4.1 Enhanced Mass Resolution

The use of a triple-quadrupole mass spectrometer with enhanced mass resolution capability has been investigated to increase the specificity and sensitivity of LC–MS/MS bioanalysis [72–75]. This is possible because of the unique feature of the mass spectrometer that allowed using a higher resolution setting without a significant decrease in response. Thus, the decrease in response was not more than 50% when the Q1 or Q3 resolution was changed from a unit mass resolution, with the mass peak full width at half its maximum height (FWHM) set at 0.70 Th, to a higher resolution, with FWHM set at 0.2 Th [69]. For a typical enhanced resolution LC–MS/MS analysis, Q1 FWHM was set at 0.2 Th and Q3 FWHM at 0.7 Th [72]. At concentrations near the limit of detection, cleaner chromatograms, with improved signal-to-noise (S/N) ratios, were obtained for plasma and urine samples when Q1 FWHM was set at 0.2 Th instead of 0.7 Th. Xu et al. [73] also reported that sensitivity for plasma assay was improved using the enhanced mass resolution mode.

Rudewicz et al. demonstrated the advantage of the enhanced resolution for mometasone analysis in the presence of polypropylene glycol (PPG) interference [75]. At unit mass resolution, the transmitted precursor ion from the first quadrupole contained not only protonated molecular ion from mometasone, but also a PPG interfering ion. At enhanced resolution, only selected mometasone ion was transmitted, and no interfering ion from PPG was detected. It was found that an enhanced resolution LC–MS/MS method requires more attention to detail than a unit resolution method [72]. For instance, the mass setting for precursor ion selection is more critical because the mass peak is narrower and hence can be affected by mass axis instability. Jemal and Ouyang proposed a set of system suitability procedures to ensure that the appropriate performance of an enhanced resolution method is maintained during the course of sample analysis [72]. It should be noted that maximum benefit in specificity and sensitivity is expected for analytes that have significant differences in mass defect relative to the elemental composition of compounds found in a biological matrix. Thus, analytes that contain one or more halogens, especially bromine and iodine, are expected to significantly benefit from the use of enhanced resolution.

8.4.2 Atmospheric Pressure Photoionization

The performance of the atmospheric pressure photoionization (APPI) technique was evaluated against several sets of standards and drug-like compounds and compared to atmospheric pressure chemical ionization and electrospray ionization (ESI). This analysis shows that APPI is a valuable tool for day-to-day usage in a pharmaceutical company setting because it is able to successfully ionize more compounds, with greater structural diversity, than the other two ionization techniques. Consequently, APPI could be considered a more universal ionization method, and therefore has great potential in high-throughput drug discovery and development applications. Atmospheric pressure photoionization serves as a complement to the more established electrospray ionization and atmospheric pressure chemical ionization techniques by expanding the range of compounds that can be analyzed [76–80]. Ionization is initiated by photons produced by a vacuum-ultraviolet lamp via a direct mode or dopant-assisted mode. The analyte molecular ion generated could be a radical ion, or protonated ion, as well as deprotonated ion. The mechanism of APPI has been described in a review article [76] and other publications [77, 78]. In the dopant-assisted mode, a large excess of a dopant, such as toluene or acetone, is added. The dopant first undergoes photoionization and then acts as a charge carrier for subsequent ionizing trace levels of the analyte. It appears that the extent of the effect of the addition of dopant on analyte ionization and hence sensitivity is dependent on the specific APPI source configuration used [78], as well as on whether a reversed phase or a normal phase mobile phase is used [79]. Because the APPI mechanism is closer to that of APCI than ESI, it has often been compared for sensitivity with APCI using different model compounds. Syage reported that APPI achieves significantly better sensitivity than APCI at flow rates below 200 μL/min [78]. They also reported that APPI is less susceptible to ion suppression than APCI and ESI. The major benefit of APPI, compared to APCI and ESI, is its ability to ionize broad classes of nonpolar compounds. APPI–LC–MS/MS has been successfully applied for the quantitative determination of drugs in biological matrices [79, 80].

8.4.3 High-Field Asymmetric Waveform Ion Mobility Spectrometry

An emerging technique known as high-field asymmetric waveform ion mobility spectrometry (FAIMS) has been demonstrated to reduce chemical noise, enhance analyte detection, and eliminate some interfering metabolites and prodrugs in LC–MS/MS bioanalysis [81–84]. FAIMS separation is based on the differences in ion mobility at high versus low electric fields, and the separation occurs in an atmospheric pressure gas-phase environment. The FAIMS device is located after the sprayer between the ion source and the orifice. Thus, the FAIMS device, when used in conjunction with LC–MS/MS, acts as a post-column, pre-MS ion filter in which only a subset of the ions formed are transmitted. Kapron et al. [82] demonstrated the advantage of using FAIMS in a situation where the selected reaction monitoring (SRM) transition used for a drug exhibited interference due to in-source conversion of its N-oxide metabolite to generate the same molecular ion as the drug. The FAIMS device removed the metabolite ion before entrance to the mass spectrometer. Thibault et al. [84] reported that the combination of FAIMS with nano LC–MS improved peptide analysis sensitivity by reducing the chemical noise associated with singly charged ions and increasing the S/N ratio by 6- to 12-folds. Review of applications of high-field asymmetric waveform ion mobility spectrometry and differential mobility spectrometry (DMS) was recently presented by Kolakowski and Mester [85]. High-field asymmetric waveform ion mobility spectrometry and differential mobility spectrometry harness differences in ion mobility in low and high electric fields to achieve a gas-phase separation of ions at atmospheric pressure. This separation is orthogonal to either chromatographic or mass spectrometric separation, thereby increasing the selectivity and specificity of analysis. The orthogonality of separation, which in some cases may obviate chromatographic separation, can be used to differentiate isomers, to reduce background, to resolve isobaric species, and to improve signal-to-noise ratios by selective ion transmission. Hatsis and Kapron published a recent review on the application of FAIMS in drug discovery [86]. The applications of ion mobility methods (e.g., ion mobility spectrometry (IMS), DMS, and FAIMS) to quality assurance and process monitoring by the pharmaceutical industry have been burgeoning. Specifically, the uses of IMS and FAIMS for cleaning verification of manufacturing equipment, direct analysis of formulations, and maintenance of worker health and safety can help save money and increase overall efficiency of production. This review discussed these ion mobility methods currently employed in pharmaceutical companies and research that could further these methods [87].

8.4.4 Electron Capture Atmospheric Pressure Chemical Ionization

Under the conditions normally used for APCI, it has been shown that suitable compounds can undergo dissociative or nondissociative electron capture due to the presence of low-energy thermal electrons arising from the interaction of the corona discharge with the nitrogen used in the APCI source [88–91]. This technique, which has been named electron capture APCI (EC-APCI), has been shown to increase sensitivity by two orders of magnitude for a number of compounds, compared with conventional APCI [88]. This is achieved by forming pentafluorobenzyl ether or ester derivatives, which undergo dissociative electron capture in the gas phase to generate negative ions through the loss of a pentafluorobenzyl radical. A number of nitroaromatic compounds were investigated for their electron-capture behavior and found to undergo either dissociative or nondissociative electron capture depending on the structure of the analytes [89]. Ultrahigh sensitivity was achieved by tagging neutral steroids with electron-capturing moieties [90]. This was achieved by using boronic acid and hydrazine derivatives having electron-capturing moieties as derivatization reagents for steroids containing 1,2-diol and carbonyl groups, respectively. LC/EC–APCI–MS methodology was developed and used to analyze some low polarity compounds. In recent studies, an EC mechanism was found to be highly efficient in the analysis of fullerenes and perfluorinated compounds when dopant-assisted negative ion-APPI was employed. In addition, a dissociative EC mechanism was also found to be highly efficient in the analysis of pentafluorobenzyl derivatives. These results suggest that LC/EC–APPI–MS may permit the analysis of nonvolatile, low polarity compounds with the sensitivity and selectivity that GC/EC–NICI–MS delivers for volatile nonpolar compounds.

8.4.5 Mobile Phase Optimization for Improved Detection and Quantitation

The effect of mobile phase on ESI efficiency is not well understood and hence the behavior of a compound under a set of LC mobile phase conditions cannot be routinely predicted. However, our practical knowledge is growing due to the active contributions by bioanalytical chemists, as evidenced by the wide coverage of a number of published papers in this arena [91–93]. In one such published study of negative ESI of a carboxylic acid compound, both formic acid and ammonium formate in a water/acetonitrile mobile phase decreased analyte response, but the ammonium formate caused the more severe decrease [91]. Acidification of the mobile phase with formic acid also had the added benefit of maintaining a reasonably high retention factor (k) for the analyte even at a relatively high acetonitrile concentration. A concentration of 1 mM formic acid in the mobile phase was found to be optimal as it achieved reproducibility, elongated retention time with only 60% loss in response. Dalton et al. investigated the effects of various mobile phase modifiers on the negative ESI response of several model compounds (all without carboxylic acid or any other strongly acidic group). They found that acetic, propionic, and butyric acid at low concentrations improved the responses of the analytes to varying degrees [92]. On the other hand, formic acid decreased response, as did neutral salts (ammonium formate and ammonium acetate) and bases (ammonium hydroxide and triethylamine). In another published study, a dramatic difference in the positive ESI response was found when acetonitrile was substituted by methanol in the mobile phase [92]. The compound in a water/acetonitrile mobile phase gave only a weak ESI response in the positive ion mode with formic acid and/or ammonium acetate. The compound in a water/methanol mobile phase, in contrast, gave a significantly higher response, approximately 25-fold, with formic acid and/or ammonium acetate. On the other hand, for the same analyte, acetonitrile and methanol mobile phases gave about the same response in the negative ion mode.

8.4.6 Anionic and Cationic Adducts as Analytical Precursor Ions

Cole et al. undertook a systematic investigation of attachment of small anions, such as halides, to neutral molecules as a means of increasing the negative ion ESI response, and presented a rationale for the formation and stability of preferred anionic adducts, $M + X^-$, of the compounds studied [94, 95]. Kumar et al. extended the investigation using different model compounds and showed that the collision-induced dissociation (CID) mass spectra of $M + X^-$ ions reflected the gas-phase basicities of both the halide ion and $M - H^-$ ion of the analyte [96]. Sheen and Her later reported an excellent application of fluoride adduct formation for the sensitive quantitation of neutral drugs in human plasma [97]. During method development, they found that the fluoride, chloride, and bromide adducts of the neutral drugs exhibited intense signals in negative ion ESI. Under CID, the major product ions of bromide and chloride adducts were the nonspecific bromide and chloride anions, respectively. In contrast, fluoride adducts produced strong $M - H^-$ ions, as well as the $M - H^-$ product ions with good intensity and reproducibility. Harvey reported on use of nitrate and other anionic adducts for the production of negative ion ESI spectra from N-linked carbohydrates [98].

Adduct formation has also been successfully used to increase ESI response in the positive ion mode. Zhao et al. evaluated the effect of mobile phases with buffers made from ammonium, hydrazine, or alkyl substituted ammonium acetate (70:30 acetonitrile-buffer, 2 mM, pH 4.5) on the positive ESI of simvastatin [99]. With the alkylammonium buffers, they showed that the alkylammonium-adducted simvastatin was observed as the only major molecular ion, while the formation of other adduct ions ($M + H^+$, $M + Na^+$, and $M + K^+$) was successfully suppressed. Methylammonium acetate provided the most favorable condition among all the buffers evaluated and improved the sensitivity several fold for the simvastatin LC–MS/MS quantitation compared with that obtained using ammonium acetate buffer. Similarly, other studies have shown that alkylammonium adducts can be used to improve analyte sensitivity by reducing the spread of the ionization of the analyte among multiple molecular ions [100–104].

8.4.7 Derivatization

Derivatization in LC–MS/MS bioanalysis has been employed to increase detection sensitivity, improve chromatographic retention or peak shape, facilitate sample clean up, and form a stable derivative for unstable analytes. Esterification with dansyl chloride of analytes containing phenolic hydroxyl groups has been found to increase the positive ESI responses of a number of compounds [105–109]. The dansylation reaction introduces ionizable basic nitrogen that enhances the ESI response. The

MS/MS of such a derivative produces a product ion at *m/z* 171, corresponding to the protonated 5-(dimethylamino) naphthalene moiety. LC–MS/MS bioanalysis based on this product ion was highly selective and sensitive for these compounds. Leavens et al. reported derivatization of alcohols using *S*-pentafluorophenyl *tris* (2,4,6-trimethoxyphenyl)phosphonium acetate bromide (TMPP-AcPFP), and derivatization of aldehydes and ketones using (4-hydrazino-4-oxobutyl) *tris* (2,4,6-trimethoxyphenyl) phosphonium bromide (TMPP-PrG) for LC–ESI/MS assay [110]. The TMPP acetyl ester and TMPP propyl hydrazone derivatives formed with their respective target analytes produced enhanced positive ESI response. *p*-Toluenesulfonyl isocyanate was reported as a novel derivatization reagent with strong nucleophilic reactivity for the derivatization of hydroxyl compounds in order to enhance detection by ESI [111].

Aldehydes and ketones were derivatized with 2,4-dinitrophenylhydrazine and detected by LC–APPI–MS [112]. After derivatization, the deprotonated molecular ions were the most abundant for the carbonyls in the negative ion mode. Since the derivative contained an easily ionizable sulfonylcarbamate moiety, it greatly improved the analyte's sensitivity in negative ESI. Evans et al. employed *tris* (2,4,6-trimethoxyphenyl)phosphonium propylamine bromide (TMPP) for the derivatization of carboxylic acid groups in fumaric, sorbic, maleic, and salicylic acids to facilitate their positive ESI LC–MS/MS assay [113]. The detection limits obtained were significantly better than those achieved for the underivatized acids using negative ESI. A paper describes a strategy for the isolation, derivatization, chromatographic separation, and detection of carnitine and acylcarnitines [114]. This is an excellent example of using derivatization of small polar molecules to enhance detection and chromatography and sample clean up. Derivatization of the two compounds was achieved using pentafluorophenacyl trifluoromethanesulfonate to form pentafluorophenacyl ester.

Sayre et al. used *N,N*-dimethyl-2,4-dinitro-5-fluorobenzylamine (DMDNFB) to derivatize amino acids for LC–ESI–MS analysis [115]. Compared to derivatization with 2,4-dinitrofluorobenzene (DNFB), the DMDNFB-derivatized amino acids exhibited larger positive ESI response, which was attributed to the introduction of the *N,N*-dimethylaminomethyl protonatable site. Derivatization with 7-fluoro-4-nitrobenzoxadiazole for enhancing capillary LC–ESI–MS/MS detection of biogenic amines was employed by Liu et al. [116]. Determination of amines in biological samples was achieved using pentafluoropropionic acid anhydride derivatives [117]. Derivatization with propionyl and benzoyl acid anhydride was employed as a means for improving reversed-phase separation and ESI responses of bases, ribosides, and intact nucleotides [118]. Due to the formation of the more hydrophobic derivatives, both retention and ESI response were increased. Benzoyl derivatization was used as a method to improve retention of hydrophilic peptides in tryptic peptide mapping [119]. Among the derivatizing reagent used, *N*-hydroxy succininamide sulfonyl benzoate was found to be a better reagent than benzoyl chloride since it was soluble in water and did not decompose in water. Yang et al. used 4-dimethylaminopyridine (DMAP) for trapping 4-fluorobezyl chloride in plasma samples prior to storage, which is very volatile and reactive and thus very unstable in human plasma [120]. The resulting derivative was a stable quaternary amine salt derivative with excellent positive ESI response. A review of different derivatization approaches for neutral steroids in order to enhance their detection in positive and negative ESI as well as APCI has recently been reported [121].

8.5 STRATEGIES FOR ENHANCING CHROMATOGRAPHY

Emerging Approaches in HT LC–MS/MS: While advances in analytical methodology have dramatically reduced analysis times, these methods are considered slow in comparison to some photometric assays which are often 100-fold faster than HT LC–MS/MS. Based on the continuing increase in the production of biological samples, the trend will always be to decrease the analysis time and increase sample throughput of the HPLC/MS system. Improvements in auto-sampler design have allowed for the simultaneous processing of up to eight samples at once, with simultaneous or staggered injection, dramatically reducing the time penalty arising from auto-sampler rinse and injection cycles. Elimination of delay due to liquid handling has reduced HT bioanalytical run times to approximately 15 s/sample. Given that this overall cycle time is likely to be close to what conventional LC–MS/MS can achieve. Further improvements in this arena will only be made through multiplexing techniques. The concept of fast serial sample introduction, where the use of MS analysis time is maximized with no "wasted" time has recently been introduced. This is accomplished by minimizing the dead time between eluting peaks when no measurement is being made. Such methods are realized either through the use of complex valve systems or parallel HPLC. In its simplest form, fast serial introduction is accomplished by linking two chromatographic systems to a single MS instrument and deconvolution of the data post-run [122]. This approach has been demonstrated for the unattended analysis of 20 individual compounds. Other approaches to fast serial sample introduction involve the use of column switching techniques. In this approach, a single HPLC system with dual columns operating with a multiple probe auto-sampler introduces analyte bands from each instrument in rapid succession, utilizing MS detection time during the equilibration phases of parallel separations. Thus, a set of chromatographic bands from the first HPLC system is being introduced while the second is equilibrating. Such approaches have been demonstrated to offer a great deal of increase in efficiency and speed. These techniques have only recently been presented, so no detailed reports have yet appeared in the literature. Through the introduction of multisprayer technology, the concept of parallel sample introduction has also begun to take shape in the form of multiple sprayer sources. To date, multisprayer sources have been used to examine ESI mechanisms [123], in improvement of detection sensitivity for high flow rates [124], and for calibrant cointroduction [125].

Applications directed toward throughput improvements have only recently emerged and not many quantitative applications have yet appeared to our knowledge.

To date, the only published application of multisprayer MS is in support of combinatorial chemistry. In this application, four HPLC streams were monitored with a single time-of-flight instrument interfaced with a four-sprayer source [126]. Termed a MUX source, this device samples each flow stream in sequence, while summing the data for each channel. The MUX source allows independent chromatographic optimization of each channel and has been demonstrated to have minimal intrachannel contamination. More detailed information on this emerging technique is available from Micromass (Waters' Corporation, Manchester, UK). Multisprayer MS sources offer two principal advantages. The first, most obvious advantage is to increase throughput. Two, four, and eight sprayers have been used in conjunction with a single MS instrument to increase the throughput of compound characterization in combinatorial chemistry. Application of multisprayer sources to the analysis of biological samples has only just recently begun and has the potential for increasing throughput for this application. Second, multisprayer sources avoid dilution and ion suppression effects that can be observed when mixing samples prior to analysis. As these sources operate by introducing two to four separate flow streams into an MS source, these issues are avoided by Hiller et al. (unpublished results).

8.5.1 Ultra-Performance Chromatography

According to chromatographic theory, the use of small-particle HPLC packings provides great benefits in reducing chromatographic time with minimum sacrifice in resolution [127–130]. Thus, reducing the particle diameter from 5.0 μm to 1.7 μm will, in principle, result in a threefold increase in efficiency, a 1.7-fold increase in resolution, a 1.7-fold increase in sensitivity, and a threefold increase in speed [130]. The other advantage of small-particle packings is the flatter nature of the van Deemter plot of linear velocity versus height equivalent to a theoretical plate (HETP), which allows analysis at flow rates much higher than the optimum with minimal loss in efficiency. However, decrease in the particle size from 5.0 μm to 1.7 μm results in a large increase in the backpressure. Thus, there is a need for an appropriately designed chromatographic system that would withstand such a high backpressure and also provide the least possible extra-column effects. The performance of one such ultra-performance system has been reported using a rat bile sample *vis-à-vis* a conventional HPLC system [130]. The comparison involved a 1.7 μm column (100 × 2.1 mm) in conjunction with an ultra-performance system against a 3.5 μm column (100 × 2.1 mm) in conjunction with a conventional system, using the same gradient time of 30 min and same flow rate of 400 μL/min. The peak widths generated on the ultra-performance system were of the order of 6 s at the base, giving a separation peak capacity of 300, more than twice that produced by the conventional system. The availability of ultra-performance systems should, in principle, allow analysts/bioanalysts to use the newly achievable chromatographic resolving power per unit time to either reduce the chromatographic time or improve the resolution by maintaining the same chromatographic time. For information on the benefits of using small particle packings in conjunction with shorter column lengths using conventional chromatographic systems, the readers are referred to an excellent book and review articles [76, 128]. Monolithic columns, although limited in scope at the present time due to unavailability of columns with desirable dimensions, may play a role in enhancing the all-important chromatographic resolving power per unit time [6, 130]. A recent application of UPLC–MS/MS was published by Tettey-Amlalo and Kanfer [129]. Typically, the concentrations of drug collected in dialysates are very low, generally in the nanograms/milliliters or even picograms/milliliters range. An additional challenge is the very low volume of sample collected at each collection time and which can range from 1 to 30 μL only. A rapid, accurate, precise, reproducible, and highly sensitive UPLC–MS/MS method was developed and validated for the quantitative analysis of ketoprofen (KET) in dialystes following the application of a topical gel product to the skin of human subjects. The Acquity-trade mark UPLC BEH C_{18} column (100 mm × 2.1 mm i.d., 1.7 μm) was used and KET was separated and analyzed in negative-ion (NI) electrospray ionization mode. The mobile phase (MP) consisted of acetonitrile:methanol:water (60:20:20, *v/v/v*) under isocratic conditions at a flow rate of 0.3 mL/min.

8.5.2 Hydrophilic Interaction Chromatography for Polar Analytes

The HILIC mode of separation is used intensively for separation of some biomolecules by differences in polarity differences, organic and some inorganic molecules. Its utility has increased due to the simplified sample preparation for biological samples, when analyzing for metabolites, since the metabolic process generally results in the addition of polar groups to enhance elimination from the cellular tissue. For the detection of polar compounds with the use of electrospray-ionization mass spectrometry as a chromatographic detector, HILIC can offer a 10-fold increase in sensitivity over reversed-phase chromatography because the organic solvent is much more volatile. Retention of polar analytes in reversed-phase chromatography often requires a highly aqueous mobile phase to achieve retention, which can cause a number of problems such as the collapse of the stationary phase due to pore dewetting and decreased sensitivity in ESI detection. The result of phase collapse is that chromatography becomes more problematic with retention loss, retention irreproducibility, increased tailing, and long gradient regeneration times [130]. Hydrophilic interaction chromatography (HILIC) is a mode of chromatography that addresses these problems with chromatography of polar analytes [131]. HILIC is conducted on polar stationary phases such as bare (nonbonded, underivatized) silica, amino, diol, and cyclodextrin-based packings [132, 133]. A high-organic, low-aqueous mobile phase is used to retain polar analytes. Retention is proportional to the polarity of the solute and inversely proportional to the polarity of the mobile phase. A number of LC–MS bioanalytical methods using HILIC with underivatized silica columns have been reported [134–136]. There are several retention mechanisms on underivatized silica. A combination of hydrophilic interaction, ion exchange, and reversed-phase retention results in unique selectivity. The HILIC mechanism involves partitioning between the adsorbed polar component of the mobile phase and the remaining hydrophobic component of the mobile phase. As the aqueous content of the mobile phase is increased, the observed retention time of the analyte will decrease due to the strong elution strength of water in HILIC mode. Thus, greater retention is achieved when the organic content of the mobile phase is high, typically greater than 70%. This high organic mobile phase is usually ideal for analyte desolvation and ionization, which leads to enhanced response in MS detection, compared with traditional reversed-phase methods. Because water is the strongest elution solvent in HILIC, samples dissolved in high organic content, such as samples extracted via SPE, can be injected directly onto the HILIC column, obviating the need for evaporation followed by reconstitution [134]. The backpressure experienced with HILIC using underivatized silica is significantly lower than that experienced with reversed-phase chromatography for the same column dimension and flow rates [137]. Hence, the use of higher flow rates is feasible, resulting in significant time saving.

8.5.3 Specialized Reversed-Phase Columns for Polar Analytes

To avoid the phase collapse that may be experienced during high-aqueous reversed-phase chromatography for polar compounds using silica-based alkyl-bonded phases, specialized phases have been developed, including polar-embedded and polar-endcapped phases [138–140]. These phases are modifications of classical alkyl-bonded silica chemistry (typically C_{18}) with the addition of a polar functional group, such as an amide or a carbamate group, within the alkyl chain itself, or with a polar functional group used as an endcapping agent. Compared with the use of the classical alkyl-bonded phases, these modifications prevent dewetting of the alkyl-bonded phases, provide increased retention for polar analytes, give good peak shapes, and allow higher sample throughput because of faster mobile phase reequilibration.

Another alternative for achieving retention and separation of polar analytes is the use of special packing material known as porous graphitic carbon (PGC) [141]. PGC generally provides markedly greater retention and selectivity for polar compounds than alkyl-bonded silica columns. Solvents and additives that are used for elution from PGC columns are similar to those used in traditional reversed-phase chromatography, with most of them being MS-compatible. Xing et al. used the PGC column to resolve very polar mixtures containing nucleosides and their mono-, di-, and triphosphates under conditions suitable for LC–MS/MS [141]. A water/acetonitrile mobile phase containing ammonium acetate and diethylamine as modifiers was used. The ammonium acetate proved to be critical for retention and diethylamine was found to improve the peak shapes of di- and triphosphates.

Utilizing the ultrahigh efficiency provided by core-shell technology columns KINETEX™ (Phenomenex, Inc.) has been emerging without heavy investment in "high-end" equipment. The performance of columns packed with the new 2.6 μm Kinetex-C_{18} shell particles was investigated in gradient elution chromatography and compared with those of the 2.7 μm Halo-C_{18} shell particles and the 1.7 μm BEH-C_{18} totally porous particles with comparability for ultrahigh efficiency chromatography. Numerous application notes have been incepted with great potentials in analytical/bioanalytical R&D programs in the years to come.

8.5.4 Ion-Pair Reversed-Phase Chromatography for Polar Analytes

Ion-pair reversed-phase chromatography using perfluorinated carboxylic acids, such as trifluoroacetic acid (TFA), pentafluoropropanoic acid (PFPA), and heptafluorobutanoic acid (HFBA), in the mobile phase has been found to be a viable alternative to straight reversed-phase chromatography for the quantitation of basic polar analytes by positive ion ESI [142–147]. The advantages of this technique include the elimination of peak tailing and the increase of the retention. However, the ESI signal may be decreased substantially, depending on the type and concentration of the ion-paring reagent used. The analyte retention generally increases with increasing hydrophobicity and concentration of the ion-pairing reagent. Apffel et al. conducted a systematic study of the effect of TFA on the positive ESI of peptides, and demonstrated that the signal suppressing effect of TFA is due to the combined effects of ion-paring, and surface tension and conductivity modifications [142]. The post-column addition of a propanoic acid-2-propanol (75:25, *v/v*) in a 1:2 proportion with the LC mobile phase counteracted the deleterious effects of TFA resulting in 10–100 improvement in S/N ratio. This "TFA fix" acted by competitively interfering with the ion-paring equilibrium between the analyte and TFA. On the other hand, in a study on the effect of TFA on in amphetamine positive ESI, Fuh et al. reported that the post-column addition of propanoic acid did not improve the ESI signal [142]. Ion pairing has also been used for the negative ESI quantification of acidic compounds, such as nucleoside mono-, di-, and triphosphates [148, 149]. Using *N,N*-dimethylhexylamine as an ion-pairing reagent, Tuytten et al. found that the retention of the analytes increased with the increase in the concentration of the reagent [148], as would be expected. However, the surprising finding was that the $M - H^-$ response also increased with the increase in the reagent concentration, which was not easily explainable.

8.6 POTENTIAL PITFALLS IN LC–MS/MS BIOANALYSIS

Although LC–MS/MS has increasingly become powerful tools in all phases of drug/life science product discovery and development, it has challenges and yet to be improved ever since. Carryover, matrix effect and potential interferences (intersource conversion and degradation, and possibly cross-talk), are probably among the most commonly encountered problems in LC–MS/MS method development and robust quantitative applications.

8.6.1 Interference from Metabolites or Prodrugs due to In-Source Conversion to Drug

Metabolites or prodrugs may produce, via in-source conversion, a molecular ion that is identical to the parent drug [150–159]. Consequently, the MS/MS transition adopted for the quantitative determination of a drug responds not only to the drug but also to its metabolite or prodrug. Thus, in the absence of adequate chromatography to separate the drug from the metabolite, the LC–MS/MS method is not specific to the drug. Typical metabolites that can potentially be formed from drugs of specified functional groups and undergo in-source conversion (Table 8.2). Additionally, isomeric metabolites, such as the *Z*-isomeric metabolite of an *E*-isomer drug [9], and epimeric metabolites [11] obviously have the same MS/MS transitions as their parent drugs. Inadequate chromatographic separation will result in over estimation of parent drug concentrations in the presence of these isomeric compounds.

8.6.2 Interference from Metabolites or Prodrugs due to Simultaneous $M + H^+$ and $M + NH_4^+$ Formation or Arising from Isotopic Distribution

8.6.2.1 Adduct Formation The adduct formation also adds to the complexity of quantitative LC–MS. Generally, ESI or APCI result in deprotonated $M - H$ molecules in the negative mode and protonated $M + H^+$ molecules in the positive ionization mode. However, several adduct ions such as $M + Na^+$, $M + K^+$, or $M + NH4^+$ were also reported in addition to $M + H^+$. Although the exact mechanism involved in adduct formation is not clearly understood, carboxyl or carbonyl ether or ester groups are believed to be responsible for binding the alkali metal ions. Sodium and potassium originate from the biological matrix or from the glass containers used. On the other hand, ammonium ions result from the addition of ammonium acetate or formate to mobile phases for the LC–MS determination of organic molecules.

Difficulties related to adduct formation arise mainly by developing a quantitative procedure because the adduct formation process is not reproducible and consequently, it is not clear what adduct ion can be used for multiple reaction monitoring. A first approach is to measure the adduct ion with the highest response. However, this results in very high (and unacceptable) analytical variations. Summation of all adduct ions can help but complicates MS/MS experiments and assumes an equal response factor for all adduct ions. A second approach involves the attempt to eliminate sodium from the ionization process, for example, by addition of alkali metal complexation products (crown ethers) and by the use of ultra pure deionized water. This is quite laborious due to the ubiquitous presence of sodium, often originating from the glassware, or as an impurity in chemicals and solvents, or due to its presence in the analyzed sample itself. The opposite approach, for example, the addition of sodium acetate to the mobile phase to enhance the formation of sodium adducts has also been described. However, due to the nonvolatile character this is not advisable. Other successful attempts to replace all adducts by one single desired adduct in view of sensitivity and reproducibility have also been reported. In this way reproducible formation of mainly one ion was achieved with the addition of dodecylamine to the eluent (in the case of paclitaxel) [157].

Addition of ammonium acetate results in $M + NH4^+$. The first step in the fragmentation of $M + NH4^+$ is the loss of the neutral NH3 or of the neutral alkylamine (in the case of alkylammonium adducts) with the formation of $M + H^+$. The latter can then fragment further. On the other hand, sodium adduct ions are much

TABLE 8.2 Metabolite or Prodrug Interference due to In-Source Conversion for Drugs of Different Functional Groups

Drug	SRM from Drug	Metabolite or Prodrug	SRM from Metabolite or Prodrug	Reference
Amine	$[M + H]^+ \rightarrow P^{+a}$	*N*-glucuronide	$[M + H + 176]^+ \rightarrow [M + H]^{+b}$ $[M + H]^+ \rightarrow P^{+a}$	[158, 159]
Amine	$[M + H]^+ \rightarrow P^{+a}$	N-oxide	$[M + H + 16]^+ \rightarrow [M + H]^{+b}$ $[M + H]^+ \rightarrow P^{+a}$	[152, 153]
Carboxylic acid	*m/z* 441 → *m/z* 423[a]	Acylglucuronide	*m/z* 617 → *m/z* 441[b]	[150, 157]
E-isomer: Methyloxime	*m/z* 327 → *m/z* 97[a]	*Z*-isomer: Methyloxime	*m/z* 441 → *m/z* 423[a] *m/z* 327 → *m/z* 97[a]	[150]
Fosinoprilat	*m/z* 436 → *m/z* 390[a]	Fosinopril	*m/z* 564 → *m/z* 436[b] *m/z* 436 → *m/z* 390[a]	[154]
Hydroxyl OH or phenolic OH	$[M + H]^+ \rightarrow P^{+a}$	*O*-glucuronide	$[M + H + 176]^+ \rightarrow [M + H]^{+b}$ $[M + H]^+ \rightarrow P^{a}$	[151, 156, 159]
Hydroxyl OH or phenolic OH	$[M + H]^+ \rightarrow P^{+a}$	*O*-sulfate	$[M + H + 80]^+ \rightarrow [M + H]^{+b}$ $[M + H]^+ \rightarrow P^{+a}$	
Lactone	*m/z* 363 → *m/z* 285[a]	Carboxylic acid	*m/z* 381 → *m/z* 363[b]	[150]

Phenol (structure: R_{VII}, $R_{VII'}$, OH)	m/z 369 → m/z 229[a]	Prodrug (structure: R_{VII}, $R_{VII'}$, O, O, O, RII")	m/z 363 → m/z 285[a] m/z 646 → m/z 369[b]	[150]
γ or δ Hydroxy carboxylic acid	$[M + H]^+ \rightarrow P^{+a}$	Lactone	m/z 369 → m/z 229[a] $[M + H-18]^+ \rightarrow [M + H]^{+b}$	[155]
Simvastatin	m/z 419 → m/z 285[a]	Simvastatin acid	$[M + H]^+ \rightarrow P^{+a}$ m/z 437 → m/z 419[b] m/z 419 → m/z 285[a]	[155]
Thiol R_{IX}-SH	m/z 407 → m/z 280[a]	Disulfide R_{IX}-S-S-R_{IX}	m/z 813 → m/z 407[b] m/z 407 → m/z 280[a]	[150]

[a] SRM adopted for drugs.
[b] In-source conversion of metabolite or prodrug to the parent drug.

more stable and yield less fragments. This could point in view of sensitivity, however, the formation of sodium adduct ions is highly affected by the sodium content of a sample. In special situations, a chromatographic separation between a drug and its nonisomeric metabolite may be necessary even in the absence of in-source conversion in order to accurately quantitate the drug. One such situation involves a lactone drug and its open-ring hydroxy acid form (or vice versa). There is only one mass unit difference between the ammonium-adduct molecular ion ($M + NH_4^+$) of a lactone compound and the protonated molecular ion ($M + H^+$) of its open-ring hydroxy acid. This has implications for the accurate quantitation of a lactone drug in the presence of its hydroxy acid form, or vice versa. This was demonstrated by using simvastatin and simvastatin acid as model lactone and hydroxy acid compounds, respectively [158]. Under the LC–MS conditions used, simvastatin produced both $M + H^+$ and $M + NH_4^+$ ions, *m/z* 419 and *m/z* 436, respectively. Similarly, simvastatin acid produced *m/z* 437 and *m/z* 454. Under such conditions, the use of the protonated lactone as the precursor ion (*m/z* 419) for the accurate LC–MS/MS quantification of the lactone compound provides selectivity against the hydroxy acid, whereas the use of the ammonium-adduct lactone as the precursor ion (*m/z* 436) does not ensure such selectivity. With the use of the protonated lactone as the precursor ion, chromatographic separation from the acid is needed only if there is in-source lactonization. On the other hand, when the ammonium-adduct lactone is used as the precursor ion, chromatographic separation of the lactone from the acid would be required if a low Q1 mass resolution is used even if there is no in-source lactonization of the acid. This is due to the fact that the lactone ammonium-adduct (*m/z* 436) is only one mass unit lower than the protonated acid (*m/z* 437), which would require at least a unit mass resolution. For the quantification of hydroxy acid, the use of the ammonium-adduct acid as the precursor ion (*m/z* 454) provides selectivity against the lactone. Using ammonium-adduct acid as the precursor ion, chromatographic separation of the hydroxy acid from the lactone is needed only if there is in-source hydrolysis of the lactone. On the other hand, when the protonated acid is used as the precursor ion, chromatographic separation of the two compounds is required, no matter what the Q1 resolution is or whether or not there is in-source hydrolysis. This is due to the fact that the protonated hydroxy acid (*m/z* 437) is only one mass unit higher than the lactone ammonium-adduct (*m/z* 436), which causes interference due to the $M + 1$ isotopic contribution. It should be noted that ammonium-adduct ions may be formed, in addition to the protonated ions, even when the LC mobile phase does not contain known sources of the ammonium salts.

Another situation involves a primary amide drug ($RCONH_2 = R + 44$) and its acid metabolite ($RCOOH = R + 45$), where there is only one mass unit difference between the two compounds. Thus, unless there is chromatographic separation between the two, the $M + 1$ isotopic contribution from the molecular ion of the amide drug will interfere with quantification of the acid metabolite. Other drug–metabolite pairs that can present similar analytical pitfalls involve a primary alcohol drug ($RCH_2OH = R + 31$) and its aldehyde metabolite ($RCHO = R + 29$) or vice versa, and a secondary alcohol drug ($RCHOHR' = R + 30 + R'$) and its ketone metabolite ($RCOR' = R + 28 + R'$) or vice versa. Hence, there is a potential for drug–metabolite interference due to the M + 2 isotopic contribution, especially for analytes that contain Cl or Br. Another situation of interest is where a lactone drug or metabolite, in the presence of ammonium hydroxide that may be used in any step of sample preparation, may produce the acid, amide, or lactam form of the compound prior to or during LC–MS/MS. The potential interference of one form over another should be evaluated under the conditions used.

8.6.3 Pitfall in Analysis of Two Interconverting Analytes due to Inappropriate Method Design

For drugs and metabolites that are unstable, with one converting to the other, conditions used during sample extraction and other steps of analysis must be optimized in order to minimize such a conversion so that accurate quantification of the drug or the metabolite can be achieved. However, even the optimal conditions adopted may not totally prevent conversion of the one to the other. It is thus essential to appropriately design bioanalytical methods, with appropriate calibration standards and quality control (QC) samples, in order to minimize the effect of such a conversion on the accuracy and precision of the method. Jemal and Xia illustrated the significance of proper method design in quantification of two compounds that undergo interconversion [159]. The important feature of the method design is the use of the appropriate concentration ratios for the QC samples vis-à-vis the concentration ratios of calibration standards. It has been demonstrated that a method validated for the quantitation of the two interconverting analytes using calibration standards with 1:1 analyte concentrations and QC samples with the same 1:1 analyte concentrations can be used for accurate measurement of the analytes in post-dose samples only if such samples contain the two analytes in the 1:1 ratio. On the other hand, if a method is to be used for quantitating interconverting analytes in pharmacokinetic study samples, where analyte concentration ratios are expected to vary from sample to sample, QC samples that cover the entire spectrum of the composition of the post-dose samples should be incorporated during validation of the method.

8.6.4 Matrix Effect

It is desirable to start from the enhanced selectivity and sensitivity of the MS/MS detector and in view of higher sample throughput misconceptions evolved about speeding up or even elimination of the sample preparation and the chromatographic separation. However, in numerous reports matrix effects and adduct formation became a major concern. It was demonstrated that these two phenomena can compromise or even invalidate both qualitative and quantitative results. The performance of LC–MS/MS methods in ESI and APCI is somehow hampered by matrix effects, which result mostly in ion suppression and sometimes in ion enhancement. Matrix effect can be defined as any change in the ionization process of an analyte due to a coeluting compound. This can affect precision, sensitivity, and accuracy of an analytical procedure. In electrospray ionization, the ionization process is taking place in the liquid phase and matrix effects in ESI are due to a competition of matrix constituents and analyte molecules for access to the droplet surface and subsequent gas-phase emission. In addition, matrix constituents can also change eluent properties such as boiling point, surface tension, and viscosity, all factors known to affect the ionization process. In atmospheric pressure chemical ionization, the ionization process is taking place in the gas phase and especially the nonvolatile matrix constituents are thought to coprecipitate with the analyte of interest, thus influencing the ionization process.

Kebarle and Tang first observed the matrix effect phenomena in ESI where the analyte response decreased in the presence of other organic compounds [160]. Recent publications describe the mechanism of the matrix effects, methodologies to detect the phenomena, and biological sample preparation procedures to minimize the effects [161–183]. In ESI, droplets with a surface excess charge are created. Enke introduced a predictive model based on competition among the ions in the solution for the limited number of excess charge sites [161]. Thus, at low concentrations of the analyte, the response–concentration relation is linear. However, at higher concentrations, the response becomes independent of the analyte concentration but highly affected by the presence of other analytes [161]. Pan and McLuckey studied the effects of small cations on the positive ESI responses of proteins at low pH. They concluded that the extent to which ions concentrate on the droplet surface is the major factor in determining the ion suppression efficiency [162]. King et al. investigated the mechanism of ion suppression in ESI and found out that the gas-phase reaction leading to the loss of net charge on the analyte is not the important process that causes ion suppression [163]. However, the presence of nonvolatile solute is much more important since it changes the droplet solution properties.

A number of approaches to assess ion suppression or enhancement have been proposed in the literature. Among them, a post-column infusion of analyte of interest while injecting the blank matrix [164] has been widely used. Use of this system allows the analyst/bioanalyst to determine the extent of the effect of endogenous components in blank plasma on the analyte response as a function of chromatographic retention time. Ion suppression or enhancement is illustrated as a valley or a hill, respectively, in the otherwise flat response–time trace [171]. Kaufmann and Butcher reported the use of segmented post-column analyte addition to visualize and compensate signal suppression/enhancement [165]. Matuszewski et al. reported a practical approach for the quantitative assessment of the absolute and relative matrix effect [166] as part of formal validation of bioanalytical LC–MS/MS methods.

The main source of the commonly observed matrix effect in LC–MS/MS bioanalysis of plasma samples is believed to be the endogenous phospholipids and proteins found in plasma [167–169]. As a zwitterion, phosphatidylcholines, a class of phospholipid molecular species, can cause ion suppression in both positive and negative ESI modes. In addition to the endogenous components in biological samples, dosing vehicles, such as PEG400, propylene glycol, Tween 80, hydroxypropyl-β-cyclodextrin, and *N,N*-dimethylacetamide, have also been found to cause "matrix effect" to varying degrees [170, 171]. The post-dose samples, especially the early time-point samples, may contain high concentrations of these dosing vehicles. Thus, the drug concentrations in such samples could be underestimated if lower analyte response is obtained due to the matrix effect caused by the dosing vehicle. Another potential source of "matrix effect" is the containers used for sample storage or sample processing [172, 173]. The use of selective sample extraction, such as LLE and SPE, adequate LC separation, and appropriate ionization mode (negative ESI or APCI) can minimize the matrix effect. It is very important to recognize that matrix effect can also be caused by coeluting internal standards, including stable isotope-labeled analogs [174, 175]. This phenomenon can be explained by competition between the analyte and internal standard for the droplet surface charge [161, 176]. It should be noted that analyte suppression is flow rate dependent and is practically absent at flow rates below 20 nL/min [177].

In the FDA guidelines for industry: bioanalytical method validation the following is advised: "*. . . In the case of LC–MS/MS based procedures, appropriate steps should be taken to ensure the lack of matrix effect throughout the application of the method.*" However, it is not stated how to evaluate the presence of matrix effect, neither how to eliminate this. Compared to the reference signal of the model analyte (in mobile phase), the trace levels of analytes from plasma treated by protein precipitation (PPT) had shown matrix suppression by hydrophilic and hydrophobic compounds. Solid phase extraction can eliminate the matrix suppression associated with salts, but still suffers from suppression by hydrophobic compounds. First of all, a reduction of the amount of matrix constituents injected can help. This can be done by injecting a smaller sample volume (with subsequent loss of sensitivity) or by applying more selective extraction techniques, for example, based on ion exchange or immunoaffinity chromatography. Second, coelution of the analyte and matrix constituents should be avoided. For this, optimization of the mobile phase can help, but the separation capacity of a liquid chromatographic system is limited as compared to a capillary GC-column. Third, the use of matrix-matched calibrators is absolutely essential but also this cannot compensate completely for each individual sample [178]. The fourth strategy, for example, the use of coeluting internal standard(s) seems ideal because it is expected that the matrix effect on the analyte and on the IS is identical. However, as already mentioned, a very high level of the analyte suppresses the ionization of the coeluting internal standard, and in addition, for multicomponent analysis labeled internal standards are not always available or are cost prohibitive [179]. Less common strategies to minimize or compensate for matrix effect include the use of a nano-splitting device [180], echo-peak injection [181], continuous post-column infusion of the internal standard [182], and standard addition to each sample [183]. The latter procedure compensates for matrix effect in each individual sample but is very labor intensive for use in routine analysis. The influence of the mobile phase composition on ionization efficiency is well known in LC–MS. However, mobile phase additives can also have an effect on matrix-induced ionization suppression or enhancement of an analyte [184]. In a LC–MS determination of endocrine disruptors in water samples we demonstrated that the addition of small amounts of ammonium formate to the HPLC eluent (1 mM) resulted in substantially better ME% (matrix effect) values (less suppression). On the other hand, higher levels or acids (formic acid, acetic acid) suppressed the signal [185].

8.7 TRENDS IN HIGH-THROUGHPUT QUANTITATION

8.7.1 System Throughput

The most fundamental consideration for any high-throughput method is that of system throughput. It is particularly appropriate to set some definitions and discuss this concept where HT LC–MS/MS is concerned, as there are fundamental differences between HT LC–MS/MS and traditional LC–MS/MS. A common misconception made in this field is that capacity and throughput is equivalent to the number of samples the auto-sampler can hold. While being fundamentally true, other considerations come in to question. First, compound diversity is an important measure. How many individual compounds can the system handle per unit time? Measuring one or two individual compounds in many samples is a completely different task from making a small number of measurements on 100 compounds. Each individual compound must have a set of conditions specific for it in terms of chromatography, ionization, and fragmentation pattern prior to sample analysis. Method development for a given analyte has long been the rate-limiting step in the process. Traditional MS/MS optimization techniques are inappropriate in the HT mode due to the time-consuming nature of the task. Most instrument manufacturers have recently incorporated into their software procedures that encompass scanning and building of MS/MS methods for each compound in an automated fashion. These procedures allow the operator flexibility in regards to the number of parameters to be optimized in the MS/MS method. Typically, generic lens conditions are applicable for most compounds entering the analysis stream, requiring only the identification of the parent and fragment ion for MS/MS analysis. While important in both discovery and development applications, the matrix and sample clean up procedures are particularly critical considerations when designing HT methodologies. The complexity of the matrix, compound recovery, and LLOQ are important factors in HT sample preparation methodology. Any measure of system throughput must take into consideration the number of individual compounds analyzed, the number of samples per individual compound, and the overall success rate of analysis. The highest efficiency of the system can be achieved by selecting generic methodologies that encompass the majority of compounds entering the HT analysis stream.

8.7.2 High-Speed HPLC

High-throughput applications have driven the separation step of the LC–MS/MS process toward reduced dimension column formats aimed at the generation of decreased analysis times. Where the purpose of more traditional HPLC is to cleanly resolve the analyte band from endogenous interferences, throughput demands in modern drug discovery and development have substantially altered requirements placed on the separation step. The selectivity of MS/MS largely eliminates the need for individually developed, compound-specific chromatographic separations, changing the focus of chromatography in high-throughput bioanalysis from resolution to speed. The use of HPLC sample introduction is particularly critical where electrospray ionization is utilized. For acceptable ionization, ESI requires the separation of the analyte of interest from matrix salts. Although separation from salt and other polar material is performed using such rapid chromatography, other endogenous material usually coelutes with the analyte. The suppression effects of these materials are typically observed as imprecision at the LLOQ and should be kept in mind when performing this type of fast chromatography [186].

The primary focus in regards to the separation step of the HPLC system is speed. Run times have been reduced from multiple (5–10) minute to subminute cycle times. This has been achieved by moving away from the standard 15 cm column length to those of 1–3 cm long. Further gains in speed are to be led by the use of reduced diameter columns operated at relatively high flow rates. Reduction of column size in both dimensions gives a linear reduction due to the length of the column bed, and allows high flow to be maintained without serious back pressure. For a given column length, halving the diameter while maintaining constant flow increases the linear velocity through the column by approximately fivefold. While this has a negative impact on resolution, the nature of high-speed quantitation allows this trade off of chromatographic performance for speed. Issues concerning the use of reduced dimension columns have been discussed elsewhere. The separation component of HT LC–MS/MS must be applicable to the diversity of compounds entering the analysis stream. The most effective means of achieving such "generic" chromatography is through the use of fast or ballistic gradient elution [187, 188]. Typically, this involves application of the sample to the column bed under aqueous conditions; after a few seconds a rapid or step gradient to highly organic is run. Systems based on this type of solvent, gradient elute structurally diverse analytes into a relatively narrow retention time window. While this generic feature is desirable, ionization suppression [186] and specificity for the analyte of interest [189] are of potential concerns to consider during method development and the establishment of reliable methodologies. While allowing application for many compounds, gradient elution also exacts a time toll in that the column bed must be regenerated after a run. Typical reequilibration time for a 10×2 mm ID column is approximately 30 s to allow 10 column volumes to thoroughly wash the packing material. As this represents a third (or more) of the total analysis time, this is a high price to pay. One particularly elegant means to circumvent this is the use of alternating column regeneration. The basic principle of the method is to employ two pumping systems to achieve regeneration of one column while the other proceeds with sample analysis. Thus, the time required for the run is the only factor in determining total run time. In some instances, the use of this technique can reduce the analysis time by up to 40%. In addition to the requirement of a relatively clean sample, the use of atmospheric pressure ionization (API) MS as a detector places limitation on the mobile phase composition. In general, mobile phase composition is limited to volatile buffers and additives that will not interfere with the ionization process. For example, mobile phase additives typically used in HPLC to improve chromatographic performance on silica-based columns (trifluoroacetic acid and triethylamine) have limited utility with MS detection due to potential ionization suppression and adduct formation [190]. For the most part, acetic acid, formic acid, and ammonium acetate are adequate for the quantitative analysis for vast majority of compounds.

8.8 TRENDS IN HYBRID COUPLING DETECTION TECHNIQUES

Quantitative analysis of drugs by a few powerful hybrid detection techniques, which can be enforced under diverse assay methods is one of the most important subjects in pharmaceutical analysis and also the primary means in quality evaluation of drugs. Complicated pretreatment that is not only time-consuming and cumbersome, but also specimen wasting and unacceptable for biological specimen. But it is needed and usually involves heating or digesting step before analysis of metal- or halogen-containing organic drugs. Although identification of drug metabolites can be executed through LC–MS, quantitative analysis of metabolites will be very difficult if it is without proper reference (usually there is no reference). Synthesis with radioactive labeling is another way for this problem, but it is money-costing and time-consuming, and not a good choice unless combining NMR (^{1}H, ^{19}F) for quantitative analysis.

ICP–MS not only offers high temperature (8000 K) ionization source for specimen provocation, but also overcomes the limitation of RI, UV, or MS for accurate quantification, which is concerned with the molecular structure of specimen. Even if the chemical structure or elemental composition is known, the response from these detectors is difficult to predict with any accuracy. In ICP–MS, compounds are atomized and ionized irrespective of the chemical structure incorporating the element of interest. Therefore, it is not necessary to choose the same or analogous molecular structure of reference as the specimen and one reference is enough when quantitative analysis is carried out with ICP–MS. Axelsson et al. [191] applied ICP–MS coupling with LC in generic detection for structurally noncorrelated organic pharmaceutical compounds with common elements such as phosphorus and iodine. They found that detection of selected elements gave a better quantification of tested "unknowns" than UV and organic mass spectrometric detection and did not introduce any measurable dead volume and preserves the separation efficiency of the system.

The wider linear range, lesser interference, higher analytical precision, shorter analytical time, and lower detection limits of ICP–MS provide enormous convenience for quantitative analysis of drug and its metabolites. One of these reports employing this approach was carried out by Nicholson et al. [192], which described the profiling and quantification of metabolites of 4-bromoaniline in rat urine. This was followed by a similar study [193] that provided the simultaneous detection of the metabolites of 2-bromo-4-trifluoromethyl-^{13}C-acetanilide in rat urine by ICP–MS. The metabolites present in the sample were separated by reversed-phase LC and introduced into ICP–MS instrument where bromine-containing metabolites were detected and quantified by ICP–MS.

8.9 TRENDS IN INTERNAL R&D AND EXTERNAL OUTSOURCING

Thanks to significant advances in discovery science, there is a dramatic increase in the number of new compounds moving from discovery into life science product development. The goal of nonclinical testing is to eliminate poor candidates while moving quality candidate drugs into clinical development as quickly as possible. At any given time, there are 3000–4000 compounds in nonclinical testing that have made the cut from discovery into development. Ongoing incremental improvements in the discovery and selection process continue to occur, which should result in a higher rate of success for compounds entering the expensive clinical development phase.

Given today's time pressures and the high cost of discovery and development, it is more important than ever that supporting function/services, such as bioanalysis, are optimized for speed and success. Bioanalysis, simply stated, is the quantitative measurement of an active drug in a biological matrix; it is required for evaluation of a therapeutic entity throughout the entire drug development cycle. In the early stages of discovery, lead selection/optimization and nonclinical development, the focus is on science: creating and refining analysis methods that will accurately and efficiently assess the drug level in a variety of species and biological matrices. As development moves from nonclinical to clinical investigations, the bioanalytical support required also undergoes a transition. In the later stages of development (Phases II–IV), the challenges of sample transport and handling from complex, multisite, often multi-country trials, along with substantial data management needs, add a requirement for logistical excellence in addition to scientific capability.

Bioanalytical sciences such as feasibility and method development support critical pharmaceutical evaluations, including toxicokinetics and pharmacokinetics, bioavailability, drug–drug interaction studies and other systematic assessments. Due to the sheer volume of work required in nonclinical and clinical product development, lead times must be carefully managed to keep projects on track. Today, many pharmaceutical and biotech companies utilize outsourcing to accomplish at least a portion of their bioanalytical needs. Outsourcing of bioanalytical projects may be done strategically or tactically and depends on business, financial and technical factors/considerations. Contract research may be described as a service sector within the scientific community that has silently contributed to modern day advancements in drug discovery and development. In generalized terms, contract research is simply the outsourcing of in-house research projects by one organization to another independent organization. Although contract research spans numerous scientific disciplines and industries, it is most commonly associated to the pharmaceutical and biotech (Pharma) industry. This industry in itself is subdivided into two main branches, those being the generic and brand name (innovator) sectors. The global number of contract research organizations (CROs) servicing the Pharma industry is always in constant flux as a result of mergers, closures, and hopeful new ventures. Never the less, one can safely say that this number steadily grows year after year. The overall quality of a current CRO has also improved dramatically over the past decade. What was once thought to be a European and North American niche market player has expanded globally. As a result, one can now find competitive CROs in other regions such as India and China. This reality will likely shape the future of contract research in the years to come.

Using outsourcing as a strategy helps manage the ebb and flow of projects entering or leaving product development due to toxicity, economics, or changing priorities. It may amortize the strengths of a particular outsourcing partner and provide additional scientific and operational excellence that can help provide information to make go/no-go decisions. Depending upon the specific type of project and compound, commercial bioanalytical laboratories can often quickly develop and validate scientifically robust methods, then smoothly integrate those methods into sample analysis within weeks or sometimes even days of the initial project inquiry. Project initiation may begin immediately for established, validated assays, depending on the laboratory's capacity. Bioanalytical outsourcing often provides sponsors with more flexibility, global reach and less investment in personnel and capital infrastructure, while expanding scope and capabilities that may not be available internally. The pharmaceutical industry has experienced a number of difficulties during recent years. Greater competition from generics, with over 60% of prescription drugs being supplied from the generic market and increased gaps in the drug pipeline resulting in acquisitions or strategic alliances has lead to an uncertainty in the bio/pharma market place. There has also been a change in the market place, with a shift from primary care to specialty drugs, the introduction of personalized medicine driving the need for biomarker/diagnostic technology and the introduction of biopharmaceutical product. With the pressures of time to market, the associated costs of discovery, the attrition rate from generics, patent expiry, and development in pharmaceutical products, pharmaceutical companies are increasing their investment in offshoring and outsourcing. Whether this is strategic or tactical outsourcing with contract research organizations, the CROs need to be proactive in responding to their sponsor's requirements for cost-effective services. For example, by offering a breadth of services, an innovative approach and a commitment to the sponsor in terms of understanding and sharing risk. As industry is constantly striving to achieve a reduction in development times, maintain quality, reduce development costs, and boost productivity, there has been an increase in outsourcing expenditure. Since early 2000s, the pharmaceutical industry has increased its financial commitment to contract clinical services by 16% annually to a staggering $6.6 billion, which is greater than the annual rate of growth in overall development costs (11%). Key reasons for outsourcing assessments are as follows:

- Strategic outsourcing allows pharmaceutical organizations to focus on its partners' core strengths and long-term goals. This route allows the sponsor to maximize the internal resource and capacities of the vendor adding value to the product development process. This also allows for a flexible service model.
- Tactical outsourcing relates to short-term projects, which cannot be handled internally due to the lack of internal capacity or resource allowing for streamlining processes, fast turnaround times, and access to additional equipment, and so on.

In general, company head counts have remained static and current challenges have altered the dynamics of the market with focus on R&D performance and strategy. With the influx of small to midsize biotech enterprises, pharmaceutical organizations have become reliant on the CROs for a full service infrastructure, experience and greater efficiencies they can provide, supporting late stage full development programs (Phase II through Phase III/IV). The criteria of choosing the right outsourcing partner have been exacerbated by the growth in the number of CRO's onto the market over the past several years. Often these organizations appear to offer similar services and returns for the customer thus making the partnering decision a bit difficult. With strategic outsourcing becoming the preferred outsourcing approach, more and more companies are looking to develop partnerships with CROs, however, this requires initial assessment, as discussed in Chapter 6 (Section 6.6). Important factors should include, the right corporate culture, an effective cost structure, sufficient and relevant technology including software infrastructure, and flexible human resources.

All of the above coupled with effective communication lead to the basis of a good CRO/Pharma partnership of which supporting elements such as bioanalysis (LC–MS/MS, LBAs, etc.) are increasingly being outsourced adding value in addition to scientific and operational capacities. Bioanalytical outsourcing adds a further level of complexity in managing a CRO but allows for added value such as rapid but regulatory compliant method validation. There are several critical factors outlining below, which should be considered in making the decision for partnering with a CRO:

- Quality and regulatory compliance.
- Scale and flexibility and capacity as well as in understanding potential delays and variations in projects.
- Commitment to the project and associated risk, a clear understanding and shared vision allows for the basis of growing an excellent partnership.
- Costing versus confidence in bioanalytical partner(s). Technical expertise, a breadth of experience in method development, validation, and sample analysis support is paramount.
- Communication, this is two-way, and allows for clear up-to-date project status and managing expectations.
- Key performance indicators, all projects vary in complexity, but performance and track record measuring processes, people, and the end result are key to understanding future project expectations.
- Contracts—clear goals and objectives established from the start of projects.
- Financial stability, ensuring your CRO partner is a reliable long-term resource during lengthy studies.

Careful evaluation and ultimate decision/selection are apparently important, since sponsors want to have reasonable return on investment (ROI) while shifting resources and expenditure externally. Risk and benefits are without doubt associated with this entire process. As a rule of thumb, five or more fundamentals need to be taken into consideration while pursuing outsourcing venue: quality, commitment, communications, balances of cost versus confidence, and lastly (but not least) managing expectations.

(1) *Quality:* Quality of work is of paramount importance to bioanalytical studies and is a measure of the systems in place within a CRO to facilitate timely delivery of high-quality data and reports. Important elements when

considering a CRO are what quality systems are in place and the principal elements should include the following:

- Whether the vendor's quality systems are adequate such as SOPs and accreditation to GLP
- A proactive inspection of the quality process, review of procedures and peer review steps, and adopting any changes in the regulatory environment
- The length of maintained accreditation and the level of sponsor audits and any feedback positive/negative to allow an informed decision
- Complete training records and an adequate training program therefore having a technically astute resource for projects.

(2) *Commitment:* With the time and financial commitment in planning and outsourcing clinical and supporting bioanalysis, commitment between the sponsor and CRO enables an effective partnership. Each project is individual and can vary in size but by implementing the following ideas; this can lead to an effective partnership.

- A shared vision—planning and forecasting future workloads
- A single point of contact within the CRO and continued project updates ensuring all parties are aware of changes in the projects and timelines
- On-going assessment of quality
- Having a shared understanding of the risk and sharing the successful outcome of a project
- Having a master services agreement in place and detailed contracts allowing the CRO to understand the exact requirements from the project outset.

(3) *Communication:* A structured and disciplined approach to communication can help deliver desired business benefits which should include the following:

- Frequent face-to-face meetings
- Effective kick off meetings stating clear expectations, milestones, and delivery targets
- Clearly define roles and responsibilities of the sponsor and CRO
- Tools, templates, and formal process for documenting sponsors expectations prior to the start of any project
- A recognized format for feedback through the course of a project, such as a project manager or status reports
- Review a project using CRO/ratings to allow timely sharing of knowledge and information to reenforce the relationship/partnership.

(4) *Balancing Cost and Confidence:* When faced with the prospect of outsourcing, you must be convinced of the viability and robustness of your decision to hand over control of a project to a third party. As the project manager, you must also be confident about getting value for money, quality, and delivery. Cost is often a key factor but is very rarely the most important. Sponsor will be

reluctant to spend even a small amount of money if not confident in the CRO's ability to deliver on time and to the right quality. Poor performance in achieving agreed targets can be very costly to both sponsor and CROs, and absorbs valuable time.

Indicators to good service provision are as follows:

- Appropriate quality standards
- CROs understanding of sponsors expectations
- The CRO analysts are enthusiastic and helpful thus indicating problems will be dealt with promptly and effectively
- Knowledge and competence of the staff, with continuous training
- Involvement of the staff in improving their knowledge and competence through networking and membership of scientific groups
- Presence of a robust training plan which includes both scientific and commercial development.

(5) *Managing Expectations through Contracts (Protocols):* Setting up contracts can sometimes be seen as a contentious issue and as a project manager may be unwilling to commit to a contractual agreement. There are advantages and disadvantages to contracts that need to be addressed. To manage expectations, it is necessary to define the types of "contract" required

- A Confidentiality Agreement is needed to ensure protection from fraudulent use of your intellectual property. Most pharmaceutical companies have a standard template that can be quickly issued to a CRO on request.
- A Technical Agreement detailing the nature of the testing and also the way in which both parties conduct business. Included in this are the official names and addresses of each company and details of quality standards, for example. There should also be a template for this available within most organizations.
- A Financial Agreement needs to be drawn up to define the payment schedules. Often the above can be seen as an obstructive series of hurdles delaying getting the work started and just additional work for the project manager. This does not have to be the case and a good CRO should take the initiative and make this process as smooth as possible by providing in-house templates and promptly signing and returning documents. Deliverables and timelines should be clearly defined within the contracts and communication associated with the project(s).

Like any business decision there are many contributing factors in the process of choosing laboratory partner(s) or collaborator(s). Collectively, they indicate the competence of an organization and more importantly, their willingness to move that little bit further to make your project a success. There will be regular dialogue between the CRO and sponsor before, during and after the project so it is extremely important that sponsor feel comfortable with the CRO and staff involved. The right combination of capacity, experience, quality, performance, communication, and other attributes must be carefully assessed for "ease of use" in the evaluation of a contract laboratory. The contracting process is also an important consideration and must be factored into the overall planning process and timelines. A vigilantly nurtured bioanalytical outsourcing relationship has many benefits, including expanded resources, enhanced communication, global reach, and the flexibility to accommodate time-critical product development. Fully understanding the scope, depth and breadth of what a commercial contract laboratory offers can provide the necessary information to make an informed decision and establish an outsourcing relationship that extends across time with a wide range of services for sponsor support. There are mainly three important elements for consideration: evolution, regulatory hurdles, and competitions in outsourcing business.

Evolution: CROs have had to evolve over the past years in an effort to survive and adapt to the competitive nature of the industry. In addition to the traditional research and development (R&D) services, CROs may also provide additional expertise and services such as regulatory compliance, product testing, manufacturing, statistics, clinical trials, analytical, bioanalytical, the list goes on. An estimated $14 billion is spent annually by the Pharma industry on outsourcing services provided by CRO's, globally (*Source*: Contract Pharma, May 2006) and this number is steadily growing. In fact, by 2010 more than 40% of Pharma R&D expenses are expected to be outsourced, up from 28% in 2006 (*Source*: Frost and Sullivan, 2005). How successful a CRO is depending entirely on how large a piece of that outsourcing pie it can grab?

Regulatory Hurdles: However, not just any organization can join this pie-grabbing contest. Due to the nature of the industry, all work realized for a Pharma company, with the exception of R&D, is governed by strict regulatory requirements. The main reason for the existence of these regulations is to guarantee the safety and efficacy of the drugs and thereby protecting the end users or consumers. Despite their extensive nature, the regulations are often summarized into the following categories: Good Laboratory Practices, Good Clinical Practices, Good Clinical Laboratory Practices, and current Good Manufacturing Practices (cGMP), each governing its specific areas of research. Each region and/or country has its own set of regulations and regulatory agencies, such as FDA (USA), HPFB (Canada), EMEA (Europe), ANVISA (Brazil) just to name a few. However, the FDA regulations and guidelines for laboratory, clinical and manufacturing practices are so well recognized that they form the foundation for many CRO standard operating procedures. Since Pharma work impacts human beings across the globe, international agencies have made some efforts to harmonize the regulations and guidelines governing this ever-growing industry. The International Conference on Harmonization is an on-going project between the regulatory authorities of Europe, Japan, and United States and experts from the Pharma industry in the three regions to discuss scientific and technical aspects of product registration. Although not directly involved, other agencies, such as Canadian regulatory authority, HPFB, observe and review all ICH discussions and documentations. Thus far, the ICH has produced several guidelines that harmonize those three region's regulations, however, to date they have not been completely adopted or accepted by all regulatory agencies, even by those involved in the ICH projects.

Competition: Regulations aside, the competition within the CRO service sector is especially fierce and challenging for existing and new CROs (or new sites opened by established CROs). Often there are many criteria (such as pricing, work quality, timelines, and communication efficiency) for the new CRO to satisfy before they are awarded any outsourced work. For the most part these criteria are predetermined by the regulatory agency to which the outsourced work will be submitted for approval. To a lesser extent, the Pharma companies outsourcing that work may add their own "flavor" to the criteria, especially to those associated to the procedures and conduct of their project. The existence of these concrete (agency) and subjective (client) criteria force many Pharma companies to adopt cautious attitudes when choosing a CRO. Some avoid the doubt altogether by mainly working with CROs that have been audited by a regulatory agency, preferably FDA.

These audits are often associated to specific studies submitted for review and approval to an agency and often include facility inspections at the various locations (clinical, analytical, etc.) where the study data were generated. Although most regulatory agencies do not provide any type of "certificate of approval" for a site upon completion of a successful inspection, simply having completed an agency audit often provides reassurance (although sometimes false) to the Pharma companies that the CRO has well-established procedures and methods. This criterion may be especially challenging for new CROs to satisfy. This reality is primarily due to the lengthy timeframes associated to the study's inception to final submission. Even if directly invited by a CRO, most regulatory agencies do not perform inspections of the CRO upon request as such it is very unlikely that any regulatory agency inspections would occur within the first few years of the CRO's lifespan.

The number of CROs has steadily increased over the past years. An estimated 88 CROs are currently providing laboratory analysis services in North America alone (*Source*: Biopharma Knowledge Publishing, The Contract Research Annual Review 2006). From the Pharma "consumer" perspective, a wider range of choice translates to lower costs for the same, if not better, quality of work. A Pharma company is no longer willing to pay premium dollar for the analysis of its projects and it also demands more aggressive study timelines and bullet-proof data quality. From the perspective of the CRO, however, this ability to "shop around" has made it increasingly more difficult to manage the operational costs against the shrinking revenues. The end result is a more fierce competitive environment between CROs, each struggling to differentiate themselves from the other through innovation, aggressive study quotes, added value services, and above all, quality. If the lifeline of a Pharma is its compound pipeline, then the same can be said for the method development schedule within a Bioanalytical CRO. The method development schedule outlines the current and projected analytical methods in R&D and/or validation. As with a Pharma's compound pipeline, this schedule is generally not divulged externally however, the analytical method list is widely (and enthusiastically) marketed. The analytical method list outlines all the validated methods immediately available at the Bioanalytical CRO without further R&D or validation. In this case, "bigger is better" since the more validated method assays a Bioanalytical CRO has available, the greater the likelihood that it may have an assay that a potential Pharma client may need. In short, an extensive analytical method list and a good reputation are the most effective ways for a Bioanalytical CRO to distinguish itself above and beyond others, especially when dealing with the generic sector.

The generic sector Pharma clients are primarily focused on bioequivalence (BE) or bioavailability (BA) studies. Therefore, their ideal goal is to have their generic equivalent approved by the regulatory agency as soon as the brand-name original is off patent, or as soon as possible thereafter. The clients are now more price conscious and demand short study timelines. When given the choice they would more likely outsource a study to a Bioanalytical CRO with an established validated method rather than incur additional costs or time due to R&D and/or validation prior to analysis. Consequently, to succeed with the generic sector the Bioanalytical CRO must be capable of anticipating their market trends, quickly determine where to focus their R&D efforts and revise their method development schedules accordingly. This type of philosophy toward managing their R&D is risky since the Bioanalytical CRO must divert resources (financial, personnel, etc.) from revenue driven activities to speculative work that may not provide immediate returns in the near future. Never the less, the Bioanalytical CRO that adopts this vision views this speculative work as an investment and stays focused directly on the generic Pharma needs. Although a strong method assay list may impress the generic sector, a Bioanalytical CRO must use a different approach when attempting to win contracts from the brand name (innovator) sector. These Pharma clients are primarily involved in proprietary new chemical entities under (or soon to be under) patent with little or no established analytical method information.

Furthermore, to outsource any type of work involves the release of vital confidential information to the CRO prior to product approval, a move that could be (potentially) detrimental to the Pharma's future. As a result, a Bioanalytical CRO wishing to deal with this sector must demonstrate scientific prowess, quality, and confidentiality. Reputation is the golden key. Of the three aforementioned qualities, scientific prowess is the most challenging to achieve. Although significant advancements in equipment, instrumentation, and automation have greatly improved the quality of the data and simplified many laboratory activities, the heart of a Bioanalytical CRO's scientific knowledge and capabilities lies within their personnel. Without experienced teams of scientists and technicians in place, the Bioanalytical CRO cannot hope to meet the R&D expectations of the brand name Pharma, even with the state-of-the-art equipment in place. The diversity in experience and education within its personnel further strengthens its capabilities to research and develop precise, accurate, and rugged methods for any new chemical entity, in any biological matrix. The secret to success is to focus on the team and not the individual. A few CROs with multisite bioanalytical facilities have taken this sense of R&D cooperation to a whole new level not attainable for other Bioanalytical CROs. Essentially, the multisite bioanalytical facilities are encouraged to communicate and discuss technical and scientific issues encountered at each site, share experiences and problem resolutions and harmonize and simplify procedures. In essence, they have created a science "think-tank" within the organization, independent of any one site, budget and above all any one "exceptional" scientist. The think-tank's goal is simply to better satisfy their client's needs and expectations, thus creating a marketable edge while dealing with the brand-name Pharma sector.

8.10 TRENDS IN LIGAND-BINDING ASSAYS AND LC–MS/MS FOR BIOMARKER ASSAY APPLICATIONS

Biomarkers are playing an increasingly important role in drug discovery and development from target identification and validation to clinical application, thereby making the overall process a more rational approach. The incorporation of biomarkers in drug development has clinical benefits that lie in the screening, diagnosing, or monitoring of the activity of diseases or in assessing therapeutic response. The development and validation of these mechanism-based biomarkers serve as novel surrogate end points in early-phase drug trials. This has created a much-appreciated environment for protein biomarker discovery efforts and the development of a biomarker pipeline that resembles the various phases of drug development. The components of the biomarker development process include discovery, qualification, verification, research assay optimization, clinical validation, and commercialization. The role of biomarkers in rational drug development has been a major focus of the Food and Drug Administration (FDA) critical path initiative and the National Institute of Health (NIH) roadmap. Although the overwhelming majority of biomarkers are proteins used as surrogate end points for drug development, diagnostic biomarkers may also prove useful for understanding the biology of the disease. Successful biomarker development depends on a series of pathway approach that originates from the discovery phase and culminates in the clinical validation of an appropriately targeted biomarker. New biomarkers can revolutionize both the development and use of therapeutics but are contingent on the establishment of a concrete validation process that addresses technology integration and method validation as well as regulatory pathways for efficient biomarker development.

A biomarker is a characteristic that is objectively measured and evaluated as an indicator of normal biological processes, pathogenic processes, or pharmacologic responses to a therapeutic agent [194]. Thus, biomarkers may enable more reliable and earlier detection of illness, enhance the selection of patients most likely to respond to targeted therapeutics and allow real-time monitoring or even prediction of efficacy to treatment. Throughout the long and costly cycle of drug development, biomarkers are seen as facilitating go/no go decision making by either accelerating the promotion of active compounds into man or rejecting early those compounds destined to fail [195]. In the future, biomarkers may pave the way for optimal therapeutic intervention on a case-by-case basis (personalized medicine), or identify those at risk of disease enabling early or preventative intervention. At least five different categories of biomarker assays have been recognized based on the level of quantitation inherent in the methodology ranging from: nominal (yes/no); ordinal (discrete, nonquantitative, arbitrary scores); quasi-quantitation; relative quantitation to absolute quantitation [196]. Thus, biomarker assays span across a wide diversity of technology platforms. The fit-for-purpose method validation is an umbrella terminology that is used in biomarker and ligand-binding assays to describe distinct stages of the validation process, including prevalidation, exploratory and advanced method validation, and in-study method validation. Method validation is thus a continuous and iterative process of assay refinement with validation criteria that is driven by the application of the biomarkers with increasing rigor at each successive validation step and focusing on method robustness, cross-validation, and documentation control. The need for a standardized pathway approach toward the biomarker validation process is becoming increasingly important given the recent surge in the biomarker development pipeline. Application of a biomarker in scientific decision making requires its validation for the given purpose. Validation of biomarkers should include two distinct activities. The first step is the technical validation of the assay procedure used for quantitative determination of the biomarker. The second step is the biological/pharmacological validation of the biomarker's selectivity and sensitivity by establishing appropriate correlations between the measured biomarker concentration and target activity; physiological, pathological, and pharmacological observations; and/or disease state and progression rate as assessed by alternative valid methods. Regarding the validation of biomarker assays, position papers for the validation of immunoanalytical or ligand-binding assays have been published [196]. As liquid chromatography–tandem mass spectrometry (LC–MS/MS) gains increasing importance as an analytical technique for the quantification of biomarkers, there is a need to describe the validation approaches that are unique to endogenous analytes. Although validation guidelines from the regulatory agencies for the analysis of xenobiotics in biological matrices are available and widely accepted by the industry for LC–MS/MS assays, guidelines specific to biomarker LC–MS/MS assays are not available at this time. Therefore, special considerations need to be taken into account when validating LC–MS/MS-based chromatographic assays for endogenous analytes. In general, the considerations should take into account the selection of standard curve matrix, definition of the lower limit of quantitation, assay selectivity, and accuracy. Approaches for the validation of tissue-based biomarker assays using these considerations were published [197, 198].

Chromatographic techniques, such as LC, coupled to mass spectrometry (MS), or MS on its own, represents one of only a limited number of analytical platforms which can claim to offer absolute quantitation and is being increasingly utilized in the quantitative analysis of biomarkers. However, the technological challenges that LC–MS/MS faces in this arena remain formidable. Serum concentrations of potential biomarkers, such as proteins, can vary by a factor of 10^8–10^{10} between high abundance species such as albumin and classic biomarkers such as prostate-specific antigen (PSA). It is conceded by many that immunoassays still offer far greater sensitivity, reproducibility, and dynamic range than LC–MS/MS [199]. In Europe, when biomarker measurements are performed on samples collected from subjects entered into clinical trials, laboratories conducting these analyses are subject to the Clinical Trials Regulations, requiring the implementation of a full quality assurance (QA) system. In order to comply with the regulations, the biomarker assay also has to undergo extensive method validation, a whole science in itself [200]. Therefore, it was considered timely important to devote a special effort on the subject of quantitative analysis of biomarkers. It has been trending attempt to span the analytical spectrum from technology developments through biomarker discovery and development of validation methodologies to informatics. At the same time, it is also imperative to incorporate a range of applications in both *in vitro* and *in vivo* settings. Due to the nature of the subject material, method validation and clinical applications are intentionally prominently featured.

Bioanalysis has evolved considerably over the past 10 years with the rapid implementation of three key technology levers: *robotics* for automated sample extraction, fast liquid chromatography coupled to mass spectrometry for *high-throughput* trace analysis, and finally Laboratory Information Management Systems (LIMS) for seamless data processing and data interpretation.

8.11 TRENDS IN STUDY DESIGN AND EVALUATION RELATING TO BIOANALYSIS

With advancement of new technologies and techniques nowadays, new approaches and assays have been increasingly developed, validated, and established over the past several years. As a result, regulatory bodies have ever scrutinized emerging techniques and approaches along with new guidances for industry for consideration and implementation. Nonclinical safety studies to support the conduct of human clinical trials for pharmaceuticals has been internationally harmonized by the ICH as outlined in ICH Topic M3: Note for Guidance on Nonclinical Safety Studies for the Conduct of Human Clinical Trials for Pharmaceuticals, Topic S7B: Note for Guidance on Safety Pharmacology Studies for assessing the Potential for Delayed Venteicular Repolarization (QT Interval Prolongation) by Human Pharmaceuticals. However, different regional requirements still exist with regard to nonclinical studies to support the first dose in humans. In April 2005, the FDA released a draft guidance for Exploratory IND studies that clarifies preclinical and clinical approaches that should be considered when planning exploratory IND studies in humans. As part of FDA's "Critical Path Initiative," this process is a new tool available to the industry that enables a faster, more cost-effective path to early clinical development. A primary application of an Exploratory IND study is Microdosing or Phase 0 Clinical Trials. Microdosing studies permit collection of human pharmacokinetic (PK) and bioavailability data earlier in the drug development process. This human data is combined with preclinical data to select the best candidates to advance to further, more expensive, and extensive clinical development. Since microdosing studies are designed not to induce pharmacological effects, the potential risk to human subjects is very limited. Therefore, these studies can be initiated with less preclinical safety data, help reduce the number of human subjects needed and require fewer resources for selecting promising drugs candidates for further development. Microdosing offers a faster and potentially less expensive approach to obtaining human *in vivo* PK data in early clinical drug development. It encompasses the use of pharmacologically inactive doses of test drug in the low microgram range along with ultrasensitive assay methods (PET, AMS) to assess human exposure in order to extrapolate the PK of higher, clinically more relevant doses, assuming linear PK. This strategy allows early evaluation of systemic clearance, oral bioavailability as well as sources of intersubject variability and questions of specific metabolite formation. It does take advantage of reduced regulatory requirements of preclinical safety studies, bulk drug synthesis (CMC requirements) and easier formulation options, for example, as part of an exploratory IND.

Microdosing for Human Dose Prediction: In the current context, the term "microdose" is defined as less than 1/100th of the dose calculated to yield a pharmacological effect of the test substance based on primary pharmacodynamic data obtained *in vitro* and *in vivo* (typically doses in, or below, the low microgram range) and at a maximum dose of less than 100 μm [201]. An example of such a clinical trial is the early characterization of a substance's pharmacokinetic/distribution properties or receptor selectivity profile using positron emission tomography (PET) imaging, accelerator mass spectrometry (AMS), or other extra sensitive analytical techniques. Nevertheless, new analytical tools and approaches may be required to achieve ultimate goals of obtaining meaningful data for regulatory review and consideration. Exploratory IND studies are intended to provide clinical information for a new drug candidate at a much earlier phase of drug development. These studies help to identify the best candidates for continued development and eliminate those lacking promise. These clinical trials occur very early in Phase I, involve very limited human exposure, and have no therapeutic intent. Exploratory IND studies are conducted prior to the traditional dose escalation, safety, and tolerance studies and provide important information on pharmacokinetics and bioavailability of a candidate drug.

Accelerator Mass Spectrometry is an ultrasensitive bioanalytical platform capable of quantifying ^{14}C-labeled compounds with attomole (10^{-18} M) sensitivity. The exceptional sensitivity of AMS enables microdosing studies performed at subpharmacologic levels. Human microdosing was endorsed in 2003 by the European Medicines Evaluation Agency (EMA) in a Position Paper on the nonclinical safety studies to support human microdosing trials. EMEA defined a preclinical toxicology package that is substantially reduced from the ICH M3 package and the US Acute Toxicology Guidance traditionally used for Phase I studies, indicating the European Agency's comfort with the safety of human microdosing. The FDA is currently reviewing 21 CFR Part 361 as to whether or not new isotopically labeled new chemical entities destined for ultimate pharmaceutical use is covered by the Regulation.

Prospect of Phase 0 Study Design (Risks Versus Benefits): It will be some time before it will be known whether Phase 0 trials have had a positive impact on the development of new drugs for cancer or possibly other indications. It is anticipated that PD-driven studies will expedite the evaluation of those agents, which directly modulate their targets. Needless to say, conducting a Phase 0 trial does not delay clinical development or divert resources from Phase I investigations; rather, the reduced preclinical requirements for an exploratory IND enabled us to reach the Phase 0 study endpoints well before the first NCI-supported Phase I. Use of the lowest efficacious dose theoretically will allow an extra margin of safety for studies of new drug candidates. It also remains to be seen whether the conduct of Phase 0 trials will result in more drugs "failing faster" than is currently the case. It is clear that the exploratory IND allows great flexibility in clinical development. Patient safety is still paramount, but the emphasis of Phase 0 first-in-human testing is on a drug's target rather than its toxicity. Determining the MTD will still be required for further clinical development, but drugs that fail proof-of-principle target inhibition studies may be discarded before reaching formal Phase I/II evaluation.

Phase 0 trials are characterized by two critical determinants: (1) patients must be willing to participate in a clinical trial which offers no possibility of direct clinical benefit and (2) serial blood (and often tumor) samples for research purposes are required. Although risks to the patient participating in a Phase 0 trial are considered low because of the limited drug exposure, they are not negligible. Every effort must still be made to minimize potential risk by selecting appropriate patients and monitoring their safety. Ethical considerations extend to ensuring that biopsy samples are only collected once the PD and PK parameters established during preclinical testing are met, and that the blood and tumor biopsy samples collected are put to the best possible research use.

The second critical determinant for Phase 0 trials is the integration of a robust and reliable assay of agent activity into the development plan. In addition to informing further clinical testing, the assay introduces the possibility of identifying biomarkers of drug efficacy in surrogate tissues. Validating the assay prior to clinical testing to measure whether the agent is or is not having its anticipated effect is labor- and resource-intensive, factors that may influence the decision to conduct a Phase 0 rather than Phase I trial. The availability of a qualified PD assay, however, is a requirement for measuring drug effect on target in tumor tissue. The development of the exploratory IND Guidance has provided researchers with the opportunity to evaluate molecularly targeted agents in a context that facilitates clinical assay development; it is anticipated that this will lead to more rational and rapid evaluation of promising anticancer therapies.

Dried Blood Spot for Bioanalytical Methodologies: Dried blood spot (DBS) approach was utilized for biological sample collections as early as in 1960s. DBS had been reported to be a convenient matrix, readily amenable to measurement of biomarkers and drugs such as the antimalarials, in blood for both clinical and preclinical studies. However, it was not routinely used in bioanalytical laboratory for sample/specimen collection as part of sample analysis until recent years. A novel approach has been developed and incorporated for the quantitative determination of circulating drug concentrations in clinical studies using DBS on paper, rather than conventional plasma samples. A quantitative bioanalytical HPLC–MS/MS assay requiring small blood volumes (15 μL) has been validated using acetaminophen as a tool compound (range 25–5000 ng/mL human blood). The assay employed simple solvent extraction of a punch taken from the DBS sample, followed by reversed-phase HPLC separation, combined with selected reaction monitoring mass spectrometric detection. In addition to performing routine experiments to establish the validity of the assay to internationally accepted criteria (precision, accuracy, linearity, sensitivity, selectivity), a number of experiments were performed to specifically demonstrate the quality of the quantitative data generated using this novel sample format, namely, stability of the analyte and metabolites in whole human blood and in DBS samples; effect of the volume of blood spotted, the device used to spot the blood, or the temperature of blood spotted. The validated DBS approach was successfully applied to a clinical study and reported by Spooner et al. [202]. This work demonstrates the capability of DBS analysis to provide high quality TK information using significantly smaller volumes of blood than are conventionally required. Further, the reduction in blood volume leads to a decrease in animal numbers used, through serial rather than composite sampling regimes, giving ethical benefits and an increase in data quality. In addition, if continued through the life cycle of a new drug, the technology offers the advantage of simpler sample collection, storage and shipment for both preclinical and clinical study samples, leading to notable ethical and financial benefits. The success of this and similar, related studies has led to the intent to apply DBS technology as the recommended analytical approach for the assessment of PK/TK data for all new oral small molecule drug candidates, which have previously demonstrated a successful bioanalytical validation.

More recently, a review article was published by Li and Tse [203]. As a less invasive sampling method, DBS offers simpler sample collection and storage and easier transfer, with reduced infection risk of various pathogens, and requires a smaller blood volume. To date, DBS–LC–MS/MS has emerged as an important method for quantitative analysis of small molecules. Despite the increasing popularity of DBS–LC–MS/MS, the method has its limitations in assay sensitivity due to the small sample size. Sample quality is often a concern. Systematic assessment on the potential impact of various blood sample properties on accurate quantification of analyte of interest is necessary. Whereas most analytes may be stable on DBS, unstable compounds present another challenge for DBS as enzyme inhibitors cannot be conveniently mixed during sample collection. Improvements on the chemistry of DBS card are desirable. In addition to capturing many representative DBS–LC–MS/MS applications, this review highlights some important aspects of developing and validating a rugged DBS–LC–MS/MS method for quantitative analysis of small molecules along with DBS sample collection, processing, and storage. In the early stage of a DBS–LC–MS/MS method development, it is important to systematically assess the possible impact of hematocrit and blood volume, and possible uneven distribution of the analyte of interest in the blood spots. To ensure the required ruggedness and suitability of an intended DBS–LC–MS/MS assay method, quality control samples should be prepared from three individual blood lots with hematocrit values ranging from low to high in the three validation runs along with the calibration standards prepared from a blood lot with a medium hematocrit value from the intended population. Including six blood lots with various hematocrit values (from low to high) in the assay selectivity and sensitivity assessment is necessary. Whenever the impact of various blood properties becomes a concern, accurate blood spotting to DBS card/paper and cutting the whole spot for analysis should be employed. With a good quality control of blood sampling (procedure standardization and adhesiveness to the established procedure), implementation of automation for DBS sample punching and extraction, improved chemistry of pretreated card/paper for enhanced stability of unstable compounds, and a higher analysis throughput (<1 min) via UPLC, DBS–LC–MS/MS is expected to play an increasingly important role in the quantitative analysis of drugs and metabolites in blood samples.

8.12 TRENDS IN APPLYING GLP TO *IN VITRO* STUDIES IN SUPPORT OF REGULATORY SUBMISSIONS

In the mid-1970s and early 1980s, the US FDA, and Environmental Protection Agency (EPA) were discovering that the toxicology data submitted to them was not as reliable as they believed. As a result of these revelations, the US Senate gave power to the agencies to set forth guidelines for laboratories to follow while conducting studies intended for submission. These guidelines, which grew out of industry "best practices," became the Good Laboratory Practices (FDA 21 CFR 58, EPA TSCA 21 CFR 792, FIFRA 21 CFR 160). Many new versions and revisions to the GLPs have been published since the 1970s, so that now, studies conducted in compliance with GLP regulations are recognized throughout the member countries of the OECD and elsewhere as promoting good science and yielding trustworthy data. The early versions of the GLPs were written with long-term, animal-based toxicology studies in mind. Today, many of these animal-based studies are being replaced by *in vitro* studies, which are being used by many companies to make safety decisions without accompanying animal data. A review paper was published to illustrate quantitative prediction of *in vivo* drug clearance and drug interaction from *in vitro* data on metabolism, together with binding and transport [204]. In order to assure that the data from these validated *in vitro* methods are universally accepted as being of high quality, the studies must be carried out in compliance with Good Laboratory Practices. Confusion in the interpretation of the regulations arises when the traditional GLPs are applied to short-term, nonanimal studies. Differences in test system handling and basic assay design require that the GLPs be supplemented with guidelines that fit the unique structure of these *in vitro* assays. Workshops were sponsored by such organizations as the European Center for the Validation of Alternative Methods (ECVAM) and IIVS where interested individuals (such as cell culture technologists, toxicologists, and quality assurance personnel from industry, academia, and government) were able to discuss the possible design of GLPs for these methods. Following these consensus building workshops, the OECD published "The Application of the GLP Principles to Short Term Studies" (1993, rev. 1999), Number 7 in their series on the Principles of GLP and Compliance Monitoring. This document was followed (November 2004) by the OECD advisory document titled "The Application of the Principles of GLP to *In Vitro* Studies" (Number 14). These documents supplement the GLPs to effectively assure that there are GLP guidelines to cover all critical aspects of *in vitro* assay design.

The monitoring of *in vitro* GLP studies requires the auditors to interpret regulations and guidance documents within a new framework. Emphasis in auditing should be placed on those areas of the *in vitro* laboratory and assay systems that are critical to control when working with *in vitro* systems. Critical phases are identified prior to performing the study based on the protocol design. Some examples of phases that would be considered "critical" are: test system receipt and handling, test system preparation, application of the test material to the test system, and collection of data. During the in-life audits of the study, auditors should review the documentation (e.g., laboratory notebook or workbook), equipment used, personnel training, and the actual performance of the chosen step. The results of the audit would be circulated to the Study Director and Facility Management. All elements of documentation combined should provide a full explanation of the conduct of the assay without the need for additional clarification. The final report should accurately reflect the conduct of the assays without any of the data gaps, fabrications, or creative penmanship that led to the establishment of the GLPs in the late 1970s. The steady increase in industry use and regulatory acceptance of *in vitro* test methods has resulted in an increased need to apply Good Laboratory Practice regulations to these systems. The original GLP regulations, developed to address the conduct of animal studies, are concerned with many special conditions that apply to animal housing and care, and the relatively long duration of animal studies that are not present in the shorter *in vitro* studies. In animal studies, for example, emphasis is placed on the isolation of species and periodic analysis of feed and water; whereas in nonanimal studies, there is increased importance on the justification of the test system. Zhang et al. [205] described a way of predicting drug–drug interaction using *in vitro* model. This article summarizes critical elements in the *in vitro* evaluation of drug interaction potential during drug development and uses a case study to highlight the impact of *in vitro* information on drug labeling from FDA perspectives.

8.13 TRENDS IN GLOBAL R&D OPERATIONS

In the past decade, drug discovery and development programs have increasingly become global. Progressively, many pharmaceutical and biotech companies have enhanced their research activities to emerging economies such as South America, Eastern Europe, and Asia Pacific. In recent years, a shift in nonclinical studies and clinical trials sponsored by the pharmaceutical/biotech industries to emerging economies, especially in Eastern European, Latin American, and Asian countries, has been noticeable. This provides pharmaceutical/biotech companies with access to diverse scientific talent, flexible capacity with a more favorable cost structure, a way to share risk, a stepping stone into new emerging markets future, a mechanism to make our scientists and our current infrastructure more productive, and a way to explore new drug-discovery and -development paradigms. These activities range from strict "fee for service work" to "risk-shared" projects [206].

In the area of clinical research outsourcing, emerging economies offer a huge patient base, diversity/uniqueness of diseases and population, drug-naive population, ease in recruiting and retaining patients, strong intellectual capital, and competitive economic advantage. There is tremendous potential for emerging regions in the realm of bioanalytical research and development. There is also an increased emphasis on rapid access to bioanalytical data during nonhuman studies and Phase I safety trials to guide dose-escalation decisions. An increase in preclinical work under GLP in Asia Pacific to support first-in-human dose-enabling studies has also been observed. With no doubt, all of these programs need bioanalytical R&D support. Bioanalytical laboratories within the geographical region can provide speed, economy, and relative simplicity in terms of sample storage, shipment, and other logistics. The gap (talent pools) is closing very rapidly because individuals with training in Western countries are moving back to these emerging countries (aka reverse brain drain) in recent years. This paradigm shift has dictated that alternative locations for sourcing bioanalysis be explored and has led to an increased investment in the number of laboratories offering bioanalytical services in these emerging regions. One indicator for this investment is almost a 100-fold increase in the number of LC–MS/MS systems in some of these countries in the past decade. Approximately 40 or more laboratories providing bioanalytical services are available in the Asia–Pacific region. Historically, these contract research organizations were providing bioanalytical services to the generic drug industry by supporting bioequivalence/bioavailability studies. There is also an abundant supply of qualified talent (e.g., China and India have almost one billion people under the age of 35). However, experience in method development, trouble shooting, and validation capabilities for regulated bioanalysis of new chemical entities (NCEs) are limited to a fraction of these CROs. This particular concern is diminishing simply because Western CROs have aggressively expanded their presence and operations to these newly growing regions.

A successful and sustained relationship between a sponsor and a CRO would need a careful evaluation (due diligence) and thoughtful selection of a partner from a number of CROs claiming to provide bioanalytical services. For regulatory bioanalysis, complete familiarity with the current regulatory environment/requirements is imperative for any CRO. Strict adherence to regulatory guidelines, such as FDA BMV Guidelines (May 2001), Crystal City Workshops (I–III), and the Incurred Sample Reanalysis white paper, should also be a prerequisite prior to evaluating and selecting the CRO. Quality of bioanalytical data must never be compromised regardless of regional differences. One way to ensure the quality is to evaluate CROs from emerging economies with the same level of rigor during due diligence and hold these to the same standard during bioanalysis as any other western CROs. Before a decision is made to support preclinical or clinical bioanalysis in emerging economies, performing proper due diligence is a must to ensure scientific competence, capability, capacity, and economic benefits. During evaluation of a bioanalytical services provider, the following aspects should be considered (but not limited to) by the sponsor(s):

- *Leadership:* Assessment of key personnel; presence/development of strong technical and logistical expertise, scientific depth of key personnel, willingness to develop infrastructure; overall approach to GLP bioanalytical support;
- *GLP Readiness:* GLP standard operating procedures, sample management group, supporting software and hardware, training, overall infrastructure, and so on;
- *GLP Experience:* Prior or current experience in developing, validating, and implementing (i.e., study sample analysis) GLP methods;
- *Scientific Depth:* Supporting scientific staff (e.g., number, expertise, or experience); assessment of technical knowledge below leadership level;
- *Assay-Development Expertise:* Ability to troubleshoot real-world bioanalytical issues and resolve complex/atypical LC–MS/MS problems;
- *Instrumentation:* Current instrument platforms; personnel's ability to understand pros and cons of each system; technical support/maintenance of systems;
- *Current Capacity:* Assessment of partners' overall capacity; includes time needed to increase capacity;
- *Expansion Capabilities:* Ability (and agility) to expand the facilities in terms of key personnel, support personnel, instrumentation, space, etc.;
- *Partnering Experience:* Preferably across geographies.

If the initial assessment is favorable, it is recommended that sponsors conduct feasibility or pilot experiments to ensure that proper skills have been demonstrated in terms of method development, validation, and sample analysis. These experiments should be designed to test all phases of execution, starting with method feasibility to reporting of final data in the proper format. Weaknesses found in any phase of these experiments can then be isolated and corrected before undertaking a real project. If a bioanalytical method needs to be transferred, proper cross-validation criteria must be in place prior to actual experiments and execution.

If the study is being supported within the same country, sample shipping and handling becomes less critical. Logistics for sample shipping and handling become vital for successful support for a global or multisite study. Knowledge of country-specific import–export laws and practices are very important to ensure smooth and timely delivery of bioanalytical data. This would include specific governmental permits, time needed to obtain these permits, any matrix-specific customs restrictions and procedures to release the samples through customs or port of entry. With proper procedures in place, it may take somewhere between 3 and 6 working days to ship samples from Northern and Southern Asian countries such as China and India. The timeline for delivery of bioanalytical data must be constructed to account for shipping time. Reliable courier services, albeit at a cost, are available to assist sponsors with sample handling and shipping logistics. It is advisable to collect samples in duplicate whenever possible as a contingency to account for any shipping failures.

For sponsors, issues such as site visits for due diligence, quality assurance audit, method development and transfer, oversight of the project, interactions for issues resolution, data management, transfer and flow are magnified many times in a global setting compared with a local or regional situation. Understanding cultural difference is essential to effective communication. The culture in emerging economies may be hierarchical, the pace may be relaxed and people may be more friendly and personal. However, in Western culture, people may be less rank-dependent and more fast-paced; people may be more private and would like to proceed quickly into business relationships. A preferred method of communication should be agreed upon in the beginning of the relationship, such as email, face-to-face meeting, videoconference, or teleconference. While emailing, excessive abbreviations should not be used and detailed messages should be used to avoid any misunderstandings. During conversations, one should listen actively, speak slowly and clearly, and avoid idioms, slang, and acronyms. Mechanisms for real-time communication must be established, particularly during method development, validation, and troubleshooting.

In summary, emerging economies provide a great opportunity to access scientific talent and infrastructure and provide a regional alternative for bioanalysis to support increasing local preclinical and clinical activities. Effective communication, operational excellence, and good relationships are all keys to success.

8.14 TRENDS IN REGULATORY IMPLEMENTATIONS

In 1990, leaders in the field of bioanalytical sciences met at Crystal City Conference Center in Arlington, VA to discuss important issues in the field of bioanalysis. A consensus paper was published from this meeting that set the industry standard for bioanalysis and provided the foundation for the FDA guidelines, which were promulgated in May 2001. A third meeting occurred in May 2006 and the working group from this 3rd AAPS/FDA Bioanalytical Workshop has published a consensus report [207], which is likely to be as standard-setting as the original report was 15 years ago. Sponsors need to ensure that the laboratories where they have bioanalytical testing performed are aware of these recommendations and are taking steps to come into compliance with them. The FDA was represented at the conference and it is reasonable to expect that the FDA will start requiring compliance with some or all of these recommendations, even if it is slow to publish its own follow-up guidelines. There are many individual recommendations published in the paper for both ligand-binding assays and chromatographic assays. Here are a few of the pivotal recommendations for chromatographic assays with regards to their implementation.

8.14.1 Calibration Range and Quality Control Samples

For studies involving pharmacokinetic profiles spanning all or most of the calibration curve, three QC samples run in duplicate (or at least 5% of the unknown samples), spaced across the standard curve as per the FDA Guidance.

8.14.1.1 Recommendation on Implementation There are six QC samples analyzed in a typical bioanalytical batch. This number is adequate for batches of up to 120 samples (6 = 5% of 120). Many methods are employing 96-well plate formats and this becomes problematic when more than one plate is employed for a batch. It has always been a good idea to document the batch size during method validation, but given this specific recommendation not only should the batch size be documented, but if multiple 96-well plates will constitute one batch, then a *set number of QC samples per plate should be assigned*—typically six QC samples per 96-well plate.

If a narrow range of analysis values is unanticipated, but observed after start of the sample analysis, it is recommended that the analysis be stopped and either the standard curve narrowed, existing QC concentrations revised, or QC samples at additional concentrations added to the original curve prior to continuing with sample analysis. It is not necessary to reanalyze samples analyzed prior to optimizing the standard curve or QC concentrations.

8.14.1.2 Recommendation on Implementation This reinforces the 483's that the FDA has issued over the past several years for bioanalytical methods used to support bioequivalence studies, cited for too great an analytical range. While laboratories like to make calibration ranges as wide as linearity permits, it is clear that the FDA expects calibration ranges to be equal to the *actual* range of concentrations found in the study samples and that the low, medium, and high QC samples represent the actual concentration range of samples analyzed within the batch. This means that labs will have to take one of two approaches: validate a series of ranges, each with three levels of QC concentrations; or validate a very wide range but with multiple QC concentrations spread over the entire range.

8.14.2 Incurred Sample Reproducibility (Duplicate Sample Analysis)

A proper evaluation of incurred sample reproducibility and accuracy needs to be performed on each species used for GLP toxicology experiments. It is not necessary for additional incurred sample investigations to be performed in toxicology species once the initial assessment has been performed. Incurred sample evaluations performed using samples from one study would be sufficient for all other studies using that same species.

8.14.2.1 Recommendation on Implementation Laboratories should plan on randomly selecting samples from each treatment arm of the first scheduled study for each species to perform this evaluation. Incurred sample-to-sample precision should be evaluated using the same criteria as the method validation. The results of all duplicate samples should be presented in the study report and evaluated in an addendum to the method validation report.

The final decision as to the extent and nature of the incurred sample testing is left to the analytical investigator, and should be based on an in-depth understanding of the method, the behavior of the drug, metabolites, and any concomitant medications in the matrices of interest. There should be some assessment of both reproducibility and accuracy of the reported concentration.

8.14.2.2 ***Recommendation on Implementation*** As a candidate compound progress through the clinical study program and more is known about metabolites and potential concomitant medications, the sponsor and the laboratory should proactively determine if and when additional duplicate sample information should be collected.

In selecting samples to be reassayed, it is encouraged that issues such as concentration, patient population, and special populations (e.g., renally impaired) be considered, depending on what is known about the drug, its metabolism, and its clearance. First in human, proof of concept in patients, special population and bioequivalence studies are examples of studies that should be considered for incurred-sample concentration verification. The study sample results obtained for establishing incurred sample reproducibility may be used for comparison purposes, and do not necessarily have to be used in calculating reported sample concentrations.

8.14.2.3 ***Recommendation on Implementation*** It will not be sufficient to just randomly select samples to be evaluated as duplicates. Samples may be randomly selected within subgroups of a study or studies to ensure that all possible treatment arms, concentrations, and subpopulations are adequately evaluated. Laboratories should use their previously established sample reanalysis SOP to determine how specific duplicate samples should be reported for a specific study, but the results of all duplicate samples should be presented in the study report and evaluated in an addendum to the method validation report.

The results of incurred sample reanalysis studies may be documented in the final bioanalytical or clinical report for the study, and/or as an addendum to the method validation report.

8.14.2.4 ***Recommendation on Implementation*** Driven by the decision to collect additional duplicate sample information, the results of this testing should be published in the individual study report (as duplicate sample reporting) and as an addendum to the method validation report as support of the method robustness over a range of conditions and matrices for the method.

Workshop Report and Follow-Up—AAPS Workshop on Current Topics in GLP Bioanalysis: Assay Reproducibility for Incurred Samples—Implications of Crystal City Recommendations was published by Fast et al. [208].

8.14.3 LIMS and Electronic Data Handling, Security, Archiving, and Submission

When the principles of good laboratory practice were drafted in 1970s by US FDA, and in 1982 by the Organization for Economic Cooperation and Development the electronic era was in its infant stages and many of the issues surrounding what may affect the environment and human health was not expected. Today, advances in technology for capturing and recording data for the reconstruction of a study are available and are being developed operating at speeds, which could not have been known or understood in years past. Since that time, the US FDA has required the conduct of additional studies in support of a drug registration in accordance with the GLP regulations. However, not all of these studies are required in other countries or may not require adherence to the principles of GLP. Companies are using computer models as virtual studies instead of in-life or bench-type regulated research. Studies are often conducted at institutions of higher learning because of the academic expertise they offer. What is the overall impact advancing technology has on the principles of GLP? Are monitoring authorities ready? The medical products field faces similar issues. Development and testing of these products and devices are being conducted similar to development and testing in other life science arena.

To garner trust in mutual acceptance of data, each participating country must adhere to practices that ensure the highest standards of quality and integrity. The GLP inspector will need to have a good understanding of the science supporting the study conducted and the electronic systems that generate process and maintain study records. The FDA expects that all computer systems, which control quality and production data must be validated and the data within must be controlled. The principle method of demonstrating control of computer systems is validation. Most observations relating to validation are due to missing or incomplete validations. Other common causes of validation observation come from companies not following their own validation procedures, or for omitting mandated technological controls such as audit trails or security.

8.14.3.1 Recommendation on Implementation The level of laboratory instrument integration depends on the complexity of the instrument itself and the functionality of the LIMS. In any case, data entry should be performed according to a validated process in which the identification of the laboratory instrument, date, and time stamp should be recorded at all times of the interaction.

The processes, such as integration and calibration, are defined by processing parameters or calibration factors and affect only the resultant data after processing, but not the acquired electronic raw data. In contrast to the acquired electronic raw data the processing parameters may be changed during data evaluation. The changed processing parameters, methods, and processed data should be identified by versioning. Only the processed data that are finally used and the corresponding process should be retained and archived in addition to the electronic raw data. Once processed data have been approved and released during the evaluation process, the processed data and the corresponding processes should also be retained, even if approval has been withdrawn. In any case, the electronic raw data, once acquired, should be retained and archived. If in justified cases exclusion of specific electronic raw data is necessary these data should be clearly marked as not used. No approved processes or processed data may be discarded.

To be competitive in regulated bioanalysis, bioanalytical laboratories should bear technical excellence along with up to date of compliance. Where automation of sample preparation will go in the future can be anticipated by the foreseeable evolution of the necessities in the analytical/bioanalytical laboratory, namely:

(1) Shortening of the time required for automated SP, as required by the growing number of samples to be analyzed. This reduction in time can be achieved by the design and commercialization of both, general and dedicated devices assisted by different sources of energy.
(2) Integration of several steps involved in common SP procedures in a single commercial device.
(3) Broadening of robotics implementation—either workstations or robotic stations, as required—for unattended development of routine analyses.
(4) Extension of laser uses, which will encompass improvement of present uses and new applications to the three physical states of samples, but particularly to solid samples.
(5) Miniaturization of automated SP devices is a present trend which will grow in the future as a result of the continuously smaller volumes required in the subsequent steps of the analytical process.

Good laboratory management is the key to success. Reducing costs while improving or maintaining quality is crucial to boost profitability. There are several ways to achieve this:

- By increasing the batch size, laboratories can analyze more samples per batch, which lowers the cost per sample.
- Labor is still the main cost factor for bioanalysis. By investing in robotic and automated equipment to perform the routine stages of sample handling, not only will sample throughput increase but laboratory technicians can then focus their time on other value-added areas.
- A systematic approach to method development and validation can shorten timelines. This should be implemented to attain consistency, method robustness, time efficiency, and a higher success rate.
- Investing in high-throughput equipment can mean a higher initial outlay in capital expenditure, but contributes to an overall reduction in costs. For example, investing in chromatography technologies, such as ultra-performance liquid chromatography (UPLC), can result in considerably faster analysis with a higher sample/hour rate.
- Using a state-of-the-art laboratory information management system to manage the entire primary process, from sample reception to report writing, saves time for laboratory technicians, project managers, and others involved in the process. Changing from an outdated LIMS system to an up-to-date system can save 10–15% of time spent on the process.
- Using LC–MS/MS or other techniques to expand into other areas of scientific research, such as microdosing studies, analyzing pharmacokinetic profiles of large molecules (proteins), and metabolic profiling research. Immunoassays, for example, used for pharmacodynamic and PK analysis of large molecules, are not subject to the price pressure seen in LC–MS/MS analysis.

8.15 TRENDS IN GLOBAL REGULATIONS AND QUALITY STANDARDS

Discovery and development of medicines have undertaken tremendous effort, investment, and safety scrutiny for regulatory review and approval. To safeguard well being of our societies and life, regulatory agencies have enforced common regulations from global perspectives (FDA GLP, OECD GLP, ICH GCP, etc.) before drug may be considered for approval. The use of drug product is widespread and touches all aspects of our daily lives without our knowledge. From aspirin to cardiac medicine and from hormones in animal feed to over-the-counter dietary supplements, these products are used every day. What they all have in common are specific government regulations known as current Good Manufacturing Practices established to ensure that the drug product is safe, pure, and effective. They intend to ensure that all aspects of manufacturing use proper science to ensure the integrity and validity of all information and data used. These regulations can all be found under their respective sections of the Code of Federal Regulations.

Routine medical laboratories involved in patient care are covered by well-defined international quality standards (ISO 15189) and national laws (42 CFR 493). In pharmaceutical research and development, only the nonclinical (not involving humans) laboratory safety studies are governed by the Good Laboratory Practices regulations (21 CFR 58). No other well-defined quality standards exist for other so-called non-GLP laboratory research. This leads to the absurd situation that in a research laboratory, for example, a blood sample from a rabbit is theoretically subject to a stricter quality standard than a human sample. The FDA expects that "sound quality principles" are applied to the processing of human samples, but these principles are nowhere defined. Because of the rapid progress of biological sciences, non-GLP laboratory research (e.g., biomarker development, exploratory, mechanistic studies, etc.) has become increasingly critical for modern drug development. Many research institutions are recognizing the need to establish solid quality standards to ensure the integrity and validity of the data they generate. But which standards to follow, and how to introduce them, is the big question.

Non-GLP biomedical research has traditionally been considered to be off-limits for formal quality systems or management. Scientists in general regard their work as a highly intellectual activity where quality is knowledge and experience is an integral part of the scientific rigor that they apply. A longstanding tradition of quality control in science has been peer review of the results, but modern pharmaceutical research has become so complex that peer review has a limited value today.

The scope and sheer dimension of modern research is moving science out of the individual scientist's domain and into a globalized team space where standards, transparency, and reproducibility have become key requirements. Scientific work that cannot be reproduced by others is a waste of valuable, limited resources. Also, biomedical research generates intellectual property, which in the pharmaceutical industry accounts for a large share of the overall company value. Biomedical research has become increasingly subject to internal and external scrutiny and is often challenged as potential litigations. The authenticity and integrity of scientific data is therefore of utmost importance. To prove the authenticity and integrity of scientific data, studies and experiments must be conducted under controlled and verifiable conditions.

In the face of serious global health challenges in developed as well as developing countries, the World Health Organization (WHO) has become increasingly concerned that time and resources in biomedical research are often wasted because of insufficient quality practices. Currently only binding quality regulations for nonclinical laboratory safety studies (GLP) and clinical trials (good clinical practice or GCP) exist. The WHO concluded that there is a pressing global need to provide minimum quality guidance for basic biomedical research not covered by GLP and/or GCP regulations.

In 2000, a scientific working group composed of independent scientific experts convened by the WHO drafted a guide outlining quality standards for biomedical research. The draft was reviewed by a group of experts from a wide range of scientific disciplines and finally published in 2005 under the title "*Handbook: Quality Practices in Basic Biomedical Research*." The handbook describes basic quality requirements ranging from personnel and training to study plans, resources, documentation, and reporting. The overall objective is to ensure the validity and credibility of the scientific data obtained.

Driven by similar concerns as the WHO, the British Association of Research Quality Assurance convened a Working Party on Quality in Non-Regulated Research and published specific quality guidelines in 2006 for biomedical research laboratories that do not fall under GLP. The group recognized that the entire landscape of drug research and development is changing rapidly and that research not (yet) governed by GLP or GCP has become increasingly vital for the success of a pharmaceutical company. Low quality in this area can have disastrous consequences ranging from loss of intellectual property rights to costly repetition of work already performed.

There are many different organizations that interface with and provide quality input to various regulatory bodies. The American Society for Quality (ASQ), as one of the largest quality organizations in the world, takes up the initiative and develop a national quality standard for biomedical laboratory research in the United States. The success of this standard could potentially lead to an international ISO standard.

The proposal for the new quality standard is based on the following:

- The 2005 WHO Handbook "*Quality Practices in Basic Biomedical Research*"
- The 2006 BARQA "*Guidelines for Quality in Non-Regulated Research*"
- The 2007 ISO Standard 15189 "Medical Laboratories—Particular Requirements for Quality and Competence"
- The 2005 ISO Standard 17025 for Uncertainty Measurement for Laboratories.

A quality standard based on statistical models and scientific design is important for the future of drug research worldwide. Implementation of a national quality standard for biomedical research would not only serve as a strong base for the entire drug development process, but also reduce the waste of valuable resources and increase business productivity. It is recognized that a number of countries are already applying the GCLP principles to the analysis of clinical trial samples. Indeed it is possible within some of these countries for clinical laboratories to be accredited by the National Monitoring Authority. Some organizations and indeed countries operate proficiency testing schemes to which laboratories subscribe. While these ensure the integrity of the analytical process they may not assure compliance with GCP. The GCLP document is intended to provide a unified framework for sample analysis to lend credibility to the data generated and facilitate the acceptance of clinical data by regulatory authorities from around the world. It is important to recognize that the framework outlined in the document will be applied across a diverse set of disciplines involved in the analysis of samples from clinical trials. It is therefore important to understand that this framework should be interpreted and applied to the work of those organizations that undertake such analyses with the objective of assuring the quality of every aspect of the work that they perform.

8.16 TRENDS IN COMPLIANCE WITH 21 CFR PART 11

In the fast moving world of today where electronic systems have become so much apart of our daily lives, it is difficult to imagine life before electronic systems and the Internet. In fact it is difficult to think of an area of modern life were computer technology and the Internet has not made a dramatic impact. More recently, scientists unveiled the world's fastest supercomputer, which reportedly has the capability of performing 1000 trillion calculations per second. Such advances may someday allow scientists to simulate highly complex toxicology and other preclinical studies without ever stepping foot into a laboratory. Technology such as this, if ever fully implemented, will certainly raise compelling questions for the world's regulators, for example, what kind of training will the GLP inspectors require and how will they go about inspecting a virtual study? Will the GLP regulations have any relevance to such a study? And if so, how can the test substance and system be defined?

Historically, all the quality documents including SOPs, laboratory notebooks, manufacturing production batch records (MPBRs), product batch records (PBRs), and log books have been maintained on paper by companies in order to comply with FDA's GLP, GCP, and cGMP. Even as companies automated their laboratory methods, procedures, process, production, and quality processes, they were still being forced to maintain and track paper records. The Code of Federal Regulations (CFR) Part 11 was implemented in 1997 to let the FDA accept electronic records and signatures in place of paper records and handwritten signatures for compliance. The regulation outlines controls for ensuring that electronic records and signatures are trustworthy, reliable, and compatible with FDA procedures and as verifiable and traceable as their paper counterparts.

Hence, 21 CFR Part 11 also specifies a number of requirements for software systems to enable trustworthy and reliable electronic records and signatures. These software requirements must be met for the resulting electronic records to comply with FDA's GLP, GCP, and cGMP. If an organization does employ electronic records and signatures, but fails to comply with these system requirements, the FDA will cite the firm for violating the underlying regulation. For example, if a drug company maintains its written complaint records, required by 21 CFR 211.198(b), in electronic form, but the agency finds for some reason that these records are unacceptable substitutes for paper records, then the FDA would charge the firm with violating 211.198(b)—"Master production records are generated from a computer as electronic records without any apparent controls to assure authenticity and integrity 21 CFR 211.186(a)."

8.16.1 21 CFR Part 11 Software Requirements

The following are the specific software requirements that are specified in Section 11.10:

- Validation of systems to ensure accuracy, reliability, consistent intended performance, and the ability to discern invalid or altered records.
- The ability to generate accurate and complete copies of records in both human readable and electronic form.
- Protection of records to enable their accurate and ready retrieval throughout the records retention period.
- Limiting system access to authorized individuals.
- Use of secure, computer-generated, time-stamped audit trails.
- Use of operational system checks to enforce permitted sequencing of steps and events.
- Use of authority checks to ensure that only authorized individuals can use the system, electronically sign a record, access the operation or computer system input or output device, alter a record, or perform the operation at hand.
- Use of device checks to determine the validity of the source of data input or operational instruction.
- Determination that persons who develop, maintain, or use electronic record/ electronic signature systems has the education, training, and experience to perform their assigned task.
- The establishment of, and adherence to, written policies that hold individuals accountable and responsible for actions initiated under their electronic signatures.
- Use of appropriate controls over systems documentation.

8.16.2 Building a Roadmap for Compliance with 21 CFR Part 11

According to analysts who track FDA regulations, the cost of Part 11 compliance could vary from $5 million to $400 million, depending on a company's size and requirements. Companies with low budgets and lots of computer systems that are not compliant with 21 CFR Part 11 must prioritize which systems to fix first. They are now beginning to use risk-based methodology to create a compliance plan for their systems.

Risk-based compliance methodology begins with an inventory of all the existing systems and carefully identifies all systems that are either paper-based or noncompliant. The approach then carefully analyzes each system to assess their risk, as well as, the cost of either converting paper-based system or upgrading/replacing a noncompliant system to comply with the regulations. A key aspect to determining risk is assessing the computer system's potential impact on affecting consumer safety. Incorporated in this assessment must be the role that system plays in the product life cycle, as well as the potential capability of the company's products to injure the consumer as a result of the use of that system. Another aspect to determining risk relates to system's potential to fail due to issues such as software code complexity, lack of good vendor support or lack of change control procedures. Finally, the company must consider the risk of intervention by FDA during an inspection, leading to a large fine or delay in drug approval or a consent decree. While calculating the cost of upgrading, one should determine if the total costs of legacy system upgrade and validation is more expensive than its replacement.

8.16.3 Low Hanging Fruits in the Roadmap for Compliance with 21 CFR Part 11

Based on research by various analysts and consulting firms, one of the low hanging fruits is upgrading quality management systems to become compliant with 21 CFR Part 11. Such systems provide a core infrastructure for electronic records for SOPs and training/certification, implement strict change control, and enable auditable corrective action processes. Hence, these systems are considered quick hits because of their high-risk (high risk of FDA intervention due to direct correlation with cGMP) and lower-cost (relatively lower cost of replacement than a manufacturing system) profile. Quality Management systems should support multiplant and multiorganization architecture, including any outsourced operations such as clinical trials, R&D, or production. Multiorganization architecture enables companies to ensure consistency of practices and processes across the entire internal supply chain leading to a reduction of overall risk of customer-safety. Since existing implementations of quality management systems do not have the architecture to support global operations, enhancements to existing legacy systems is more expensive than implementing a new solution with a global architecture.

Capabilities addressed by Quality Management Systems include the following:

- Document Management and Control (for SOPs)
- Audit Management
- Out-of-Specifications/Noncompliance Tracking
- Corrective action and preventive action (CAPA)
- Change Control
- Training and continued improvement
- Equipment Maintenance and Calibrations

Leading pharmaceutical, drug discovery and development companies are aggressively investing in quality management systems through initiatives that

- establish and monitor company wide quality programs;
- assure compliance with company and regulatory procedures and guidelines;
- provide release and approval of all GLP, GCP, and cGMP documentation, including standard operating procedures (SOPs), data management, and batch records; and
- enable auditing of
  - chemical development, medicinal chemistry, and analytical departments;
  - manufacturing and packaging facilities;
  - analytical chemistry laboratories;
  - drug formulation facilities;
  - raw material supplier audits; and
  - contract research organization.

The regulatory process begins and ends with regulators' responsibilities to protect human health and public interests in well being. The important question for the regulators is: what should the government regulate and how? With the amazing speed at which new technology is introduced, it will be more important than ever for the world's regulatory agencies to recognize the value of staying ahead of modern innovation. For good regulations to be passed and work collaboratively and successfully, there must be a logical balance between the interests of the producer and the needs and concerns of the consumer. It is time for countries around the world to work together in an effort to broaden the scope of their respective GLP programs and realize the great value of global harmonization. Nations whose boarders were historically closed to foreign interaction are now beginning to realize the benefits of policies that promote openness, providing them access to vast experience from experts around the world, while promoting a spirit of cooperation and partnership. Despite these changes, there still exist several challenges that lie ahead. In the modern age of rapid technological advancement, people have become accustomed to hearing about new and exciting scientific discoveries. However, when a new innovation has the potential to affect human health and the environment, as it has been seen with simulated (virtual) studies, genetically modified organisms (GMO)'s or nanotechnology, regulatory review, and oversight must follow, compelling governments to commit significant resources to training their regulatory personnel.

The GLP inspector of the twenty-first century must have the appropriate qualification and training, not only in the basic science, but also in the new technologies. He/she must first have a basic knowledge of the study design, before he/she can inspect the facilities and audit the new technology studies. Along this line, there remains disharmonization among nations regarding appropriate levels of government oversight and testing required for making regulatory decisions on the risks and benefits of pesticide and pharmaceutical products. These inconsistencies have created data gaps and uncertainty about the integrity of foreign data, resulting in the loss of valuable research time and money. Still another challenge facing receiving and regulatory officials within the US Regulators is an increased trend of data submissions consisting of only cited study literature. This information is sometimes very old and, in most cases, not performed in accordance with the GLP regulations. In some situations the laboratory identified in the literature as having generated the data is unknown or may no longer exist. Challenges such as these examples will be solved in time and organizations such as the OECD are a logical place in which to discuss them. OECD's mutual acceptance of data (MAD) program has been highly successful in translating the principles of GLP and procedures, while promoting better data consistency throughout the world. These programs have helped governments establish their own procedures for monitoring GLP compliance through regulatory inspections, global training, and study audits. The author's believe this is the best way to solve the most pressing issues in new millennia, while fostering friendship, respect, and harmonization between nations, and assuring the quality and integrity of data in regulatory decision making.

Laboratory information management systems integration with ERP and other information systems permits incorporation of scientific data in business decisions, optimization of the entire organization, and better information utilization among different corporate divisions to ensure smooth and seamless operation of business processes. Apart from globalization imperatives and the need for enterprise-level solutions, standardized business practices and regulatory compliance have catapulted the LIMS market into the growth stage. Standardized business processes are the primary advantage of an integrated LIMS solution, which ultimately increases productivity and traceability not only in the laboratories but also at the organizational level. End users are now increasingly turning to specialist LIMS vendors since in-house development has proven to be expensive and has failed to deliver global implementations that standardize processes across multiple locations, languages, time zones, and data formats. The pharmaceutical sector remains the largest end-user segment for LIMS, primarily because of Food and Drug Administration and 21 CFR regulations mandating electronic documentation. Here, LIMS is expected to facilitate compliance and ensure a smooth transition to an entirely automated and paperless lab. Following this trend, biotech companies are expected to increase LIMS purchases. LIMS vendors must understand evolving customer needs to achieve long-term success. Strategic Anticipation® is a technique that helps identify potential opportunities and create customer-centric solutions. Various stakeholder groups have different needs. For instance, while selecting a LIMS vendor, management executives emphasize standardization and organizational control through enterprise integrations. Conversely, enterprise integrations are not high priority for scientists. LIMS vendors must, therefore, create different presentations for each stakeholder highlighting those features that are most important for each group. "Flexible, highly configurable, low maintenance, easy-to-use systems with reduced costs of implementation and ownership are some of the common demands of stakeholder and end-user groups." However, LIMS faces a distorted value perception with end users considering the standard 18 month implementation process tedious and lengthy. Intense marketing efforts that reiterate the LIMS' shift from a mere quality assurance and control tool to a total R&D solution could offset such negative perceptions. Value-added services such as maintenance contracts could further ease adoption and generate additional revenues. A complete and successfully designed LIMS solution would need to incorporate instrument interfacing provisions. With genomics and proteomics gaining popularity in the drug discovery field, end users are increasingly demanding LIMS interfacing with DNA sequencing, mass spectrometers, and protein array instruments. Workflow analysis and specific recommendations could smoothen the implementation and simplify the integration of LIMS with laboratory instruments.

As LIMS has become an enterprise system as opposed to an isolated laboratory solution, several barriers confront new entrants. However, alternative technologies such as SAP QM that have large customer bases remain a significant threat to traditional LIMS vendors. Also challenging LIMS vendors is the reduced customer base due to rampant consolidation among pharmaceutical companies. Fortunately, such consolidation is stimulating the need for standardized integrated solutions with global reach, thereby driving LIMS implementations. As the market becomes competitive, top LIMS vendors are partnering with scientific data management system providers to renew vertical-market focus, offer complete, differentiated solutions, and enhance the value of LIMS integrations.

8.17 SUMMARY

From technical point of view, the need for some degree of sample preparation prior to LC–MS/MS analysis will likely continue. Current trends appear to focus on minor improvements to existing technologies. The manner in which we approach this topic in the future, however, should continually be evaluated. Although there is a move toward parallel sample processing and miniaturization, more dramatic changes will likely be required for the future. It is believed that the sample preparation and analysis scheme of the future will likely include micro-device designed for nanoliter quantities of sample in possibly chip-based reservoirs connected via nanofabricated closed channels, which may alleviate potential carryover and cross contamination. This miniaturized sample preparation, separation, and analysis system will be an integrated unit, making our current concepts appear very old-fashioned. Miniaturization affords challenges and benefits. The application of nanoliter quantities of samples and materials to chip-based analysis systems will require nano-pipetters capable of dispensing large numbers of samples in parallel. The nanofabricated devices will be very inexpensive and disposable and will preclude current problems associated with contamination or carryover between samples because each sample track will be used only once. This strategy will also demand continued improvements in LC–MS/MS sensitivity because very small quantities of sample will be utilized. Although this may seem a challenge, modern techniques often use only 50 µL of plasma, for example, whereas a decade ago, 2–5 mL of biological sample was used.

Typically, an LC–MS/MS experiment involves serial analysis of an auto-sampler tray with sample throughput of up to 384 samples per 24 h period. The analytical demands of modern drug discovery practices, such as drug mixture dosing to single animals and analysis of samples from high-throughput screening of combinatorial chemical syntheses that generate very large numbers of samples, will demand much improved sample preparation and analysis capabilities. High-throughput bioanalysis for regulated studies has also become vitally important for speedy and accurate data generation. Author strongly believes one key to improving sample throughput is to prepare and analyze samples in parallel. The so-called "massively parallel" concept imposes some interesting challenges to the way things are currently undertaken. A massively parallel system has two orthogonal parallel axes in simultaneous operation. Author envisions technologies that will allow simultaneous, parallel sample preparation and chromatographic separation and detection of, for example, 96 samples. Thus, multiples of the current 96-well plates may be miniaturized to integrated analysis formats. This vision would allow the simultaneous, parallel preparation of 96 biological samples, which could be transferred directly and simultaneously to 96 HPLC or electrophoretic separation channels.

After simultaneous chromatographic separation of the 96 sample extracts, the effluent from these separations would be transferred directly to a mass spectrometer, which could monitor the effluent from all 96 columns simultaneously. Thus, such an instrument, which does not currently exist commercially, would have 96 parallel ion paths or mass analyzers in one vacuum system. Detection of parallel ion beams and data handling of 96 separate signals would impose new demands on the mass spectrometer data system, but these obstacles must be overcome. Of course, the high volume of data generated by this approach will also challenge information technology and quality assurance programs. The goal will be to develop integrated, miniaturized chip-based or alike devices that form a system capable of routine LC–MS and LC–MS/MS operations, which we have come to expect from modern commercial instrumentation. Automation and integration of everything from sample preparation to detection and data analysis will be important components of future developments in miniaturized LC–MS and LC–MS/MS systems. Parallel concepts should be used whenever possible because so much more may be accomplished in this mode. The future is bright for technique and instrumentation developments that can revolutionize the way we currently approach our analytical and bioanalytical problems. LC–MS/MS concepts will be pivotal to future developments, and these systems will likely be even more common than other hyphenated MS/MS instruments that we have today.

The sensitive and accurate analysis of biological samples remains rewarding and challenging. Improvements in the automation of sample preparation, and enhancements in the sensitivity and robustness of methods have been achieved in past years. In most cases, various off-line and on-line strategies can be applied to support studies in drug discovery and development. Parallelism in chromatography is one of the most efficient ways to improve sample throughput and many different configurations have been applied. The challenge of the analyst is to select the right tools for the right problems and right needs. Again, high-throughput analysis with LC–MS/MS is a very dynamic field, and many new ideas to improve the bioanalytical process or to solve new challenges are emerging on a daily basis. From analytical chemistry standpoint, there is a continuous trend toward miniaturization to reduce costs associated with laboratory operations, and simplify systems for accuracy, reliability, and ruggedness.

From regulatory perspective, the FDA has implemented a GLP Working group, which is in the process of soliciting inputs from the various FDA Centers, as well as industry. The group will then begin the process of how to modernize the GLPs and other related regulations or GxPs. FDA will publish preliminary information to the public and, after a comment period, new GLP/GxP regulations may emerge. At this time, there is no firm timetable for the activities to be completed. As one of the examples in recent years, the FDA has implemented a review of regulation and has revised the Good Manufacturing Practices (GMP) to more accurately reflect current practice and technology. These revisions have placed an increased emphasis upon risk-based approaches and upon quality systems. As part of this ongoing regulatory review process, the GLPs are also being considered for potential future revision and comments from the industry have been solicited by FDA. The Society of Quality Assurance (SQA) and the Pharmaceutical Research and Manufacturers of America (PhRMA) are two groups which have implemented working groups and provided comments for suggested changes and improvements.

REFERENCES

1. Lipinski CA, Lombardo F, Dominy BW, Feeney PJ. Experimental and computational approaches to estimate solubility and permeability in drug discovery and development settings. *Adv Drug Deliv Rev* 1997;23:3–25.
2. Khan S, Teitz DS, Jemal M. *Anal Chem* 1998;70:1622–1628.
3. Shipkova M, Armstrong VW, Oellerich M, Wieland E. *Ther Drug Monit* 2003;25:1–16.
4. Jemal M, Ouyang Z, Chen BH, Teitz D. *Rapid Commun Mass Spectrom* 1999;13:1003–1015.
5. Jemal M, Rao S, Salahudeen I, Chen BH, Kates R. *J Chromatogr B* 1999;736:19–41.
6. Ong VS, Stamm GE, Menacherry ES, Chu S-Y. *J Chromatogr B* 1998;710:173–182.
7. Fura A, Harper TW, Zhang H, Fung L, Shyu WC. *J Pharm Biomed Anal* 2003;32:513–522.
8. Boink ABTJ, Buckley BM, Christiansen TF, Covington AK, Maas AHJ, Müller-Plathe O, Sachs C, Siggaard-Anderson O. *Clin Chim Acta* 1991;202:S13–S22.
9. Xia Y-Q, Whigan DB, Jemal M. *Rapid Commun Mass Spectrom* 1999;13:1611–1621.
10. Wang CJ, Pao LH, Hsiong CH, Wu CY, Whang-Peng JJK, Hu OYP. *J Chromatogr B* 2003;796:283–291.
11. Testa B, Carrupt PA, Gal J. *Chirality* 1993;5:105–111.
12. Won CM. *Pharm Res* 1994;11:165–170.
13. Gilbert HF. *Meth Enzymol* 1995;251:8–29.
14. Evans MJ, Livesey JH, Ellis MJ, Yandle TG. *Clin Biochem* 2001;34:107–112.
15. Testa B, Mayer JM. *Hydrolysis in Drug and Prodrug Metabolism*. Wiley-VCH Verlag GmbH, Zurich, 2003.
16. Jemal M, Khan S, Teitz DS, McCafferty JA, Hawthorne DJ. *Anal Chem* 2001;73:5450–5456.

17. Redinbo MR, Potter PM. *Drug Discov Today* 2005;10:313–325.
18. Satoh T, Taylor P, Bosron WF, Sanghani SP, Hosokawa M, La Du BN. *Drug Metab Dispos* 2002;30:488–493.
19. Watt AP, Morrison D, Locker KL, Evans DC. *Anal Chem* 2000;72:979–984.
20. Berna M, Murphy AT, Wilken B, Ackermann B. *Anal Chem* 2002;74:1197–1201.
21. Sadagopan NP, Li W, Cook JA, Galvan B, Weller DL, Fountain ST, Cohen LH. *Rapid Commun Mass Spectrom* 2003;17:1065–1070.
22. Teitz DS, Khan S, Powell ML, Jemal M. *J Biochem Biophys Methods* 2000;45: 193–204.
23. Heath TG, McLaughlin K, Knotts K. Precipitation of plasma proteins with a trichloroacetic acid solution for direct quantitative analysis by LC–MS/MS. The 46th ASMS Conference on Mass Spectrometry and Allied Topics, Orlando, FL, USA, 1998, p. 1400.
24. Blanchard J. Evaluation of the relative efficacy of various techniques for deproteinizing plasma samples prior to HPLC analysis. *J Chromatogr B* 1981;226:455–460.
25. Kaye B, Herron WJ, Macrae PV, Robinson S, Stopher DA, Venn RF, Wild W. Rapid, solid phase extraction technique for the high-throughput assay of darifenacin in human plasma. *Anal Chem* 1996;68:1658–1660.
26. Janiszewski JS, Schneider RP, Hoffmaster K, Swyden M, Wells D, Fouda H. Automated sample preparation using membrane microtiter extraction for bioanalytical mass spectrometry. *Rapid Commun Mass Spectrom* 1997;11:1033–1037.
27. Steinborner S, Henion J. Liquid–liquid extraction in the 96-well plate format with SRM LC/MS quantitative determination of methotrexate and its major metabolite in human plasma. *Anal Chem* 1999;71:2340–2345.
28. Zweigenbaum J, Heinig K, Steinborner S, Wachs T, Henion J. High-throughput bioanalytical LC–MS/MS determination of benzodiazepines in human urine: 1000 samples per 12 hours. *Anal Chem* 1999;71:2294–2300.
29. Hendriks G, Uges DR, Franke JP. *J Chromatogr B* 2007;853 (1–2): 234–241.
30. Needham SR, Cole MJ, Fouda HG. Direct plasma injection for HPLC/MS quantitation of the anxiolytic agent CP-93,393. *J Chromatogr B* 1998;718:87–94.
31. Jemal M, Qing Y, Whigan DB. The use of high-flow HPLC coupled with positive and negative ion electrospray tandem mass spectrometry for quantitative bioanalysis via direct injection of the plasma/serum samples. *Rapid Commun Mass Spectrom* 1998;12:1389–1399.
32. Ayrton J, Dear GJ, Leavens WJ, Mallett DN, Plumb RS. Optimization and routine use of generic ultra-high flow rate HPLC with mass spectrometric detection for the direct on-line analysis of pharmaceuticals in plasma. *J Chromatogr A* 1998;828:199–207.
33. Jemal M, Teitz D, Ouyang Z, Khan S. *J Chromatogr B* 1999;732:501.
34. Wells DA. *High Throughput Bioanalytical Sample Preparation: Methods and Automation Strategies*. Elsevier, Amsterdam, 2003.
35. Ouyang Z, Khan S, Jemal M. Proceedings of the 53rd ASMS Conference on Mass Spectrometry and Allied Topics, San Antonio, Texas, USA, 2005.
36. Mallet CR, Lu Z, Fisk R, Mazzeo JR, Neue UD. *Rapid Commun Mass Spectrom* 2003;17:163–170.
37. AbuRuz S, Millership J, McElnay J. *J Chromatogr B* 2003;798:203–209.
38. Wachs T, Henion J. *Anal Chem* 2003;75:1769–1775.

39. Zhao JJ, Xie IH, Yang AY, Roadcap BA, Rogers JD. *J Mass Spectrom* 2000;35:1133–1143.
40. Bolden RD, Hokell, SH, Eichhold TH, McCauley-Myers DL, Wehmeyer KR. *J Chromatogr B* 2002;772:1–10.
41. Eerkes A, Shou WZ, Naidong W. *J Pharm Biomed Anal* 2003;31:917–928.
42. Xu N, Kim GE, Gregg H, Wagdy A, Swaine BA, Chang MS, El-Shourbagy TA. *J Pharm Biomed Anal* 2004;36:189–195.
43. Ji QC, Reimer MT, El-Shourbagy TA. *J Chromatogr B* 2004;805:67–75.
44. Xue Y-J, Pursley J, Arnold ME. *J Pharm Biomed Anal* 2004;34:369–378.
45. Zhang N, Yang A, Rogers JD, Zhao JJ. *J Pharm Biomed Anal* 2004;34:175–187.
46. O'Connor D, Clarke DE, Morrison D, Watt AP. *Rapid Commun Mass Spectrom* 2002;16:1065–1071.
47. Polson C, Sarkar P, Incledon B, Raguvaran V, Grant R. *J Chromatogr B* 2003;785:263–275.
48. Ayrton J, Dear GJ, Leavens WJ, Mallett DN, Plumb RS. *Rapid Commun Mass Spectrom* 1997;11:1953–1958.
49. Ayrton J, Dear GJ, Leavens WJ, Mallett DN, Plumb RS. *J Chromatogr A* 1998;828:199–207.
50. Souverain S, Rudaz S, Veuthey J-L. *J Chromatogr B* 2004;801:141–156.
51. Jemal M, Xia Y-Q, Whigan DB. *Rapid Commun Mass Spectrom* 1998;12:1389–1399.
52. Jemal M, Ouyang Z, Xia Y-Q, Powell ML. *Rapid Commun Mass Spectrom* 1999;13:1462–1477.
53. Xia Y-Q, Hop CECA, Liu DQ, Vincent SH, Chiu S-HL. *Rapid Commun Mass Spectrom* 2001;15:2135–2144.
54. Herman JL. *Rapid Commun Mass Spectrom* 2002;16:421–426.
55. Herman JL. *Rapid Commun Mass Spectrom* 2005;19:696–700.
56. Chiap P, Rbeida O, Christiaens B, Hubert Ph, Lubda D, Boos K-S, Crommen J. *J Chromatogr A* 2002;975:145–155.
57. Papp R, Mullett WM, Kwong E. *J Pharm Biomed Anal* 2004;36:457–464.
58. Hsieh Y, Brisson J-M, Ng K, Korfmacher WA. *J Pharm Biomed Anal* 2002;27:285–293.
59. Xia Y-Q, Liu DQ, Bakhtiar R. *Chirality* 2002;14:742–749.
60. Xia Y-Q, Bakhtiar R, Franklin RB. *J Chromatogr B* 2003;788:317–329.
61. Wu ST, Xing J, Apedo A, Wang-Iverson DB, Olah TV, Tymiak AA, Zhao N. *Rapid Commun Mass Spectrom* 2004;18:2531–2536.
62. Cassiano NM, Barreriro JC, Moraes MC, Cliverira RY, Cass QB. *Bioanalysis* 2009;1 (3): 577–594.
63. Plumb R, Dear G, Mallett D, Ayrton J. *Rapid Commun Mass Spectrom* 2001;15:986–993.
64. Hsieh Y, Wang G, Wang Y, Chackalamannil S, Korfmacher WA. *Anal Chem* 2003;75:1812–1818.
65. Zeng H, Deng Y, Wu J-T. *J Chromatogr B* 2003;788:331–337.
66. Gilbert JD, Olah TV, McLoughlin DA. HPLC with atmospheric pressure ionization tandem mass spectrometry as a tool in quantitative bioanalytical chemistry. In: Snyder AP, editor. *Biological and Biotechnological Applications of Electrospray Ionization MS*, The American Chemical Society, New York, 1996, pp. 330–350.

67. Bruins AP. Atmospheric pressure ionization mass spectrometry I. Instrumentation and ionization techniques. *Trends Anal Chem* 1994;13:37–43.
68. Bruins AP. Atmospheric pressure ionization mass spectrometry II. Applications in pharmacy, biochemistry and general chemistry. *Trends Anal Chem* 1994;13:81–90.
69. Covey T. Analytical characteristics of the electrospray ionization process. In: Snyder AP, editor. *Biochemical and Biotechnological Applications of Electrospray Ionization MS. The American*, Chemical Society, Washington, DC, 1996, pp. 21–59.
70. Hiller DL, Zuzel TJ, Williams JA, Cole RO. Rapid scanning technique for the determination of optimal tandem MS conditions for quantitative analysis. *Rapid Commun Mass Spectrom* 1997;11:593–597.
71. Wang T, Zeng L, Strader T, Burton L, Kassel DB. A new ultra-high throughput method for characterizing combinatorial libraries incorporating a multiple probe autosampler coupled with flow injection mass spectrometry analysis. *Rapid Commun Mass Spectrom* 1998;12:1123–1129.
72. Jemal M, Ouyang Z. *Rapid Commun Mass Spectrom* 2003;17:24–38.
73. Xu X, Veals J, Korfmacher WA. *Rapid Commun Mass Spectrom* 2003;17:832–837.
74. Hughes N, Winnik W, Dunyach J-J, Amad M, Splendore M, Paul G. *J Mass Spectrom* 2003;38:743–751.
75. Yang L, Amad MA, Winnik WM, Schoen AE, Schweingruber H, Mylchreest I, Rudewicz PJ. *Rapid Commun Mass Spectrom* 2002;16:2060–2066.
76. Raffaelli A, Saba A. *Mass Spectrom Rev* 2003;22:318–331.
77. Kauppila TS, Kuuranne T, Meurer EC, Eberlin MN, Kotiaho T, Kostiainen R. *Anal Chem* 2002;74:5470–5479.
78. Hanold KA, Fisher SM, Cormia PH, Miller CE, Syage JA. *Anal Chem* 2004;76:2842–2851.
79. Wang G, Hsieh Y, Korfmacher WA. *Anal Chem* 2005;77:541–548.
80. Hsieh Y, Merkle K, Wang G, Brisson J-M, And Korfmacher WA. *Anal Chem* 2003;75:3122–3127.
81. Guevremont R. *J Chromatogr A* 2004;1058:3–19.
82. Kapron JT, Jemal M, Duncan G, Kolakowski B, Purves R. *Rapid Commun Mass Spectrom* 2005;19:1979–1983.
83. Shvartsburg AA, Tang K, Smith RD. *J Am Soc Mass Spectrom* 2005;16:2–12.
84. Venne K, Bonneil E, Eng K, Thibault P. *Anal Chem* 2005;77:2176–2186.
85. Kolakowski BM, Mester Z. *Analyst* 2007;132 (9): 842–864.
86. Hatsis P, Kapron JT. *Rapid Commun Mass Spectrom* 2008;22 (5): 735–738.
87. O'Donnell RM, Sun XB, Harrington P. *Trac-Trends Anal Chem* 2008;27 (1): 44–53.
88. Singh G, Gutierrez A, Xu K, Blair IA. *Anal Chem* 2000;72:3007–3013.
89. Hayen H, Jachmann N, Vogel M, Karst U. *Analyst* 2002;127:1027–1030.
90. Higashi T, Takido N, Yamauchi A, Shimada K. *Anal Sci* 2002;18:1301–1307.
91. Jemal M, Almond R, Ouyang Z, Teitz D. *J Chromatogr B* 1997;703:167–175.
92. Wu Z, Gao W, Phelps MA, Wu D, Miller DD, Dalton JT. *Anal Chem* 2004;76: 839–847.
93. Jemal M, Hawthorne DJ. *Rapid Commun Mass Spectrom* 1999;13:61–66.
94. Cai Y, Cole RB. *Anal Chem* 2002;74:985–991.

95. Vai Y, Concha MC, Murray JS, Cole RB. *J Am Soc Mass Spectrom* 2002;13:1360–1369.
96. Kumar MR, Prabhakar S, Kumar MK, Reddy TJ, Vairamani M. *Rapid Commun Mass Spectrom* 2004;18:1109–1115.
97. Sheen JF, Her GR. *Rapid Commun Mass Spectrom* 2004;18:1911–1918.
98. Harvey DJ. *J Am Soc Mass Spectrom* 2005;16:622–630.
99. Zhao JJ, Yang AY, Rogers JD. *J Mass Spectrom* 2002;37:421–433.
100. Miao X-S, Metcalfe CD. *J Chromatogr A* 2003;998:133–141.
101. Teshima K, Kondo T, Maeda C, Oda T, Hagimoto T, Tsukuda R, Yoshimura Y. *J Mass Spectrom* 2002;37:631–638.
102. Miao X-S, Metcalfe CD. *J Chromatogr A* 2003;998:133–141.
103. Mortier JA, Zhang G-F, Peteghem CHV, Lambert WE. *J Am Soc Mass Spectrom* 2004;15:585–592.
104. Dams R, Benijts T, Günther W, Lambert W, De Leenheer *Rapid Commun Mass Spectrom* 2002;16:1072–1077.
105. Anari R, Bakhtiar R, Zhu B, Huskey S, Franklin RB, Evans DC. *Anal Chem* 2002;74:4136–4144.
106. Zhang F, Bartels MJ, Brodeur JC, McClymont EL, Woodburn KB. *Rapid Commun Mass Spectrom* 2004;18:2739–2742.
107. Xia Y-Q, Chang SW, Patel S, Bakhtiar R, Karanam B, Evans DC. *Rapid Commun Mass Spectrom* 2004;18:1621–1628.
108. Nelson RE, Grebe SK, Okane DJ, Singh RJ. *Clin Chem* 2004;50:373–384.
109. Shou WZ, Jiang X, Naidong W. *Biomed Chromatogr* 2004;18:414–421.
110. Barry SJ, Carr RM, Lane SJ, Leavens WJ, Manning CO, Monté S, Waterhouse I. *Rapid Commun Mass Spectrom* 2003;17:484–497.
111. Zuo M, Gao M-J, Liu Z, Cai L, Duan G-L. *J Chromatogr B* 2005;814:331–337.
112. van Leeuwen SM, Hendriksen L, Karst U. *J Chromatogr A* 2004;1058:107–112.
113. Cartwright AJ, Jones P, Wolff J-C, Evans EH. *Rapid Commun Mass Spectrom* 2005;19:1058–1062.
114. Minkler PE, Ingalls ST, Hoppel CL. *Anal Chem* 2005;77:1448–1457.
115. Liu Z, Minkler PE, Lin D, Sayre LM. *Rapid Commun Mass Spectrom* 2004;18:1059–1065.
116. Song Y, Quan Z, Evans JL, Byrd EA, Liu Y-M. *Rapid Commun Mass Spectrom* 2004;18:989–994.
117. Marand , Karlsson D, Dalene M, Skarping G. *Analyst* 2004;129:522–528.
118. Nordström A, Tarkowski P, Tarkowska D, Dolezal K, stot C, Sandberg G, Moritz T. *Anal Chem* 2004;76:2869–2877.
119. Julka S, Regnier FE. *Anal Chem* 2004;76:5799–5806.
120. Yang E, Wang S, Bowen C, Kratz J, Cyronak MJ, Dunbar JR. *Rapid Commun Mass Spectrom* 2005;19:759–766.
121. Higashi T, Shimada K. *Anal Bioanal Chem* 2004;378:875–882.
122. Korfmacher WA, Veals J, Dunn-Meynell K, Zhang X, Tucker G, Cox KA, Lin C. Demonstration of the capabilities of a parallel LC–MS/MS system for use in the analysis of drug discovery plasma samples. *Rapid Commun Mass Spectrom* 1999;13:1991–1998.

123. Kostianen R, Bruins AP. Effect of multiple sprayers on dynamic range and flow rate limitations in electrospray and ionspray MS. *Rapid Commun Mass Spectrom* 1994;8:549–558.
124. Shia J, Wang C. Applications of multiple channel electrospray ionization sources for biological sample analysis. *J Mass Spectrom* 1997;32:247–250.
125. Andrien BA, Whitehouse C, Sansone MA. Multiple inlet probes for electrospray and APCI sources. The 46th ASMS Conference on Mass Spectrometry and Allied Topics, Orlando, FL, USA, 1998, p. 889.
126. de Biasi V, Haskins N, Organ A, Bateman R, Giles K, Jarvis S. High throughput liquid/mass spectrometric analysis using a novel multiplexed electrospray interface. *Rapid Commun Mass Spectrom* 1999;13:1165–1168.
127. Plumb R, Castro-Perez J, Granger J, Beattie I, Joncour K, Wright A. *Rapid Commun Mass Spectrom* 2004;18:2331–2337.
128. Neue UD. *HPLC Columns: Theory Technology and Practice*. Wiley-VCH Verlag GmbH, New York, 1997.
129. Tettey-Amlalo RN, Kanfer I. *J Pharm Biomed Anal* 2009;50 (4): 580–586.
130. Wehr T. *LCGC North Am* 2002;20 (1): 41–47.
131. Przybyciel M, Majors RE. *LCGC North Am* 2002;20 (6): 516–523.
132. Guo Y, Gaiki S. *J Chromatogr A* 2005;1074:71–80.
133. Grumbach ES, Wagrowski-Diehl DM, Mazzeo JR, Alden B, Iraneta PC. *LCGC North Am* 2004;22 (10): 1010–1023.
134. Li AC, Junga H, Shou WZ, Bryant MS, Jiang X-Y, Naidong W. *Rapid Commun Mass Spectrom* 2004;18:2343–2350.
135. Pan J, Song Q, Shi H, King M, Junga H, Zhou S, Naidong W. *Rapid Commun Mass Spectrom* 2004;18:2549–2557.
136. Brown SD, White CA, Bartlett MG. *Rapid Commun Mass Spectrom* 2002;16:1871–1876.
137. Oertel R, Neumeister V, Kirch W. *J Chromatogr A* 2004;1058:197–201.
138. Majors RE, Przybyciel M. *LCGC North Am* 2002;20 (7): 584–593.
139. Layne J. *J Chromatogr A* 2002;957:149–164.
140. Walter TH, Iraneta P, Capparella M. *J Chromatogr A* 2005;1075:177–183.
141. Xing J, Apedo A, Tymiak A, Zhao N. *Rapid Commun Mass Spectrom* 2004;18: 1599–1606.
142. Apffel A, Fischer S, Goldberg G, Goodley PC, Kuhlmann FE. *J Chromatogr A* 1995;712: 177–190.
143. Fuh M-R, Huang C-H, Wu T-U, Lin S-L, Pan WHT. *Rapid Commun Mass Spectrom* 2004;18:1711–1714.
144. Gustavsson SA, Samskog J, Markides KE, Långström B. *J Chromatogr A* 2001;937: 41–47.
145. Petritis K, Brussaux S, Guenu S, Elfakir C, Dreux M. *J Chromatogr A* 2002;957:173–185.
146. McCalley DV. *J Chromatogr A* 2004;1038:77–84.
147. Wu T-U, Fuh M-R. *Rapid Commun Mass Spectrom* 2005;19:775–780.
148. Tuytten R, Lemière F, Dongen WV, Esmans EL, Slegers H. *Rapid Commun Mass Spectrom* 2002;16:1205–1215.

149. Pruvost A, Becher F, Bardouille P, Guerrero C, Creminon C, Dekfraissy JF, Goujard C, Grassi J, Benech H. *Rapid Commun Mass Spectrom* 2001;15:1401–1408.
150. Jemal M, Xia Y-Q. *Rapid Commun Mass Spectrom* 1999;13:97–106.
151. Weng Naidong, Lee JW, Jiang X, Wehling M, Hulse JD, Lin PP. *J Chromatogr B* 1999;735:255–269.
152. Jemal M, Huang M, Mao Y, Whigan D, Schuster A. *Rapid Commun Mass Spectrom* 2000;14:1023–1028.
153. Jemal M, Ouyang Z, Powell M. *J Pharm Biomed Anal* 2000;23:323–340.
154. Weng, Naidong W, Jiang X, Newland K, Coe R, Lin P, Lee J. *J Pharm Biomed Anal* 2000;23:697–704.
155. Liu DQ, Pereira T. *Rapid Commun Mass Spectrom* 2002;16:142–146.
156. Ayrton J, Clare RA, Dear GJ, Mallett DN, Plumb RS. *Rapid Commun Mass Spectrom* 1999;13:1657–1662.
157. Yan Z, Caldwell GW, Jones WJ, Masucci JA. *Rapid Commun Mass Spectrom* 2003;17:1433–1442.
158. Jemal M, Ouyang Z. *Rapid Commun Mass Spectrom* 2000;14:1757–1765.
159. Jemal M, Xia Y-Q. *J Pharm Biomed Anal* 2000;22:813–827.
160. Kebarle P, Tang L. *Anal Chem* 1993;65:972A–986A.
161. Enke CG. *Anal Chem* 1997;69:4885–4893.
162. Pan P, McLuckey SA. *Anal Chem* 2003;75:5468–5474.
163. King R, Bonfiglio R, Fernandez-Metzler C, Miller-Stein C, Olah T. *J Am Soc Mass Spectrom* 2000;11:942–950.
164. Bonfiglio R, King RC, Olah TV, Merkle K. *Rapid Commun Mass Spectrom* 1999;13:1175–1185.
165. Kaufmann A, Butcher P. *Rapid Commun Mass Spectrom* 2005;19:611–617.
166. Matuszewski BK, Constanzer ML, Chavez-Eng CM. *Anal Chem* 2003;75:3019–3030.
167. Ahnoff M, Wurzer A, Lindmark B, Jussila R. Proceedings of the 51st ASMS Conference on Mass Spectrometry and Allied Topics, Montreal, Canada, 2003.
168. Bennett PK, Meng M, Van Horne KC. Proceedings of the 53rd ASMS Conference on Mass Spectrometry and Allied Topics, San Antonio, Texas, USA, 2005.
169. Shen JX, Motyka RJ, Roach JP, Hayes RN. *J Pharm Biomed Anal* 2005;37:359–367.
170. Tong XS, Wang J, Zheng S, Pivnichny JV, Grifin PR, Shen X, Donnelly M, Vakerich K, Nunes C, Fenyk-Melody J. *Anal Chem* 2002;74:6305–6313.
171. Shou WZ, Naidong W. *Rapid Commun Mass Spectrom* 2003;17:589–597.
172. Mei H, Hsieh Y, Nardo C, Xu X, Wang S, Ng K, Korfmacher WA. *Rapid Commun Mass Spectrom* 2003;17:97–103.
173. Xia Y-Q, Patel S, Bakhtiar R, Franklin RB, Doss GA. *J Am Soc Mass Spectrom* 2005;16:417–421.
174. Liang HR, Foltz RL, Meng M, Bennett P. *Rapid Commun Mass Spectrom* 2003;17:2815–2821.
175. Sojo LE, Lum G, Chee P. *Analyst* 2003;128:51–54.
176. Cech NB, Enke CG. *Anal Chem* 2001;73:4632–4639.
177. Schmidt A, Karas M, Dülcks T. *J Am Soc Mass Spectrom* 2003;14:492–500.

178. Sancho JV, Pozo OJ, et al. *Rapid Commun Mass Spectrom* 2002;16:639–645.
179. Liang HR, Foltz RL, et al. *Rapid Commun Mass Spectrom* 2003;17:2815–2821.
180. Gangl ET, Annan M, et al. *Anal Chem* 2001;73:5635–5644.
181. Zrostlikova J, Hajslova J, et al. *J Chromatogr A* 2002;973:13–26.
182. Choi BK, Gusev AI, Hercules DM. *Anal Chem* 1999;71:4107–4110.
183. Ito S, Tsukada K. *J Chromatogr A* 2002;943:39–46.
184. Choi BK, Hercules DM, Gusev AI Fres. *J Anal Chem* 2001;369:370–377.
185. Benijts T, Dams R, et al. *J Chromatogr A* 2004;1029:153–159.
186. Matuszewski BK, Constanzer ML, Chavez-Eng CM. *Anal Chem* 1998;70:882–889.
187. Ayrton J, Dear GJ, Leavens WJ, Mallett DN, Plumb RS. *J Chromatogr B* 1998;709:243–254.
188. Cole RO, Laws KA, Hiller DL, Kiplinger JP, Ware RS. *Am Lab* 1998;30:15–20.
189. Jemal M, Xia Y. *Rapid Commun Mass Spectrom* 1999;13:97–106.
190. Voyksner RD. Combining liquid chromatography with electrospray MS. In: Cole R, editor. *Electrospray Ionization Mass Spectrometry: Fundamentals, Instrumentation and Applications*, Wiley, New York, 1996, pp. 323–341.
191. Axelsson BO, Jornten-Karlsson M, Michelsen P, Abou-Shakra F. *Rapid Commun Mass Spectrom* 2001;15:375–385.
192. Nicholson JK, Lindon JC, Scarfe G, Wilson ID, Abou-Shakra F, Castro-Perez J, Eaton A, Preece S. *Analyst* 2000;125:235–236.
193. Nicholson JK, Lindon JC, Scarfe GB, Wilson ID, Abou-Shakra F, Sage AB, Castro-Perez J. *Anal Chem* 2001;73:1491–1494.
194. NIH. *Clin Pharmacol Ther* 2001;69:89–95.
195. Sarker D, Workman P. *Adv Cancer Res* 2007;96:213–268.
196. Lee JW, Devanarayan V, Barrett YC, Weiner R, Allinson J, Fountain S, Keller S, Weinryb I, Green M, Duan L, Rogers JA, Millham R, O'Brien PJ, Sailstad J, Khan M, Ray C, Wagner JA. *Pharm Res* 2006;23:312–328.
197. Szekely-Klepser G, Wade K, Woolson D, Brown R, Fountain S, Kindt E. *J Chromatogr B* 2005;826:31–40.
198. Kindt E, Shum Y, Badura L, Snyder PJ, Brant A, Fountain S, Szekely-Klepser G. *Anal Chem* 2004;76:4901–4908.
199. van der Merwe DE, Oikonomopoulou K, Marshall J, Diamandis EP. *Adv Cancer Res* 2007;96:23–33.
200. Lee JW, Weiner RS, Sailstad JM, Bowsher RR, Knuth DW, O'Brien PJ, Fourcroy JL, Dixit R, Pandite L, Pietrusko RG, Soares HD, Quarmby V, Vesterqvist OL, Potter DM, Witliff JL, Fritche HA, O'Leary T, Perlee L, Kadam S, Wagner JA. *Pharm Res* 2005;22:499–511.
201. Mats Bergstro, Anders Grahne, Bengt Langstrom. *Eur J Clin Pharmacol* 2003;59:357–366.
202. Spooner N, Barfield M, Lad R, Parry S, Fowles S. *J Chromatogr B* 2008;870:32–37.
203. Li W, Tse FLS. *Biomed Chromatogr* 2010;24:49–65.
204. Iwatsubo T, Ito K, Kanamitsu S, Nakajima Y, Sugiyama Y. *Ann Rev Pharmacol Toxicol* 1998;38:461–499.
205. Zhang L, Zhang YD, Zhao P, Huang S-M. *AAPS J* 2009;11 (2): 300–306.

206. Chaudhary AK. Bioanalysis going global. *Bioanalysis* 2009;1 (3): 503–505.

207. Viswanathan CT, Bansal S, Booth B, DeStefano AJ, Rose MJ, Sailstad J, Shah VP, Skelly JP, Swann PG, Weiner R. *AAPS J* 2007;9 (1): Article 4.

208. Fast D, Kelley M, Viswanathan CT, O'Shaughnessy J, Peter King S, Chaudhary A, Weiner R, DeStefano AJ, Tang D. Workshop Report and Follow-Up—AAPS Workshop on Current Topics in GLP Bioanalysis: Assay Reproducibility for Incurred Samples—Implications of Crystal City Recommendations. *AAPS J* 2009;11 (2): 238–241.

9

GENERAL TERMINOLOGIES OF GxP AND BIOANALYTICAL LABORATORIES

This is author's intent to provide a broad scope of terminologies as possible within GxP and bioanalytical functions (chromatographic and ligand-binding assays) along with other useful definitions to facilitate better understanding within the areas of interest for our readers.

9.1 GENERAL TERMINOLOGIES FOR GxP AND BIOANALYTICAL LABORATORIES

Acceptance criteria: For specimens or samples: Procedures for acceptance or rejection of specimens or samples arriving at the analytical laboratory. Such procedures are focused on assessing the adequacy of the chain of custody (also see *Method Acceptance Criteria*).

Accreditation: Procedure by which an accreditation body formally recognizes that a laboratory or person is competent to carry out specific tasks.

Accreditation body: Independent science-based organization that has the authority to grant *accreditation*.

Accuracy: Ability to get the true result. For quantitative tests the accuracy expresses the closeness of agreement between the true value and the value obtained by applying the test procedure a number of times. It is affected by systematic and random errors.

Regulated Bioanalytical Laboratories: Technical and Regulatory Aspects from Global Perspectives,
By Michael Zhou

Aliquot: Portion of a liquid sample or solution; alternative hypothesis (see *Hypothesis testing*).

Analysis: See *Test* analysis of variance (ANOVA): Statistical technique that can be used to separate and estimate the different causes of variation.

Analyte or target analyte: Substance to be identified or measured. Surrogate analyte: Well-characterized substance that is taken as representative of the analyte.

Analytical batch or run: Group of one or more specimens or samples that are analyzed under conditions approaching repeatability. Usually it should contain calibrators and quality control specimens or samples in addition to the samples to be analyzed (also see *Batch*).

Analytical method: See *Test*.

Antibody: The functional component of antiserum, often referred to collectively as a population of molecules, each member of which is capable of reacting with a specific antigenic determinant. An antibody molecule is, by definition, monospecific but might also be "idiospecific," "heterospecific," "polyspecific," or of "unwanted specificity." It cannot be "nonspecific" except in the sense of nonimmunochemical binding. These proteins are immunoglobulins and bind by means of specific binding sites to a specific antigenic determinant.

Antigen: Classically, a substance that will elicit the formation of antibodies in a suitable host. A more recent connotation defines an antigen as a substance that will combine with antibody through its antibody-binding sites.

Antigenic determinant: Part of the structure of an antigen molecule that is responsible for specific interaction with antibody molecules evoked by the same or a similar antigen.

Antiserum: A serum-containing antibodies.

Apoenzyme: The protein part of an enzyme without the cofactor necessary for catalysis. The cofactor can be a metal ion, an organic molecule (coenzyme), or a combination of both.

Archive: Collection of documents and records purposefully stored for a defined period of time.

Arithmetic mean or average: Sum of the individual values in a set divided by the number of values.

Assay: Quantitative measurement of an *Analyte*.

Assigned value: See *Value*.

Average: See *Arithmetic mean*.

BARQA: British Association of Research Quality Assurance.

Batch or analytical batch: Group of one or more specimens or samples that are analyzed under conditions approaching repeatability. Usually it should contain calibrators and quality control specimens or samples in addition to the samples to be analyzed.

Best fit: See *Goodness-of-fit*.

Bias: Difference between the expectation of the test result and an accepted reference value. There may be one or more systematic error components contributing to the bias.

Binding capacity: The capacity of a receptor to bind a ligand, expressed in operational units, unlike the quantitative mass units of the affinity constant.

Binomial distribution: See *Distribution.*

Bioavailability (in pharmacokinetics): Ratio of the systemic *exposure* from extravascular (e.v.) exposure to that following intravenous (i.v.) exposure as described by the equation:

$F = A_{e.v.} D_{i.v.} / B_{i.v.} D_{e.v.}$ where F is the bioavailability, A and B are areas under the (plasma) concentration–time curve following extravascular and intravenous administration, respectively, and $D_{e.v.}$ and $D_{i.v.}$ are the administered extravascular and intravenous doses.

Biodegradation: Breakdown of a substance catalyzed by enzymes *in vitro* or *in vivo*.

Biomarker (biological marker): A characteristic that is objectively measured and evaluated as an indicator of normal biological processes, pathogenic processes, or pharmacologic responses to a therapeutic agent.

Biotransformation: Any chemical conversion of substances that is mediated by living organisms or *enzyme* preparations derived therefrom.

Blank: Specimen or sample not containing the analyte.

Blind specimen or sample: Specimen or sample that is analyzed by an operator who is unaware at the time of the analysis that the sample is for control purposes.

Blood: The fluid that circulates through the heart, arteries, capillaries, and veins. Blood is composed of plasma, the fluid portion, and cells, the particles suspended in the plasma.

Blood cell: Any cellular element of the blood including erythrocytes, leukocytes, and platelets.

Blood cell count: The number of red blood cells and white blood cells per unit volume in a specimen of venous blood. In some instances, blood cell count may also include measurement of the hematocrit, hemoglobin, and various computed erythrocyte indices (mean corpuscular volume; mean corpuscular hemoglobin; mean corpuscular hemoglobin concentration).

Blood group: Red cell phenotypes classified by their antigenic structural characteristics, which are under the control of various allelic genes. The cell membrane properties that provide the specific antigenicity of the blood groups are called agglutinogens as they agglutinate or clump in the presence of their specific antibody.

Cord blood: The blood contained in the vessels of the umbilical cord at the time of birth.

Occult blood: Blood present in such small amounts that its presence can be ascertained only by chemical analysis or by spectroscopic or microscopic examination; particularly, the blood found in stools.

Peripheral blood: Blood obtained from parts of the body that are located at some distance from the heart. Examples are blood drawn from the earlobe, fingertip, or heel pad.

Plasma: A clear, yellowish fluid that accounts for about 55% of the total volume of blood. Plasma is obtained by centrifuging a whole blood sample that has had an anticoagulant added to it. Plasma from which fibrinogen and related coagulation proteins have been removed is called serum (a).

Serum: The clear, yellowish fluid that separates from blood when it is allowed to clot. It closely resembles plasma except T or the absence of some coagulation factors.

Protein-free filtrate: A sample of blood, serum, or plasma from which all proteins have been removed by chemical or physical denaturation, dialysis, ultrafiltration, or solvent extraction.

Blunder: Big mistake, especially one that seems to be the result of carelessness or stupidity (see *Outlier*).

Calibration: Set of operations that establish, under specified conditions, the relationship between values indicated by a measuring instrument or measuring system, or values represented by a material measure, and the corresponding known values of a measurand.

Calibration curve: Relationship between the signal response of the instrument and various concentrations of analyte in a suitable solvent or matrix.

Calibration laboratory: Laboratory that performs calibrations.

Calibration method: Defined technical procedure for performing a calibration.

Calibrator: Pure analyte in a suitable solvent or matrix, used to prepare the calibration curve.

Carryover: A process by which materials are carried into a reaction mixture in which they do not belong. This material can be either parts of a sample or reagents including diluent or wash solution. In this case, carryover means transfer of material from one container or from one reaction mixture to another. It can be either unidirectional (backward or forward) or bidirectional in a series of samples or assays.

Catalyst: A catalyst is a substance that increases the rate of a reaction without modifying the overall standard Gibbs energy change in the reaction; the process is called catalysis, and a reaction in which a catalyst is involved is known as a catalyzed reaction.

Certification: *Procedure* by which a certifying body formally recognizes that a body, person, or product complies with given specifications.

Certified reference material (*CRM*): A reference material one or more of whose property values have been certified by a technical procedure, accompanied by or traceable to a certificate or other documentation that has been issued by a certifying body.

Certifying body: Independent science-based organization that has the competence to grant certifications.

Chain of custody: Procedures and documents that account for the integrity of a specimen or sample by tracking its handling and storage from its point of collection to its final disposition.

Chemical ionization (in mass spectrometry): This concerns the process whereby new ionization species are formed when gaseous molecules interact with ions. The process may involve transfer of an electron, proton, or other charged species between the reactants. When a positive ion results from chemical ionization, the term may be used without qualification. When a negative ion is involved, the term negative ion chemical ionization should be used; note that negative ion formation by attachment of a free electron does not fall within this definition. Chemical ionization and chemi-ionization are two terms, which should not be used interchangeably.

Chirality: Property of a molecule not superimposable on its mirror image. Due to asymmetry in their structures, chiral molecules can exist as different isomers and will have special optical and biological properties.

Chiral mobile phase: A mobile phase containing a *chiral selector*.

Chiral selector: The chiral component of the separation system capable of interacting enantioselectively with the enantiomers to be separated.

Chiral stationary phase (in liquid chromatography): A *stationary phase* which incorporates a chiral selector. If not a constituent of the stationary phase as a whole, the chiral selector can be chemically bonded to (chiral-bonded stationary phase) or immobilized onto the surface of a solid support or column wall (chiral-coated stationary phase), or simply dissolved in the liquid stationary phase.

Chi-square distribution: See *Distribution*.

Chromatograph (noun): The assembly of apparatus for carrying out chromatographic separation.

Chromatography: A process that separates a chemical mixture into its component parts for subsequent identification and quantification.

Clerical error: See *Error*.

CLIA: Clinical Laboratory Improvement Amendments.

Cluster ion (in mass spectrometry): An ion formed by the combination of more ions or atoms or molecules of a chemical species often in association with a second species. For example, $(H_2O)nH^+$ is a cluster ion.

Cochran Test: See *Outlier*.

Code of Federal Regulations (CFR): The categorized set of regulations that implement federal statutes. Regulations that pertain to EPA are at 40 CFR.

Coefficient of variation or relative standard deviation: Measure used to compare the dispersion or variation in groups of measurements. It is the ratio of the standard deviation to the mean, multiplied by 100 to convert it to a percentage of the average.

Coenzyme: The dissociable, low-relative-molecular-mass active group of an enzyme, which transfers chemical groups, hydrogen, or electrons. A coenzyme binds with its associated protein (apoenzyme) to form the active enzyme (holoenzyme).

Collaborative studies or interlaboratory test comparisons: Organization, performance, and evaluation of tests on the same or similar items or materials by two or more different laboratories in accordance with predetermined conditions. The main purpose is validation of analytical methods or establishment of reference methods.

Colorimetric analysis: A method of chemical analysis by which the concentration of a compound in solution can be determined by measuring the strength of its color by visual or photometric methods.

Comparison-of-means test: See *Significance test.*

Competitive binding assay: An *assay* based on the competition between a *labeled* and an unlabelled *ligand* in the reaction with a receptor-binding agent (e.g., *antibody*, receptor, transport protein).

Competitive protein binding assay: A type of radioligand assay in which the binding protein is a transport protein or enzyme.

Concentration: Amount of a substance, expressed in mass or molar units, in a unit volume of fluid.

Confidence coefficient: See *Confidence level.*

Confidence interval: Range of values that contains the true value at a given level of probability. This level of probability is called the confidence level.

Confidence level or confidence coefficient: Measure of probability, a, associated with a confidence interval, expressing the probability of the truth of a statement that the interval will include the parameter value.

Confidence limits: The extreme values or end values in a confidence interval (see *Limit*).

Confirmatory test: Second test by an alternative chemical method for unambiguous identification of a drug or metabolite.

Conjugate: A material produced by attaching two or more substances together. Conjugates of antibody with fluorochromes, radioactive isotopes, or enzymes are often used in immunoassays.

Conjugate solutions: Two solutions that coexist in equilibrium at a given temperature and pressure and, at constant pressure (temperature), change their compositions and relative proportions with a variation of temperature (pressure). The term usually refers to two immiscible liquids, but it is also applicable to two immiscible solid solutions.

Consensus value: See *Value.*

Contract Research Organization (*CRO*): A scientific organization (commercial, academic, or other) to which a sponsor may transfer some of its tasks and obligations. Any such transfer should be defined in writing.

Contractor: *Organization or individual* who provides a service under contractual conditions. It should be ensured that the contractor provides services in line with specified criteria of competence.

Control chart: Plot of test results with respect to time or sequence of measurements, with *limits* drawn within which results are expected to lie when the analytical scheme is in a state of statistical control.

Cusum chart: In a cusum chart each result is compared with a reference, usually the assigned or target value. The differences from the reference are then accumulated, respecting the sign, to give a cumulative sum of differences from the standard. The cusum chart has the advantage of identifying small persistent changes in the analytical scheme faster than the Shewhart chart.

Shewhart chart: Chart where the variable of interest is plotted against batch or time. The observed values are compared with the expected or true value. Lines corresponding to the mean value obtained from replicate analysis of reference material and warning and action limits are inserted to provide objective criteria for the interpretation of the chart.

Control limit: See *Limit*.

Controls: Specimens or samples used to determine the validity of the calibration, that is, the linearity and stability of a quantitative test or determination over time. Controls are either prepared from the reference material (separately from the calibrators, that is, weighed or measured separately), purchased, or obtained from a pool of previously analyzed specimens or samples. Where possible, controls should be man-matched to specimens or samples and calibrators.

Positive control: Control that contains the analyte at a concentration above a specified limit.

Negative control: Control that contains the analyte at a concentration below a specified *limit*. Usually a drug-free specimen or sample (blank) is used as a negative control.

Corrective action: Action taken to eliminate the causes of an existing deviation, defect, or other undesirable situation in order to prevent recurrence.

Correlation coefficient: Number showing the degree to which two variables are related. Correlation coefficients range from 0 (no correlation) to -1 or $+1$ (perfect correlation).

Cross-reacting substance: In immunoassays, a substance that reacts with antiserum produced for the target analyte.

Cross reactivity: The reaction of an antibody with an antigen other than that which elicited the formation due to the presence of related determinants.

Cusum chart: See *Control chart*.

Cut-off concentration: Concentration of a drug in a specimen or sample used to determine whether the specimen or sample is considered positive or negative. In some circumstances, it is recommended that the cut-off concentration should be set equal to the limit of detection (see *Threshold*).

Cytochrome P450: Member of a superfamily of heme-containing monooxygenases involved in *xenobiotic* metabolism, cholesterol biosynthesis, and steroidogenesis, in eukaryotic organisms found mainly in the endoplasmic reticulum and inner mitochondrial membrane of cells. "P450" refers to a feature in the carbon monoxide absorption difference spectrum at 450 nm caused by the presence of a thiolate in the axial position of the heme opposite to the carbon monoxide ligand.

Daughter ion (in mass spectrometry): An electrically charged product of reaction of a particular parent ion. In general, such ions have a direct relationship to a particular precursor ion and indeed may relate to a unique state of the precursor ion. The reaction need not necessarily involve fragmentation. It could, for example, involve a change in the number of charges carried. Thus, all fragment ions are daughter ions, but not all daughter ions are necessarily fragment ions.

Dead-(void) volume (in chromatography): This term is used to express the *extra-column volume*. Strictly speaking, the term "dead-volume" refers to volumes within the chromatographic system, which are not swept by the mobile phase. On the other hand, mobile phase is flowing through most of the extracolumn volumes. Due to this ambiguity the use of term "dead-volume" is discouraged.

Degrees of freedom: Number of independent comparisons that can be made between the members of a sample.

Denaturation: The partial or total alteration of the structure of a protein without change in covalent structure by the action of certain physical procedures (heating, agitation) or chemical agents. Denaturation is the result of the disruption of tertiary bonding, which causes the opening of the folded structure of a protein and the loss of characteristic physiologic, enzymatic, or physico-chemical properties; it can be either reversible or irreversible.

Desorption: The converse of *adsorption*, for example, the decrease in the amount of adsorbed substance.

Detection limit: Smallest measured content from which it is possible to deduce the presence of the analyte with reasonable statistical certainty (also see *Limit*).

Deviation: Departure from what is considered normal (see *Standard deviation*).

Dialysis: The process of separating a colloidal *sol* from a colloid-free solution by a membrane permeable to all components of the system except the *colloidal* ones, and allowing the exchange of the components of small molar mass to proceed for a certain time.

Difference absorption spectroscopy: A highly concentrated analyte in an analytical sample can be determined with better precision by replacing the *blank (reference) cell* by one containing a solution of the analyte or other absorber of known concentration; this is known as difference absorption spectroscopy. Difference spectra can also be obtained by computer or other subtraction methods.

Differential detector (in chromatography): A device that measures the instantaneous difference in the composition of the column effluent.

Diluent (in solvent extraction): The liquid or homogeneous mixture of liquids in which extractant(s) and possible modifier(s) may be dissolved to form the solvent phase.

Dilution linearity: A test to demonstrate that the analyte of interest, when present in concentrations above the range of quantification, can be diluted to bring the analyte concentrations into the validated range for analysis by the method. Samples used for this test are, in general, the ones containing high concentrations of spiked analyte, not endogenous analyte.

Discrimination: Ability to recognize and understand the differences between two things.

Distribution: A ranking, from lowest to highest, of the values of a variable and the resulting pattern of measures or scores when they are plotted on a graph. A frequency distribution, for example, gives the possible values of a parameter versus the number of times each value occurred in the sample or population. In many instances, it refers to the spread of the individual values of a sample or population around the mean.

Binomial distribution: Based on the idea that if only one of two possible outcomes can occur on any one occasion, then the theoretical distribution of the different combinations of outcomes that could occur can be worked out if the number of occasions is known. One characteristic of this distribution is that it consists of a limited or finite number of events, n. When n becomes very large, tending to infinity, the binomial distribution becomes the normal distribution.

Chi-square distribution: This distribution may be considered as that of the sum of squares of v independent normal variates in standard form. The parameter v is known as the number of degrees of freedom.

F-distribution: Theoretical distribution used to study population variances. It is the distribution of the ratio of two independent variables each of which has been divided by its degrees of freedom.

Normal distribution: Purely theoretical continuous probability distribution in which the horizontal axis represents all possible values of a variable and the vertical axis represents the probability of those values occurring. The scores on the variable are clustered around the mean in a symmetrical, unimodal pattern known as the bell-shaped (normal) curve. In a normal distribution, the mean, the median, and the mode are all the same. The normal distribution is obtained when the number of events in the binomial distribution, n, becomes very large, tending to infinity.

Probability distribution: Distribution giving the probability of a value of x as a function of x or, more generally, the probability of joint occurrence of a set of variates $x_p, \ldots, x_p$ as a function of those quantities.

t-Distribution: Theoretical probability distribution used in hypothesis testing. Like the normal distribution, the t-distribution is unimodal, symmetrical, and bell-shaped.

Theoretical probability distribution: Number of times it can be expected to get a particular number of successes in a large number of trials. Important theoretical probability distributions are the normal t-, chi-square, and F-distributions.

Z-distribution: Normal distribution in which the scores are the z-scores.

Distribution function: The distribution function $F(x)$ of a variate x is the total frequency of members with variate values less than or equal to x. As a general rule, the total frequency is taken to be unity, in which case the distribution function is the proportion of members bearing values less than or equal to.

Dixon test: See *Outlier*.

Dose–response and dose–effect relationships: The graph of the relation between dose and the proportion of individuals responding with an all-or-none effect; it is essentially the graph of the probability of an occurrence (or the proportion of a population exhibiting an effect) against *dose*. Typical examples of such all-or-none effects are mortality or the incidence of cancer. The dose–effect curve is the graph of the relation between dose and the magnitude of the biological change produced, measured in appropriate units. It applies to measurable changes giving a graded response to increasing doses of a drug or *xenobiotic*. It represents the effect on an individual animal or person, when biological variation is taken into account. An example is the increased effect of lead on the heme synthesis, for example, on activity of the enzyme 6-amino levulinic acid dehydratase in blood serum or coproporphyrin levels in urine.

Double-beam spectrometer (for luminescence spectroscopy): Double (spectral) beam spectrometers are used where two samples are to be excited by two different wavelengths. A double- (synchronous) beam spectrometer is a luminescence spectrometer in which both the excitation and emission monochromators scan the excitation and emission spectra simultaneously, usually with a fixed wavelength difference between excitation and emission.

Double blind procedure: Means of reducing bias in an experiment. In the clinical context, for example, such a procedure ensures that both those who administer a treatment and those who receive it do not know (are blind to) which subjects are in the control group and which are in the experimental group, that is, who is and is not receiving the treatment.

Double-focusing mass spectrometer: An instrument which uses both direction and velocity focusing, and therefore an ion beam of a given mass/charge is brought to a focus when the ion beam is initially diverging and contains ions of the same mass and charge with different translational energies. The ion beam is measured electrically.

Drug design: Drug design includes not only ligand design, but also pharmacokinetics and toxicity, which are mostly beyond the possibilities of structure- and/or computer-aided design. Nevertheless, appropriate chemometric tools, including experimental design and multivariate statistics, can be of value in the planning and evaluation of *pharmacokinetic* and toxicological experiments and results. Drug design is most often used instead of the correct term "ligand design."

Duplicate samples or specimens: Two aliquots of a sample or specimen analyzed at the same time.

Dynamic range: Range over which a relationship exists between analyte concentration and *assay* response. A distinction may be made between the linear dynamic range, where the response is directly proportional to concentration, and the dynamic range where the response may be nonlinear, especially at higher concentrations.

Effective theoretical plate number (of a chromatographic column): A number indicative of column performance when resolution is taken into account: $N = 16Rs^2(1 - \alpha^2)$where Rs is the *peak resolution*, and α is the *separation factor*.

Effluent (in chromatography): The mobile phase leaving the column.

Electrodialysis: *Dialysis* conducted in the presence of an electric field across the membrane(s).

Electron ionization (in mass spectrometry): This is the term used to describe ionization of any species by electrons. The process may, for example, be written: $M + e \rightarrow M+ + 2e$ for atoms or molecules. $M + e \rightarrow M+ + 2e$ for radicals.

Electrophoresis: The motion of *colloidal* particles in an electric field.

Eluent: The liquid or gas entering a chromatographic bed and used to effect a separation by *elution*.

Endoenzymes: *Enzymes* that cut internal bonds of a polymer. Endonucleases are able to cleave phosphodiester bonds within a *nucleic acid* chain by hydrolysis either randomly or at specific base sequences (refer to: *Restriction enzymes*).

Enzyme: An enzyme is a protein that acts as a catalyst.

Enzyme conjugate: Designates a inatcrial that has an enzyme bound covalently.

Enzyme immunoassay (EIA): A generic term for an immunoassay in which the analyte content of the sample is estimated by measuring the catalytic activity of a specific enzyme conjugate on a substrate.

Enzyme-linked immunosorbent assay (ELISA): A heterogeneous enzyme immunoassay method where an antigen or antibody is firmly attached to a solid support.

Error: Something done that is considered to be incorrect or wrong.

Absolute error: Difference between the analytical result and the true value.

Clerical error: Mistake made during routine jobs in an office or laboratory, for example, a transcription error, a *specimen* misidentification or a filing error.

Maximum tolerable error: Extreme values of an error permitted by specifications, regulations, etc. for a given determination.

Random error: Component of the total error of a measurement that varies in an unpredictable way. This causes the individual results to fall on both sides of the average value.

Relative error: *Absolute error* of a measurement divided by the assigned value of the analyte (see *Coefficient of variation* and *Relative standard deviation*).

Systematic error: Component of the total error of a measurement that varies in a constant way. This causes all the results to be in error in the same sense.

Total error: Sum of random and systematic errors.

Type I error: Error made by wrongly rejecting a true null hypothesis. If the null hypothesis is that the sample should be negative, a type I error will generate a false positive result.

Type II error: Error made by wrongly accepting a false null hypothesis. If the null hypothesis is that the sample should be negative, a type II error will generate a false negative result.

Estimate value: See *Value*.

Evaluation: Systematic examination of the extent to which a product, process, or service fulfils specified requirements.

Expert witness: Knowledgeable person, for example, a forensic scientist, familiar with the testing and the interpretation of test results and able to give an expert opinion based on scientific fact or evidence, for example, in court or at a hearing.

Expiration date: Date after which the specified characteristics of a reagent, solution, specimen, control, etc. can no longer be guaranteed.

Extractability (in solvent extraction): A property that qualitatively indicates the degree to which a substance is extracted. The term is imprecise and generally used in a qualitative sense. It is not a synonym for fraction extracted.

Extraction: *Distribution* and *partition* are often used as synonyms for the general phenomenon of extraction where appropriate (see *Liquid–liquid extraction*).

False negative: Test result that states that no drug or metabolite is present when, in fact, such a drug or metabolite is present in an amount greater than a threshold or designated cut-off concentration.

False positive: Test result that states that a drug or metabolite is present when, in fact, it is not present or is present in an amount less than a threshold or designated cut-off concentration.

F-distribution: See *Distribution.*

Federal Food, Drug, and Cosmetic Act (FFDCA): FFDCA regulates, among other things, the use of drugs (human and veterinary), and chemicals in cosmetics and human and animal foods. It includes the legal requirement that tolerances (maximum residue limits) be established for pesticide residues in and on raw agricultural commodities, processed food, and feed items (see Sections 408 and 409). These tolerances are established by EPA.

Federal Insecticide, Fungicide, and Rodenticide Act (FIFRA): FIFRA sets forth regulations for the sale, distribution, and use of pesticides in the United States.

Filtration: The process of segregation of phases; for example, the separation of suspended solids from a liquid or gas, usually by forcing a carrier gas or liquid through a porous medium.

Flow rate (in chromatography): The volume of mobile phase passing through the column in unit time. The flow rate is usually measured at column outlet, at ambient pressure (p_a) and temperature (T_a, in K); this value is indicated with the symbol F. If a water-containing flowmeter was used for the measurement (e.g., the so-called soap bubble flowmeter) then F must be connected to dry gas conditions in order to obtain the mobile phase flow rate at ambient temperature (F_a): $F_a = F(1 - p_w/p_a)$ where p_w is the partial pressure of water vapor at ambient temperature. In order to specify chromatographic conditions in column chromatography, the flow rate (mobile phase flow rate at column temperature, F_c) must be expressed at T_c (kelvin), the column temperature: $F_c = F_a(T_c/T_a)$.

Fluorescence: *Luminescence* that occurs essentially only during the *irradiation* of a substance by electromagnetic radiation.

Fluorescence immunoassay (FIa): A generic term for an immunoassay in which the analyte is measured by fluorescence. This type of assay is carried out by conjugating fluorescent compounds to the antigen or antibody and then measuring the fluorescence in the antigen–antibody reaction.

Federal Register (FR): A daily government publication where all federal regulatory actions, including proposed rules, final rules, and notices, are published.

Food and Drug Administration (FDA): The federal agency responsible for carrying out the provisions of the Federal Food, Drug, and Cosmetic Act, which includes pesticide tolerance enforcement (see also *Federal Food, Drug, and Cosmetic Act*).

Food Quality Protection Act (FQPA): The Food Quality Protection Act, passed by Congress in 1996, amends prior pesticide legislation to establish a more consistent protective regulatory scheme, based on sound science. It mandates a single, health-based standard for all pesticides in all foods; special protections for infants and children; expedites approval of safer pesticides; and creates incentives for the development and maintenance of effective crop protection tools for American farmers. It also requires periodic reevaluation of pesticide registration will remain up to date in the future.

Fourier transform-ion cyclotron resonance (FT-ICR) mass spectrometer: A high-frequency mass spectrometer in which the cyclotron motion of ions, having different mass/charge ratios, in a constant magnetic field is excited essentially simultaneously and coherently by a pulse or a radio frequency electric field applied perpendicular to the magnetic field. The excited cyclotron motion of the ions is subsequently detected on so-called receiver plates as a time domain signal that contains all the cyclotron frequencies that have been excited. Fourier transformation of the time domain signal results in the frequency domain FT-ICR signal which, on the basis of the inverse proportionality between frequency and the mass/charge ratio, can be converted into a mass spectrum. The term is sometimes contracted to Fourier transform-mass spectrometer (FT-MS) (see also *ion cyclotron resonance (ICR) mass spectrometer*).

Fourier transform spectrometer: A scanning interferometer, containing no principal dispersive element, which first splits a beam into two or more components, then recombines these with a phase difference. The spectrum is obtained by a Fourier transformation of the output of the interferometer.

Fragment ion (in mass spectrometry): An electrically charged dissociation product of an ionic fragmentation. Such an ion may dissociate further to form other electrically charged molecular or atomic moieties of successively lower formula weight (See also *Daughter ion*).

F-test: See *Significance test*.

Gas chromatography (GC): A separation technique in which the *mobile phase* is a gas. Gas chromatography is always carried out in a column.

Gas–liquid chromatography: Comprises all gas-chromatographic methods in which the *stationary phase* is a liquid dispersed on a solid support. Separation is achieved by partition of the components of a sample between the phases.

Gas–solid chromatography: Comprises all gas chromatographic methods in which the *stationary phase* is an active solid (e.g., charcoal, molecular sieves). Separation is achieved by adsorption of the components of a sample. In gas chromatography, the distinction between gas–liquid and gas–solid may be obscure because liquids are used to modify solid stationary phases, and because the solid supports for liquid stationary phases affect the chromatographic process. For classification by the phases used, the term relating to the predominant effect should be chosen.

Geometric mean: See *Mean*.

Good clinical laboratory practices (GCLP)*:* The GCLP standards were developed with the objective of providing a single, unified document that encompasses IND sponsor requirements to guide the conduct of laboratory testing, which possesses a unique quality, as it embraces both the research and the clinical aspects of GLP. The development of GCLP standards encompasses applicable portions of 21 CFR Part 58 (GLP) and 42 CFR Part 493 (Clinical Laboratory Improvement Amendments, CLIA) for human clinical trials.

Good clinical practices (GCPs): GCP is an international ethical and scientific quality standard for designing, conducting, recording, and reporting trials that involve the participation of human subjects. Compliance with this standard provides public assurance that the rights, safety, and well being of trial subjects are protected, consistent with the principles that have their origin in the Declaration of Helsinki, and that the clinical trial data are credible.

Good laboratory practices (GLPs): *Organizational* process and conditions under which laboratory studies are planned, performed, monitored, recorded, and reported. Includes a system of protocols (standard operating procedures) recommended to be followed so as to avoid the production of unreliable and erroneous data.

GLP Compliance Review: A review of a facility's current procedures and practices to determine that the Good Laboratory Practice Standards regulations of TSCA or FIFRA are being observed for pertinent studies being conducted at the facility. One or more ongoing TSCA- or FIFRA-regulated studies will normally be selected from the master schedule to serve as a partial basis for the review.

GLP Inspection: An inspection or investigation conducted at a testing facility, which consists of a GLP compliance review and/or a study audit of one or more final reports. It is possible that a GLP inspection may consist only of study audits, without the conduct of a formal facility GLP compliance review. Generally, an inspection will be conducted as a result of the neutral scheme for targeting facilities. An investigation differs from an inspection in that it will consist of a more detailed review of facility policies, procedures, or data as—a result of suspected data deficiencies and/or violations of the FIFRA or TSCA regulations.

Good manufacturing practices (GMP or cGMP): cGMP is that part of quality assurance which ensures that products are consistently produced and controlled to the quality standards appropriate to their intended use and as required by the Marketing Authorization of product specifications. cGMP is concerned with both production and quality control.

Goodness-of-fit: How well a model, a theoretical distribution, or an equation matches actual data.

Gradient elution (in chromatography): A procedure in which the composition of the *mobile phase* is changed continuously or stepwise during the elution process.

Grubbs test: See *Outlier*.

Half-life, $t_{1/2}$: For a given reaction, the half-life $t_{1/2}$ of a *reactant* is the time required for its concentration to reach a value that is the arithmetic mean of its initial and final (equilibrium) values. For a reactant that is entirely consumed, it is the time taken for the reactant concentration to fall to one-half of its initial value: The half-life of a reaction has meaning only in the following special cases:

(1) For a first-order reaction, the half-life of the reactant may be called the half-life of the reaction.
(2) For a reaction involving more than one reactant, with the concentrations of the reactants in their stoichiometric ratios, the half-life of each reactant is the same, and may be called the half-life of the reaction. If the concentrations of reactants are not in their stoichiometric ratios, there are different half-lives for different reactants, and one cannot speak of the half-life of the reaction.

Harmonization: Bringing about agreement on terminology, concepts, etc. so that different entities can interact based on the same terms of reference.

Heterogeneous immunoassay: An immunoassay that requires the physical separation of free-labeled antigen (or antibody) from labeled antigen (or antibody) bound in an immune complex, prior to measurement of the quantity of label.

Homogeneous immunoassay: An immunoassay in which no physical separation is performed. The specific activity of the label or the signal is modulated according to the analyte content of the sample.

Holoenzyme: An active enzyme consisting of the apoenzyme and coenzyme.

Hypothesis test: See *Significance test.*

Hypothesis testing or significance testing: Process of assessing the statistical significance of a finding. It involves comparing empirically observed sample findings with theoretically expected findings, expected if the null hypothesis is true (see *Significance test*). This comparison allows one to compute the probability that the observed outcomes could have been due to chance alone (see *Nonparametric test*).

Alternative hypothesis: Hypothesis that must be accepted if the null hypothesis is rejected.

Null hypothesis: Any hypothesis to be tested. The term null implies that there is no difference between the observed and known values other than that which can be attributed to random variation.

Imprecision: See *Precision.*

Immobilized enzymes: Soluble enzymes bound to an insoluble organic or inorganic matrix, or encapsulated within a membrane in order to increase their stability and make possible their repeated or continued use.

Immunoassay: A ligand-binding assay that uses a specific antigen or antibody capable of binding to the analyte.

Immunogen: A substance that elicits a cellular immune response and/or antibody production.

Immunogenicity: The ability of an immunogen to elicit an immune response.

Immunoglobulin: A glycoprotein found in serum or other body fluids possessing antibody activity.

Independent test result: Result obtained in a manner not influenced by any previous results on the same or similar material.

Induction: In enzymology, induction is a biological process, which results in an increased biosynthesis of an enzyme thereby increasing its apparent activity. It results from the presence of an inducer.

Inhibition: An inhibitor is a substance that diminishes the rate of a chemical reaction; the process is called inhibition.

Influence quantity: Quantity, for example, an environmental condition that is not the subject of measurement but that influences the result.

In-house reference material: See *Reference material.*

Initial test: See *Screening test.*

Inspector: The single individual with overall responsibility for the conduct of the inspection. The inspector is the leader of the inspection team, is responsible for presenting inspection credentials and the Notice of Inspection to the responsible facility representative, and for preparing the Receipt for Samples and Documents, and TSCA Confidentiality Claim. The inspector is responsible for the collection and quality of all evidence necessary to support any potential enforcement action. The inspector must be present at the facility whenever other inspection team members are present. If he/she cannot be present at the facility, the inspector may delegate one of the team members to temporarily assume the responsibilities of the inspector. However, the delegated person must also have inspector credentials.

Instrument linearity: Straight-line relationship between concentrations of analyte and instrument response, in which a change in concentration causes a proportional change in response.

Interfering substance: Substance other than the analyte that gives a similar analytical response or alters the analytical result.

Interlaboratory test comparisons: See *Collaborative studies.*

Intermediate precision: Precision of repeated measurements within-laboratories taking into account all relevant sources of variation affecting the results (e.g., day, analyst, or batch). ICH is also referred to as interbatch, interassay, and interrun precision.

Internal standard: Addition of a fixed amount of a known substance that is not already present as a constituent of the specimen or sample in order to identify or quantify other components. The physicochemical characteristics of the internal standard should be as close as possible to those of the analyte.

Interpretation: Explanation of what analytical results mean based on chemical, pharmacological, toxicological, and statistical principles.

Intralaboratory test comparisons: Organization, performance, and evaluation of tests on the same or similar items or materials within the same laboratory in accordance with predetermined conditions.

In vitro: Testing or occurring outside an organism (e.g., in a test tube or a culture dish).

In vivo: Testing or occurring inside an organism.

Ion-exchange chromatography: Chromatography in which separation is based mainly on differences in the ion-exchange affinities of the sample components. Present day ion-exchange chromatography on small particle high-efficiency columns and usually utilizing conductometric or spectroscopic detectors is often referred to as ion chromatography (IC).

Ion trap mass spectrometer: An arrangement in which ions with a desired range of quotients mass/charge are first made to describe stable paths under the effect of a high-frequency electric quadrupole field, and are then separated and presented to a detector by adjusting the field so as to selectivity induce path instability according to their respective mass/charge ratios.

Isocratic analysis (in chromatography): A procedure in which the composition of the mobile phase remains constant during the elution process.

Isoenzyme: One of a group of related enzymes catalyzing the same reaction but having different molecular structures and characterized by varying physical, biochemical, and immunological properties.

Isotope dilution analysis: A kind of quantitative analysis based on the measurement of the isotopic abundance of a *nuclide* after *isotope dilution* with the test portion.

Isotopically labeled: Describes a mixture of an *isotopically unmodified* compound with one or more analogous *isotopically substituted* compound(s).

Isotopic labeling: *Labeling* in which the resulting product is only different from the initial one by its isotopic composition.

Laboratory: Facilities where analyses are performed by qualified personnel using adequate equipment.

Laboratory Information Management System: See *LIMS*.

Laser ionization (in mass spectrometry): Occurs when a sample is irradiated with a *laser* beam. In the irradiation of gaseous samples, ionization occurs *via* a single- or multiphoton process. In the case of solid samples, ionization occurs *via* a thermal process.

Laser micro mass spectrometry (LAMMS): Any technique in which a specimen is bombarded with a finely focused *laser* beam (diameter less than 10 μm) in the ultraviolet or visible range under conditions of vaporization and ionization of sample material and in which the ions generated are recorded with a *time-of-flight mass spectrometer*.

Laser Raman microanalysis (LRMA): Any technique in which a specimen is bombarded with a finely focused *laser* beam (diameter less than 10 μm) in the ultraviolet or visible range and the intensity versus wavelength function of the Raman radiation is recorded yielding information about vibrational states of the excited substance and therefore also about functional groups and chemical bonding.

Least-squares: Statistical method of determining a regression equation, that is, the equation that best represents the relationship among the variables.

Lethal dose: Amount of a substance or physical agent (radiation) that causes death when taken into the body by a single absorption (denoted by LD).

Level of significance: Probability that a result would be produced by chance alone, for example, the probability of incorrectly rejecting the null hypothesis. It is, therefore, the probability of making a type I error.

Ligand: A substance or part of a substance that binds to a specific receptor.

Light-scattering immunoassay: A type of immunoassay that involves the detection of the antigen–antibody complex formation in an immune reaction by changes in turbidity (turbidimetry) or light scattering (nephelometry) in a fluid medium.

Limit: Prescribed or specified maximum or minimum amount, quantity, or number.

- *Action limit*: Corresponds to a ± 3 standard deviation from the mean. If an observed value falls outside the action limit, the cause must be identified immediately and remedial action taken.
- *Confidence limit*: The limits of the confidence interval.
- *Control limit*: The limits, on a control chart, that are used as criteria for action or for judging whether a set of data does or does not indicate lack of statistical control.
- *Detection limit*: Smallest measured content from which it is possible to deduce the presence of the *analyte* with reasonable statistical certainty.

Quantitation limit: The smallest measured content from which it is possible to quantitate the *analyte* with an acceptable level of accuracy and precision.

Warning limit: Corresponds to a ±2 standard deviation from the mean. Even if the method is under statistical control, approximately 5% of results may be expected to fall outside the warning limits.

Limit of quantification (LOQ):

Lower limit of quantification (LLOQ): The lowest amount of an analyte in a sample that can be quantitatively determined with suitable precision and accuracy.

Upper limit of quantification (ULOQ): The highest amount of an analyte in a sample that can be quantitatively determined with precision and accuracy.

LIMS (Laboratory Information Management System): Software package for collating, calculating, controlling, and disseminating analytical data. It can perform a variety of functions, from specimen or sample registration, and tracking to processing captured data, quality control, financial control, and report generation.

Linear range: Concentration range over which the intensity of the signal obtained is directly proportional to the concentration of the species producing the signal.

Linear regression: Method of describing the relationship between two or more variables by calculating a best-fitting straight line or graph.

Liquid chromatography (LC): A separation technique in which the mobile phase is a liquid. Liquid chromatography can be carried out either in a column or on a plane. Present-day liquid chromatography generally utilizing very small particles and a relatively high inlet pressure is often characterized by the term high-performance (or high-pressure) liquid chromatography, and the acronym HPLC.

Liquid–liquid extraction: This term may be used in place of liquid–liquid distribution when the emphasis is on the analyte(s) being distributed (or extracted). *Note*: The distinction between the distribution constant (K_D) and the partition constant (K_D o) or the concentration distribution ratio (D_c) is reaffirmed and it is recommended that the terms partition constant, partition coefficient, and extinction constant should not be used as synonyms for the (analytical) *distribution ratio*, D_c. A distinction is drawn between the terms solvent and diluent, and the term extractant is now restricted to the active substance in the solvent (e.g., the homogeneous "organic phase" which comprises the extractant, the diluent, and/or the modifier), which is primarily responsible for the transfer of solute from the "aqueous" to the "organic" phase.

Logbook: Book that records laboratory activities, for example, instrumentation, maintenance of instrumentation, sample preparation, and reagents.

Lowest-observed-adverse-effect-level (LOAEL): Lowest concentration or amount of a substance, found by experiment or observation, which causes an adverse alteration of morphology, functional capacity, growth, development, or life span of a target organism distinguishable from normal (control) organisms of the same species and strain under defined conditions of exposure.

Lowest-observed-effect-level (LOEL): Lowest concentration or amount of a substance, found by experiment or observation, that causes any alteration in morphology, functional capacity, growth, development, or life span of target organisms distinguishable from normal (control) organisms of the same species and strain under the same defined conditions of exposure.

Luminescence spectrometer: The instrument used to measure luminescence emission spectra.

Maintenance: Activity of keeping something such as facilities, machines, or instrumentation in good condition by regularly checking it and doing necessary repairs.

Mass analysis (in mass spectrometry): A process by which a mixture of ionic or neutral species is identified according to the mass-to-charge (*m*/*z*) ratios (ions) or their aggregate atomic masses (neutrals). The analysis may be qualitative and/or quantitative.

Mass range (in mass spectrometry): The range of mass numbers that can be characterized by a *mass spectrometer* with sufficient resolution to differentiate adjacent peaks.

Mass Spectrometer: An instrument in which beams of ions are separated (analyzed) according to the quotient mass/charge, and in which the ions are measured electrically. This term should also be used when a scintillation detector is employed.

Mass spectrometer (operating on the linear accelerator principle): A mass spectrometer in which the ions to be separated absorb maximum energy through the effect of alternating electric fields which are parallel to the path of the ions. These ions are then separated from other ions with different mass/charge by an additional electric field.

Mass-to-charge ratio (in mass spectrometry), m/z: The abbreviation *m*/*z* is used to denote the dimensionless quantity formed by dividing the mass number of an ion by its charge number. It has long been called the mass-to-charge ratio although *m* is not the ionic mass nor is *z* a multiple or the elementary (electronic) charge, *e*. The abbreviation *m*/*e* is, therefore, not recommended. Thus, for example, for the ion $C_7H_{7^2}^{+}$, *m*/*z* equals 45.5.

Matrix: Material that contains the analyte, for example, urine or blood.

Matrix effect: The direct or indirect alteration or interference in response due to the presence of unintended analytes (for analysis) or other interfering substances in the sample.

Maximum tolerable dose (MTD): Highest amount of a substance that, when introduced into the body, does not kill test animals (denoted by LD0).

Maximum tolerable error: See *Error*.

Maximum tolerable exposure level (MTEL): Maximum amount or concentration of a substance to which an organism can be exposed without leading to an adverse effect after prolonged exposure time.

Mean: When not otherwise specified, refers to *Arithmetic mean.*

Geometric mean: The *n*th root of the product of *n* individual values.

Median: Middle value of a ranked set of data.

Metabolism: The process by which chemicals are transformed and stored in an organism—plant, animal, and human.

Metabolite: Any substance produced by metabolism (see *Metabolism*).

Miscellaneous body fluids:

Amniotic fluid: The fluid that surrounds the fetus in the amniotic sac. A specimen is obtained by a technique called amniocentesis in which a long needle is inserted into the amniotic sac through the abdominal wall through which fluid is withdrawn.

Cerebrospinal fluid (CSF): A clear, colorless fluid that fills spaces within and around the central nervous system. It is formed from plasma by a biological ultrafiltration process. Specimens are obtained by a lumbar puncture (a spinal tap).

Lymph: A yellowish, slightly basic fluid derived from tissue fluid. Lymph is collected from peripheral tissues throughout the body and is carried in lymph vessels to the circulatory system via the thoracic duct and the right lymphatic duct.

Saliva: The clear, viscous secretion from the parotid, submaxillary, sublingual, and smaller mucous glands in the cavity of the mouth.

Metabolite: Compound produced in the body as a result of biochemical processes.

Method: Detailed (defined) procedure for performing an analysis (see *Test procedure*).

Method Acceptance Criteria: A set of parameters has been set forth for evaluation and establishment for intended use and purposes, which may include (but not limited to): precision, accuracy, linearity/range, stability, recovery, and other parameters of interests for testing specifications.

Method traceability: Property of a method whose measurements give results that can be related, with a given *uncertainty,* to a particular reference, usually a national or international standard, through an unbroken chain of comparisons.

Mobile phase (in chromatography): A fluid that percolates through or along the stationary bed, in a definite direction. It may be a liquid (*liquid chromatography*) or a gas (*gas chromatography*) or a supercritical fluid (*supercritical-fluid chromatography*). In gas chromatography, the expression carrier gas may be used for the mobile phase. In elution chromatography, the expression "eluent" is also used for the mobile phase.

Mode: In statistics, the value or values occurring most frequently in a set of data.

Molecular ion (in mass spectrometry): An ion formed by the removal from (positive ions) or addition to (negative ions) a molecule of one or more electrons without fragmentation of the molecular structure. The mass of this ion corresponds to the sum of the masses of the most abundant naturally occurring isotopes of the various atoms that make up the molecule (with a correction for the masses of the electron(s) lost or gained).

Monoclonal: Arising from a single clone of cells, in the case of immunoglobulin, refers to its origin; usually the monoclonal antibody is of a single immunoglobulin class containing only one light chain type of either the K or L variety. Also refers to all antibody molecules having identical physical–chemical characteristics and antibody specificity. Monoclonal antibodies have very restricted structural diversity and are homogeneous compared with polyclonal antibodies.

Monoclonal antibodies (MAbs): A single species of *immunoglobulin* molecules produced by culturing a single *clone* of a *hybridoma* cell. MAbs recognize only one chemical structure, that is, they are directed against a single *epitope* of the antigenic substance used to raise the antibody.

Monoisotopic mass spectrum: A spectrum containing only ions made up of the principal isotopes of atoms making up the original molecule.

Monospecificity: Monospecificity is functionally defined as the immunoreactivity of an antiserum with its designated antigen (e.g., antihuman IgG, antihuman IgG Fc piece, human IgG3 Fc piece, etc.). In practice, true monospecificity to naturally occurring antigens does not occur in antisera produced by the immunization of the intact animal. An attempt is made to reduce the level of unwanted specificities below that which will interfere with the intended use of a particular immunochemical test.

Multienzyme: A protein possessing more than one catalytic function contributed by distinct parts of a polypeptide chain ("domains"), or by distinct subunits, or both.

Multienzyme complex: A *multienzyme* with catalytic domains on more than one type of polypeptide chain.

Multiplex spectrometer: A *spectrometer* in which a single photodetector simultaneously receives signals from different spectral bands, which are specifically encoded. In the case of frequency multiplexing, each spectral band is modulated at a specific frequency. Decoding is achieved by filtering out, by electronic means, the corresponding signals. Frequency multiplexing may be realized by changing the path difference between the two interfering beams at a uniform rate. Fourier transform of the interferogram so obtained yields the spectrum. This method is called *Fourier transform spectrometry* (*FTS*).

Mutagenesis: The introduction of permanent heritable changes, for example, *mutations* into the *DNA* of an organism.

Mutation: A heritable change in the *nucleotide* sequence of genomic *DNA* (or *RNA* in RNA viruses), or in the number of *genes* or *chromosomes* in a cell, which may occur spontaneously or be brought about by chemical mutagens or by radiation (induced mutation).

Necrosis: Sum of morphological changes resulting from cell death by lysis and/or enzymatic degradation, usually affecting groups of cells in a tissue.

Negative: Indicates that the analyte is absent or below a designated cut-off concentration. "Not detected" is sometimes used as a synonym for negative, although this is not recommended.

Negative control: See *Control.*

Negative ion (in mass spectrometry): An atom, radical, molecule, or molecular moiety in the vapor phase, which has gained one or more electrons thereby acquiring an electrically negative charge. The use of the term anion as an alternative is not recommended because of its connotations in solution chemistry.

NOAEL (No-observed-adverse-effect-level): Greatest concentration or amount of a substance, found by experiment or observation, which causes no detectable adverse alteration of morphology, functional capacity, growth, development, or life span of the target organism under defined conditions.

NOEL (No-observed-effect-level): Greatest concentration or amount of a substance, found by experiment or observation, that causes no alterations of morphology, functional capacity, growth, development, or life span of target organisms distinguishable from those observed in normal (control) organisms of the same species and strain under the same defined conditions of exposure.

Noise: The random fluctuations occurring in a signal that are inherent in the combination of instrument and method.

Nominal linear flow (in chromatography), F: The volumetric flow rate of the *mobile phase* divided by the area of the cross section of the column (cm/min), for example, the linear flow rate in a part of the column not containing *packing*.

Normalization: A mathematical procedure which ensures that the integral of the square of modulus of a wave-function over all space equals 1. The constant required to ensure that a wave-function is normalized is termed the normalization constant.

Nonradioisotopic immunoassay: A type of immunoassay in which the antigen–antibody reaction is measured through the light-scattering properties of immune complexes or through the use of marker molecules attached to constituents of the immune reaction.

None detected: Indicates the absence of an analyte within the specifications of the test(s) performed.

Nonparametric test: Statistical method that makes no assumptions about the distribution of the population from which the sample data are taken.

One-tail test: *Hypothesis test* stated so that the chances of making a *type I error* are located entirely in one tail of a *probability distribution*, for example, it is applicable if we wish to test only whether method A is more precise than method B and not whether method B is more precise than method A.

Two-tail test: Statistical test in which the critical region (the region of rejection of the null hypothesis) is divided into two areas at the tails of the sampling distribution, for example, it is applicable if we wish to test whether methods A and B differ in their precision.

Normal distribution: See *Distribution.*

Not detected: The use of this term as a synonym for negative is not recommended, null hypothesis (see *Hypothesis testing*).

Nucleic acids: Macromolecules, the major organic matter of the nuclei of biological cells, made up of *nucleotide* units, and hydrolyzable into certain *pyrimidine* or *purine bases* (usually adenine, cytosine, guanine, thymine, and uracil), D-ribose or 2-deoxy-D-ribose and phosphoric acid.

OECD Guidelines: Testing guidelines prepared by the Organization of Economic and Cooperative Development of the United Nations. These guidelines assist in the preparation of protocols for toxicological, environmental fate, and so on studies.

One-point calibration: Simplified calibration procedure using a single calibrator and a blank.

One-tail test: See *Nonparametric test*.

Organization: Company, corporation, or institute (or part thereof, e.g., a laboratory), private or public that has its own functions and administration. Some of the international organizations dealing with quality assurance are the International Association of Forensic Toxicologists (TIAFT), the International Federation of Clinical Chemistry (IFCC), the International Olympic Committee (IOC), the International Organization for Standardization (ISO), the International Programme on Chemical Safety (IPCS), the International Union of Pure and Applied Chemistry (IUPAC), and the Organization for Economic Cooperation and Development (OECD).

Outlier: Result that appears to differ unreasonably from the population of the other results. Tests for outliers include the following:

Cochran test: Compares the largest of a set of variances with the other variances in the set.

Dixon test: Compares the difference between a measurement and the one nearest to it in size with the difference between the highest and lowest measurements in the set.

Grubbs test: Now recommended to replace the Dixon test or to be used sequentially after the Dixon test. The single Grubbs test statistic is calculated as the percentage decrease in the standard deviation of a set of results following the removal of either the highest or lowest value in the set, whichever gives the largest decrease in the standard deviation. The pair Grubbs test statistic is calculated in an analogous manner by removing the two highest, two lowest, or else both the highest and the lowest values in the original set of results, whichever gives the lowest standard deviation. The presence of an outlier or a Grubbs outlier pair is indicated if the Grubbs statistic exceeds a critical value, which depends on the number of results in the set and which is given by a reference table.

Parallelism: Relative accuracy from recovery tests on the biological matrix, incurred study samples, or diluted matrix against the calibrator in a substitute matrix. It is commonly assessed with multiple dilutions of actual study samples or samples that represent the same matrix and analyte combination of the study samples.

Parametric test: Statistical techniques designed for use when data have certain characteristics, usually when they approximate a normal distribution and are measurable.

Parent ion (in mass spectrometry): An electrically charged molecular moiety, which may dissociate to form fragments, one or more of which may be electrically charged, and one or more neutral species. A parent ion may be a molecular ion or an electrically charged fragment of a molecular ion.

Partition: This term is often used as a synonym for distribution and extraction. However, an essential difference exists by definition between *distribution constant* or *partition ratio* and *partition constant*. The term partition should be, but is not invariably, applied to the distribution of a single definite chemical species between the two phases.

Partition chromatography: Chromatography in which separation is based mainly on differences between the solubility of the sample components in the stationary phase (*gas chromatography*), or on differences between the solubilities of the components in the mobile and stationary phases (*liquid chromatography*).

Partition coefficient: This term is not recommended and should not be used as a synonym for *partition constant*, *partition ratio*, or *distribution ratio*.

Peak elution volume (time) (in column chromatography), VR, tR: The volume of mobile phase entering the column between the start of the elution and the emergence of the peak maximum, or the corresponding time. In most of the cases, this is equal to the total retention volume (time). There are, however, cases when the elution process does not start immediately at sample introduction. For example, in liquid chromatography, sometimes the column is washed with a liquid after the application of the sample to displace certain components, which are of no interest and during this treatment the sample does not move along the column. In gas chromatography, there are also cases when a liquid sample is applied to the top of the column but its elution starts only after a given period. This term is useful in such cases.

Peak widths (in chromatography): Peak widths represent retention dimensions (time or volume) parallel to the baseline. If the baseline is not parallel to the axis representing time or volume, then the peak widths are to be drawn parallel to this axis. Three peak-width values are commonly used in chromatography. Peak width at base (w_b) is the segment of the peak base intercepted by the tangents drawn to the inflection points on either side of the peak. Peak width at half height (w_h) is the length of the line parallel to the peak base at 50% of the peak height that terminates at the intersection with the two limbs of the peak. Peak width at inflection points (w_i) is the length of the line drawn between the inflection points parallel to the peak base. The peak width at base may be called the "base width." However, the peak width at half height must never be called the "half width" because that has a completely different meaning. Also, the symbol $w_{1/2}$ should never be used instead of w_h.

Personnel: Persons qualified, by virtue of training and experience to carry out their assigned functions.

Pharmacology: Study of the interactions of drugs with living systems.

Pharmacodynamics: Study of pharmacological actions on living systems, including the reactions with and binding to cell constituents, and the biochemical and physiological consequences of these actions.

Pharmacokinetics: Process of the uptake of *drugs* by the body, the biotransformation they undergo, the distribution of the drugs and their metabolites in the tissues, and the elimination of the drugs and their metabolites from the body. Both the amounts and the concentrations of the drugs and their metabolites are studied. The term has essentially the same meaning as toxicokinetics, but the latter term should be restricted to the study of substances other than drugs.

Phase I reaction (of biotransformation): Enzymic modification of a substance by oxidation, reduction, hydrolysis, hydration, dehydrochlorination, or other reactions catalyzed by enzymes of the cytosol, of the endoplasmic reticulum (microsomal enzymes) or of other cell organelles.

Phase II reaction (of biotransformation): Binding of a substance, or its *metabolites* from a phase I reaction, with endogenous molecules (conjugation), making more water-soluble derivatives that may be excreted in the urine or bile.

Plasma desorption ionization (in mass spectrometry): The ionization of any species by interaction with heavy particles (which may be ions or neutral atoms) formed as a result of the fission of a suitable nuclide adjacent to a target supporting the sample.

Plasmid: An extrachromosomal genetic element consisting generally of a circular duplex of *DNA*, which can replicate independently of chromosomal DNA. R-plasmids are responsible for the mutual transfer of antibiotic resistance among microbes. Plasmids are used as *vectors* for cloning DNA in bacteria or yeast host cells.

Plot: Representation of data on or by a graph.

Polyclonal: Arising from different clones. A typical antiserum obtained from a conventional immunization is polyclonal.

Positive ion (in mass spectrometry): This is an atom, radical, molecule, or molecular moiety, which has lost one of more electrons thereby attaining an electrically positive charge. The use of the term cation as an alternative is not recommended. The use of mass ion is not recommended.

Post-column derivatization (in chromatography): A version of reaction chromatography in which the separated components eluting from the column are derivatized prior to entering the detector. The derivatization process is generally carried out "on-the-fly," for example, during transfer of the sample components from the column to the detector. Derivatization may also be carried out before the sample enters the column or the planar medium; this is precolumn (preliminary) derivatization.

Potency: The characteristic of an antibody representing the concentration (titre) of antibody and the acidity for a given substrate (antigen) in the defined method.

Population or universe: (Theoretical) entity defined as an entire group of people, things, or events that have at least one trait in common.

Population statistics: Statistical descriptors of the population, for example, mean, median, mode, or standard deviation.

Positive: Indicates that the analyte is present at a level above a designated cut-off concentration.

Positive control: See *Control.*

Power of test: *Probability* of rejecting the null hypothesis when it is false.

Precision: Closeness of agreement between independent *test* results obtained under prescribed conditions. It is generally dependent on analyte concentration, and this dependence should be determined and documented. The measure of precision is usually expressed in terms of imprecision and computed as a standard deviation of the test results. Higher imprecision is reflected by a larger standard deviation. Independent test results refer to results obtained in a manner not influenced by any previous results on the same or similar material. Precision covers repeatability and reproducibility.

Precursor ion (in mass spectrometry): Synonymous with *Parent ion.*

Presumptive: Describes things that are based on presumptions about what is probably true rather than on certainty.

Presumptive negative: *Specimen* or *sample* that has been flagged as negative by screening. Usually no further tests are carried out, so there is no certainty about its content.

Presumptive positive: *Specimen* or *sample* that has been flagged as positive by screening but that has not yet been confirmed by an adequately sensitive alternate chemical method.

Presumptive test: See *Screening test.*

Preventive action: Action taken to eliminate the causes of a potential deviation or other undesirable situation in order to prevent occurrence.

Principal ion (in mass spectrometry): A molecular or fragment ion that is made up of the most abundant isotopes of each of its atomic constituents. In the case of compounds that have been artificially isotopically enriched in one or more positions such as CH_3, $^{13}CH_3$, or CH_2D_2 the principal ion may be defined by treating the heavy isotopes as new atomic species. Thus, in the above two examples, the principal ions would be of masses 31 and 18, respectively.

Probability: Mathematical measurement of how likely it is that something will happen, expressed as a fraction or percentage. Values for statistical probability range from 1% or 100% (always) to 0 or 0% (never). The relative frequency obtained after a long run of measurements or results will give good approximations to the true probability. It is also understood in other ways: as expressing in some indefinable way a "degree of belief," or as the limiting frequency of an occurrence in an infinite random series.

Probability distribution: See *Distribution*.

Probability function: Function of a discrete variate that gives the probability that a specified value will occur.

Procedure: Specified way to perform an activity. For quality assurance purposes, procedures should be written.

Test procedure: Total operation necessary to perform the analysis, for example, the preparation of the specimen or sample, of the reference materials, or of the reagents, the use of instruments and of formulas for the calculations (when the test is quantitative), the preparation and use of calibration curves, and the determination of the number of replicates.

Processed data: Raw data that have been acted upon to make them clearer or more readily usable.

Proficiency testing: Ongoing process in which a series of proficiency specimens or samples, the characteristics of which are not known to the participants, are sent to laboratories on a regular basis. Each laboratory is tested for its accuracy in identifying the presence (or concentration) of the drug using its usual procedures. An accreditation body may specify participation in a particular proficiency-testing scheme as a requirement of accreditation.

Progenitor ion (in mass spectrometry): Synonymous with *parent or precursor ion (in mass spectrometry)*.

Quadrupole mass analyzer: An arrangement in which ions with a desired quotient mass/charge are made to describe a stable path under the effect of a static and a high-frequency electric quadrupole field, and are then detected. Ions with a different mass/charge are separated from the detected ions because of their unstable paths.

Qualitative test: Analysis in which substances are identified or classified on the basis of their chemical or physical properties, such as chemical reactivity, solubility, molecular weight, melting point, radiative properties (emission, absorption), mass spectra, nuclear half-life, and so on. Test that determines the presence or absence of specific drugs or metabolites in the specimen or sample.

Quality assessment: Overall system of activities whose purpose is to provide assurance that the overall quality control job is being done effectively. It involves a continuing evaluation of the products produced and of the performance of the production system.

Quality assurance (QA): System of activities whose purpose is to provide to the producer or user of a product or a service the assurance that it meets defined standards of quality with a stated level of confidence.

Quality assurance management: All activities of the overall management function that determines and implements quality policy, objectives, and responsibilities.

Quality assurance program: Internal control system designed to ascertain that the studies are in compliance with the principles of good laboratory practices.

Quality audit: Systematic and independent examination to determine whether quality activities and related results comply with planned arrangements and whether these arrangements are implemented effectively and are suitable to achieve the objectives.

Quality control: Overall system of activities whose purpose is to control the quality of a product or service so that it meets the needs of users. The aim is to provide quality that is satisfactory, adequate, dependable, and economic.

External quality control: See *Proficiency testing*.

Internal quality control: Set of procedures undertaken by a laboratory for continuous monitoring of operations and results in order to decide whether the results are reliable enough to be released. Quality control of analytical data primarily monitors the batchwise trueness of results on quality control specimens or samples and precision on independent replicate analysis of test materials.

Quality Control (QC) Sample: A spiked sample used to monitor the performance of a bioanalytical method and to assess the integrity and validity of the results of the unknown samples analyzed in an individual batch.

Quality management: That aspect of the overall management function that determines and implements the quality policy.

Quality manual: Document stating the general quality policies, procedures, and practices of an organization.

Quality policy: Statement by top management regarding the laboratory's adherence to principles of quality. It may set forth codes of practice or ethics.

Quality system: The organizational structure, responsibilities, procedures, processes, and resources for implementing quality management. In a laboratory, it refers to the total features and activities of a laboratory aimed at producing accurate work and a high-quality product.

Quantification range: The range of concentration, including ULOQ and LLOQ, that can be reliably and reproducibly quantified with accuracy and precision through the use of a concentration–response relationship.

Quantitation limit: See *Limit*.

Quantitative analysis: Analyses in which the amount or concentration of an analyte may be determined (estimated) and expressed as a numerical value in appropriate units. *Qualitative analysis* may take place without quantitative analysis, but quantitative analysis requires the identification (qualification) of the analytes for which numerical estimates are given.

Quantitative structure–activity relationships (QSAR): The building of structure–biological activity models by using regression analysis with physicochemical constants, indicator variables, or theoretical calculations. The term has been extended by some authors to include chemical reactivity, that is, activity is regarded as synonymous with reactivity. This extension is, however, discouraged.

Quantitative structure–activity relationship (in drug design): Quantitative structure–activity relationships are mathematical relationships linking chemical structure and pharmacological activity in a quantitative manner for a series of compounds. Methods that can be used in QSAR include various regression and pattern recognition techniques. QSAR is often taken to be equivalent to *chemometrics* or multivariate statistical data analysis. It is sometimes used in a more limited sense as equivalent to Hansch analysis. QSAR is a subset of the more general term SPC.

Quantitative test: Test to determine the quantity of drug or metabolite present in a specimen.

Radioligand assay: A technique in which unlabeled and radioactive labeled molecules of the same species compete for a limited number of binding sites on a specific binding protein. The binding protein may be an antibody, transport protein, hormone receptor, or any other cell-associated receptor or tissue component. The unlabeled ligand is the analyte. In the procedure, after a suitable reaction period, the bound ligand (both labeled and unlabeled) is separated from the free ligand, and the radioactivity of either fraction is measured. Calibration reference materials are included in the assay, and the concentration of unlabeled ligand can be estimated from the calibration curve or computed after application of a suitable curve fitting routine.

Radioimmunoassay (RIA): A type of radioligand assay in which the binding protein is an antibody.

Radioreceptor assay: A type of radioligand assay in which the binding protein is a hormone receptor.

Random error: See *Error*.

Random sample: Sample taken in such a way that all the members of the population have an equal chance of being included, that is, each is chosen entirely by chance.

Range: *Concentration* interval for which acceptable accuracy and precision can be achieved. Statistically, it is the difference between the minimum and the maximum values of a set of measurements.

Raw data: Data that are in their original state and have not been processed.

Receptor: A specific molecule on the surface of a neutron or target cell that specifically binds to a specific molecule such as a neurotransmitter, hormone, antigen complement component, lymphokine, and so on.

Record: Document that furnishes objective evidence of activities performed or results achieved.

Recovery: Percentage of the drug, metabolite, or internal standard originally in the specimen or sample that reaches the end of the procedure. In toxicology, the process leading to partial or complete restoration of a cell, tissue, organ, or organism following its damage from exposure to a harmful substance or agent.

Reference material: Material or substance one or more properties of which are sufficiently well established to be used for calibrating an apparatus, assessing a measurement method, or assigning values to materials.

In-house reference material: Material whose composition has been established by the user *laboratory* by several means, by a reference method or in collaboration with other laboratories.

Reference method or standard consensus method: Method developed by *organizations* or groups that use collaborative studies or similar approaches to validate it. Its value depends on the authority of the organizations that sponsor it.

Reference preparation: Processed reference material.

Reference standard: A standard, generally of the highest quality available at a given location, from which measurements made at that location are derived.

Regression analysis: Method of explaining or predicting the variability of a dependent variable using information about one or more independent variables. Also, techniques for establishing regression equations.

Regression curve: Curve that comes closest to approximating a distribution of points in a scatter diagram.

Relative error: See *Error*.

Relative frequency: Number calculated by dividing the number of values with a certain characteristic by the total number of values. Also, the frequency of an event that would occur in the long run given the probability of the event.

Relative standard deviation: See *Coefficient of variation*.

Reliability: Extent to which an experiment, test, or measuring procedure yields accurate results in repeated trials.

Repeatability: Closeness of the agreement between the results of successive measurements of the same *analyte* made under repeatable conditions, for example, same method, same material, same operator, and same laboratory and carried out in a narrow time period. Results should be expressed in terms of the repeatability standard deviation, the repeatability coefficient of variation, or the confidence interval of the mean value.

Replicate analysis: Multiple analyses of separate portions of a test material using the same test method under the same conditions, for example, same operator, same apparatus, same laboratory.

Report: Document containing a formal statement of results of tests carried out by a laboratory. It should include the information necessary for the interpretation of the test results.

Representative sample: Statistically, a sample that is similar to the population from which it was drawn. When a sample is representative, it can be used to make inferences about the population. The most effective way to get a representative sample is to use random methods to draw it. Analytically, it is a specimen or sample that is a portion of the original material selected in such a way that it is possible to relate the analytical results obtained from it to the properties of the original material.

Reproducibility: Closeness of the agreement between the results of successive measurements of the same analyte in identical material made by the same method under different conditions, for example, different operators and different laboratories and considerably separated in time. Results should be expressed in terms of the reproducibility standard deviation, the reproducibility coefficient of variation, or the confidence interval of the mean value.

Resolution: Ability to distinguish meaningfully between closely adjacent values.

Resolution (in mass spectroscopy): (1) Energy—By analogy with the peak width definition for mass resolution, a peak showing the number of ions as a function of their translational energy should be used to give a value for the energy resolution; (2) 10% valley definition—Let two peaks of equal height in a mass spectrum at masses m and $m - \Delta m$ be separated by a valley which at its lowest point is just 10% of the height of either peak. For similar peaks at a mass exceeding m, let the height of the valley at its lowest point be more (by any amount) than 10% of either peak height. Then the resolution (10% valley definition) is $m/\Delta m$. It is usually a function of m. The ratio $m/\Delta m$ should be given for a number of values of m; (3) Peak width definition—For a single peak made up of singly charged ions at mass m in a mass spectrum, the resolution may be expressed as $m/\Delta m$, where Δm is the width of the peak at a height which is a specified fraction of the maximum peak height. It is recommended that one of three values 50, 5, or 0.5% should always be used. For an isolated symmetrical peak recorded with a system, which is linear in the range between 5% and 10% levels of the peak, the 5% peak width definition is technically equivalent to the 10% valley definition. A common standard is the definition of resolution based upon Δm being full width of the peak at half its maximum height, sometimes abbreviated "FWHM." This acronym should preferably be defined the first time it is used.

Restriction enzymes: Endonucleases which recognize specific base sequences within a *DNA* helix, creating a double-strand break of DNA. Type I restriction enzymes bind to these *recognition sites* but subsequently cut the DNA at different sites. Type II restriction enzymes both bind and cut within their recognition or target sites.

Result: Information obtained from a test or series of tests. Usually it refers to processed data.

Result (in analysis): The final value reported for a measured or computed quantity, after performing a measuring procedure including all subprocedures and evaluations.

Retention sample or specimen: Amount of material equivalent in quantity to the *assay specimen* or *sample* and taken from the consignment in a manner similar to that used to assay the sample or specimen. It should be stored under specified conditions.

Reversed-phase chromatography: An elution procedure used in liquid chromatography in which the mobile phase is significantly more polar than the stationary phase, for example, a microporous silica-based material with chemically bonded alkyl chains. The term "reverse phase" is an incorrect expression to be avoided.

Reverse transcriptases: *Enzymes* found in retroviruses that can synthesize complementary single strands of *DNA* from an *m*RNA sequence as *template*. They are used in genetic engineering to produce specific *cDNA* molecules from purified preparations of *m*RNA.

Review: Evaluation of laboratory results to ensure that they have been correctly interpreted.

Risk assessment: Identification and quantification of the risk resulting from a specific use or occurrence of an agent, taking into account possible harmful effects on individuals exposed to the agent in the amount and manner proposed and all the possible routes of exposure.

Robustness or ruggedness: Capacity of a test to remain unaffected by small variations in the procedures. It is measured by deliberately introducing small changes to the method and examining the consequences (see *Influence quantity*).

Sample: Analytically equivalent to specimen, it is a representative portion of the whole material to be tested. Statistically, it is a set of data obtained from a population.

Sample handling (in analysis): Any action applied to the sample before the analytical procedure. Such actions include the addition of preservatives, separation procedures, storage at low temperature, protection against light, irradiation and degradation, loading, and chain of custody.

Sample injector (in chromatography): A device by which a liquid or gaseous sample is introduced into the apparatus. The sample can be introduced directly into the carrier-gas stream, or into a chamber temporarily isolated from the system by values, which can be changed so as to make an instantaneous switch of the gas stream through the chamber. The latter is a *by-pass injector*.

Sample statistics: Statistical descriptors of the sample, for example, mean, median, mode, standard deviation, range, or size.

Sampling: Analytically, the whole set of operations needed to obtain a sample or specimen, including planning, collecting, recording, labeling, sealing, and shipping. Statistically it is the process of determining properties of the whole population by collecting and analyzing data from a representative segment of it.

Sandwich immunoassay: An immunoassay using the chemical or immunochemical binding of the analyte to a solid phase and the immunochemical binding of a second (labeled) reagent to the analyte.

Scatter diagram or scatter plot or scattergram: Pattern of points that results from plotting two variables on a graph. Each point or dot represents one subject or unit of analysis and is formed by the intersection of the values of the two variables.

Screening test or initial test or presumptive test: First test carried out on a specimen or sample for the purpose of determining a presumption of a positive or negative assay. Usually, presumptive positives are followed by a confirmatory test.

Selectivity: Extent to which a method can determine particular analyte(s) in a complex mixture without interference from the other components in the mixture. A method that is perfectly selective for an analyte or group of analytes is said to be specific.

Sensitivity: Difference in analyte concentration corresponding to the smallest detectable difference in the response of the method. It is represented by the slope of the calibration curve. Sometimes it is used, erroneously, to mean detection limit.

Shewhart chart: See *Control chart*.

Significance test or hypothesis test: Statistical test whose purpose is to draw a conclusion about a population using data from a sample. It is used to determine the likelihood that observed characteristics of samples have occurred by chance alone in the population from which the samples were selected. Frequently used significance tests include the following:

Comparison-of-means test or t-test: Compares the mean of the results from one sample taken from a given population with the mean of the results from a second sample taken from the same population, with the two sets of results having, for example, been produced by different analytical methods. It answers the question, "Are the two means significantly different?" The null hypothesis is that the means are not significantly different and that the samples are therefore part of the same population. It is assumed that the variances of the samples are the same and that the samples are representative of the whole population. The larger the number of results for each sample, the more likely this is to be true. Statistical comparison of the means will indicate whether any differences between the samples could have arisen by chance alone. The *t*-test is used under particular circumstances, for example, when the size of the samples is small (usually less than 20) or when a single sample is taken from a population for which the variance is unknown.

Variance-ratio test or F-test: In the comparison-of-means test, it is assumed that the variance of each sample is the same. A variance-ratio test is used to check if this assumption is reasonable.

Significant figures: Number of figures that are consistent with the precision of the test.

Size-exclusion chromatography (SEC): A separation technique in which separation mainly according to the hydrodynamic volume of the molecules or particles takes place in a porous nonadsorbing material with pores of approximately the same size as the effective dimensions in solution of the molecules to be separated.

Skewness: Said of measures or scores that are bunched on one side of a central tendency parameter (mean, median, and mode) and trail out on the other. The more skewness in a distribution, the more variability in the scores. Also used to refer to asymmetry in, for example, a chromatographic peak shape ("tailing" and "fronting").

Solid phase antibody radioimmunoassay: A kind of *radioimmunoassay* employing an *antibody* bound to a solid phase.

Solid phase extraction (SPE): This is a separation process that is used to remove compounds from a mixture, using their physical and chemical properties on a sorbent (stationary phase) to retain analytes of interest and remove as much as other "unwanted" components. Analytical laboratories use solid phase extraction to concentrate and purify samples for analysis. Solid phase extraction can be used to isolate analytes of interest from a wide variety of matrices, including urine, blood, water, beverages, soil, and animal tissue.

Solubility: The analytical composition of a saturated solution, expressed in terms of the proportion of a designated solute in a designated solvent, is the solubility of that solute. The solubility may be expressed as a concentration, molality, mole fraction, and mole ratio.

Solution: Liquid in which a solid substance or a gas has been dissolved.

Solvent extraction: The process of transferring a substance from any matrix to an appropriate liquid phase. If the substance is initially present as a solute in an immiscible liquid phase the process is synonymous with *liquid–liquid extraction. Notes*: (1) If the extractable material is present in a solid (such as a crushed mineral or an ore) the term leaching may be more appropriate. The extractable material may also be a liquid entrapped within or adsorbed on a solid phase; (2) Common usage has established this term as a synonym for liquid–liquid distribution. This is acceptable provided that no danger of confusion with extraction from solid phases exists in a given context.

Specification: Statement of requirements, usually in written form.

Specificity: See *Selectivity*.

Specimen: Analytically, equivalent to sample. In the context of this Glossary, any biological material for examination, study, or analysis.

Spiked sample: A test material containing a known addition of analyte.

Split-level model: Statistical model that splits the study sample according to a predetermined assumption so that only a portion of the cases falls into the category of interest, for example, so that only some of the specimens that were positive for group A of drugs will be positive for group B.

Split specimen or sample: Practice of dividing a specimen or sample. A urine specimen, for example, may be divided into two portions, one of which may be submitted for analysis and the other preserved by freezing for confirmatory analysis or reanalysis.

Stable ion (in mass spectrometry): An ion which is not sufficiently excited to dissociate spontaneously into a daughter/product ion and associated neutral fragment(s) or to react further in any other way within the time scale of the experiment, for example, until hitting the detector.

Stability: Resistance to decomposition or other chemical changes, or to physical disintegration.

Standard addition: The addition of a known amount of a pure component supposed to be present as a constituent of the specimen or sample in order to verify and quantitate this component. Operationally, a measurement is made on the specimen or sample, a known amount of the desired constituent is added, the modified specimen or sample is remeasured, and the amount of the constituent originally present is determined by proportionation.

Standard consensus method: See *Reference method.*

Standard deviation: A statistic that shows the spread or dispersion of scores in a distribution of scores. It is calculated by taking the square root of the variance. It is applicable to all kinds of repeated measurements, for example, between batch, within batch, repeatability, and reproducibility.

Standard operating procedures (SOPs): Written *procedures* that describe how to perform certain laboratory activities.

Standard solution: Solution of known concentration prepared from characterized material.

Statistical control: A procedure is in statistical control when results consistently fall within established control limits, that is, when they have constant mean and variance. Statistical control should be monitored graphically with control charts.

Statistical correlation: Extent to which two or more things are related to one another. This is usually expressed as a correlation coefficient.

Statistical significance: Said of a value or measure of a variable when it is larger or smaller than would be expected by chance alone. Statistical significance does not necessarily imply practical significance.

Stock solution: Concentrated standard solution used to prepare calibrators.

Study: Experiment or set of experiments performed to obtain information on a particular subject.

Study audit: A review of a completed study, interim report, supplemental data, or other document which has been submitted to the Agency under the appropriate sections of TSCA or FIFRA. This is normally accomplished by reviewing available raw data, records and reports, interviewing study personnel or other facility officials, and reviewing laboratory operations. The study audit should be based on completeness of the raw data; conformance of the raw data to study report findings and conclusions; adherence to GLP Standards regulations, and to available and appropriate standard operating procedures and protocols.

Substrate: A substrate is a reactant (other than the catalyst itself) in a catalyzed reaction.

Surrogate analyte: See *Analyte.*

Surveillance: Monitoring of certain activities to ensure that specified requirements have been fulfilled.

Survey: Study conducted among organizations to collect information on their activities or performance.

Systematic error: See *Error*.

Target analyte: See *Analyte*.

Target value: See *Value*.

T-distribution: See *Distribution*.

Test: Technical operation to determine one or more characteristics of or to evaluate the performance of a given product, material, equipment, organism, physical phenomenon, process, or service according to a specified procedure.

Testing facility: A person (including an individual, partnership, corporation, association, scientific or academic establishment, government agency, or any other legal entity) who actually conducts a study, encompassing those operational units that are being, or have been used to conduct studies. In elaborate studies, the testing facility may be comprised of one or more test sites, typically field sites and laboratories, in addition to offices and administrative units, computer laboratories and other facilities for analyzing data, and storage/archiving sites, any of which may be subcontracted.

Test linearity: Ability within a given range to obtain test results directly proportional to the concentration (amount) of analyte in the specimen or sample.

Test procedure: See *Procedure*.

Theoretical probability distribution: See *Distribution*.

Threshold: A particular, significant amount, level or limit, at which something begins to happen or take effect. See *Cut-off concentration*.

Total error: See *Error*.

Toxicodynamics: Study of toxic actions on living systems, including the reactions with and binding to cell constituents, and the biochemical and physiological consequences of these actions.

Toxicokinetics: Process of the uptake of potentially toxic substances by the body, the biotransformation they undergo, the distribution of the substances and their metabolites in the tissues, and the elimination of the substances and their metabolites from the body. Both the amounts and the concentrations of the substances and their metabolites are studied. The term has essentially the same meaning as pharmacokinetics, but the latter term should be restricted to the study of pharmaceutical substances. See also *Pharmacokinetics*.

Toxicology: Scientific discipline involving the study of the actual or potential danger presented by the harmful effects of substances (poisons) on living organisms and ecosystems, of the relationship of such harmful effects to exposure, and of the mechanisms of action, diagnosis, prevention, and treatment of intoxications.

Traceability: Ability to trace the history, application, or location of an entity by means of recorded identification. See also *Chain of custody*.

Transferability: Transferability assumes invariance of properties, associated conceptually with an atom or a fragment present in a variety of molecules. The property, such as *electronegativity, nucleophilicity*, and *NMR chemical shift,* is held as retaining a similar value in all these occurrences.

Transformation: The conversion of a *substrate* into a particular product, irrespective of reagents or *mechanisms* involved. For example, the transformation of aniline ($C_6H_5NH_2$) into *N*-phenylacetamide ($C_6H_5NHCOCH_3$) may be effected by use of acetyl chloride or acetic anhydride or ketene. A transformation is distinct from a reaction, the full description of which would state or imply all the reactants and all the products.

Trueness: Closeness of agreement between the average value obtained from a large series of *test* results and an accepted reference or true value.

True value: See *Value.*

T-test: See *Significance test.*

Type I errors and type II errors: See *Error.*

Uncertainty: A parameter, associated with the result of a measurement that characterizes the dispersion of the values that could reasonably be attributed to the analyte.

Universe: See *Population.*

Unknown: A biological sample that is the subject of the analysis.

Urine: A fluid, containing water and metabolic products, that is excreted by the kidneys, stored in the bladder, and normally discharged by way of the urethra (u).

Validated method: Method whose performance characteristics meet the specifications required by the intended use of the analytical results. Some of the performance characteristics to be evaluated are limit of detection, limit of quantitation, linearity, precision, range, ruggedness, selectivity and specificity, and trueness.

Validation: Confirmation by examination and provision of objective evidence that the particular requirements for a specific intended use are fulfilled.

Full validation: Establishment of all validation parameters to apply to sample analysis for the bioanalytical method for each analyte.

Partial validation: Modification of validated bioanalytical methods that do not necessarily call for full revalidation.

Cross-validation: Comparison validation parameters of two bioanalytical methods.

Value: The expression of a quantity in terms of a number and an appropriate unit of measurement.

Assigned value: Best available estimate of the true value.

Consensus value: Value produced by a group of experts or referee laboratories using the best possible methods. It is an estimate of the true value.

Estimate value (statistical): Value(s) of population characteristic(s) obtained from sample data.

Target value: Numerical value of a measurement result that has been designated as a goal for measurement quality.

True value: Value that characterizes a quantity perfectly defined in the conditions that exist when that quantity is considered. The true value of a quantity is an ideal concept and, in general, cannot be known exactly.

Variability: Spread or dispersion of scores in a group of scores; the tendency of each score to be unlike the others.

Variable: Generally, any quantity that varies. More precisely, a quantity that may take any one of a specified set of values.

Variance: Statistic that shows the spread or dispersion of scores in a distribution of scores. It is calculated as the sum of the squares of the differences between the individual values of a set and the arithmetic mean of the set, divided by one less than the number of values.

Variance-ratio test: See *Significance test.*

Variate: In contradistinction to a variable, a variate is a quantity that may take any of the values of a specified set with a specified relative frequency or probability. It is often known as a random variable.

Verification: Confirmation by examination and provision of objective evidence that specified requirements has been fulfilled.

Warning limit: See *Limit.*

Working standard solutions: Standard solutions prepared by diluting the stock solution containing the concentrations used to establish the calibration curve.

Z-distribution: See *Distribution.*

Z-score: Number of standard derivation units that separate a value from its mean.

9.2 GLP BASIC CONCEPTS AND IMPLEMENTATION

What are the criteria that a facility must follow?

The facility must first comply with the OECD Series on Principles of Good Laboratory Practice. Other compliance documents include the following:

a. Quality Assurance and GLP
b. Compliance with Laboratory Suppliers with GLP Principles
c. The Application of the GLP Principles to Field Studies
d. The Application of GLP Principles to Short-Term Studies
e. The Role and Responsibilities of the Study Director in GLP Studies
f. The Application of Principles of GLP to Computerized Systems
g. The Application of the OECD Principles of GLP to the Organization and Management of Multisite Studies
h. The Application of the Principles of GLP to *In Vitro* Studies

To understand some of the major differences in the way that GLP is organized it is important to look at the key roles and responsibilities defined in the regulations. There are five key roles, which are as follows.

Study director (SD): This person has the sole responsibility for the design, conduct, and reporting of the study including compliance with the regulations (in pharmaceutical manufacturing an equivalent role is the qualified person). As most GLP studies are multidisciplinary the study director needs to coordinate with scientists in other organizational departments to conduct the study effectively.

Principal investigator (PI): These are the individuals who act on behalf of the study director and have defined responsibilities for the delegated phases of a study. They are usually scientists who have a functional responsibility for the work conducted, so for the bioanalytical portion of a protocol there will be a bioanalytical PI who is identified in the protocol and report, similarly there will be a PI responsible for the pharmacokinetic or toxicokinetic analysis. Note that the responsibilities of the study director for the overall conduct of the study cannot be delegated to the PIs including approval of the study protocol, any amendments, the final report, and GLP compliance. Note that the PI concept is only found in OECD GLP regulations and not FDA GLP.

However, this is a practical way of dividing the multidisciplinary work found in a nonclinical study. In a European GLP report, the PIs sign for the scientific content of their portion of the study.

Test facility manager: This person has the authority and formal responsibility for the organization and functioning of the GLP test facility. Identification and documentation of the responsibilities of senior management in the GLP regulations contrasts with the omission in the pharmaceutical Good Manufacturing Practice regulations.

Members of quality assurance unit (QAU) or quality assurance program: These are independent individuals responsible to the test facility management for ensuring that studies were conducted according to the protocol and GLP. The key word is independent; they must not conduct any study work nor report to management responsible for conducting the study.

Archivist: This individual manages the GLP archive—a controlled area for storing records and regulatory documents generated during the execution of study protocols as well as regulatory documents written by the test facility. This will be discussed in the following section.

9.2.1 The Study Protocol

Each GLP study is conducted to a formal plan that must be preapproved before execution and formally reported after completion. This is called either a study plan or protocol and is used to describe:

- How the study will be set up?
- Doses that each group of animals will get and for how long?

- What is the route of administration?
- When (and how) the samples will be taken and analyzed (e.g., blood for the measurement of the clinical chemistry parameters such as cell counts, liver function, and hematology or for the measurement of the experimental compound under development)?
- How the data will be analyzed: the comparison of the measured blood chemistry values against reference ranges to see if the experimental drug has altered anything, or the pharmacokinetic analysis of the concentration profiles of the experimental drug?
- How the study data and results will be collated and reported?

The plan will be written by the study director with input from the various principal investigators involved and then reviewed by management and the QAU before being released for use.

During the laboratory phase of the study, changes to the protocol must be made by issuing protocol amendments that, similar to the approval of the protocol, must be formally issued and documented. Protocol changes come in two forms:

1. Amendments are planned changes from the study plan and occur before they occur.
2. As life cannot always be planned, study deviations document changes that were unplanned and are written and authorized after the event.

9.2.2 Raw Data

Raw data is one of the requirements of both the FDA and OECD GLP regulations. It is a frequently used term but what does it really mean?

OECD regulations state

- *Raw data means all original test facility records and documentation, or verified copies thereof, which are the result of the original observations and activities in a study.*
- *Raw data may also include, for example, photographs, microfilm or microfiche copies, computer readable media, dictated observations, recorded data from automated instruments, or any other data storage medium that has been recognized as capable of providing secure storage of information for the retention period of the study.*

FDA regulations state

- *Raw data means any laboratory worksheets, records, memoranda, notes, or exact copies there of, that are the result of original observations and activities of a non-clinical laboratory study and are necessary for the reconstruction and evaluation of the report of that study.*

- *In the event that exact transcripts of raw data have been prepared (e.g., tapes which have been transcribed verbatim, dated and verified accurate by signature), the exact copy or exact transcript may be substituted for the original source as raw data.*
- *Raw data may include photographs, microfilm or microfiche copies, computer printouts, magnetic media, including dictated observations, and recorded data from automated instruments.*

Raw data are generated during the course of the execution of a protocol and they are the original observations. These can either be written in a laboratory notebook, printed from a chromatograph, LIMS, or other computer system or be electronic records generated by a computer. This topic has been written about in earlier "Questions of Quality" columns in 1996 and 2000.

In the FDA GLP A758.130, part (e) states

- *All data generated during the conduct of a non-clinical laboratory study, except those that are generated by automated data collection systems, shall be recorded directly, promptly and legibly in ink.*
- *All data entries shall be dated on the date of entry and signed or initialed by the person entering the data.*
- *Any change in these entries shall be made so as not to obscure the original entry, shall indicate the reason for such change and shall be dated and signed or identified at the time of the change.*
- *In automated data collection systems, the individual responsible for direct data input shall be identified at the time of the data input. Any change in automated data entries shall be made so as not to obscure the original entry, shall indicate the reason for the change, shall be dated and the responsible individual shall be identified.*

Hence, changes to GLP raw data both for paper and automated records have the following information associated with them:

- Original data must still visible.
- Corrections to the data added.
- Initialed by the individual.
- Dated (contemporaneously—at the time the change was made).
- The reason for the change must be documented.

9.2.3 The GLP Archive and the Archivist

One of the major differences between GLP and other quality systems is the concept of the archive. This is a secure storage facility (i.e., locked, with controlled environmental conditions, and fireproof) designed to protect the records that are under the direction of an archivist. At the completion of a study all the raw data, protocols, amendments, analytical data, and specimens are transferred to the archive.

There needs to be formal procedures for the transfer of study records into the archive: a log of accession describing what the records consist of and where they are located. If material is removed from the archive, there needs to be records of this and when it will be replaced. In the latter instance when the material is returned it must be checked to ensure that all records are returned. Many GLP archive facilities also have a log of visitors to the archive: they have to sign into the archive in a log, state the reason for the entry and sign out again at the end of the visit, and be escorted by the archivist or a deputy.

9.2.4 Expansion of GLP Scope

The scope of GLP has expanded outside the original concept of nonclinical studies into drug metabolism, clinical bioanalysis, as well as agrochemical and environmental studies.

9.2.4.1 Bioanalysis and Drug Metabolism Expansion of the scope of GLP to include classic drug metabolism (excretion balance and metabolite identification) and bioanalytical studies occurred in the early 1980s. As many of the samples analyzed in these laboratories were taken from animals during the course of toxicology studies, it was logical to expand the laboratories to include those in drug metabolism and pharmacokinetic (DMPK) departments, so LC and LC–MS–MS techniques have been included.

Samples taken for bioanalysis are related to the dose groups of the animals in the study and, after LC or LC–MS analysis, are interpreted by pharmacokinetic analysis. The nearest equivalent in more conventional laboratories is a stability study, where samples are taken over time and must relate to the product batch from which they originate.

This difference in approach to GLP study samples has important considerations when trying to automate a bioanalytical laboratory by implementing a laboratory information management system. Instead of the LIMS being sample-driven and the results generated being compared with a development or product specification, there needs to be a protocol driver that allows the study to be defined electronically and the samples to be taken in the various animal dose groups used in the protocol. This difference in the way the LIMS works is so important that any organization trying to implement a sample-driven LIMS for a bioanalytical laboratory is heading for failure before the ink is dry on the project proposal.

9.2.5 OECD GLP

In addition to the FDA's GLP regulations, Europeans use a slightly different variant developed and maintained by the OECD. Based in Paris this organization appears at first a little interesting as the originator of GLP regulations, but it is the work of the Environmental Directorate that produces the regulation. The original scope of toxicology studies remains but it also extends to include environmental studies such as ecotoxicology, environmental behavior and bioaccumulation, residue analysis, and the impact on natural ecosystems. OECD defines GLP *as a quality system concerned with the organizational process and the conditions under which non-clinical health and environmental safety studies are planned, performed, monitored, recorded, archived and reported.* So agrochemical and environmental studies are covered under these GLP regulations.

The OECD has also produced a number of guidance documents, which are listed in Table 2.1 and discussed in the next section.

9.3 GLP GUIDANCE DOCUMENTS

To help and interpret GLP regulations a number of organizations (FDA, OECD, and the Swiss AGIT group) have published several guidance documents that are freely available on the Internet. These are listed in Table 2.1 and some key ones discussed below.

9.3.1 FDA Guidance for Industry on Bioanalytical Method Validation

This document was published in May 2001 and is an update of an earlier consensus conference held in 1990 and published in 1992. The guidance is intended validation and application of bioanalytical methods used to generate data from both nonclinical and human studies used in regulatory submissions. It covers the criteria to be applied for quantitative chromatographic method validation; for example, precision, accuracy, and selectivity, as well as the stability of the analyte to be measured in biological matrices such as whole blood, plasma, or urine.

In addition, what is expected when a validated method is applied to the analysis of GLP samples including reporting of the results. For example, instead of selecting the best chromatograms, the study plan needs to define which ones will be included in the report before the analysis starts so that a representative selection is used.

9.3.2 OECD GLP Guidance Documents

As you can see from Table 2.1, the OECD have produced a large number of guidance documents to help organizations interpret GLP within the areas of chromatographic and ligand-binding assays.

Application of the Principles of GLP to Computerized Systems: This provides an overview of the validation of computerized systems in a GLP environment. It is a consensus document that was published in 1995 as a collaboration between industry and the regulators. Please note that there is no mention of the need for a requirements specification, which is recognized as the key for any system validation. It contains some good elements that are useful but look at the Swiss AGIT document for a better validation guide.

Establishment and Control of Archives that Operate in Compliance with the Principles of GLP: Issued in 2007 it looks at both electronic and paper archives, in a Part 11 world it is a useful document.

9.3.3 Swiss GLP Guidance Documents

In Switzerland, collaboration between a group of GLP inspectors and industry experts has resulted in the formation of the Working Group on Information Technology (AGIT in German) who has published a series of guidance documents over the past 8 years. These documents are very practical and valuable additions to the GLP literature and can also be used outside of the GLP subject area. All the guidance documents are all listed in Table 2.1 but three of these are worth highlighting:

Validation of Computerized Systems: A more focused and up-to-date guidance on the validation of GLP computerized systems, the section on change control in the operational environment is very practical and recommended even outside of GLP.

Electronic Raw Data Archiving: Regulations always lag technological advances so the regulations look at paper-based archives when electronic records are created in abundance within GLP laboratories. These guidelines entitled "Swiss AGIT: *Guidelines for the Archiving of Electronic Raw Data in a GLP Environment*" will help readers understand more about the importance of e-data archiving.

Acquisition and Processing of Electronic Raw Data: The guidelines "Swiss AGIT: *Guidelines for the Acquisition and Processing of Electronic Raw Data in a GLP Environment*" is very informative while dealing with chromatography and complying with 21 CFR 11 (Electronic Records and Electronic Signatures regulations) because it has an excellent discussion when electronic records are created in chromatographic analysis.

REFERENCES AND SOURCES FOR ABOVE TERMINOLOGIES

- United Nations Drug Control Programme. *Report of the Consultative Meeting on Quality Assurance and Good Laboratory Practices*. Glasgow, 2–6 November 1992.
- National Institute for Drug Abuse *Urine Testing for Drugs of Abuse*, Research Monograph 73. Department of Health and Human Services, Rockville, MD, 1986.
- Miller JC, Miller N, *Statistics for Analytical Chemistry*. Ellis Horwood, Chichester, 1984.
- Wennig R. *Practical Compendium for Health Professionals: Drugs of Abuse Currently Used in Europe*, Publication CEC/V/E/I/Lux 92. Commission of the European Communities, Health and Safety Directorate, Luxembourg, 1992.
- International Olympic Committee/Reference Materials Committee of the International Organization for, Standardization. Quality control of analytical data produced in chemical laboratories, Publication 271, draft protocol presented to the *Fifth International Symposium on the Harmonization of Internal Quality Assurance Schemes for Analytical Laboratories*, Washington, DC, July 23, 1993.

- *PONS English Language Dictionary*. Collins Cobuild, London, 1987.
- Garfield FM. *Quality Assurance Principles for Analytical Laboratories*. Association of Official Analytical Chemists, Arlington, VA, 1991.
- *American Society of Crime Laboratory Directors, Laboratory Accreditation Manual*. American Society of Crime Laboratory Directors, Norfolk, VA, 1992.
- *International Organization for Standardization/International Electrotechnical Commission*. Guide 25: *General Requirements for the Competence of Calibration and Testing Laboratories*, Geneva, 1990.
- Voight WP. *Dictionary of Statistics and Methodology: A Non-Technical Guide for the Social Sciences*. Sage Publications, Thousand Oaks, CA, 1993.
- *International Organization for Standardization/International Electrotechnical Commission*. Guide 2: General Terms and Their Definitions Concerning Standardization and Related Activities, Geneva, 1991.
- Kendall MG, Buckland WR. *A Dictionary of Statistical Terms*. Longman Group, London, 1976.
- Visher C. *A Comparison of Urinalysis Technologies for Drug Testing in Criminal Justice*. National Institute of Justice and Bureau of Justice Assistance, Washington, DC, 1991.
- Society of Forensic Toxicologists, Inc./American Academy of Forensic Sciences, *Forensic Toxicology Laboratory Guidelines*, 1991.
- Eckschlager K. *Errors, Measurement and Results in Chemical Analysis*. Van Nostrand Reinhold, London, 1969.
- International Union of Pure and Applied Chemistry, *Spectrochemica Acta*, 1978.
- International Organization for Standardization, *International Vocabulary of Basic and General Terms Used in Metrology*, Geneva, 1984.
- National Measurement Accreditation Service, NIS 46, *Accreditation for Forensic Analysis and Examination*. Teddington, Middlesex, 1992.
- Recommendations (Alle Rechte Documents, Nils-Eric Saris, cd.). *International Federation of Clinical Chemistry*, Vol. 1. Walter de Gruyter, New York, 1978–1983.
- Giacomo P. *International Vocabulary of Basic and General Terms in Metrology, Metrology, 1993, 1984*, 2nd ed. International Organization for Standardization, Geneva, 1993. ISBN 92-67-01075-1.
- Horwitz, W. Nomenclature for sampling in analytical chemistry: recommendations (1990) *Pure Appl Chem* 1990;62:1193–1208.
- Bennington JL. *Saunders Dictionary and Encyclopedia of Laboratory Medicine and Technology*. W. B. Saunders Co., New York, 1984.
- Haeckel, R. Recommendations for definition and determination of carry-over effects. *J Autom Chem* 1988;10:181–183.
- Freiser H, Nancollas GH. *Compendium of Analytical Nomenclature—Definitive Rules 1987. International Union of Pure and Applied Chemistry—Analytical Chemistry Division*. Blackwell Scientific Publications, Oxford, 1987.
- *Development of Definitive Methods for the Notion Til Reference System for the Clinical Laboratory*; Approved Guideline, NCCLS Document No. NRSCCL-A, National Committee for Clinical Laboratory Standards, Villanova, PA, 1991.

- *Precision of Test Methods—Determination of Repeatibility and Reproducibility by Interlaboratory Tests*, ISO 5725-1981, Proposed Revision Part 4 (1985), International Organization for Standardization, Geneva, Switzerland.
- *Development of Certified Reference Materials for the National Reference System for the Clinical Laboratory*; Approved Guideline, NCCLS Document No. NRSCL3-A, National Committee for Clinical Laboratory Standards, Villanova, PA, 1991.
- *Tentative Guidelines for Calibration Materials in Clinical Chemistry*, NCCLS Publication 2(17). National Committee for Clinical Laboratory Standards, Villanova, PA, 1982, pp. 499–526.
- *Tentative Guidelines for Control Materials in Clinical Chemistry*, NCCLS Publication 2(18). National Committee for Clinical Laboratory Standards, Villanova, PA, 1982, pp. 527–554.
- Guilbault, GG. and Hjelm, M. Nomenclature for automated and mechanized analysis. *Pure Appl Chem*, 1989;61:1657–1664.
- compiled by Gold V, Loening KL, McNaught AD, Schmi P, *Compendium of Chemical Terminology, IUPAC Recommendations*, Blackwell, 1987.
- Brewer M, Scott T. *Concise Encyclopedia of Biochemistry*, Walter de Gruyter, New York, 1983.
- Bechtler G, Haeckel R, Horder M, Kiiffer H, Porth AJ. *Guidelines (1980) for Classification Calculation and Evaluation of Conversion Rates in Clinical Chemistry*, IFCC Document 1984, Stage 2, Draft 1. Enzyme Nomenclature Recommendations (1961) of the I.U.B. Enzyme Commission. *J Clin Chem Clin Biochem*, 1985;23:493–503. *J Autoin Chem*, 1987;9:105–112.
- Units of Enzyme Activity, Recommendations, Nomenclature Committee of the International Union of Biochemistry, Eur. J. Biochem., 97319–320, 1979. Correction 104, 1 (1980).
- Approved Recommendation (1978), Quantity and Units in Clinical Chemistry. IUPAC Section on Clinical Chemistry and IFCC Committee on Standards. Clin Chim Acta, 1979;96:F155–F204.
- Symbolism and Terminology in Enzyme Kinetics, Recommendations 1981. Nomenclature Committee of the International Union of Biochemistry (NC-IUB). Arch Biochem Biophys 224:732–740, 1983; Biochem J 21356–21371, 1983; Eur J Biochem 128281–128291, 1982.
- Svehla G. Nomenclature for kinetic methods of analysis. *Pure Appl Chem*, 1993;652291–652298.
- *Kinetics Analysis of Enzyme Reactions*, Tentative Guideline C7T, Vol. 2(9), pp. 271–328, National Committee for Clinical Laboratory Standards, Villanova, PA, 1982.
- Mills IM, Cvitas T, Kallay N, Homann K, Kuchitsu K,editors. *Quantities, Units and Symbolism in Physical Chemistry, International Union of Pure and Applied Chemistry*, 2nd edition, Blackwell Scientific Publications, Oxford, 1988; 1993, ISBN 0-632-03583-8.
- *Glossary and Guidelines for Immunodiagnostic Procedures, Reagents and Reference Materials*, 2nd ed. Approved Guideline NCCLS Document DII-A2. National Committee for Clinical Laboratory Standards, Villanova, PA, 1992.

Appendix A

GENERIC CHECKLIST FOR GLP/GxP INSPECTIONS/AUDITS

Example of an Audit Checklist

Organization and Personnel	Yes	No	NA	Observations/ Recommendations
Organizational chart exists and accurately represents the organization?				
Is the laboratory affiliated with other organizations? Identify the organizations				
Are training records available? List the components of the training record				
Are there personnel curricula (training matrix/plan) established and documented for each individual?				
Does the training program include new hire training and requalification training for personnel?				

Regulated Bioanalytical Laboratories: Technical and Regulatory Aspects from Global Perspectives,
By Michael Zhou

(*Continued*)

Organization and Personnel	Yes	No	NA	Observations/ Recommendations
Has personnel been appropriately trained to perform functions required by job descriptions?				
Is there a procedure to assess and document personnel competency on an annual basis?				
Is there an escalation process by which personnel that do not pass competency are retrained etc.?				
Has the Laboratory manager received GCLP training? If so, by what organization and when?				
Has personnel received regulatory training? GCLP □ GCP □ GLP □ cGMP □ Other				
Has personnel received health/safety training? List safety training provided				
Is there a system in place for personnel to report any safety concerns or incidents?				
Does the laboratory have sufficient qualified personnel to perform functions that support the GCLP clinical trial?				
Is there a list of consultants and is their qualifications maintained?				
Are external contractors/vendors utilized? Are they qualified/ approved for use?				
Is there an SOP that outlines this process?				

(*continued*)

(*Continued*)

Organization and Personnel	**Yes**	**No**	**NA**	**Observations/ Recommendations**
Is there a Quality Assurance Unit? If so, what are the roles of the Quality Control and the Quality Assurance group?				
Does the Quality Assurance Unit perform audits, trend metrics, and report the results to the Laboratory Management?				
Is the Quality Assurance Unit independent from the personnel engaged in the direction or conduct of a clinical trial?				
Certifications/Licenses	**Yes**	**No**	**NA**	**Observations/ Recommendations**
Does the laboratory maintain any certifications/licenses? CAP ☐ Date CLIA ☐ Date Other				
Are there copies of certifications/ licenses available?				
Standard Operating Procedures/ Methods	**Yes**	**No**	**NA**	**Observations/ Recommendations**
Is there a governing SOP that outlines the creation, revision, approval, distribution, document control, and retirement of SOPs?				
Are SOPs in compliance with the current version governing SOPs?				
Is there a current index listing of the SOPs available?				
Is there a schedule for review of the SOPs?				
Are the SOPs in locations where they are used?				

(*continued*)

(*Continued*)

Standard Operating Procedures/ Methods	**Yes**	**No**	**NA**	**Observations/ Recommendations**
Is there a system for documenting and handling SOP/method deviations and CAPAs?				
Is there a change control system for SOP/methods?				
Facility	**Yes**	**No**	**NA**	**Observations/ Recommendations**
Security and confidentiality is adequate to prevent unauthorized access to records/test samples and a procedure to report unauthorized access exists?				
Is there sufficient space to store materials, archive records, equipment to function properly, and conduct laboratory testing?				
Is the work flow designed to prevent contaminations and mixups of test samples				
Is the facility maintained and clean?				
Is there safety equipment (e.g., showers, eyewash stations) available? Is the equipment maintained?				
Are updated Material Safety Data sheets and Certificate of Analysis available?				
Are there environmental controls within laboratory and are the controls monitored?				
Are personnel wearing appropriate garmenting for designated areas?				
Is there an SOP detailing the designated routes and methods available for waste disposal?				
Is the biohazardous and hazardous chemical waste disposal described?				

(*continued*)

(*Continued*)

Facility	**Yes**	**No**	**NA**	**Observations/ Recommendations**
Is there a sanitation or cleaning procedure established and is being followed and documented?				
Are facilities maintenance procedures established and being followed and documented?				
Does the Laboratory have a disaster recovery plan that covers all areas of the facility including computer systems and equipment?				
Are generators utilized at the facility?				
Does the lab have an SOP for the testing and maintenance of generators? Request to review generator logs				
Does the lab use a water purifying system? Are there logs to show maintenance of the system? Grade of water utilized?				
Equipment	**Yes**	**No**	**NA**	**Observations/ Recommendations**
Is equipment used for GCLP studies readily distinguishable from equipment used for non-GCLP Studies?				
Is there a Master Equipment Inventory present?				
Are there site-specific SOPs detailing equipment use, maintenance, and calibration?				
Is the equipment utilized in the lab suitable to perform the required operations?				
Are user logs utilized for equipment and do they include a chronological record of use?				

(*continued*)

(*Continued*)

Equipment	**Yes**	**No**	**NA**	**Observations/ Recommendations**
Log entries show the date, time, name of person performing and checking the work, as appropriate				
Equipment calibration and/or preventative maintenance schedules have been established and are being followed and documented				
Records of equipment calibration or maintenance are maintained in the laboratory and archived				
Is there a system for moving or removing a piece of equipment from service or tagging the equipment? Documentation?				
Calibration of the equipment is traceable to NIST or another recognized standards institution				
Are there established tolerance limits for the equipment? Who established the tolerance ranges?				
Are the equipment manuals available?				
Is there a written equipment qualification/validation program?				
Laboratory Controls	**Yes**	**No**	**NA**	**Observations/ Recommendations**
Are there an assay validation, revalidation, and limited validation process outlined in a SOP?				
Is there a written procedure for repeat testing or invalidating lab data? Is there a repeat decision tree?				
How are results that fail specifications investigated or nonconformances investigated?				

(*continued*)

(*Continued*)

Laboratory Controls	**Yes**	**No**	**NA**	**Observations/ Recommendations**
Are there validated methods and acceptance criteria for each test method?				
Is there a SOP for significant figures?				
Is there a SOP that outlines good documentation practices?				
Reagent and Solution Labeling and Qualification	**Yes**	**No**	**NA**	**Observations/ Recommendations**
Is there an SOP that outlines how reagents are labeled, how expiration dates are established?				
Are reagents qualified for use? Is parallel testing of reagents performed?				
Is there a current inventory of all reagents and solutions?				
Sample Shipment, Receipt, and Storage	**Yes**	**No**	**NA**	**Observations/ Recommendations**
Is there a SOP for sample receipt, shipment, and storage of materials and test samples?				
Does the SOP contain a chain of custody procedure?				
Is the sample receipt area maintained separate from the sample processing area?				
Data-Handling Procedures and Computer Validation	**Yes**	**No**	**NA**	**Observations/ Recommendations**
Is access to computers limited by an individual username and password system (lab members cannot share a username)?				
How is the computer network and computer systems maintained, if applicable?				

(*continued*)

(*Continued*)

Data-Handling Procedures and Computer Validation	**Yes**	**No**	**NA**	**Observations/ Recommendations**
Are there a computer validation master plan and/or SOPs?				
List computers systems and software utilized. Validated?				
Are changes to computer systems controlled and documented?				
Are records of computer system errors maintained and investigated?				
Are records of hardware maintenance and repairs maintained?				
Are computers backed up routinely to prevent loss of data? Is there a backup log?				
Is there a preventative maintenance program for computer systems?				

Records and Reports	**Yes**	**No**	**NA**	**Observations/ Recommendations**
A documentation control system exists and is functional				
Is raw laboratory data recorded in lab notebooks, electronically, or controlled data sheets?				
Are laboratory final reports generated for clinical studies? Who reviews the reports?				
Is there a SOP that outlines the content of the final report?				
Is there a SOP or a system for the retention, storage, and destruction of records?				
How does the site ensure the sponsor's proprietary information is not disclosed to unauthorized personnel or external organizations?				

(*continued*)

Record Retention and Archival	**Yes**	**No**	**NA**	**Observations/ Recommendations**
Is there a dedicated facility/area for the archival of records?				
Is there control access to the archival facility?				
Is the environment of the facility monitored and controlled?				
Is the procedure for archiving records outlined in an SOP?				
Is the retention time for records stated in the SOP?				
Is there a method of electronic data archive?				

Appendix B

GENERAL TEMPLATE FOR SOP

GLP Standard Operating Procedures	
SOP number here	Title: Insert SOP title here
Print ID number	
Original effective date	
Current version effective date	Page _1_ of _??_

Insert Title Here

1.0 Purpose

The purpose of this procedure is to define the ________ of all GLP studies at Group/Department/Division/Company.

2.0 Application/Scope

This procedure shall be applied to ________ in support of GLP studies on the subject/study.

Regulated Bioanalytical Laboratories: Technical and Regulatory Aspects from Global Perspectives,
By Michael Zhou

3.0 Definitions

Insert key definitions here (examples listed can be used or deleted).
GLP = (FDA) Good Laboratory Practice as defined by 21 CFR 58.
QAU = Quality Assurance Unit.
Management = GLP Management.

Designated Director = Director of a study or unit such as Study Director, Technical Director, or QAU Lead.

4.0 References

Insert key references here.
Example: FDA 21 CFR 58, Revision April 1, 2004.

5.0 Associated SOPs
5.1 XXX-000-0000-v00
5.2 XXX-000-0000-v00

Or

There are no SOPs associated with this SOP.

6.0 Responsibilities
6.1 Insert Job Title
6.1.1 Insert Responsibilities
6.2 Insert Job Title
6.2.1 Insert Responsibilities
6.3 Insert Job Title
6.3.1 Insert Responsibilities

GLP Standard Operating Procedures	
SOP number here	Title: Insert SOP title here
Print ID number	
Original effective date	
Current version effective date	Page _1_ of _??_

7.0 Materials and Equipment
7.1 Example: Centrifuge
7.2 Example: LC–MS/MS Instrument
7.3 ELISA Scanner

Or

There are no specialized materials or equipment associated with this SOP.

8.0 Procedure

8.1 Start List here

8.1.1 More here
8.1.2 More here
8.1.3 More here

8.1.3.1 More here

8.2 More here

8.2.1 More here
8.2.2 More here, etc.

9.0 Controlled Forms

9.1 The following forms are available in Appendix __ of the GLP SOP Manual.

9.1.1 Title of Controlled Form here
9.1.2 Title of Controlled From here

Or

There are no controlled forms associated with this SOP.

10.0 Safety Precaution

Or

There are no safety concerns associated with this SOP—N/A.

11.0 Revision History

SOP Number including Revision Number	**Effective Date**
SOP number here	

Appendix C

TYPICAL SOPs FOR GLP/REGULATED BIOANALYTICAL LABORATORY

QUALITY ASSURANCE—GLP

1. Quality Assurance Unit—Roles, responsibilities, authorities, and accountabilities
2. Training Requirements—QA Unit Personnel
3. Master Schedule—QA Unit (GLP Lab)
4. QA Reports and QA Statement (GLP Lab)
5. Deviations: Notification, Handling, and Investigation including laboratory OOS investigations and production deviation/nonconformance investigations
6. Management notification
7. Change Management—tracking, managing, and documenting of evolving knowledge of processes, formulations, methods, specifications, and contractors.
8. Metrics to monitor performance
9. Quality Agreement
10. Drug Regulatory Inspection
11. QA Audit:
    a. Critical Phase (in-study) Inspection audit (GLP Lab)
    b. Vendor/contractors qualification and audits

Regulated Bioanalytical Laboratories: Technical and Regulatory Aspects from Global Perspectives,
By Michael Zhou

c. Internal audits
d. Audit of bioanalytical raw data and reports (GLP Lab)
e. Audit of contractors raw data and reports
f. etc.

BIOANALYTICAL—GLP LABORATORIES

1. Organization, roles, and responsibilities of GLP Lab
2. Study Director role and responsibilities
3. Procedure for the control of incoming test specimen—receiving, storage, inventory, login/logout, and disposal—chain-of-custody check
4. Procedure for the control, qualification, calibration, and PM of laboratory freezers
5. Control, documentation, and control of control articles—qualification of reference standards
6. Control, documentation, and control of blank biological matrices
7. Control, storage, expiry dates, and destruction of chemicals, reagents
8. Procedure to Label test specimens, chemicals, and solutions
9. Preparation and control of Project Binder
10. Preparation of Standard and Quality Control samples
11. Preparation and qualification of stock standard solutions
12. Procedure for validation of bioanalytical methods—content, review, approval, and change control
13. Sample analysis plan—in study control
14. Chromatography Acceptance Criteria—Procedure to review and accept HPLC. LC–MS/MS chromatographic data—peak integration, baseline review, and acceptance criteria
15. Calibration curves, quality control, and analytical runs acceptance criteria
16. Sample reassay procedure, acceptance, and reporting
17. Significant figures round off procedure
18. System suitability tests requirements
19. Documentation and record-keeping and record-retention practices—procedure to enter raw data
20. Laboratory verification, review and approval of analytical raw data, calculation, derived data, and reports
21. Procedure to determine and reassay of PK outliers
22. Procedure to investigate and resolve of significant recurring events/deviations—SOP for Event Investigation and Resolution
23. Backup and archiving of study data
24. Testing of incurred sample reanalysis (ISR)

25. Procedure to record deviations
26. Procedure to issue analytical report
27. Procedure to evaluate the stability of analyte in stock solution and biological samples
28. Procedure to evaluate recovery of analytes in biological fluid
29. Procedure for the cleaning of laboratory glassware
30. Instrument qualification, calibration, and PM program
31. Operation, maintenance, and calibration of various lab instruments:
    a. HPLC, LC–MS/MS, centrifuge, balances, autopipette, and robotic multiprobe dilutors
32. Computer system validation (CSV) and related CSV SOPs
33. Training and qualification requirements for bioanalytical personnel
34. etc.

Appendix D

BASIC EQUIPMENT/APPARATUS FOR BIOANALYTICAL LABORATORY

Sample Receiving and Sample Login: Watson, Item Track softwares, Bar code reader, computerized system for sample login, tracking, chain of custody, and so on.

Sample Storage/Monitoring: Refrigerators, Freezers (− 20 and − 70°C), Temperature monitoring device (chart paper recorder, Amega, or alike environment monitoring system), and power backup (UPS or other generator).

Sample Preparation: Analytical balance, Liquid handler, pipette, sonicate, vortex, vacuum manifold, TurboVac, centrifuge, tissue (grinder, homogenizer, mill, etc.), and automated system (robots, etc.).

Wet Chemistry/Conventional Analysis: Laboratory glassware, general suppliers, HPLC, and UV spectrometer/spectrophotometer/plate reader.

Instrumental Analysis: HPLC–UV/HPLC–MS/HPLC–MS/MS and other types of hypernated systems.

Data Acquisition and Analysis: Computer or computerized system with vendor softwares (Analyst, ChemStation, X-Caliber, etc.).

Data Storage/Retention: LIMS (Watson, Neugenesis, etc.), archive storage facility, and so on.

Regulated Bioanalytical Laboratories: Technical and Regulatory Aspects from Global Perspectives,
By Michael Zhou

Appendix E

WEBSITE LINKAGES FOR REGULATED BIOANALYSIS

1. **Regulatory Documents and Information**

   Comparison of Definitions and Interpretation of GLP (FDA, EPA, OECD, etc.) http://www.fda.gov/ora/compliance_ref/bimo/comparison_chart/definitions.html

   FDA Compliance Program Guidance Manual http://www.fda.gov/ora/cpgm/default.htm

   FDA/ORA Compliance Guides Manual http://www.fda.gov/ora/compliance_ref/cpg/default.htm

   Title 21 Code of Federal Regulations (21 CFR Part 11) Electronic Records; Electronic Signatures http://www.fda.gov/ora/compliance_ref/part11/

   FDA's Electronic Freedom of Information Reading Room—Warning Letters and Responses http://www.fda.gov/foi/warning.htm

   21 CFR Part 58 (GLP Regulations) http://www.accessdata.fda.gov/scripts/cdrh/cfdocs/cfcfr/CFRSearch.cfm

   FDA Guidance Documents http://www.fda.gov/opacom/morechoices/industry/guidedc.htm

   FDA ORA Lab Manual http://www.fda.gov/ora/science_ref/lm/default.htm

   Irish GLP Compliance Program Manual http://www.inab.ie/media/GLP%20Manual.pdf

   Comparison Charts for GLP re FDA_EPA_OECD http://www.fda.gov/ora/compliance_ref/bimo/comparison_chart/FDA-EPA-OECD2.pdf

Regulated Bioanalytical Laboratories: Technical and Regulatory Aspects from Global Perspectives,
By Michael Zhou

2. **Representative Regulatory Affairs around the World**

European Medicines Agency (EMEA) http://www.emea.europa.eu/
Food and Drug Administration (FDA) http://www.fda.gov/
Heads of Medicines Agencies (HMA) http://www.hma.eu/
International Conference on Harmonization of Technical Requirements for Registration of Pharmaceuticals for Human Use (ICH) http://www.ich.org/cache/compo/276-254-1.html
International Cooperation on Harmonization of Technical Requirements for Registration of Veterinary Products (VICH) http://www.vichsec.org/en/guidelines.htm
State Food and Drug Administration, P.R. China (SFDA) http://eng.sfda.gov.cn/eng/
Medicines Healthcare products Regulatory Agency (MHRA) http://www.mhra.gov.uk/index.htm
Pharmaceuticals and Medical Devices Agency, Japan (PMDA) http://www.pmda.go.jp/index.html

3. **Representative Pharmacopoeias around the World**

British Pharmacopoeia (BP) http://www.pharmacopoeia.co.uk/
European Pharmacopoeia (EP) http://www.edqm.eu/en/Homepage-628.html
Japanese Pharmacopoeia (JP) http://jpdb.nihs.go.jp/jp14e/
United States Pharmacopoeia (USP) http://www.usp.org/

4. **Representative Organizations around the World**

Active Pharmaceutical Ingredients Committee (APIC) http://apic.cefic.org/
Association of European Self-Medication Industry (AESGP) http://www.aesgp.be/
Belgian Association for Bioindustries (BIO.BE) http://www.bio.be/
Biotechnology Industry Organization (BIO) http://www.bio.org/
Clinical Research Cluster (ARESA) http://www.aresa.be/
European Association for Bioindustries (EuropaBio) http://www.europabio.org/
European Association of Pharma Biotechnology (EAPB) http://www.eapb.de/
European Biopharmaceutical Enterprises (EBE) http://www.ebe-biopharma.org/
European Commission—Enterprise and Industry, Pharmaceuticals (ECDG) http://ec.europa.eu/enterprise/pharmaceuticals/index_en.htm
European Directorate for the Quality of Medicines Healthcare (EDQM) http://www.edqm.eu/en/Homepage-628.html
European Federation of Biotechnology (EFB) http://www.efb-central.org/
European Federation of Pharmaceutical Industries and Associations (EFPIA) http://www.efpia.org/
European Generic Medicines Association (EGA) http://www.egagenerics.com/
European Patent Office (EPO) http://www.epo.org/
European Vaccine Manufacturers (EVM) http://www.evm-vaccines.org/

French Society of Pharmaceutical Science and Technology (SFSTP) http://www.sfstp.org/
International Federation for Animal Health (IFAH) http://www.ifahsec.org/
International Federation for Animal Health—Europe (IFAH-Europe) http://www.fedesa.be/
International Pharmaceutical Federation (FIP) http://www.fip.org/www/
International Federation of Pharmaceutical Manufacturers and Associations (IFPMA) http://www.ifpma.org/
Metabolism and Pharmacokinetic Group (GMP) http://www.gmp.asso.fr/
Organization for Economic Cooperation and Development (OECD) http://www.oecd.org/
Pharmaceutical Group of the European Union (PGEU) http://www.pgeu.org/
Plasma Protein Therapeutics Association (PPTA) http://www.pptaglobal.org/
World Health Organization (WHO) http://www.who.int/en/
World Organization for Animal Health (OIE) http://www.oie.int/eng/en_index.htm
World Self-Medication Industry (WSMI) http://www.wsmi.org/

5. **Representative Technical Organization/Associations around the World**

Analytical and Mass Spectrometry Organizations
ACS Division of Analytical Chemistry Homepage http://www.analyticalsciences.org/
American Physical Society http://www.aps.org/
The American Society for Mass Spectrometry http://www.asms.org/
Australian and New Zealand Society for Mass Spectrometry (ANZAMS) http://www.anzsms.org/index.htm
Belgian Society for Mass Spectrometry (BSMS) http://www.bsms.be/
British Mass Spectrometry Society (BMSS) http://www.bmss.org.uk/
Canadian Society for Mass Spectrometry (CSMS) http://www.csms.inter.ab.ca/
European Society for Mass Spectrometry (ESMS) http://www.bmb.leeds.ac.uk/esms/
International Mass Spectrometry Society/Foundation (IMSS/IMSF) http://www.imss.nl/
Indian Society of Mass Spectrometry (ISMS) http://www.ismas.org/massspectrometers.htm
Mass Spectrometry Society of Japan (MSSJ) http://www.mssj.jp/
Polish Mass Spectrometry Society (PMSS) http://ptsm.ibch.poznan.pl/index_-eng.php
South African Association for Mass Spectrometry (SAAMS) http://www.saams.up.ac.za/
Swiss Group for Mass Spectrometry (SGMS) http://www.sgms.ch/

6. **Representative Local Mass Spectrometry Discussion Groups around the World**

Atlanta–Athens Mass Spectrometry Discussion Group (AAMSDG) http://www.aamsdg.org/

B.C. Regional Mass Spectrometry Discussion Group http://www.csms.inter.ab.ca/bc.htm

Colorado Biological Mass Spec Society http://cbmss.org/

Delaware Valley Mass Spectrometry Discussion Group http://science.widener.edu/svb/msdg/

East Tennessee Mass Spectrometry Discussion Group. http://www.chem.utk.edu/~etmsdg/

Greater Boston Mass Spectrometry Discussion Group (GBMSDG) http://www.gbmsdg.org/

Madison–Chicago–Milwaukee Mass Spectrometry Discussion Group (MCM–MSDG). http://www.chem.uic.edu/mcm/

Canadian LC–MS Group http://www.canadianlcmsgroup.com/

Minnesota Mass Spectrometry Discussion Group (MMSDG) http://www.minnmass.org/MinnMass/Welcome.html

Montreal Mass Spectrometry Discussion Group http://www.csms.inter.ab.ca/montreal.htm

National Institutes of Health Mass Spectrometry Discussion Group http://dir.niehs.nih.gov/dirlsb/msshome.htm

NCI-FCRDC Mass Spectrometry Interest Group or Mass Spectrometry Interest Group of the NCI at Frederick http://msig.ncifcrf.gov/

North Jersey Mass Spectrometry Discussion Group http://www.njacs.org/msdg/index.html

Pacific Northwest Mass Spectrometry Group http://www.pacmass.org/

The Pittsburgh Mass Spectrometry Discussion Group http://chemed.chem.pitt.edu/ssp-msdg/

San Francisco Bay Area Mass Spectrometry Discussion Group http://www.bams.org/

The San Diego Mass Analysis Network (aka Scripps Center for Mass Spectrometry) http://masspec.scripps.edu/sandman/sandman.php

The Triangle Area Mass Spectrometry Discussion Group http://biochem.ncsu.edu/TAMS

Washington–Baltimore Mass Spectrometry Discussion Group http://chemistry.nrl.navy.mil/6110/6111/

INDEX

AAPS, 80, 85, 127, 129, 137, 168, 169, 207, 213, 223, 224, 226, 227, 228, 276, 277, 291, 294, 295, 409, 411, 430, 431
abbreviations, 32, 204, 409
ability, 8, 20, 24, 59, 64, 68, 69, 72, 78, 113, 128, 148, 151, 155, 160, 162, 182, 198, 225, 236, 247, 260, 280, 283, 357, 365, 367, 371, 395, 397, 408, 417, 433, 441, 448, 464, 469
absorbance, 171
absorption, 145, 171, 172, 197, 276, 325, 360, 440, 450, 460
abundance, 400, 449, 477
accelerator, 402, 452
acceptability, 17, 23, 54, 69, 107, 134, 178, 190, 214, 226, 263, 267, 313, 326, 327, 334
acceptance, 7, 14, 16, 19, 24, 25, 27, 29, 30, 32, 42, 43, 55, 59, 62, 63, 66, 75, 78, 80, 82, 83, 84, 110, 127, 128, 136, 137, 139, 144, 153, 168, 175, 179, 180, 184, 191, 192, 200, 201, 204, 207, 208, 209, 210, 212, 213, 214, 215, 216, 219, 220, 223, 224, 225, 226, 227, 231, 257, 260, 263, 269, 274, 290, 291, 300, 301, 303, 304, 305, 311, 324, 325, 328, 331, 332, 333, 334, 335, 336, 337, 352, 356, 358, 406, 412, 416, 420, 433, 453, 487, 496
accessibility, 67, 367
accordance, 14, 29, 37, 39, 43, 50, 53, 56, 59, 67, 68, 69, 88, 95, 98, 107, 131, 139, 141, 143, 194, 202, 206, 304, 305, 310, 316, 412, 420, 438, 449
accountability, 15, 19, 39, 49, 62, 71, 101, 128, 155, 182, 265, 314, 315, 323, 325, 327, 335
accreditation, 7, 74, 75, 76, 77, 84, 88, 89, 95, 134, 394, 433, 460, 478
accuracy, 9, 13, 17, 20, 21, 23, 24, 25, 26, 27, 29, 31, 39, 61, 68, 82, 94, 98, 99, 142, 144, 145, 147, 148, 149, 166, 168, 175, 177, 178, 179, 180, 184, 186, 187, 190, 193, 198, 201, 206, 208, 212, 213, 214, 216, 217, 218, 219, 225, 226, 227, 237, 239, 241, 250, 256, 257, 260, 261, 263, 264, 265, 267, 269, 270, 283, 287, 299, 301, 303, 314, 315, 320, 323, 324, 325, 326, 327, 328, 329, 330, 331, 332, 333, 334, 335, 336, 338, 361, 385, 390, 400, 404, 410, 411, 417, 423, 433, 451, 453, 456, 460, 461, 462, 476

Regulated Bioanalytical Laboratories: Technical and Regulatory Aspects from Global Perspectives,
By Michael Zhou

acquisition, 137, 161, 189, 195, 196, 234, 245, 252, 259, 277, 286, 314, 319, 356, 477, 499
ACS, 503
Acta, 292, 423, 478, 479
activity, 42, 56, 69, 105, 121, 122, 146, 157, 170, 171, 191, 194, 203, 206, 211, 224, 259, 339, 348, 399, 400, 403, 414, 442, 443, 447, 448, 452, 460, 461, 462, 479
addendum, 218, 221, 360, 410, 411
additions, 64, 118, 120, 196, 202, 317, 477
adduct, 374, 381, 384, 385, 389
adequacy, 57, 109, 110, 116, 148, 186, 303, 307, 312, 338, 345, 433
adherence, 37, 40, 41, 44, 46, 67, 69, 75, 99, 149, 186, 202, 206, 303, 304, 305, 313, 338, 344, 407, 412, 417, 461, 468
ADME, 132, 171, 175, 360, 361
administration, 10, 35, 65, 72, 74, 82, 84, 98, 117, 118, 125, 128, 129, 136, 152, 158, 172, 178, 185, 202, 227, 228, 230, 265, 266, 267, 269, 294, 295, 299, 300, 302, 303, 304, 315, 319, 323, 336, 342, 352, 353, 354, 399, 420, 435, 445, 456, 473, 502
adoption, 83, 92, 161, 191, 421
adsorption, 258, 367, 440, 446
advancement, 116, 265, 358, 401, 419
advantage, 1, 125, 232, 252, 367, 369, 371, 372, 377, 402, 404, 407, 420, 439
adverse, 46, 113, 186, 187, 270, 271, 300, 301, 302, 325, 326, 330, 332, 333, 334, 451, 452, 455
advisory, 45, 406
affairs, 37, 128, 129, 345, 502
affinity, 178, 211, 212, 213, 289, 290, 435
agency, 10, 12, 13, 29, 35, 36, 37, 38, 39, 40, 54, 55, 66, 67, 68, 74, 76, 82, 95, 97, 99, 107, 109, 138, 147, 173, 183, 228, 282, 299, 300, 301, 302, 305, 306, 317, 320, 339, 344, 345, 347, 351, 352, 358, 397, 398, 402, 405, 417, 445, 468, 469, 502
agent, 176, 255, 258, 364, 379, 399, 403, 424, 435, 438, 450, 462, 465
agreement, 23, 26, 59, 75, 79, 225, 285, 303, 394, 395, 433, 447, 459, 463, 464, 470, 495
agrochemical, 8, 475, 476
algorithms, 160, 200
aliquot, 232, 237, 434
allocation, 56, 114, 217
allowance, 114, 118, 277
ambient, 25, 122, 179, 181, 192, 227, 444, 445
ambiguity, 147, 357, 440
amendment, 38, 43, 58, 63, 83, 110, 118, 119, 202, 204, 205
analog, 212, 222, 254, 257
analysts, 13, 15, 18, 20, 31, 67, 94, 148, 150, 179, 182, 184, 189, 194, 200, 224, 232, 261, 268, 269, 281, 287, 337, 365, 377, 395, 417, 418
analyte, 4, 8, 18, 19, 20, 21, 22, 23, 24, 25, 26, 27, 28, 30, 94, 132, 145, 151, 152, 169, 175, 176, 177, 178, 179, 198, 199, 210, 212, 213, 214, 216, 217, 218, 219, 221, 222, 223, 224, 226, 227, 233, 235, 237, 238, 239, 240, 241, 246, 248, 251, 252, 255, 256, 257, 258, 260, 261, 263, 267, 274, 287, 324, 325, 328, 329, 330, 337, 365, 366, 367, 370, 371, 372, 373, 374, 375, 376, 379, 380, 385, 386, 387, 388, 389, 404, 434, 435, 436, 439, 440, 441, 443, 445, 447, 448, 449, 450, 451, 452, 454, 455, 456, 459, 461, 462, 463, 464, 465, 466, 467, 468, 469, 470, 476, 497
analyzer, 256, 288, 368, 460
ANDA, 81, 131, 143, 202, 220, 323
antibody, 6, 27, 132, 221, 259, 261, 262, 290, 434, 435, 438, 439, 443, 445, 447, 448, 450, 454, 455, 459, 462, 467
anticancer, 359, 403
anticoagulant, 18, 220, 223, 224, 363, 436
antidepressant, 174, 249
antigen, 400, 434, 439, 443, 445, 447, 448, 450, 454, 455, 459, 462
antiserum, 434, 439, 454, 458
AOAC, 84, 85, 228, 353
APCI, 236, 254, 369, 370, 371, 372, 373, 376, 381, 385, 387, 428
API, 162, 163, 224, 247, 369, 389
apparatus, 11, 13, 50, 61, 90, 93, 145, 197, 234, 241, 257, 359, 364, 437, 463, 465, 499
appearance, 13, 54, 107
appendix, 128, 199, 344, 481, 482, 483, 484, 485, 486, 487, 488, 489, 491, 492, 493, 495, 496, 497, 499, 501, 502, 503, 504
APPI, 371, 372, 373, 375
applicability, 35, 42, 79, 109, 169, 190, 207, 313

application, 9, 19, 23, 25, 27, 29, 31, 38, 40, 43, 45, 46, 47, 61, 64, 65, 66, 72, 73, 76, 78, 81, 82, 83, 84, 87, 99, 143, 157, 160, 162, 166, 172, 173, 188, 191, 193, 194, 197, 205, 206, 218, 228, 245, 246, 250, 257, 264, 266, 272, 275, 277, 287, 288, 289, 290, 295, 300, 301, 304, 320, 337, 344, 356, 365, 367, 368, 370, 372, 374, 377, 378, 379, 387, 389, 399, 400, 401, 406, 421, 457, 462, 469, 471, 476, 491
appropriateness, 140, 212
approval, 7, 8, 9, 12, 16, 17, 35, 38, 39, 42, 47, 54, 55, 58, 63, 68, 74, 80, 81, 82, 83, 89, 100, 108, 118, 120, 121, 128, 131, 140, 143, 148, 149, 161, 164, 166, 173, 174, 175, 176, 186, 187, 188, 201, 202, 203, 204, 206, 220, 255, 266, 270, 273, 274, 300, 301, 302, 303, 311, 316, 323, 324, 326, 337, 342, 355, 356, 397, 398, 412, 414, 418, 419, 445, 472, 473, 483, 496
archival, 47, 55, 109, 128, 134, 135, 136, 142, 144, 155, 182, 489
archivist, 142, 472, 474, 475
arena, 163, 207, 373, 376, 400, 412
Aria, 239, 240, 242, 244
array, 44, 197, 231, 421
arrival, 62, 65, 121, 122, 126, 152
article, 38, 39, 53, 85, 106, 110, 121, 122, 123, 125, 129, 142, 190, 201, 227, 228, 240, 264, 289, 291, 295, 300, 301, 303, 308, 310, 311, 313, 314, 315, 319, 320, 321, 330, 367, 371, 404, 406, 431
ASMS, 294, 424, 428, 429, 503
assays, 4, 5, 6, 25, 26, 64, 66, 80, 85, 127, 129, 132, 145, 151, 157, 159, 160, 166, 169, 171, 180, 207, 208, 209, 210, 211, 212, 214, 215, 219, 222, 225, 227, 231, 232, 234, 239, 245, 250, 252, 258, 259, 260, 261, 262, 263, 279, 287, 288, 289, 291, 296, 303, 335, 338, 358, 360, 361, 376, 392, 398, 399, 400, 401, 405, 406, 409, 433, 436, 476
assessment, 3, 37, 39, 40, 43, 67, 73, 75, 76, 77, 81, 91, 121, 132, 135, 136, 166, 171, 182, 187, 191, 205, 211, 212, 216, 218, 262, 271, 275, 278, 280, 283, 304, 305, 306, 327, 335, 359, 386, 393, 394, 404, 405, 408, 410, 411, 418, 460, 465
association, 1, 7, 37, 70, 78, 84, 150, 276, 289, 295, 353, 354, 357, 415, 434, 437, 456, 469, 478, 502, 503
assumption, 134, 257, 260, 466, 467
attachment, 95, 204, 346, 374, 437
attrition, 287, 289, 392
auditor, 94, 95, 99, 135, 137, 138, 183, 184, 234, 340, 341
authenticity, 67, 68, 70, 72, 101, 156, 414, 415, 417
authority, 7, 11, 32, 41, 45, 53, 54, 55, 56, 58, 67, 71, 80, 102, 107, 108, 134, 155, 165, 185, 203, 269, 270, 290, 307, 338, 339, 341, 347, 350, 396, 416, 417, 433, 463, 472
authorization, 48, 49, 98, 120, 141, 148, 185, 269, 306, 312, 321, 337, 359, 447
automation, 4, 5, 83, 132, 149, 161, 170, 176, 207, 233, 234, 241, 244, 245, 251, 263, 283, 353, 355, 364, 366, 398, 405, 412, 422, 424
autosampler, 222, 232, 257, 376, 426
availability, 11, 23, 57, 120, 139, 174, 257, 261, 262, 287, 290, 311, 321, 359, 368, 377, 403

BA, 16, 17, 19, 73, 83, 168, 187, 201, 202, 203, 204, 228, 253, 265, 266, 267, 268, 269, 270, 271, 272, 273, 291, 299, 300, 323, 326, 327, 329, 330, 331, 333, 335, 337, 398, 425, 428
barcode, 158, 160, 163, 244
BARQA, 7, 78, 79, 83, 357, 415, 434
baseline, 37, 70, 192, 200, 250, 274, 276, 334, 336, 457, 496
basis, 3, 6, 9, 10, 13, 16, 21, 22, 28, 30, 36, 40, 44, 76, 77, 97, 99, 100, 101, 103, 110, 113, 116, 127, 134, 144, 159, 162, 169, 182, 184, 185, 211, 212, 215, 262, 264, 268, 269, 271, 321, 327, 328, 329, 330, 331, 333, 343, 344, 347, 348, 357, 393, 399, 422, 445, 447, 460, 482
batches, 180, 366, 410
benchmark, 207, 208, 212, 270
benchtop, 181, 216
bidirectional, 158, 160, 244, 436
BIMO, 36, 336, 353, 501
binder, 120, 194, 196, 197, 274, 307, 496
bioanalysis, 4, 7, 32, 34, 73, 80, 81, 89, 100, 126, 127, 131, 132, 133, 135, 137, 138, 151, 153, 157, 166, 168, 169, 170, 171, 175, 176,

178, 182, 183, 198, 200, 204, 205, 206, 207, 212, 231, 232, 233, 234, 235, 244, 246, 248, 250, 252, 254, 255, 256, 257, 258, 260, 262, 264, 265, 278, 286, 288, 290, 291, 296, 355, 362, 363, 365, 366, 368, 369, 370, 372, 374, 375, 380, 381, 385, 386, 387, 388, 391, 393, 394, 401, 403, 407, 409, 411, 412, 413, 422, 424, 425, 430, 431, 475, 501, 502, 503, 504
bioanalyst, 135, 386
bioavailability, 3, 73, 80, 81, 83, 84, 89, 121, 129, 166, 168, 174, 198, 206, 227, 228, 259, 265, 266, 278, 294, 299, 303, 304, 323, 353, 354, 391, 398, 401, 402, 407, 435
biochemistry, 4, 8, 79, 426, 479
bioequivalence, 16, 31, 73, 80, 81, 83, 84, 89, 129, 166, 168, 198, 205, 206, 207, 218, 220, 227, 228, 245, 265, 266, 294, 295, 299, 303, 304, 322, 323, 328, 353, 354, 398, 407, 410, 411
bioinformatics, 170, 207
biologics, 337, 359
biology, 131, 289, 399
biomarker, 5, 6, 7, 263, 287, 288, 289, 290, 392, 399, 400, 401, 414, 435
biomolecules, 166, 231, 378
biopharmaceutical, 4, 5, 207, 337, 356, 392, 502
bioresearch, 36, 37, 128, 152, 268, 295, 300, 302, 306, 320, 328, 336, 353
biosafety, 104, 105
biosynthesis, 439, 448
biotechnology, 5, 69, 207, 231, 285, 305, 337, 356, 360, 502
biotransformation, 435, 458, 469
biowaivers, 323
blood, 4, 17, 74, 79, 151, 177, 202, 207, 212, 216, 235, 236, 237, 242, 248, 290, 319, 344, 358, 362, 363, 403, 404, 405, 414, 435, 436, 442, 452, 467, 473, 476
BMV, 16, 23, 29
bottlenecks, 95, 235, 280, 282, 283, 286, 368

calculation, 31, 126, 158, 160, 274, 365, 479, 496
calibrations, 76, 77, 88, 151, 188, 272, 418, 436
calibrator, 26, 27, 216, 260, 287, 436, 456
cancer, 287, 295, 296, 306, 403, 430, 442
CAPA, 141, 186, 270, 322, 324, 326, 340, 341, 418
capability, 61, 94, 160, 244, 248, 278, 328, 370, 391, 404, 408, 416, 418
capacity, 23, 113, 132, 135, 141, 152, 236, 246, 258, 259, 262, 275, 279, 280, 281, 286, 311, 317, 361, 377, 387, 388, 392, 393, 395, 407, 408, 435, 451, 452, 455, 465
carcinogenicity, 174, 306, 322
carryover, 216, 217, 248, 380, 421, 436
cartridge, 236, 237, 238, 239, 240, 241, 243, 253, 364
categories, 90, 93, 133, 151, 195, 197, 260, 277, 289, 307, 345, 396, 399
CBER, 129, 302
CDER, 84, 166, 217, 228, 295, 302, 322, 354
cell, 64, 65, 211, 240, 242, 259, 290, 296, 369, 405, 435, 440, 454, 458, 462, 469, 473
centrifugation, 144, 151, 176, 237, 364, 366
cerebrospinal, 212, 453
certificates, 19, 54, 108, 164, 223, 302
certification, 12, 13, 14, 47, 53, 54, 55, 77, 82, 96, 107, 108, 116, 127, 201, 418, 436
cGMP, 3, 42, 47, 48, 50, 51, 52, 72, 73, 74, 75, 77, 82, 83, 105, 135, 138, 144, 184, 185, 187, 188, 199, 200, 201, 202, 204, 206, 266, 267, 268, 269, 270, 272, 273, 300, 302, 323, 330, 334, 336, 352, 353, 359, 360, 396, 416, 418, 419, 447, 482
characterization, 10, 15, 16, 37, 38, 39, 46, 49, 64, 65, 110, 117, 121, 122, 199, 201, 217, 287, 288, 289, 314, 315, 321, 377, 402
checklist, 92, 99, 100, 135, 136, 182, 481, 482, 483, 484, 485, 486, 487, 488, 489
chemiluminescence, 171
chromatograms, 31, 135, 137, 147, 182, 204, 216, 219, 220, 333, 337, 370, 476
chromatography, 4, 17, 28, 127, 133, 137, 145, 151, 157, 161, 169, 176, 177, 178, 193, 200, 207, 212, 213, 223, 231, 232, 239, 240, 246, 247, 248, 250, 251, 252, 254, 264, 274, 293, 332, 334, 358, 361, 365, 375, 376, 377, 378, 379, 380, 387, 388, 389, 400, 401, 413, 422, 430, 437, 440, 443, 444, 445, 446, 447, 449, 451, 453, 455, 457, 458, 465, 466, 477, 496
citation, 34, 298, 432, 480
clarification, 207, 406
classification, 196, 275, 277, 323, 446, 479

clearance, 152, 218, 402, 405, 411
clinical, 2, 3, 4, 6, 7, 8, 10, 14, 17, 23, 29, 30, 31, 32, 33, 36, 41, 42, 48, 73, 74, 78, 79, 80, 81, 82, 83, 84, 85, 88, 89, 99, 117, 128, 131, 132, 133, 139, 143, 144, 152, 153, 154, 156, 160, 169, 171, 172, 173, 174, 175, 184, 185, 205, 206, 211, 217, 218, 230, 231, 233, 234, 244, 259, 260, 262, 266, 267, 268, 271, 278, 279, 282, 285, 287, 288, 289, 290, 296, 299, 300, 302, 303, 304, 307, 319, 322, 323, 325, 326, 335, 336, 338, 341, 352, 356, 357, 358, 359, 360, 361, 364, 366, 390, 391, 392, 394, 396, 397, 399, 401, 402, 403, 404, 406, 407, 409, 411, 415, 416, 418, 437, 442, 446, 456, 473, 474, 475, 476, 478, 479, 482, 483, 488, 502
clinical pharmacology, 16
closeness, 20, 23, 433, 459, 463, 464, 470
CMC, 144, 322, 359, 360, 402
coagulation, 436
coefficient, 20, 23, 24, 215, 223, 256, 437, 438, 439, 443, 451, 457, 463, 464, 468
coelute, 8, 145, 257
coenzyme, 434, 437, 447
coextractives, 241
Cohesive Technologies, 239
collaboration, 160, 264, 463, 476, 477
collaborators, 259, 264
collection, 6, 13, 22, 23, 24, 31, 32, 53, 80, 89, 97, 106, 144, 151, 152, 153, 163, 172, 178, 181, 189, 191, 200, 203, 206, 212, 213, 216, 226, 233, 234, 239, 258, 273, 283, 284, 287, 288, 313, 315, 316, 319, 321, 362, 363, 365, 367, 370, 378, 401, 403, 404, 406, 434, 437, 448, 474
collision, 369, 370, 374
column, 8, 75, 133, 193, 231, 237, 238, 239, 240, 241, 242, 243, 245, 246, 247, 249, 250, 251, 252, 253, 254, 256, 258, 264, 275, 365, 366, 367, 368, 372, 376, 377, 378, 379, 380, 386, 387, 388, 389, 437, 440, 443, 444, 445, 446, 455, 457, 458
column switching, 264
combination, 26, 184, 249, 268, 284, 286, 323, 366, 367, 372, 395, 434, 437, 455, 456
combinatorial, 170, 355, 360, 377, 422, 426
commercialization, 241, 279, 288, 302, 399, 413
commitment, 102, 105, 114, 119, 121, 273, 279, 284, 286, 392, 393, 394
committee, 7, 73, 78, 79, 105, 206, 228, 314, 342, 456, 477, 478, 479, 502
communication, 56, 57, 58, 59, 60, 82, 91, 102, 105, 121, 259, 260, 264, 279, 280, 283, 284, 286, 393, 394, 395, 396, 397, 409
community, 41, 47, 68, 169, 191, 217, 357, 391
comparator, 18
comparison, 16, 18, 47, 53, 89, 106, 107, 110, 123, 145, 159, 216, 218, 221, 241, 245, 248, 250, 251, 318, 346, 358, 376, 377, 411, 438, 448, 466, 470, 473, 478, 501
compartments, 289
compatibility, 161, 176
Compendial, 19, 84, 228, 353
competence, 13, 37, 74, 77, 84, 88, 104, 109, 115, 116, 395, 408, 415, 436, 438, 478
competition, 174, 386, 387, 392, 397, 438
competitors, 1, 139, 184, 351
complaint, 74, 156, 323, 336, 417
complementary, 286, 465
complexity, 56, 127, 158, 159, 169, 191, 260, 277, 289, 329, 330, 333, 335, 361, 363, 368, 381, 388, 393, 412, 418
compliance, 3, 7, 12, 14, 15, 16, 32, 33, 37, 38, 39, 41, 44, 45, 46, 47, 48, 49, 51, 53, 54, 55, 56, 57, 58, 59, 60, 62, 64, 66, 67, 68, 69, 70, 71, 72, 73, 74, 75, 76, 77, 78, 79, 80, 82, 83, 87, 88, 95, 96, 98, 99, 101, 102, 103, 104, 107, 108, 109, 110, 111, 114, 118, 119, 127, 128, 129, 133, 134, 140, 143, 144, 147, 148, 149, 152, 154, 158, 159, 160, 161, 162, 163, 164, 186, 187, 188, 204, 205, 206, 207, 228, 233, 265, 267, 268, 270, 272, 273, 280, 282, 283, 285, 295, 299, 300, 301, 302, 303, 305, 306, 307, 309, 312, 315, 317, 318, 321, 322, 323, 324, 326, 327, 328, 329, 330, 331, 333, 334, 336, 338, 339, 341, 342, 347, 351, 353, 354, 357, 358, 359, 360, 393, 396, 405, 406, 409, 412, 416, 417, 418, 419, 420, 446, 447, 461, 471, 472, 477, 483, 501
components, 20, 68, 69, 71, 175, 176, 177, 185, 187, 190, 193, 222, 232, 237, 255, 260, 266, 269, 288, 304, 348, 361, 386, 399, 422, 435, 440, 445, 446, 449, 457, 458, 466, 467, 481

computer, 13, 14, 94, 100, 104, 113, 128, 145, 147, 153, 154, 155, 156, 158, 185, 186, 195, 201, 233, 269, 274, 276, 277, 278, 308, 310, 311, 312, 313, 316, 317, 324, 337, 339, 412, 416, 417, 418, 440, 442, 469, 473, 474, 485, 487, 488, 497, 499
concentration, 18, 20, 21, 24, 25, 26, 27, 28, 30, 113, 124, 125, 142, 146, 148, 169, 175, 176, 179, 180, 200, 207, 209, 211, 212, 214, 215, 216, 217, 218, 220, 221, 227, 230, 236, 246, 249, 255, 257, 258, 260, 266, 290, 313, 315, 319, 331, 334, 362, 369, 373, 380, 385, 386, 400, 410, 411, 435, 438, 439, 440, 443, 444, 447, 449, 451, 452, 454, 455, 459, 460, 461, 462, 466, 467, 468, 469, 473
concept, 3, 32, 37, 46, 73, 79, 88, 90, 91, 218, 252, 288, 355, 356, 357, 365, 376, 388, 411, 422, 471, 472, 474, 475
concurrence, 344
confidence, 12, 23, 24, 26, 30, 89, 90, 158, 165, 190, 198, 218, 263, 284, 288, 329, 347, 349, 358, 393, 394, 438, 450, 460, 463, 464
confidentiality, 2, 57, 105, 115, 139, 156, 286, 346, 398, 448, 484
configuration, 56, 132, 163, 216, 252, 255, 277, 367, 371
confirmation, 76, 77, 125, 126, 175, 215, 320, 470, 471
confirmatory, 123, 126, 438, 466, 467
conformance, 7, 12, 89, 185, 188, 197, 269, 270, 272, 276, 468
conformity, 43, 51, 78, 197, 316, 319, 320
confrontations, 349, 351
conjugation, 258, 458
connection, 7, 14, 41, 136, 157, 182, 217, 289
consensus, 41, 43, 47, 53, 61, 64, 126, 127, 137, 169, 223, 224, 225, 226, 260, 263, 406, 409, 438, 463, 468, 470, 476
consequence, 205, 234
consistency, 11, 76, 101, 199, 200, 214, 225, 234, 237, 273, 283, 334, 344, 413, 418, 420
consolidation, 158, 170, 299, 421
consultants, 117, 150, 166, 194, 204, 308, 352, 357, 482
consultation, 56, 57, 58, 110, 345
consumer, 10, 74, 82, 397, 418, 419
containers, 65, 122, 123, 124, 125, 126, 146, 152, 315, 381, 387
contingency, 284, 409
continuity, 121, 322
contractor, 97, 101, 189, 272, 279, 281, 283, 284, 314, 348, 438
contribution, 60, 79, 217, 384, 385
contributions, 58, 59, 60, 373
contributor, 214, 256
controversial, 343, 345, 347
convention, 228, 233
cooperation, 2, 40, 43, 47, 74, 84, 87, 95, 99, 134, 205, 290, 304, 344, 349, 352, 357, 398, 411, 419, 456, 502, 503
corrections, 101, 138, 183, 202, 316, 317, 321, 350, 474
correlation, 132, 215, 233, 418, 439, 468
cortisol, 236, 241, 243
cosmetics, 9, 35, 37, 128, 159, 444
council, 36, 43, 44, 75, 83, 84, 138, 183, 304, 305
coverage, 60, 95, 100, 119, 306, 325, 336, 373
credibility, 76, 80, 114, 149, 415, 416
criterion, 145, 213, 215, 263, 397
crossreactivity, 25
curriculum, 115, 184, 268, 310
customer, 75, 78, 105, 282, 393, 418, 420, 421
cyclosporin, 235, 242

database, 154, 157, 160, 188, 189, 195, 234, 245, 272
DDI, 73, 175, 267
deadlines, 88, 283, 284
decision, 10, 43, 44, 56, 57, 83, 94, 131, 140, 148, 160, 169, 171, 172, 173, 175, 193, 218, 279, 284, 285, 286, 287, 289, 305, 328, 329, 348, 352, 358, 362, 393, 394, 395, 396, 399, 400, 403, 407, 410, 411, 420, 486
declaration, 48, 53, 79, 107, 446
decomposition, 20, 468
deficiencies, 36, 95, 96, 97, 308, 309, 310, 312, 313, 314, 315, 316, 317, 320, 324, 342, 344, 351, 447
definition, 39, 47, 70, 90, 114, 115, 134, 175, 200, 205, 276, 400, 434, 437, 457, 464, 478
degradation, 24, 224, 225, 260, 325, 362, 380, 454, 465
degree, 19, 32, 111, 112, 182, 211, 224, 225, 246, 277, 286, 365, 368, 421, 439, 444, 460
delegation, 115, 317

demonstration, 7, 18, 89, 121, 288, 427
denaturation, 176, 225, 436, 440
department, 33, 37, 40, 49, 56, 83, 84, 90, 92, 97, 103, 126, 138, 149, 157, 161, 166, 203, 227, 228, 271, 281, 294, 295, 307, 342, 343, 353, 354, 477, 491
dependence, 345, 459
derivatization, 373, 374, 375, 376, 458
description, 5, 11, 23, 31, 46, 53, 59, 107, 109, 122, 139, 158, 178, 191, 198, 202, 214, 220, 226, 309, 317, 322, 341, 470
designation, 232, 306, 321, 324
designees, 193
desolvation, 379
desorption, 440, 458
detection, 5, 8, 18, 25, 93, 96, 145, 168, 176, 177, 208, 211, 212, 238, 241, 244, 246, 247, 251, 252, 254, 258, 261, 264, 275, 361, 364, 365, 368, 369, 370, 371, 372, 373, 374, 375, 376, 378, 379, 389, 390, 399, 404, 422, 424, 439, 440, 450, 466, 470
detector, 20, 177, 193, 378, 385, 389, 440, 449, 452, 458, 467
deterioration, 251, 255, 314, 315, 317
determination, 4, 12, 16, 17, 18, 20, 22, 25, 38, 67, 140, 142, 162, 168, 169, 176, 198, 201, 207, 209, 211, 213, 217, 225, 226, 235, 236, 237, 238, 241, 247, 248, 249, 251, 255, 325, 327, 328, 329, 360, 361, 362, 365, 372, 375, 380, 381, 387, 400, 403, 417, 424, 426, 439, 443, 460, 478, 479
deviation, 20, 21, 23, 30, 31, 121, 124, 140, 143, 186, 188, 201, 203, 213, 262, 270, 333, 335, 339, 437, 439, 440, 443, 450, 451, 456, 459, 463, 464, 465, 468, 495
device, 67, 69, 71, 72, 83, 128, 143, 155, 234, 237, 241, 290, 303, 342, 348, 372, 377, 387, 404, 413, 417, 421, 440, 465, 499
diagnosis, 6, 65, 258, 287, 288, 289, 469
dialysis, 4, 178, 436, 440, 443
diligence, 407, 408, 409
diluent, 436, 440, 451
dimensions, 238, 264, 368, 378, 389, 457, 466
directives, 37, 83, 119
directorate, 90, 92, 228, 475, 477, 502
disaster, 70, 322, 485
discrepancy, 123, 139, 203
discretion, 66, 101, 148
discussion, 7, 12, 127, 133, 137, 143, 151, 169, 173, 198, 218, 221, 223, 246, 256, 263, 288, 304, 327, 346, 477, 503, 504
discussions, 7, 40, 66, 191, 232, 361, 396
diseases, 85, 128, 166, 288, 290, 319, 399, 407
disintegration, 359, 468
dispatch, 111, 122
disposal, 62, 63, 93, 104, 105, 111, 112, 123, 146, 148, 152, 233, 274, 484, 496
disposition, 39, 121, 321, 335, 437
dissociation, 132, 256, 374, 446
dissolution, 176, 359
distinction, 13, 182, 443, 446, 451, 480
district, 299, 304, 321, 322, 326, 344, 350, 351
divergent, 91, 127, 169, 212, 213, 263
diversity, 364, 367, 371, 388, 389, 398, 400, 407, 454
division, 129, 267, 295, 299, 300, 321, 326, 336, 354, 478, 491, 503
DMPK, 131, 157, 175, 231, 475
DNA, 4, 166, 454, 458, 464, 465
documentary, 72
dosage, 171, 172, 173, 174, 202, 266, 303, 321
dose, 3, 36, 45, 101, 111, 112, 118, 121, 123, 124, 125, 126, 138, 182, 183, 266, 271, 287, 288, 289, 300, 304, 314, 319, 385, 386, 401, 402, 403, 407, 442, 450, 452, 475
droplet, 386, 387
dryness, 176, 251
duplicate, 9, 26, 27, 32, 209, 210, 215, 346, 349, 409, 410, 411, 442
duplication, 43, 120, 146
duties, 48, 57, 102, 118, 140, 184, 268, 307, 309, 317, 340
dynamics, 293, 392

ecosystems, 469, 476
ecotoxicology, 476
edition, 344, 357, 479
editor, 425, 426, 430
efficacy, 3, 8, 35, 79, 172, 173, 174, 178, 182, 230, 266, 267, 287, 288, 356, 359, 396, 399, 403, 424
efficiencies, 72, 159, 264, 280, 283, 392
electrodes, 254
electrodialysis, 443

electrophoresis, 289, 360, 443
electrospray, 251, 252, 254, 371, 378, 385, 389, 424, 425, 426, 428, 430
elimination, 171, 172, 255, 325, 376, 378, 380, 385, 458, 469
ELISA, 4, 5, 132, 145, 153, 158, 171, 211, 443, 492
elucidation, 171, 175
eluent, 252, 290, 381, 386, 387, 443, 453
elution, 177, 236, 237, 238, 239, 241, 251, 252, 253, 254, 258, 364, 367, 379, 389, 443, 447, 449, 453, 457, 465
EMEA, 80, 89, 305, 337, 396, 402, 502
encryption, 72, 156
encyclopedia, 478, 479
endogenous, 20, 24, 26, 27, 166, 178, 211, 212, 213, 214, 226, 232, 235, 237, 241, 254, 256, 364, 368, 369, 386, 388, 389, 400, 441, 458
endorsement, 3, 140
endpoint, 360, 361
energy, 105, 241, 256, 370, 372, 413, 436, 452, 464
enforcement, 35, 66, 67, 81, 82, 92, 98, 143, 326, 445, 448
enhancement, 222, 235, 249, 255, 256, 257, 326, 327, 385, 386, 387
entities, 39, 161, 207, 211, 279, 304, 357, 398, 402, 407, 447
entity, 17, 161, 182, 185, 230, 269, 278, 391, 398, 459, 469
enzyme, 132, 171, 404, 434, 435, 437, 438, 442, 443, 447, 448, 479
EPA, 10, 12, 37, 41, 47, 53, 54, 55, 82, 83, 84, 87, 106, 107, 108, 109, 127, 305, 352, 405, 437, 444, 501
equation, 21, 222, 435, 447, 450
equilibrium, 4, 177, 178, 238, 380, 438, 447
equipment, 8, 11, 12, 15, 19, 20, 29, 31, 36, 46, 50, 51, 57, 61, 65, 72, 73, 74, 76, 77, 84, 89, 90, 91, 92, 93, 97, 104, 106, 109, 111, 112, 113, 114, 120, 124, 134, 135, 137, 141, 144, 145, 179, 185, 187, 188, 189, 190, 200, 201, 204, 234, 236, 261, 263, 265, 269, 270, 271, 272, 275, 276, 277, 278, 280, 281, 284, 286, 295, 301, 302, 307, 308, 310, 311, 312, 313, 314, 315, 316, 317, 318, 323, 324, 326, 338, 341, 349, 353, 372, 379, 392, 398, 406, 413, 418, 450, 469, 484, 485, 486, 492, 493, 499
ESI, 249, 251, 254, 256, 369, 371, 373, 374, 375, 376, 378, 380, 381, 385, 386, 387, 389
establishment, 6, 8, 17, 19, 23, 26, 29, 31, 36, 44, 45, 47, 67, 69, 143, 170, 259, 260, 264, 307, 308, 348, 349, 351, 389, 399, 406, 417, 438, 453, 469, 470, 477
estimation, 23, 178, 219, 226, 381
EU, 37, 40, 73, 78, 83, 502
evaluation, 12, 16, 17, 22, 23, 24, 38, 41, 45, 55, 75, 76, 81, 91, 99, 104, 114, 128, 135, 154, 157, 165, 168, 170, 171, 172, 173, 174, 175, 179, 190, 198, 207, 211, 216, 217, 218, 220, 221, 222, 227, 228, 230, 250, 252, 257, 260, 263, 278, 279, 284, 285, 286, 294, 295, 299, 305, 311, 312, 322, 326, 327, 329, 330, 335, 339, 352, 353, 354, 359, 360, 390, 391, 393, 396, 401, 402, 403, 406, 407, 408, 410, 412, 424, 438, 442, 444, 449, 453, 460, 465, 473, 479
evaporation, 236, 251, 252, 364, 379
evidence, 10, 19, 24, 31, 54, 60, 62, 98, 101, 108, 120, 145, 173, 179, 190, 198, 201, 205, 226, 266, 300, 308, 324, 342, 444, 448, 462, 470, 471
examination, 41, 97, 319, 320, 435, 444, 461, 467, 470, 471, 478
exception, 80, 89, 97, 213, 214, 276, 346, 396
excipients, 255
excretion, 360, 475
expansion, 175, 265, 408, 475
expectation, 98, 135, 186, 189, 200, 201, 206, 265, 270, 323, 331, 334, 358, 435
expiration, 13, 19, 72, 142, 146, 200, 201, 216, 221, 223, 313, 444, 487
expiry, 49, 93, 120, 122, 142, 162, 274, 324, 392, 496
extractability, 198, 444
extractant, 440, 451

facilitators, 340
facility based, 100
falsification, 68, 69, 155
familiarity, 191, 285, 313, 407
faulty, 13, 113
feces, 4, 17, 202
filtration, 176, 178, 364, 444

findings, 32, 59, 75, 95, 101, 149, 265, 271, 299, 310, 318, 319, 325, 339, 343, 448, 468
firmware, 189, 195, 196, 277
flexibility, 8, 38, 39, 87, 98, 131, 148, 158, 163, 279, 284, 286, 388, 392, 393, 396, 403
flowchart, 27
fluorescence, 132, 198, 445
fluoroimmunoassay, 132
forms, 77, 91, 95, 115, 144, 147, 153, 190, 203, 300, 312, 313, 318, 324, 348, 473, 493
formulation, 8, 118, 121, 123, 124, 125, 126, 172, 175, 255, 259, 261, 266, 271, 303, 322, 325, 402, 419
forum, 40, 226, 295, 304
foundation, 127, 182, 289, 396, 409, 503
fragmentation, 256, 381, 388, 440, 446, 453
framework, 7, 9, 47, 79, 80, 102, 144, 160, 163, 178, 285, 305, 340, 406, 416
functionality, 47, 146, 156, 157, 159, 160, 192, 196, 412
fungicide, 82, 84, 352, 444

GC, 5, 7, 17, 93, 145, 157, 187, 272, 296, 373, 387, 446
GCLP, 33, 79, 80, 83, 85, 89, 105, 128, 159, 160, 267, 302, 357, 358, 416, 446, 482, 485
GCP, 2, 3, 7, 10, 42, 47, 48, 50, 51, 52, 67, 72, 78, 79, 80, 82, 83, 105, 139, 144, 159, 160, 268, 302, 336, 353, 357, 414, 415, 416, 419, 446, 482
genes, 435, 454
glassware, 93, 274, 381, 497, 499
globalization, 357, 420
GLP, 1, 2, 3, 4, 6, 7, 8, 9, 10, 11, 12, 13, 14, 15, 16, 19, 31, 32, 33, 35, 36, 37, 38, 39, 40, 41, 42, 43, 44, 45, 46, 47, 48, 49, 50, 52, 53, 54, 55, 56, 57, 58, 59, 60, 61, 62, 63, 64, 65, 66, 67, 69, 71, 72, 73, 74, 75, 76, 77, 79, 80, 81, 82, 83, 84, 87, 88, 89, 90, 91, 92, 93, 94, 95, 96, 97, 98, 99, 100, 101, 102, 103, 105, 106, 107, 108, 109, 110, 111, 112, 113, 114, 115, 116, 117, 118, 119, 120, 121, 122, 123, 124, 125, 126, 127, 128, 131, 132, 133, 134, 135, 136, 138, 139, 140, 141, 142, 143, 144, 146, 147, 149, 150, 154, 157, 158, 159, 160, 169, 172, 182, 183, 184, 185, 186, 187, 188, 199, 200, 201, 202, 204, 205, 206, 217, 228, 266, 267, 268, 269, 271, 272, 273, 274, 282, 295, 299, 300, 301, 302, 303, 304, 305, 306, 307, 308, 309, 310, 311, 312, 313, 314, 316, 317, 318, 320, 321, 322, 323, 324, 326, 328, 330, 332, 334, 335, 336, 337, 338, 339, 340, 341, 342, 343, 344, 345, 346, 347, 348, 349, 350, 351, 352, 353, 354, 356, 357, 394, 405, 406, 407, 408, 410, 411, 412, 414, 415, 416, 419, 420, 423, 431, 446, 447, 468, 471, 472, 473, 474, 475, 476, 477, 480, 481, 482, 483, 484, 485, 486, 487, 488, 489, 491, 492, 493, 495, 496, 497, 501
government, 35, 55, 64, 109, 305, 306, 406, 414, 419, 445, 469
guidances, 56, 72, 73, 135, 183, 206, 323, 401
guidelines, 2, 6, 7, 9, 10, 13, 14, 23, 29, 32, 40, 41, 43, 46, 51, 64, 66, 67, 69, 70, 72, 78, 79, 80, 81, 82, 83, 88, 89, 90, 91, 103, 117, 120, 127, 128, 131, 139, 147, 150, 158, 178, 205, 207, 212, 228, 259, 260, 266, 295, 302, 303, 304, 305, 338, 352, 353, 357, 387, 396, 397, 400, 405, 406, 407, 409, 415, 419, 456, 477, 478, 479, 502
GxP, 3, 83, 96, 103, 128, 143, 144, 145, 147, 149, 151, 153, 155, 157, 159, 160, 161, 163, 164, 165, 267, 273, 300, 305, 322, 336, 337, 339, 341, 343, 345, 347, 349, 351, 352, 353, 356, 423, 433, 434, 435, 436, 437, 438, 439, 440, 441, 442, 443, 444, 445, 446, 447, 448, 449, 450, 451, 452, 453, 454, 455, 456, 457, 458, 459, 460, 461, 462, 463, 464, 465, 466, 467, 468, 469, 470, 472, 474, 476, 478, 481, 482, 483, 484, 485, 486, 487, 488, 489

hardware, 135, 157, 189, 195, 408, 488
harmonization, 33, 40, 55, 73, 75, 82, 84, 89, 127, 178, 228, 352, 357, 359, 360, 396, 419,
headquarters, 158, 310
healthcare, 6, 228, 305, 502
hematology, 79, 319, 473
hemoglobin, 435
heparin, 223, 363
hepatitis, 145
high throughput, 261, 355, 401
HILIC, 231, 250, 251, 252, 378, 379
horseradish, 221
HPLC, 5, 8, 17, 93, 145, 157, 176, 237, 238, 241, 247, 248, 249, 251, 252, 253, 254, 265, 274, 316, 329, 361, 369, 376, 377, 387, 388, 389, 404, 422, 424, 425, 451, 496, 497, 499

HTLC, 232, 242
humidity, 112, 163, 192, 311
husbandry, 95, 117
HVAC, 145, 338
hydrolysis, 172, 362, 384, 423, 443, 458
hydrophobicity, 380
hygiene, 12, 99, 120
hyphenated, 25, 231, 287, 289, 422
hypothesis, 434, 441, 444, 448, 450, 455, 459, 466

illustration, 111, 197
immunization, 454, 458
immunoassay, 26, 132, 443, 445, 447, 448, 450, 455, 465
immunogenicity, 207, 259, 260, 262, 360, 448
immunosorbent, 4, 132, 171, 443
importance, 15, 38, 43, 44, 46, 56, 57, 62, 65, 93, 98, 103, 109, 116, 124, 128, 159, 182, 189, 233, 272, 282, 290, 301, 352, 393, 400, 406, 415, 477
incorporation, 26, 58, 62, 213, 399, 420
incubation, 144, 147
IND, 89, 172, 175, 259, 304, 322, 356, 358, 401, 402, 403, 446
indicator, 256, 399, 407, 435, 461
ingredient, 11, 35, 123, 237, 266, 267
inhibition, 403, 448
inhibitor, 216, 448
initiative, 81, 82, 102, 198, 395, 399, 401, 415
injection, 178, 181, 216, 232, 237, 239, 240, 241, 244, 248, 250, 251, 252, 255, 365, 366, 367, 369, 370, 376, 387, 424, 426
injector, 216, 239, 254, 259, 465
inlet, 428, 451
innovation, 66, 87, 295, 397, 419
innovator, 288, 303, 323, 391, 398
in process, 52, 310
inquiry, 279, 308, 392
insecticide, 82, 84, 352, 444
inspection, 7, 29, 37, 38, 39, 41, 45, 51, 54, 58, 59, 60, 61, 66, 75, 77, 84, 94, 95, 96, 97, 98, 99, 100, 108, 115, 118, 135, 141, 142, 143, 148, 155, 161, 187, 201, 202, 268, 272, 299, 301, 303, 306, 307, 309, 310, 311, 318, 320, 322, 323, 324, 326, 327, 331, 333, 336, 337, 338, 339, 340, 341, 342, 343, 344, 345, 346, 347, 348, 349, 350, 351, 352, 353, 357, 394, 397, 418, 447, 448, 495
inspector, 2, 99, 100, 143, 342, 343, 345, 346, 347, 348, 349, 350, 351, 412, 419, 448, 449
instability, 218, 233, 362, 363, 371, 449
installation, 73, 128, 134, 159, 165, 188, 191, 192, 193, 194, 195, 196, 197, 272, 276, 324
institution, 96, 97, 116, 185, 188, 189, 202, 265, 267, 269, 272, 300, 486
instruction, 40, 71, 90, 116, 119, 125, 312, 417
instrumentation, 9, 12, 13, 19, 50, 73, 83, 134, 154, 161, 179, 182, 184, 186, 188, 190, 192, 194, 196, 198, 200, 202, 204, 234, 265, 268, 276, 323, 353, 368, 369, 370, 398, 408, 422, 426, 430, 451, 452
instruments, 9, 13, 14, 15, 18, 25, 50, 90, 92, 93, 136, 139, 142, 143, 145, 149, 150, 157, 158, 160, 161, 162, 164, 189, 191, 192, 194, 195, 196, 197, 198, 223, 224, 228, 234, 244, 273, 274, 276, 277, 280, 287, 295, 311, 361, 368, 370, 421, 422, 460, 473, 474, 497
intact, 126, 375, 454
integrations, 137, 332, 334, 420, 421
intensity, 171, 245, 374, 450, 451
interaction, 19, 73, 173, 191, 212, 231, 240, 250, 251, 265, 267, 278, 280, 289, 300, 323, 367, 372, 378, 391, 405, 406, 412, 419, 434, 458
interassay, 31, 449
interbatch, 20, 180, 214, 449
interface, 157, 158, 160, 161, 162, 244, 246, 249, 254, 258, 415, 428
interference, 20, 22, 25, 180, 213, 216, 219, 251, 254, 258, 328, 332, 334, 346, 368, 371, 372, 380, 381, 382, 384, 385, 390, 452, 466
intersection, 457, 465
interval, 23, 24, 26, 30, 55, 108, 109, 114, 115, 141, 188, 193, 272, 322, 401, 438, 450, 462, 463, 464
intervention, 233, 290, 399, 418
intrabatch, 20, 214
intraday, 29
intralaboratory, 449
intravenous, 435
invalid, 36, 68, 94, 327, 417
investigation, 23, 80, 81, 82, 89, 90, 112, 174, 186, 200, 203, 206, 228, 255, 257, 260, 263, 267, 270, 274, 280, 289, 300, 303, 308, 326,

327, 332, 333, 334, 335, 342, 343, 374, 447, 495, 496
investigator, 56, 57, 58, 59, 60, 62, 63, 78, 79, 95, 98, 154, 204, 205, 218, 222, 281, 302, 303, 305, 306, 307, 310, 318, 320, 321, 322, 339, 342, 345, 346, 347, 349, 350, 410, 472
investment, 102, 128, 164, 173, 262, 279, 379, 392, 393, 398, 407, 414
involvement, 64, 114, 187, 308, 309, 395
ionization, 176, 217, 222, 236, 246, 251, 254, 255, 256, 257, 330, 361, 369, 371, 372, 374, 378, 379, 381, 385, 386, 387, 388, 389, 390, 425, 426, 428, 430, 437, 443, 450, 458
ions, 222, 223, 246, 254, 256, 372, 373, 374, 375, 381, 384, 386, 437, 440, 442, 445, 449, 450, 452, 453, 454, 458, 459, 460, 464
ionspray, 249, 428
irradiation, 445, 450, 465
irreproducibility, 246, 378
ISO, 74, 75, 76, 77, 78, 84, 88, 89, 91, 305, 357, 414, 415
isoenzyme, 449
isolation, 9, 146, 171, 236, 319, 375, 406
isotope, 212, 222, 223, 236, 256, 257, 258, 289, 387, 449
isotopic, 213, 257, 381, 384, 385, 449, 450
ISR, 23, 159, 176, 274, 496
issues, 2, 3, 10, 12, 26, 31, 36, 43, 47, 53, 56, 57, 60, 74, 81, 82, 87, 107, 120, 126, 127, 153, 163, 169, 171, 175, 188, 198, 205, 207, 210, 217, 218, 226, 255, 256, 257, 258, 259, 260, 262, 263, 264, 272, 276, 287, 300, 324, 326, 327, 328, 331, 332, 333, 334, 335, 336, 338, 339, 343, 344, 345, 347, 351, 365, 377, 389, 399, 408, 409, 411, 412, 418, 420
IUPAC, 456, 479

justification, 117, 158, 324, 329, 332, 333, 334, 342, 406

laminar, 145, 240
LC, 4, 5, 17, 19, 23, 25, 93, 127, 132, 133, 151, 153, 157, 169, 180, 187, 212, 213, 218, 224, 225, 230, 231, 232, 235, 236, 237, 238, 239, 240, 241, 244, 245, 246, 247, 248, 249, 250, 251, 252, 254, 255, 256, 257, 258, 264, 272, 274, 289, 290, 296, 316, 323, 326, 329, 330, 333, 334, 354, 358, 360, 361, 366, 368, 369, 370, 371, 372, 373, 374, 375, 376, 378, 379, 380, 381, 384, 385, 386, 387, 388, 389, 390, 393, 399, 400, 404, 405, 407, 408, 413, 421, 422, 424, 427, 451, 475, 492, 496, 497, 504
leadership, 90, 186, 408
lead quality assurance, 57
legislation, 43, 57, 62, 73, 445
ligand, 4, 5, 6, 25, 80, 85, 127, 129, 132, 151, 168, 169, 207, 208, 209, 211, 214, 225, 226, 227, 231, 261, 263, 287, 289, 399, 400, 409, 433, 435, 438, 440, 442, 448, 450, 462, 476
ligand binding, 4, 127, 169, 290
limitation, 330, 389, 390
LIMS, 92, 94, 132, 154, 156, 157, 158, 159, 160, 161, 162, 163, 164, 165, 166, 189, 190, 196, 234, 244, 271, 283, 401, 411, 412, 413, 420, 421, 450, 451, 474, 475, 499
linearity, 4, 26, 178, 193, 198, 239, 325, 404, 410, 439, 441, 449, 453, 469, 470
lipids, 237, 287
liquids, 438, 440, 446
location, 43, 57, 62, 89, 111, 117, 122, 146, 148, 152, 153, 187, 256, 262, 263, 271, 275, 312, 317, 345, 358, 463, 469
logbook, 141, 194, 197, 203, 312, 451
login, 157, 232, 233, 274, 496, 499
logistics, 407, 408, 409
logout, 274, 496
luminescence, 442, 445, 452

macromolecule, 207, 211, 212, 214, 215, 216, 263
macropores, 246, 249
magnitude, 94, 212, 217, 225, 239, 248, 334, 373, 442
maintenance, 9, 11, 14, 15, 36, 50, 61, 65, 66, 75, 93, 102, 104, 109, 113, 114, 115, 120, 135, 136, 139, 141, 144, 145, 146, 155, 159, 165, 166, 187, 188, 194, 201, 234, 271, 272, 274, 275, 277, 309, 311, 312, 344, 372, 408, 418, 420, 421, 445, 451, 452, 485, 486, 488, 497
malfunction, 312, 318, 334
malpractice, 10
manager, 30, 36, 76, 81, 94, 100, 102, 107, 133, 137, 149, 285, 346, 348, 350, 394, 395, 472, 482
mandatory, 49, 51, 118, 141, 158, 159, 204, 233, 273

manifestation, 68
manifold, 499
manipulation, 161, 239
manuals, 9, 120, 192, 300, 311, 344, 486
manufacturer, 9, 35, 46, 48, 66, 72, 121, 122, 135, 146, 187, 191, 192, 195, 222, 272, 312
margin, 120, 138, 182, 183, 403
markers, 172, 287, 295
matrices, 4, 5, 8, 17, 18, 21, 23, 25, 152, 168, 169, 198, 211, 212, 218, 231, 232, 235, 238, 255, 260, 262, 263, 264, 274, 278, 361, 362, 363, 368, 372, 391, 400, 411, 467, 476, 496
mechanism, 57, 165, 246, 287, 288, 289, 367, 371, 373, 378, 381, 386, 399, 407
median, 441, 453, 459, 465, 466
medication, 18, 502, 503
membership, 47, 116, 304, 395
membrane, 364, 424, 435, 439, 440, 443, 448
memorandum, 143, 271, 319
memos, 49, 119, 309
merits, 98, 192
metabolism, 1, 131, 171, 172, 218, 231, 259, 360, 405, 411, 423, 439, 453, 475, 503
metabolite, 4, 24, 26, 31, 166, 169, 175, 217, 237, 241, 242, 249, 250, 252, 254, 257, 266, 267, 288, 304, 362, 363, 372, 380, 381, 382, 384, 385, 402, 424, 438, 444, 453, 462, 475
metabolomics, 289
meters, 93, 145, 197
methodologies, 4, 5, 132, 136, 179, 211, 225, 226, 231, 232, 261, 287, 288, 290, 360, 361, 366, 386, 388, 389, 401, 403
methylcellulose, 240
metrics, 81, 135, 182, 282, 283, 483, 495
metrology, 100, 102, 478
mg, 174, 236, 362, 428, 478
micro, 197, 239, 247, 421, 450
microanalysis, 450
microarray, 289
microbial, 349
microbiology, 359
microbore, 264
microdose, 402
microdosing, 401, 402, 413
microextraction, 239
microfilm, 473, 474
microgram, 402
Micromass, 247, 252, 377
Micromass MUX, 254
micronization, 338
microplate, 170
microplates, 176
microporous, 465
microscopic, 435
microsomal, 458
microtiter, 361, 424
migration, 70, 164
milestones, 262, 338, 394
milligram, 113, 203
milliliter, 203, 238
millisecond, 244
miniaturization, 244, 413, 421, 422
misconception, 7, 89, 388
misconceptions, 385
misconduct, 322
misidentification, 443
misinterpretation, 339
misunderstandings, 350, 409
mobility, 244, 254, 372
mock, 336, 337, 338
mode, 219, 234, 236, 237, 238, 240, 246, 248, 253, 254, 367, 368, 370, 371, 373, 374, 375, 378, 379, 381, 387, 388, 422, 441, 453, 459, 465, 466
modification, 71, 120, 146, 211, 219, 312, 329, 458, 470
moiety, 266, 267, 375, 455, 457, 458
monochromators, 442
monoclonal, 211, 231, 454
monograph, 45, 84, 477
monolithic, 178, 231, 232, 237, 239, 240, 241, 246, 248, 249, 250, 254, 264, 368, 378
monospecificity, 454
movement, 58, 112
multianalyte, 132, 219
multicenter, 3
multicolumn, 24
multicomponent, 387
multidisciplinary, 103, 359, 472
multinational, 61, 75
multiplex, 132, 289, 454
multiplexing, 24, 244, 252, 376, 454
MultiProbe, 5, 274, 497
multisite, 47, 53, 55, 56, 57, 60, 61, 62, 63, 151, 154, 278, 391, 398, 399, 408, 471
multisprayer, 376, 377
multistudy, 65
multivariate, 171, 442, 462

mutagenesis, 454
mutagenicity, 39, 202, 307

nanogram, 238
nanoliter, 421
nanotechnology, 419
NDA, 31, 81, 83, 131, 143, 173, 174, 175, 202, 220, 322, 323, 337
necessity, 1, 2, 136, 182
necropsy, 112, 319
nomenclature, 136, 256, 478, 479
nonanchor, 209, 215
nonanimal, 64, 405, 406
nonchromatographic, 168, 207
nonclinical, 2, 3, 4, 6, 7, 17, 19, 24, 32, 36, 37, 38, 39, 41, 43, 44, 45, 48, 56, 72, 74, 75, 80, 82, 84, 88, 89, 95, 96, 97, 98, 99, 128, 139, 143, 160, 184, 185, 187, 188, 201, 202, 205, 230, 233, 234, 265, 267, 268, 269, 270, 271, 272, 278, 279, 282, 294, 299, 300, 301, 302, 303, 309, 316, 320, 322, 326, 336, 338, 352, 354, 356, 359, 390, 391, 401, 402, 406, 414, 415, 472, 475, 476
noncompliance, 96, 101, 114, 138, 149, 183, 325, 357, 418
nonhuman, 205, 207, 407
nonlinear, 24, 26, 160, 211, 212, 214, 215, 258, 443
nonpolar, 250, 371, 373
nonquantitative, 400
nonradioisotopic, 455
nonregulated, 171, 233
nonspecific, 26, 364, 374, 434
nonzero, 21, 26, 180, 208
norm, 83, 224, 353
normalization, 455
notebook, 145, 406, 474
notification, 55, 109, 185, 269, 495

OASIS, 236, 237, 242, 243
OASIS MCX, 236
objectives, 1, 2, 4, 6, 8, 9, 10, 12, 14, 16, 27, 32, 79, 90, 98, 103, 104, 111, 116, 128, 150, 176, 202, 207, 268, 288, 322, 355, 393, 460, 461
observation, 36, 96, 143, 176, 314, 318, 319, 327, 328, 329, 330, 331, 350, 351, 412, 451, 452, 455
occurrence, 441, 442, 459, 460, 465
OECD, 2, 3, 7, 10, 33, 40, 41, 43, 44, 45, 46, 47, 51, 53, 54, 55, 56, 57, 59, 62, 64, 66, 75, 79, 80, 83, 84, 87, 88, 89, 91, 100, 101, 102, 106, 107, 108, 109, 111, 113, 115, 117, 119, 121, 123, 125, 127, 128, 131, 137, 172, 205, 302, 304, 305, 352, 353, 357, 405, 406, 414, 420, 456, 471, 472, 473, 475, 476, 480, 501, 503
OOS, 52, 103, 339, 495
operation, 9, 13, 28, 33, 42, 50, 60, 73, 77, 90, 97, 103, 111, 124, 128, 134, 135, 136, 139, 141, 158, 183, 187, 188, 189, 193, 194, 195, 196, 199, 201, 233, 249, 270, 272, 273, 274, 275, 307, 308, 313, 314, 317, 346, 347, 368, 417, 420, 422, 460, 469, 497
optimization, 4, 6, 131, 132, 171, 189, 239, 244, 245, 278, 329, 364, 365, 370, 373, 377, 387, 388, 391, 399, 420, 424
option, 77, 148, 200, 277, 289
organism, 170, 171, 449, 451, 452, 453, 454, 455, 462, 469
orifice, 254, 372
origin, 41, 42, 48, 55, 65, 79, 127, 324, 446, 454
outcome, 2, 43, 75, 164, 168, 173, 263, 288, 339, 347, 360, 394
outlier, 332, 335, 436, 442, 447, 456
outline, 110, 261
overestimation, 258, 362
oversight, 182, 185, 186, 269, 270, 326, 340, 409, 419
overview, 119, 340, 476
ownership, 48, 159, 186, 420

paperless, 161, 420
paradigm, 287, 407
parallelism, 26, 422, 456
parameter, 26, 103, 179, 180, 181, 215, 230, 320, 370, 438, 441, 466, 470
paramount, 57, 148, 282, 285, 344, 352, 393, 403
participation, 48, 75, 79, 304, 309, 352, 446, 460
partition, 444, 446, 451, 457, 480
partnership, 279, 393, 394, 419, 469
password, 69, 71, 72, 147, 487
patent, 173, 264, 392, 398, 502
path, 246, 280, 295, 399, 401, 449, 452, 454, 460

pathologists, 6, 117, 357
pathology, 6, 319, 320
pathways, 6, 288, 289, 399
patients, 3, 10, 161, 173, 174, 218, 287, 288, 290, 304, 359, 399, 403, 407, 411
peptide, 132, 207, 288, 372, 375
perception, 333, 420
performance, 11, 14, 17, 23, 26, 32, 36, 47, 50, 55, 57, 61, 64, 68, 73, 89, 91, 92, 93, 94, 95, 97, 100, 113, 116, 134, 137, 144, 145, 150, 157, 159, 162, 175, 176, 177, 178, 179, 187, 188, 191, 193, 194, 197, 198, 199, 207, 214, 215, 217, 218, 220, 222, 223, 226, 231, 232, 233, 234, 240, 241, 244, 246, 247, 248, 249, 254, 260, 261, 262, 263, 264, 267, 272, 273, 276, 282, 286, 288, 290, 303, 305, 309, 324, 326, 333, 338, 358, 364, 366, 370, 371, 377, 379, 385, 389, 392, 393, 395, 406, 413, 417, 438, 443, 449, 451, 460, 461, 469, 470, 495
permeability, 240, 242, 249, 423
permission, 3, 41, 59, 90
permits, 37, 43, 74, 98, 188, 267, 272, 299, 348, 408, 410, 420
perspective, 6, 67, 160, 226, 265, 266, 275, 276, 304, 397, 423
pesticides, 1, 37, 74, 143, 159, 444, 445
pharma, 356, 391, 392, 393, 396, 397, 398, 399, 502
pharmaceutics, 172
pharmacodynamic, 46, 198, 263, 267, 287, 300, 402, 413
pharmacogenetic, 79
pharmacokinetics, 3, 4, 79, 131, 166, 169, 172, 173, 176, 217, 230, 231, 265, 278, 299, 353, 359, 391, 402, 435, 442, 458, 469
pharmacology, 3, 17, 19, 46, 73, 131, 132, 171, 172, 259, 265, 299, 322, 359, 401, 458
pharmacopoeial, 228, 359, 360
pharmacovigilance, 3
pharmacy, 111, 112, 342, 426
Phenomenex, 240, 249, 379
phenomenon, 176, 222, 255, 258, 365, 387, 444, 469
photodetector, 454
photographs, 143, 346, 473, 474
photoionization, 371
photometric, 132, 376, 438
phototoxicity, 46
physician, 174
pipeline, 189, 230, 234, 392, 397, 399, 400
pipettes, 93, 145, 197, 276
pitfalls, 264, 339, 380, 381, 384, 385, 387
pivotal, 23, 31, 46, 101, 109, 110, 288, 322, 323, 409, 422
PK, 8, 16, 17, 19, 73, 132, 157, 158, 168, 175, 187, 207, 212, 213, 220, 230, 232, 244, 265, 267, 268, 274, 289, 299, 300, 303, 323, 332, 338, 401, 402, 403, 404, 429, 496
placement, 24, 29, 208, 215, 222, 226, 281
plasma, 4, 17, 18, 20, 132, 151, 176, 177, 197, 210, 212, 213, 216, 223, 224, 230, 232, 235, 236, 237, 238, 239, 240, 241, 242, 243, 247, 248, 249, 250, 251, 252, 254, 255, 256, 266, 304, 362, 363, 365, 367, 370, 374, 375, 386, 387, 404, 421, 424, 427, 435, 436, 453, 458, 476, 503
platform, 213, 216, 242, 243, 287, 365, 402
polarity, 176, 251, 373, 378
policies, 67, 68, 69, 70, 71, 72, 83, 116, 120, 153, 155, 281, 308, 343, 344, 347, 349, 352, 353, 417, 419, 447, 461
polyclonal, 213, 454, 458
polypeptide, 454
popularity, 404, 421
population, 173, 174, 218, 262, 287, 327, 332, 405, 407, 411, 434, 441, 442, 455, 456, 459, 462, 463, 465, 466, 470
porosity, 246, 248, 249
possibilities, 91, 365, 442
possibility, 15, 40, 43, 44, 58, 110, 121, 124, 126, 152, 233, 333, 403
postacquisition, 195, 245
postapproval, 3, 36
postaudit, 343
postcollection, 212
postcolumn, 249, 251, 256, 380
postconference, 36, 96, 97, 99
postevaluation, 14
postextraction, 255, 256
postmarketing, 4, 169, 174, 288
poststudy, 175, 176
posttranslational, 211
postvalidation, 221
potency, 6, 146, 170, 207, 211, 459
potent, 166, 207, 264
PPT, 4, 232, 235, 236, 237, 245, 258, 295, 354, 387

practitioners, 207, 264
preamble, 38, 88, 96, 97, 98, 136, 138, 182, 183
preapproval, 148, 202, 264, 336, 338
preassay, 212
precaution, 113, 493
precautions, 90, 126, 142, 216, 308
precipitation, 4, 5, 176, 178, 225, 232, 234, 235, 237, 240, 251, 363, 364, 366, 387, 424
precision, 6, 13, 17, 18, 20, 21, 23, 24, 25, 26, 27, 29, 30, 76, 80, 89, 94, 113, 145, 149, 168, 175, 177, 178, 179, 180, 184, 190, 193, 198, 210, 213, 214, 215, 216, 219, 225, 226, 227, 237, 239, 241, 250, 255, 256, 257, 261, 263, 264, 266, 267, 269, 287, 325, 327, 328, 329, 330, 331, 361, 385, 390, 404, 410, 440, 449, 451, 453, 455, 459, 461, 462, 466, 470, 476, 479
preclinical, 4, 7, 16, 29, 30, 31, 32, 36, 42, 44, 73, 74, 101, 110, 111, 131, 132, 133, 152, 154, 156, 160, 169, 171, 172, 175, 178, 205, 211, 230, 231, 244, 259, 260, 287, 288, 289, 290, 299, 304, 326, 356, 360, 401, 402, 403, 404, 407, 409, 416
preclude, 57, 421
precolumn, 458
preconcentration, 176, 237
precursor, 371, 374, 384, 440, 459, 460
prediction, 399, 402, 405
predominant, 446
predose, 332, 334
preference, 97, 206
premises, 50, 51, 76, 104, 187, 270
premium, 397
preparation, 4, 19, 25, 37, 38, 45, 52, 59, 90, 93, 101, 111, 115, 118, 121, 123, 124, 125, 144, 146, 169, 176, 177, 178, 179, 181, 186, 187, 189, 200, 201, 204, 208, 212, 213, 215, 216, 218, 219, 220, 223, 226, 227, 231, 232, 234, 235, 236, 237, 239, 241, 244, 250, 251, 258, 261, 264, 265, 267, 271, 273, 274, 290, 300, 310, 311, 315, 316, 319, 321, 323, 333, 334, 336, 337, 338, 339, 340, 341, 343, 348, 352, 358, 362, 363, 364, 365, 366, 367, 368, 378, 385, 386, 388, 406, 412, 421, 422, 424, 451, 456, 460, 463, 496, 499
prerequisite, 16, 41, 43, 94, 119, 164, 273, 407
prereview, 342
prestudy, 28, 174, 175, 201, 213, 214, 215, 216, 234, 286, 327, 335, 337
pretreatment, 239, 254, 257, 390
prevalidation, 65, 400
prevention, 6, 33, 469
principles, 2, 6, 7, 9, 10, 11, 13, 15, 16, 19, 23, 25, 32, 33, 36, 37, 40, 41, 42, 43, 44, 45, 46, 47, 48, 51, 53, 54, 55, 56, 57, 59, 60, 61, 62, 63, 64, 66, 67, 73, 75, 79, 80, 81, 82, 83, 84, 87, 89, 91, 101, 102, 106, 107, 108, 109, 111, 113, 115, 116, 117, 119, 121, 123, 125, 127, 128, 131, 134, 149, 154, 158, 159, 168, 169, 178, 187, 197, 204, 205, 206, 207, 224, 226, 228, 247, 265, 272, 293, 295, 302, 304, 305, 323, 356, 357, 366, 406, 411, 412, 414, 416, 420, 446, 449, 461, 471, 476, 477, 478
priority, 133, 217, 225, 245, 275, 306, 420
probability, 178, 275, 438, 441, 442, 448, 450, 455, 459, 460, 463, 469, 471
probe, 254, 370, 376, 426
problems, 8, 28, 36, 38, 56, 96, 98, 100, 109, 118, 119, 121, 141, 185, 200, 212, 216, 219, 236, 241, 269, 277, 284, 309, 310, 311, 321, 322, 323, 325, 336, 339, 341, 342, 347, 378, 380, 395, 408, 421, 422
procedure, 6, 12, 19, 20, 21, 26, 39, 69, 90, 93, 94, 100, 115, 116, 119, 120, 121, 122, 124, 125, 126, 144, 145, 151, 165, 176, 185, 190, 192, 194, 196, 199, 200, 201, 203, 204, 206, 209, 210, 224, 225, 237, 240, 241, 245, 251, 262, 270, 273, 274, 305, 307, 311, 317, 319, 340, 341, 343, 363, 366, 381, 385, 387, 400, 405, 433, 436, 442, 447, 449, 453, 455, 456, 460, 462, 463, 464, 465, 468, 469, 482, 484, 485, 486, 487, 489, 491, 493, 496, 497
proceeding, 153, 182, 185, 192, 269
producer, 55, 76, 77, 109, 419, 460
production, 1, 32, 48, 74, 76, 90, 93, 144, 185, 269, 275, 277, 280, 281, 282, 285, 286, 290, 326, 331, 372, 374, 376, 412, 416, 417, 418, 446, 447, 448, 460, 495
productivity, 4, 161, 233, 282, 392, 415, 420
professionalism, 343, 349
professionals, 64, 202, 477
proficiency, 38, 80, 301, 416, 460, 461
profile, 102, 110, 172, 174, 232, 257, 266, 300, 313, 402, 418
profitability, 72, 413

prognosis, 287, 288, 289
programs, 3, 4, 5, 7, 8, 15, 24, 42, 44, 47, 53, 54, 55, 56, 60, 66, 67, 69, 71, 73, 85, 87, 88, 93, 94, 98, 107, 108, 127, 128, 132, 133, 134, 136, 141, 143, 157, 160, 169, 171, 172, 184, 204, 233, 244, 250, 252, 259, 260, 268, 285, 286, 290, 295, 302, 303, 313, 338, 352, 354, 355, 362, 364, 369, 379, 392, 406, 407, 419, 420, 422
promulgated, 36, 87, 99, 409
proof, 3, 13, 14, 41, 113, 132, 162, 218, 397, 403, 411
properties, 16, 21, 44, 110, 175, 211, 212, 216, 230, 238, 256, 360, 364, 370, 386, 402, 404, 405, 435, 437, 440, 449, 455, 460, 463, 464, 465, 467, 470
property, 146, 211, 395, 414, 415, 436, 437, 444, 453, 470
proportion, 257, 380, 442, 467
proposals, 43, 92, 164, 165, 288
proprietary, 73, 158, 206, 222, 228, 242, 285, 346, 351, 359, 398, 488
prospective, 10, 92, 119
protease, 212, 216, 236, 239
protection, 10, 37, 38, 43, 74, 75, 82, 105, 147, 228, 285, 295, 300, 302, 303, 304, 305, 306, 338, 352, 354, 395, 405, 417, 445, 465
protections, 445
protein, 4, 5, 132, 137, 171, 176, 178, 207, 213, 217, 225, 232, 234, 235, 237, 240, 241, 249, 251, 258, 287, 288, 289, 290, 363, 364, 366, 367, 387, 399, 421, 434, 436, 438, 440, 443, 454, 462, 503
proteomics, 161, 170, 207, 288, 296, 355, 421
protocols, 11, 15, 31, 46, 49, 52, 72, 73, 74, 91, 92, 95, 96, 98, 105, 106, 110, 118, 121, 140, 141, 142, 145, 158, 160, 164, 173, 185, 186, 189, 202, 203, 205, 234, 269, 270, 273, 284, 301, 308, 309, 310, 316, 317, 337, 338, 342, 343, 395, 446, 456, 468, 472, 475
provisions, 38, 39, 66, 75, 301, 421, 445
publication, 47, 81, 84, 87, 207, 353, 445, 477, 479
purity, 6, 19, 31, 65, 110, 123, 142, 172, 201, 211, 219, 223, 237, 261, 308, 315
purpose, 3, 6, 7, 17, 32, 36, 44, 51, 61, 62, 65, 73, 74, 96, 111, 113, 115, 116, 117, 125, 132, 134, 143, 145, 146, 155, 158, 159, 160, 164, 166, 169, 171, 174, 175, 177, 190, 198, 212, 230, 239, 260, 263, 275, 287, 288, 289, 291, 300, 304, 307, 309, 310, 311, 312, 313, 314, 315, 316, 324, 338, 344, 345, 346, 348, 349, 351, 388, 400, 438, 460, 461, 466, 491

QA, 12, 13, 14, 37, 48, 49, 90, 99, 101, 103, 111, 114, 115, 118, 119, 121, 144, 147, 149, 157, 161, 182, 185, 186, 187, 189, 194, 202, 203, 204, 232, 267, 269, 270, 271, 282, 285, 296, 300, 302, 306, 307, 309, 310, 312, 313, 316, 317, 321, 322, 335, 337, 401, 460, 495
QAU, 9, 76, 81, 95, 96, 97, 98, 99, 100, 101, 103, 106, 116, 130, 134, 135, 136, 138, 140, 141, 142, 182, 183, 185, 186, 187, 188, 203, 204, 269, 270, 273, 308, 309, 310, 318, 472, 473, 492
QC 181, 182, 184, 185, 187, 189, 190, 200, 201, 204, 208, 209, 210, 213, 215, 216, 217, 219, 220, 223, 226, 227, 232, 235, 248, 269, 271, 282, 291, 313, 328, 330, 331, 332, 333, 334, 337, 385, 409, 410, 425, 461
QCS, 27, 29, 30, 175, 180, 186, 208, 209, 210, 214, 215, 217, 220, 260
QCU, 81, 185, 269
QM, 90, 161, 421
QSAR, 461, 462
quadrupole, 245, 254, 256, 361, 368, 370, 371, 449, 460
qualification, 6, 11, 12, 13, 15, 47, 73, 74, 96, 104, 108, 109, 132, 134, 135, 139, 141, 142, 149, 150, 155, 165, 179, 182, 184, 186, 187, 188, 189, 190, 191, 192, 193, 194, 195, 196, 197, 198, 200, 201, 202, 204, 224, 228, 261, 263, 268, 269, 271, 272, 273, 274, 275, 276, 277, 278, 287, 288, 289, 295, 302, 324, 326, 337, 356, 399, 419, 437, 461, 486, 487, 495, 496, 497
quantification, 17, 20, 21, 25, 26, 31, 166, 207, 211, 212, 213, 214, 215, 231, 238, 239, 248, 290, 324, 367, 380, 384, 385, 390, 400, 404, 437, 441, 451, 461, 465
quantitation, 4, 30, 168, 213, 214, 216, 231, 235, 236, 239, 240, 241, 242, 243, 252, 256, 257, 258, 260, 267, 334, 358, 361, 373, 374, 380, 384, 385, 388, 389, 400, 424, 451, 461, 470
quantitative, 4, 5, 12, 16, 17, 80, 85, 129, 132, 160, 166, 168, 169, 175, 177, 190, 198, 207,

211, 222, 225, 227, 231, 232, 235, 236, 239, 244, 245, 246, 248, 250, 254, 255, 256, 257, 258, 263, 264, 278, 280, 288, 289, 361, 362, 366, 368, 372, 376, 378, 380, 381, 385, 386, 389, 390, 391, 400, 401, 403, 404, 405, 424, 425, 426, 433, 434, 435, 439, 449, 452, 460, 461, 462, 476
quarantine, 49, 65, 112, 348

radiation, 445, 450, 454
radioactivity, 462
radioimmunoassay, 213, 462, 467
randomization, 118, 216, 316, 325
rat, 18, 113, 239, 242, 243, 248, 249, 252, 377, 390
rates, 224, 240, 245, 246, 249, 250, 362, 365, 369, 371, 376, 377, 379, 387, 389, 479
ratio, 222, 247, 251, 252, 257, 365, 368, 372, 380, 385, 435, 437, 441, 445, 451, 452, 457, 466, 467, 471
rationale, 29, 46, 58, 212, 275, 328, 333, 347, 374
reaction, 244, 250, 252, 290, 372, 374, 381, 386, 404, 436, 438, 439, 440, 445, 447, 448, 449, 450, 455, 458, 462, 468, 470
readiness, 333, 336, 408
reagents, 11, 13, 15, 19, 20, 27, 90, 93, 142, 144, 145, 146, 147, 160, 162, 200, 201, 212, 216, 222, 251, 259, 261, 262, 263, 265, 274, 289, 301, 302, 313, 323, 324, 361, 363, 373, 436, 451, 460, 470, 479, 487, 496
reality, 109, 391, 397
realm, 203, 273, 407
reanalysis, 23, 30, 81, 137, 159, 200, 218, 263, 274, 313, 407, 411, 467, 496
receipt, 39, 49, 65, 100, 111, 112, 121, 122, 132, 144, 149, 152, 153, 161, 164, 186, 187, 189, 199, 201, 204, 232, 233, 271, 282, 283, 310, 313, 314, 315, 319, 324, 325, 337, 349, 358, 362, 406, 448, 487
receptionist, 340, 348
receptor, 171, 172, 211, 290, 402, 435, 438, 450, 462
recertification, 54, 108, 219, 223
recipient, 76
recognition, 1, 32, 40, 88, 89, 171, 223, 250, 462, 464
recombinant, 207, 212, 231
recommendations, 17, 18, 24, 26, 28, 40, 61, 95, 101, 127, 129, 135, 136, 154, 169, 207, 215, 227, 336, 350, 357, 409, 411, 421, 431, 478, 479, 481, 482, 483, 484, 485, 486, 487, 488, 489
reconciliation, 49, 202, 315
reconstitution, 236, 251, 252, 364, 379
reconstructability, 44, 110
reconstruction, 89, 110, 116, 120, 121, 218, 358, 411, 473
recovery, 20, 21, 26, 31, 70, 145, 175, 179, 198, 213, 217, 218, 227, 237, 240, 241, 257, 260, 274, 328, 330, 334, 364, 365, 388, 453, 456, 462, 485, 497
recurrence, 186, 270, 326, 439
reduction, 82, 94, 241, 248, 283, 387, 389, 392, 404, 413, 418, 458
reevaluation, 445
refinement, 289, 400
reflection, 9, 110, 286
refresher, 92, 347
refreshment, 112
refrigerators, 113, 145, 187, 204, 271, 337, 499
refusal, 95, 320, 345
regeneration, 378, 389
regimen, 174, 202
region, 256, 362, 396, 397, 407, 455
registration, 7, 10, 36, 41, 45, 46, 55, 82, 89, 90, 94, 95, 105, 109, 131, 143, 149, 305, 337, 352, 396, 412, 445, 451, 502
regression, 21, 29, 157, 160, 215, 240, 251, 331, 332, 450, 451, 461, 462, 463
regulated, 1, 3, 6, 7, 14, 24, 29, 35, 37, 38, 41, 42, 43, 44, 72, 73, 74, 82, 83, 87, 90, 98, 103, 104, 128, 131, 133, 135, 138, 140, 141, 142, 150, 153, 154, 168, 169, 172, 174, 182, 183, 187, 189, 193, 194, 198, 228, 230, 233, 244, 255, 256, 258, 260, 262, 264, 265, 271, 277, 299, 300, 301, 303, 305, 306, 336, 352, 355, 357, 361, 407, 412, 415, 422, 433, 446, 481, 491, 495, 496, 497, 499, 501, 502, 503, 504
regulations, 1, 2, 4, 6, 7, 8, 9, 10, 11, 12, 13, 14, 15, 16, 32, 35, 36, 37, 38, 39, 41, 42, 43, 45, 46, 47, 53, 54, 55, 62, 64, 66, 71, 72, 73, 74, 75, 76, 77, 80, 81, 82, 83, 84, 87, 88, 89, 92, 95, 96, 97, 98, 99, 105, 106, 107, 108, 109, 111, 113, 115, 117, 119, 121, 123, 125,

128, 129, 131, 134, 135, 136, 139, 140, 141, 143, 149, 150, 156, 158, 159, 164, 172, 173, 182, 184, 185, 187, 188, 194, 199, 200, 201, 204, 206, 259, 265, 266, 267, 268, 269, 270, 272, 273, 299, 300, 301, 302, 303, 305, 307, 322, 323, 336, 338, 344, 347, 348, 351, 352, 353, 356, 357, 396, 397, 401, 405, 406, 412, 414, 415, 416, 417, 418, 419, 420, 423, 437, 443, 444, 446, 447, 468, 472, 473, 475, 476, 477, 501
regulators, 1, 89, 114, 158, 416, 419, 420, 476
reinjection, 25, 181, 222
reintegration, 29, 30, 32, 137, 200, 219, 336, 337
rejection, 25, 32, 149, 179, 227, 331, 332, 334, 433, 455
relation, 35, 59, 66, 67, 69, 71, 73, 89, 90, 125, 215, 323, 386, 442
relationship, 3, 21, 24, 26, 28, 59, 179, 214, 215, 227, 284, 285, 286, 287, 349, 352, 394, 396, 407, 409, 436, 440, 443, 449, 450, 451, 461, 462, 469
relevance, 12, 347, 416
reliability, 2, 7, 19, 44, 50, 67, 68, 72, 76, 78, 79, 83, 113, 166, 176, 179, 187, 198, 202, 206, 226, 262, 265, 267, 270, 290, 324, 333, 336, 417, 423, 463
reliance, 238, 289, 361
relieve, 174, 305
remedial, 189, 219, 272, 318, 450
removal, 4, 176, 177, 178, 317, 335, 364, 366, 368, 453, 456
renal, 173, 359
repeatability, 20, 198, 434, 459, 463, 468
repetition, 43, 415
replacement, 58, 63, 335, 418
replicates, 26, 29, 209, 460
representation, 101, 458
representatives, 12, 40, 150, 263, 345
reproducibility, 17, 20, 24, 25, 26, 144, 168, 175, 177, 178, 181, 200, 217, 218, 222, 226, 249, 263, 325, 364, 366, 373, 374, 381, 400, 410, 411, 414, 431, 459, 464, 468, 479
reputation, 91, 398
requalification, 193, 481
requestor, 30
researchers, 159, 178, 240, 403
residual, 208, 246
residue, 35, 55, 125, 127, 307, 444, 476
resistance, 246, 249, 458, 468
resolution, 40, 81, 102, 171, 186, 188, 201, 231, 232, 246, 247, 248, 250, 271, 274, 326, 327, 333, 335, 361, 367, 370, 371, 377, 384, 388, 389, 409, 443, 452, 464, 496
resonance, 171, 445
resources, 15, 32, 46, 56, 58, 97, 102, 105, 109, 111, 114, 125, 133, 135, 136, 138, 158, 161, 163, 166, 182, 183, 233, 259, 262, 279, 280, 281, 286, 288, 308, 342, 368, 393, 396, 398, 401, 403, 414, 415, 419, 461
respondents, 165, 213
response, 6, 9, 10, 20, 21, 22, 24, 26, 28, 35, 81, 97, 98, 99, 159, 166, 170, 174, 175, 176, 178, 179, 180, 199, 211, 212, 214, 215, 216, 217, 219, 222, 225, 226, 227, 245, 256, 257, 258, 260, 262, 267, 271, 289, 313, 319, 332, 344, 350, 351, 369, 370, 373, 374, 375, 376, 379, 380, 381, 386, 387, 390, 399, 436, 442, 443, 448, 449, 452, 461, 466
responsibilities, 8, 11, 15, 44, 45, 47, 48, 53, 56, 57, 60, 62, 76, 90, 92, 96, 103, 105, 107, 109, 114, 115, 134, 136, 139, 140, 141, 150, 168, 170, 172, 174, 176, 178, 182, 184, 185, 187, 188, 194, 205, 268, 269, 270, 274, 302, 303, 307, 308, 338, 346, 347, 394, 419, 449, 460, 461, 471, 472, 492, 495, 496
responsiveness, 303, 338, 350
restriction, 443, 464
resultant, 81, 138, 183, 412
retention, 11, 38, 51, 63, 70, 72, 94, 136, 142, 143, 155, 158, 186, 188, 189, 193, 202, 233, 246, 248, 249, 251, 255, 257, 258, 274, 286, 307, 317, 332, 334, 337, 366, 373, 374, 375, 378, 379, 380, 386, 389, 417, 457, 464, 473, 488, 489, 496, 499
retrieval, 16, 96, 111, 135, 142, 144, 155, 187, 200, 202, 271, 307, 312, 317, 322, 345, 348, 358, 417
revalidation, 30, 224, 329, 470, 486
review board, 303
reviewer, 69, 97
revision, 63, 70, 75, 97, 103, 120, 140, 155, 196, 423, 479, 483, 492, 493
revisit, 127, 129, 166, 169, 227, 353
rhinitis, 174
rigor, 189, 288, 289, 304, 400, 407, 414
roadmap, 399, 417, 418

robotic, 132, 232, 234, 235, 236, 244, 274, 333, 363, 366, 413, 497
robust, 23, 84, 122, 137, 160, 186, 189, 202, 204, 261, 263, 270, 273, 279, 283, 285, 286, 353, 365, 380, 392, 395, 403
robustness, 17, 23, 94, 175, 176, 240, 248, 252, 261, 262, 264, 394, 400, 411, 413, 422, 465
rodenticide, 82, 84, 352, 444
roles, 44, 47, 48, 57, 69, 76, 78, 81, 92, 109, 118, 134, 150, 162, 168, 185, 187, 194, 198, 205, 265, 269, 270, 274, 302, 307, 343, 394, 472, 483, 495, 496
rootcause, 324
roots, 35, 191, 235
routes, 465, 484
ruggedness, 17, 23, 94, 193, 331, 405, 423, 465, 470

SaaS, 159
safeguard, 10, 414
sampler, 22, 181, 237, 238, 239, 240, 241, 253, 254, 259, 370, 376, 388, 421
sanitation, 74, 350, 485
satisfaction, 78, 182, 282
satisfactory, 13, 43, 138, 173, 183, 461
scanner, 233, 492
scenario, 135, 157, 183, 206
schematic, 4, 5, 253
scheme, 4, 54, 108, 172, 421, 439, 445, 447, 460
scientists, 2, 9, 10, 32, 37, 81, 102, 117, 136, 148, 150, 179, 182, 189, 202, 260, 263, 265, 276, 281, 285, 290, 295, 356, 361, 369, 398, 407, 414, 416, 420, 472
Sciex, 224
scintillation, 171, 452
scrap, 203, 325
scripts, 195, 501
scrutiny, 81, 82, 160, 169, 182, 414
sector, 1, 104, 105, 391, 397, 398, 399, 420
security, 15, 19, 40, 61, 68, 69, 70, 73, 100, 120, 128, 151, 152, 155, 158, 162, 165, 189, 265, 275, 316, 321, 322, 323, 411, 412, 484
seizure, 35
selection, 21, 54, 57, 58, 63, 108, 117, 132, 133, 164, 165, 191, 204, 212, 239, 278, 280, 284, 287, 288, 307, 319, 352, 365, 371, 391, 393, 399, 400, 407, 476
selector, 437
sensitivity, 4, 6, 17, 20, 23, 25, 132, 152, 168, 175, 177, 178, 198, 215, 226, 230, 231, 232, 237, 245, 246, 247, 248, 250, 251, 252, 254, 255, 258, 264, 266, 288, 323, 325, 328, 330, 334, 361, 369, 370, 371, 372, 373, 374, 375, 376, 377, 378, 381, 384, 385, 387, 400, 402, 404, 405, 421, 422, 466
separation, 5, 8, 26, 50, 57, 62, 75, 97, 111, 112, 138, 146, 176, 177, 183, 212, 231, 232, 236, 240, 241, 244, 245, 246, 247, 248, 249, 250, 251, 252, 253, 254, 255, 257, 258, 264, 265, 310, 361, 365, 367, 368, 372, 375, 377, 378, 379, 381, 384, 385, 387, 388, 389, 390, 404, 421, 422, 437, 443, 444, 446, 447, 449, 457, 465, 466, 467
series, 33, 44, 45, 64, 69, 71, 73, 89, 90, 91, 173, 193, 196, 214, 233, 247, 249, 360, 364, 369, 395, 399, 406, 410, 436, 460, 462, 464, 470, 471, 477
serum, 4, 17, 176, 177, 212, 213, 216, 238, 243, 266, 287, 304, 400, 424, 434, 436, 442, 448
server, 68, 105, 155, 163, 317
setup, 92, 253
SFDA, 502
shipment, 31, 62, 152, 157, 189, 221, 233, 244, 362, 404, 407, 487
shortfall, 103
signature, 59, 68, 69, 70, 71, 72, 84, 108, 110, 118, 125, 143, 146, 155, 203, 204, 316, 317, 318, 417, 474
significance, 158, 351, 385, 438, 446, 448, 450, 466, 468, 470, 471, 480
silica, 177, 231, 236, 242, 246, 249, 250, 251, 252, 255, 264, 367, 368, 378, 379, 389, 465
similarities, 53, 106
similarity, 47, 48, 53, 74, 75, 77
simplicity, 235, 285, 291, 364, 367, 407
situation, 14, 15, 18, 57, 61, 63, 94, 116, 124, 125, 203, 235, 259, 260, 263, 321, 325, 346, 349, 364, 369, 372, 384, 385, 409, 414, 439, 459
society, 12, 13, 84, 90, 150, 295, 353, 415, 423, 425, 426, 478, 503, 504
software, 9, 18, 69, 134, 135, 136, 137, 150, 155, 156, 157, 158, 159, 161, 162, 188, 189, 191, 192, 195, 196, 200, 228, 244, 245, 272,

275, 277, 278, 289, 290, 295, 321, 324, 388, 393, 408, 416, 417, 418, 451, 488
solubility, 137, 176, 330, 423, 457, 460, 467
soluble, 177, 375, 448, 458
solute, 378, 386, 451, 467
solution, 13, 22, 29, 93, 100, 125, 137, 148, 157, 159, 163, 164, 165, 181, 200, 208, 221, 225, 234, 237, 238, 239, 256, 261, 274, 324, 328, 363, 364, 365, 386, 418, 420, 421, 424, 434, 436, 438, 440, 444, 455, 466, 467, 468, 471, 487, 497
solvent, 22, 176, 177, 236, 237, 238, 239, 241, 246, 249, 250, 251, 258, 265, 330, 331, 363, 364, 365, 378, 379, 389, 404, 436, 440, 444, 451, 467
sorbent, 177, 236, 237, 240, 366, 367, 467
spacious, 109, 187, 271
spans, 88, 340, 391
spatulas, 124, 197
SPE, 176, 177, 210, 231, 232, 235, 236, 237, 238, 239, 240, 241, 242, 245, 264, 364, 365, 366, 379, 387, 467
specialist, 95, 182, 276, 420
specialization, 138, 156, 183
species, 5, 18, 24, 39, 65, 112, 117, 146, 171, 217, 218, 224, 255, 278, 306, 314, 319, 372, 386, 391, 400, 406, 410, 437, 443, 451, 452, 454, 455, 457, 458, 459, 462
specification, 103, 113, 147, 157, 163, 189, 194, 200, 276, 278, 312, 334, 467, 475, 476
specificity, 4, 6, 20, 25, 145, 178, 179, 198, 211, 227, 231, 267, 287, 288, 325, 361, 370, 371, 372, 389, 434, 454, 467, 470
specimen, 12, 13, 14, 89, 184, 188, 269, 274, 327, 328, 330, 335, 337, 358, 390, 403, 435, 437, 439, 442, 443, 444, 449, 450, 451, 453, 459, 460, 462, 463, 464, 465, 466, 467, 468, 469, 496
spectra, 204, 361, 374, 440, 442, 452, 460
spectrometer, 177, 244, 252, 253, 254, 264, 277, 361, 366, 367, 370, 372, 422, 442, 445, 449, 450, 452, 454, 499
spectrometric, 17, 217, 222, 231, 238, 246, 252, 258, 287, 288, 358, 361, 368, 369, 371, 372, 373, 375, 390, 404, 424, 428
spectrophotometer, 93, 499
spectroscopy, 5, 171, 440, 442, 464
spectrum, 277, 385, 401, 440, 445, 446, 454, 464
spillage, 123
spinal, 453
splitter, 249
sprayer, 254, 372, 376, 377
spreadsheet, 313
stability, 8, 17, 20, 21, 22, 23, 24, 27, 29, 31, 39, 62, 110, 117, 121, 122, 124, 125, 142, 144, 146, 152, 156, 162, 168, 172, 175, 177, 178, 179, 181, 198, 201, 202, 203, 206, 208, 211, 212, 216, 219, 220, 221, 222, 223, 224, 225, 226, 227, 235, 237, 241, 257, 260, 267, 274, 284, 287, 308, 314, 315, 319, 321, 328, 329, 334, 335, 337, 338, 362, 363, 374, 393, 404, 405, 439, 448, 453, 468, 475, 476, 497
standardization, 6, 32, 82, 84, 141, 159, 199, 200, 311, 357, 405, 420, 456, 477, 478, 479
standards, 1, 2, 3, 6, 7, 12, 13, 19, 20, 21, 22, 23, 24, 25, 26, 27, 29, 31, 32, 35, 38, 41, 43, 46, 48, 50, 64, 70, 73, 78, 81, 82, 84, 91, 113, 119, 120, 134, 142, 146, 157, 159, 161, 162, 175, 179, 180, 181, 184, 185, 186, 187, 189, 199, 201, 204, 207, 208, 209, 211, 212, 213, 215, 216, 217, 220, 221, 222, 223, 225, 226, 227, 228, 235, 248, 257, 258, 267, 269, 270, 271, 272, 274, 278, 284, 290, 301, 303, 311, 324, 325, 326, 327, 328, 329, 330, 331, 332, 333, 334, 336, 337, 338, 357, 358, 371, 385, 387, 395, 405, 412, 414, 415, 446, 447, 460, 468, 478, 479, 486, 496
standpoint, 233, 345, 365, 422
statement, 11, 14, 40, 53, 54, 60, 62, 66, 84, 90, 92, 97, 98, 101, 107, 109, 110, 117, 140, 141, 147, 154, 186, 188, 202, 204, 270, 276, 282, 310, 318, 438, 461, 463, 467, 495
statements, 60, 138, 174, 183, 206, 218, 275, 341, 349
stationary, 228, 249, 264, 365, 366, 368, 378, 437, 446, 453, 457, 465, 467
statistics, 31, 214, 396, 442, 453, 459, 465, 477, 478
status, 40, 56, 58, 110, 133, 159, 171, 310, 314, 348, 351, 393, 394
statutes, 92, 437
stepwise, 305, 447
sterility, 104, 360
sterilization, 146
sterilizers, 348
steroids, 237, 373, 376
stimulant, 249

stimulus, 127, 169
storage, 6, 11, 12, 14, 15, 16, 21, 22, 24, 31, 32, 36, 47, 55, 62, 63, 64, 65, 70, 75, 77, 93, 94, 96, 104, 108, 111, 112, 113, 118, 121, 122, 123, 125, 128, 132, 134, 139, 142, 144, 146, 147, 151, 152, 154, 162, 174, 179, 181, 187, 188, 189, 192, 199, 200, 201, 202, 204, 212, 216, 218, 219, 220, 221, 222, 223, 225, 227, 233, 241, 258, 261, 271, 272, 274, 276, 307, 308, 310, 311, 313, 314, 315, 317, 321, 324, 325, 328, 329, 335, 348, 358, 362, 363, 365, 367, 375, 387, 404, 407, 437, 465, 469, 473, 474, 487, 488, 496, 499
strategy, 70, 164, 206, 213, 234, 241, 261, 263, 279, 285, 290, 336, 357, 366, 375, 387, 391, 392, 402, 421
stratification, 288
streamline, 162, 279, 283, 370
strength, 58, 142, 174, 201, 258, 279, 308, 315, 364, 379, 438
stringent, 47, 128, 156, 163, 356
structural, 169, 211, 248, 249, 254, 364, 371, 435, 454
study plan, 54, 61
subcontract, 46, 133
subcontractor, 154, 317
subdirectory, 136
subdiscipline, 166
submissions, 7, 19, 31, 54, 64, 79, 81, 87, 107, 116, 169, 265, 282, 288, 405, 420, 476
submitter, 53, 107
subpart, 69, 97, 173
subpopulations, 411
substance, 24, 45, 53, 106, 146, 172, 179, 226, 402, 416, 434, 435, 436, 438, 439, 440, 444, 445, 448, 449, 450, 451, 452, 453, 454, 455, 458, 462, 463, 467
substances, 20, 24, 26, 35, 44, 82, 142, 144, 145, 146, 166, 178, 226, 247, 352, 435, 438, 452, 458, 460, 469
substrate, 132, 443, 459, 468, 470
subsystems, 341
success, 2, 7, 8, 14, 166, 169, 217, 223, 233, 240, 260, 261, 264, 278, 283, 289, 339, 343, 365, 369, 388, 391, 395, 398, 404, 409, 413, 415, 420
succession, 376
suitability, 28, 65, 94, 113, 124, 150, 187, 190, 191, 193, 194, 198, 199, 201, 223, 226, 261, 264, 272, 274, 328, 334, 337, 371, 405, 496
summaries, 47, 308, 340
summation, 381
supercomputer, 416
supervision, 60, 154, 184, 200, 268, 325
supervisors, 40, 179, 182, 194, 268
supervisory, 202, 203
supplement, 3, 64, 194, 231, 343, 356, 406
suppliers, 42, 45, 74, 99, 101, 134, 142, 144, 259, 261, 262, 338, 471, 499
suppression, 176, 222, 235, 255, 256, 257, 258, 371, 377, 385, 386, 387, 389
surfactant, 256
surrogate, 172, 175, 214, 262, 399, 403, 434, 468
surveillance, 39, 104, 136, 182, 288, 301, 306, 321, 468
suspension, 40, 125, 359
Symbiosis, 231, 238, 239, 243
symbol, 233, 445, 457, 480
symptoms, 174
synonym, 444, 454, 455, 457, 467
synopsis, 179, 180
synthesis, 8, 142, 170, 172, 211, 338, 360, 390, 402, 442

talent, 279, 280, 281, 282, 286, 407, 409
tandem, 4, 127, 133, 156, 169, 231, 238, 246, 248, 251, 258, 358, 364, 368, 369, 400, 424, 425, 426
techniques, 4, 5, 7, 8, 13, 16, 19, 36, 39, 65, 90, 92, 95, 99, 110, 119, 120, 121, 124, 169, 170, 171, 177, 178, 184, 231, 232, 233, 234, 235, 236, 237, 238, 240, 244, 245, 264, 265, 268, 273, 289, 290, 291, 301, 325, 339, 358, 361, 363, 365, 366, 368, 371, 376, 387, 388, 390, 400, 401, 402, 413, 421, 424, 426, 457, 462, 463, 475
technology, 6, 66, 68, 69, 70, 71, 87, 121, 127, 128, 136, 157, 163, 166, 169, 171, 189, 207, 214, 232, 233, 234, 236, 244, 246, 247, 254, 281, 283, 286, 289, 356, 357, 376, 379, 392, 393, 399, 400, 401, 404, 411, 412, 416, 419, 422, 423, 428, 477, 478, 503
teleconference, 409
temperature, 22, 25, 94, 112, 113, 122, 142, 146, 148, 151, 152, 163, 179, 181, 193, 213,

219, 221, 222, 224, 225, 227, 311, 362, 390, 404, 438, 444, 445, 465, 499
template, 90, 395, 465, 491, 492, 493
tendency, 153, 366, 466, 471
tension, 380, 386
teratogenicity, 45
termination, 55, 109
terminologies, 47, 206, 277, 433, 434, 435, 436, 437, 438, 439, 440, 441, 442, 443, 444, 445, 446, 447, 448, 449, 450, 451, 452, 453, 454, 455, 456, 457, 458, 459, 460, 461, 462, 463, 464, 465, 466, 467, 468, 469, 470, 472, 474, 476, 478
theme, 169
theory, 293, 377, 428
therapeutic, 4, 5, 6, 132, 169, 171, 182, 207, 211, 212, 216, 231, 264, 266, 267, 278, 360, 391, 399, 402, 435
therapeutics, 6, 132, 207, 211, 212, 230, 231, 259, 260, 399, 503
therapies, 127, 169, 171, 173, 403
thermometers, 93, 197, 276
threshold, 182, 262, 439, 444, 469
throughput, 4, 24, 131, 132, 160, 161, 170, 171, 189, 231, 232, 234, 235, 236, 237, 241, 244, 245, 246, 247, 248, 249, 250, 251, 252, 253, 254, 255, 259, 261, 264, 265, 355, 358, 360, 361, 363, 364, 365, 368, 369, 370, 371, 376, 377, 379, 385, 388, 389, 405, 413, 421, 422, 424, 426, 428
timelines, 88, 105, 131, 261, 285, 286, 342, 394, 395, 396, 397, 398, 413
tissue, 17, 65, 126, 170, 172, 212, 230, 236, 237, 307, 319, 321, 378, 400, 403, 453, 454, 462, 467, 499
TK, 16, 158, 168, 175, 187, 265, 291, 404
tolerance, 3, 27, 46, 93, 241, 402, 445, 486
tomography, 402
Tomtec, 5, 232
topics, 73, 133, 207, 223, 294, 351, 411, 424, 428, 429, 431
toxicity, 8, 45, 73, 171, 172, 279, 287, 300, 306, 307, 322, 391, 403, 442
toxicodynamics, 469
toxicokinetic, 16, 32, 131, 153, 168, 198, 205, 211, 231, 290, 472
toxicokinetics, 46, 73, 84, 166, 265, 278, 322, 391, 458, 469
toxicologists, 64, 405, 456, 478
toxicology, 1, 3, 10, 19, 24, 36, 38, 39, 42, 64, 65, 73, 84, 131, 133, 151, 171, 172, 174, 175, 204, 217, 259, 265, 271, 299, 301, 303, 306, 322, 338, 361, 364, 402, 405, 410, 416, 462, 469, 475, 476, 478
traceability, 2, 44, 75, 105, 120, 128, 152, 159, 199, 288, 420, 453, 469
trainer, 184, 268
transaction, 72
transcriptase, 239
transcriptases, 465
transcription, 443
transcripts, 474
transduction, 234
transferability, 262, 470
transfers, 18, 233, 241, 249, 259, 260, 262, 264, 303, 437
transformation, 360, 445, 446, 470
translation, 290, 344
translational, 288, 290, 442, 464
transparency, 82, 414
transportation, 62
treatment, 6, 65, 174, 204, 234, 288, 314, 370, 399, 410, 411, 442, 457, 469
triangle, 190, 504
triplicate, 25, 27, 29, 138
trivial, 259, 263
troubleshoot, 408
trueness, 94, 461, 470
trustworthy, 66, 405, 416
tumor, 287, 290, 403
turbidimetry, 450
turbidity, 450
turboflow, 244
turbulent, 178, 231, 239, 240, 250, 252, 365
turnaround, 131, 132, 138, 183, 232, 281, 283, 286, 392
turnover, 121, 281

ultra, 231, 246, 247, 377, 381, 413, 424, 426
ultrafast, 246
ultrafiltration, 4, 178, 436, 453
ultrasensitive, 402
ultraviolet, 371, 450
uncertainty, 392, 415, 419, 453, 470
uniformity, 207, 308, 315, 319
unit, 8, 9, 15, 47, 48, 76, 81, 82, 106, 116, 134, 135, 139, 141, 154, 158, 165, 182, 183, 185, 193, 237, 238, 268, 269, 270, 308, 309, 314,

329, 337, 361, 370, 371, 377, 378, 384, 388, 421, 435, 438, 444, 465, 470, 472, 483, 492, 495
universe, 459, 470
unknowns, 175, 215, 390
updates, 165, 344, 394
UPLC, 231, 232, 246, 247, 248, 264, 378, 405, 413
uptake, 458, 469
urgency, 135, 182
urinalysis, 319, 478
urine, 4, 17, 18, 20, 171, 176, 177, 202, 212, 241, 242, 243, 258, 266, 287, 304, 370, 390, 424, 442, 452, 458, 467, 470, 476, 477
usage, 53, 106, 203, 264, 371, 467
user, 44, 68, 69, 71, 74, 84, 128, 148, 156, 159, 160, 161, 191, 192, 193, 195, 196, 197, 198, 203, 244, 245, 262, 275, 276, 278, 313, 353, 420, 460, 463, 485
USFDA, 7, 33, 80, 172
USP, 19, 84, 178, 276, 353, 502
utilities, 100, 109, 112, 187, 192, 271, 338

vaccine, 207, 502
validation, 4, 6, 8, 9, 11, 12, 13, 14, 16, 17, 18, 19, 21, 22, 23, 24, 25, 26, 27, 28, 29, 30, 31, 32, 33, 51, 64, 65, 68, 70, 71, 72, 73, 74, 76, 80, 83, 84, 85, 94, 104, 109, 127, 129, 132, 133, 135, 137, 141, 145, 149, 150, 152, 153, 154, 155, 159, 160, 163, 165, 166, 168, 169, 174, 175, 176, 177, 178, 179, 180, 181, 185, 186, 190, 191, 194, 195, 196, 198, 199, 201, 204, 205, 206, 207, 209, 210, 211, 212, 213, 214, 215, 216, 217, 218, 219, 220, 221, 222, 223, 224, 225, 226, 227, 228, 229, 231, 234, 239, 248, 250, 252, 255, 257, 259, 260, 261, 262, 263, 265, 267, 269, 270, 273, 274, 276, 277, 278, 280, 287, 288, 289, 290, 291, 294, 295, 296, 308, 311, 313, 316, 322, 323, 324, 325, 326, 327, 328, 329, 330, 331, 332, 335, 337, 338, 339, 343, 353, 354, 385, 386, 387, 393, 397, 398, 399, 400, 401, 404, 405, 407, 408, 409, 410, 411, 412, 413, 417, 418, 438, 470, 476, 477, 486, 487, 488, 496, 497
validity, 13, 17, 23, 29, 59, 62, 71, 79, 111, 121, 123, 149, 155, 162, 186, 195, 218, 265, 270, 299, 323, 326, 327, 329, 332, 333, 334, 335, 404, 414, 415, 417, 439, 461
valve, 238, 253, 376
vaporization, 450
variability, 6, 20, 27, 120, 145, 151, 205, 209, 211, 212, 214, 218, 222, 223, 225, 232, 256, 257, 261, 263, 289, 328, 329, 330, 333, 334, 335, 402, 463, 467, 471
variance, 214, 215, 434, 466, 468, 471
variation, 20, 23, 24, 179, 223, 227, 235, 256, 366, 434, 437, 438, 442, 443, 448, 449, 463, 464
variety, 8, 36, 87, 153, 156, 163, 169, 211, 278, 351, 363, 364, 368, 391, 451, 454, 467, 470
vehicle, 110, 117, 124, 125, 387
velocity, 239, 361, 377, 389, 442
vendor, 68, 72, 74, 156, 165, 186, 191, 192, 193, 276, 279, 311, 316, 392, 394, 418, 420, 495, 499
ventilation, 12, 112, 311
verbiage, 47, 87, 286
verification, 19, 60, 63, 71, 74, 75, 89, 102, 114, 128, 158, 187, 192, 201, 204, 218, 265, 270, 271, 274, 275, 313, 323, 372, 399, 411, 471, 496
version, 44, 63, 91, 110, 118, 120, 136, 148, 149, 174, 188, 191, 250, 295, 306, 322, 344, 458, 483, 491, 492
versions, 61, 63, 64, 115, 147, 148, 149, 195, 199, 273, 303, 405
vessels, 435
viability, 56, 394
vial, 213, 239
vibrational, 450
village, 157
violation, 95, 98
violations, 36, 96, 447
virus, 236
viscometers, 197
viscosity, 177, 386
vitro, 3, 44, 45, 47, 64, 65, 66, 137, 171, 172, 287, 323, 361, 368, 401, 402, 405, 406, 435, 449, 471
vivo, 3, 65, 137, 171, 172, 266, 268, 295, 303, 320, 322, 328, 353, 401, 402, 405, 435, 449
volume, 18, 20, 22, 29, 123, 124, 125, 126, 132, 134, 138, 148, 176, 236, 237, 239, 279, 285, 364, 365, 369, 378, 387, 390, 391, 404, 422, 435, 436, 438, 440, 444, 457, 466
vortex, 125, 176, 197, 263, 276, 499

warrants, 196, 350
washers, 261
waterhouse, 427
wavelength, 93, 193, 442, 450
weaknesses, 338, 339, 408
welfare, 126, 228, 267, 302, 305, 322, 325
withdrawal, 46, 88, 120, 149
workflow, 71, 82, 159, 160, 161, 189, 231, 232, 270, 271, 279, 421
workload, 105, 115, 308
workplace, 110
worksheets, 11, 153, 200, 203, 337, 473
workshops, 2, 43, 64, 80, 207, 327, 405, 406, 407
workspace, 152
workstations, 5, 112, 235, 244, 312, 413
worldwide, 2, 15, 32, 40, 41, 76, 299, 305, 415

xenobiotic, 231, 439, 442
xenobiotics, 20, 166, 211, 400

yields, 176, 251, 454, 463